T0295357

Indian Summer Monsoon Variability

El Niño-Teleconnections and Beyond

Indian Summer Monsoon Variability

El Niño-Teleconnections and Beyond

Edited by

Jasti S. Chowdary
Indian Institute of Tropical Meteorology,
(IITM-MoES), Pune, India

Anant Parekh
Indian Institute of Tropical Meteorology,
Ministry of Earth Sciences, Pune, Maharashtra, India

C. Gnanaseelan
Indian Institute of Tropical Meteorology,
Ministry of Earth Sciences, Pune, Maharashtra, India

ELSEVIER

British Library Cataloguing-in-Publication Data
A catalogue record for this book is available from the British Library

Library of Congress Cataloging-in-Publication Data
A catalog record for this book is available from the Library of Congress

ISBN: 978-0-12-822402-1

For Information on all Elsevier publications visit our website at
https://www.elsevier.com/books-and-journals

Publisher: Candice Janco
Acquisitions Editor: Amy Shapiro
Editorial Project Manager: Grace Lander
Production Project Manager: Debasish Ghosh
Cover Designer: Victoria Pearson Esser

Typeset by Aptara, New Delhi, India

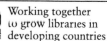

Working together
to grow libraries in
developing countries

www.elsevier.com • www.bookaid.org

Contents

7. **Teleconnections between the Indian summer monsoon and climate variability: a proxy perspective**

S. Chakraborty, Aasif M. Lone, Anant Parekh, P.M. Mohan

Part II
Indian and Atlantic Ocean – Indian Summer Monsoon teleconnections

8. **Indian Ocean Dipole influence on Indian summer monsoon and ENSO: A review**

Annalisa Cherchi, Pascal Terray, Satyaban B. Ratna, Syam Sankar, K P Sooraj, Swadhin Behera

9. **Influence of South Tropical Indian Ocean dynamics on the Indian summer monsoon**

Yan Du, Zesheng Chen, Ying Zhang, Kaiming Hu, Xiaotong Zheng, Weidong Yu

Part III
Subtropical and Extratropical teleconnections to Indian Summer Monsoon

18. The Atlantic Multidecadal Oscillation and Indian summer monsoon variability: a revisit

Lea Svendsen

19. Indian summer monsoon and its teleconnection with Pacific decadal variability

Manish K. Joshi, Fred Kucharski, Archana Rai, Ashwini Kulkarni

Part IV
Climate change and Monsoon teleconnections

20. Future changes of the ENSO–Indian summer monsoon teleconnection

June-Yi Lee, Tamás Bódai

Contributors

N.K. Agarwal   Indian Institute of Tropical Meteorology, Pune, Maharashtra, India

Sebastian Anila   Indian Institute of Tropical Meteorology, Ministry of Earth Sciences, Pune, Maharashtra, India

Karumuri Ashok   Centre for Earth, Ocean, and Atmospheric Sciences, University of Hyderabad, Hyderabad, India; Phule Pune University, Pune, India

Swadhin Behera   Application Laboratory, Research Institute for Value-Added-Information Generation, Japan Agency for Marine-Earth Science and Technology, Yokohama, Japan

Tamás Bódai   Institute for Basic Science (IBS), Center for Climate Physics (ICCP), Busan, South Korea; Research Center for Climate Sciences and Department of Climate System, Pusan National University, Busan, South Korea

Wenju Cai   Key Laboratory of Physical Oceanography–Institute, for Advanced Ocean Studies, Ocean University of China, and Qingdao National Laboratory for Marine Science and Technology, Qingdao, China; Center for Southern Hemisphere Oceans Research (CSHOR), CSIRO Oceans and Atmosphere, Hobart, Australia

S. Chakraborty   Indian Institute of Tropical Meteorology, Ministry of Earth Sciences, Pune, Maharashtra, India

Zesheng Chen   State Key Laboratory of Tropical Oceanography, South China Sea Institute of Oceanology, Chinese Academy of Sciences, Guangzhou, China; Southern Marine Science and Engineering Guangdong Laboratory, Guangzhou, China

Annalisa Cherchi   National Research Council of Italy, Institute of Atmospheric Sciences and Climate (CNR-ISAC), Bologna, Italy, Istituto Nazionale di Geofisica e Vulcanologia, Bologna, Italy

Jasti S. Chowdary   Indian Institute of Tropical Meteorology, (IITM-MoES), Pune, India

P. Darshana   Indian Institute of Tropical Meteorology, Ministry of Earth Sciences (IITM-MoES), Pune, India; Savitribai Phule Pune University, Pune, India

S. De   Indian Institute of Tropical Meteorology, Pune, Maharashtra, India

Yan Du   State Key Laboratory of Tropical Oceanography, South China Sea Institute of Oceanology, Chinese Academy of Sciences, Guangzhou, China; Southern Marine Science and Engineering Guangdong Laboratory, Guangzhou, China; University of Chinese Academy of Sciences, Beijing, China

G. Purna Durga   Atmospheric Science Research Laboratory, Department of Physics, SRM Institute of Science and Technology, Tamilnadu, India

F. Feba Centre for Earth, Ocean, and Atmospheric Sciences, University of Hyderabad, Hyderabad, India

P.A. Francis Indian National Centre for Ocean Information Services, Ministry of Earth Sciences, Hyderabad, India

Sulochana Gadgil Centre for Atmospheric and Oceanic Sciences, Indian Institute of Science, Bengaluru, India

C. Gnanaseelan Indian Institute of Tropical Meteorology, Ministry of Earth Sciences, Pune, Maharashtra, India

D. Govardhan Centre for Earth, Ocean, and Atmospheric Sciences, University of Hyderabad, Hyderabad, India

Kaiming Hu Center for Monsoon System Research, Institute of Atmospheric Physics, Chinese Academy of Sciences, Beijing, China

Manish K. Joshi Indian Institute of Tropical Meteorology, Ministry of Earth Sciences, Pune, Maharashtra, India

Yu Kosaka Research Center for Advanced Science and Technology, The University of Tokyo, Tokyo, Japan

Ramesh Kripalani Indian Institute of Tropical Meteorology (MoES), Pune, India (retired)

Fred Kucharski Earth System Physics Section, Abdus Salam International Centre for Theoretical Physics, Trieste, Italy; Center of Excellence for Climate Change Research/Department of Meteorology, King Abdulaziz University, Jeddah, Saudi Arabia

Ashwini Kulkarni Indian Institute of Tropical Meteorology, Ministry of Earth Sciences (IITM-MoES), Pune, Maharashtra, India

T.V. Lakshmi Kumar Atmospheric Science Research Laboratory, Department of Physics, SRM Institute of Science and Technology, Tamilnadu, India

June-Yi Lee Institute for Basic Science (IBS), Center for Climate Physics (ICCP), Busan, South Korea; Research Center for Climate Sciences and Department of Climate System, Pusan National University, Busan, South Korea

Ziguang Li Key Laboratory of Physical Oceanography–Institute, for Advanced Ocean Studies, Ocean University of China, and Qingdao National Laboratory for Marine Science and Technology, Qingdao, China

Won-il Lim Department of Atmospheric Sciences, Pusan National University, Busan, South Korea

Aasif M. Lone Indian Institute of Science Education and Research, Bhopal, India

Mengmeng Lu School of Atmospheric Sciences and Guangdong Province Key Laboratory for Climate Change and Natural Disaster Studies, Sun Yat-sen University, Guangzhou, China

R.K. Mall DST Mahamana Center of Excellence in Climate Change Research, Banaras Hindu University, Varanasi, India

P.M. Mohan Pondicherry University, Port Blair, India

Harini Nagendra   Centre for Climate Change and Sustainability, Azim Premji University, Bengaluru, India

Ravi S. Nanjundiah   Indian Institute of Tropical Meteorology (IITM-MoES), Pune, India; Centre for Atmospheric and Oceanic Sciences, Indian Institute of Science, Bengaluru, India

Benjamin Ng   Center for Southern Hemisphere Oceans Research (CSHOR), CSIRO Oceans and Atmosphere, Hobart, Australia

Anant Parekh   Indian Institute of Tropical Meteorology, Ministry of Earth Sciences, Pune, Maharashtra, India

Hyo-Seok Park   Department of Ocean Science and Technology, Hanyang University, Ansan, South Korea

Archana Rai   Indian Institute of Tropical Meteorology, Ministry of Earth Sciences (IITM-MoES), Pune, Maharashtra, India

K. Koteswara Rao   Centre for Climate Change and Sustainability, Azim Premji University, Bengaluru, India

Satyaban B. Ratna   Climatic Research Unit, School of Environmental Sciences, University of East Anglia, Norwich, United Kingdom, Climate Research and Services, India Meteorological Department, Pune, India

Indrani Roy   University College London (UCL), IRDR, London, United Kingdom

S. Sajani   CSIR-Fourth Paradigm Institute, Bengaluru, India

Syam Sankar   Advanced Centre for Atmospheric Radar Research (ACARR), Cochin University of Science and Technology, Kochi, India

Kyong-Hwan Seo   Department of Atmospheric Sciences, Pusan National University, Busan, South Korea

K P Sooraj   Centre for Climate Change Research, Indian Institute of Tropical Meteorology, Ministry of Earth Sciences (IITM-MoES), Pune, Maharashtra, India

Lea Svendsen   Geophysical Institute, University of Bergen and the Bjerknes Centre for Climate, Research, Bergen, Norway

C.T. Tejavath   Centre for Earth, Ocean, and Atmospheric Sciences, University of Hyderabad, Hyderabad, India

Pascal Terray   LOCEAN/IPSL, Sorbonne Universités (UPMC, Univ Paris 06)-CNRS-IRD-MNHN LOCEAN Laboratory, Paris, France

P.N. Vinayachandran   Centre for Atmospheric and Oceanic Sciences, Indian Institute of Science, Bengaluru, India

Guojian Wang   Key Laboratory of Physical Oceanography–Institute, for Advanced Ocean Studies, Ocean University of China, and Qingdao National Laboratory for Marine Science and Technology, Qingdao, China; Center for Southern Hemisphere Oceans Research (CSHOR), CSIRO Oceans and Atmosphere, Hobart, Australia

Hai Wang   Department of Marine Meteorology, College of Oceanic and Atmospheric Sciences, Ocean University of China, Qingdao, China

Lin Wang   Center for Monsoon System Research, Institute of Atmospheric Physics, Chinese Academy of Sciences, Beijing, China

Wei Wei   School of Atmospheric Sciences, Sun Yat-sen University, and Southern Marine Science and Engineering Guangdong Laboratory (Zhuhai), Zhuhai, China; Guangdong Province Key Laboratory for Climate Change and Natural Disaster Studies, Sun Yat-sen University, Zhuhai, China

Renguang Wu   Southern Marine Science and Engineering Guangdong Laboratory (Zhuhai), Guangdong, China; Key Laboratory of Geoscience Big Data and Deep Resource of Zhejiang Province, School of Earth Sciences, Zhejiang University, Hangzhou, China

Shang-Ping Xie   Scripps Institution of Oceanography, University of California San Diego, La Jolla, CA, United States

Peiqiang Xu   Center for Monsoon System Research, Institute of Atmospheric Physics, Chinese Academy of Sciences, Beijing, China

Ramesh Kumar Yadav   Indian Institute of Tropical Meteorology, Pune, India

Kai Yang   Center for Southern Hemisphere Oceans Research (CSHOR), CSIRO Oceans and Atmosphere, Hobart, Australia; State Key Laboratory of Numerical Modeling for Atmospheric Sciences and Geophysical Fluid Dynamics, Institute of Atmospheric Physics, Chinese Academy of Sciences, Beijing, China

Song Yang   School of Atmospheric Sciences and Guangdong Province Key Laboratory for Climate Change and Natural Disaster Studies, Sun Yat-sen University, Guangzhou, China; Southern Marine Science and Engineering Guangdong Laboratory (Zhuhai), Guangdong, China

Weidong Yu   School of Atmosphere Sciences, Sun Yat-Sen University, Guangzhou, China

Ying Zhang   State Key Laboratory of Tropical Oceanography, South China Sea Institute of Oceanology, Chinese Academy of Sciences, Guangzhou, China; Southern Marine Science and Engineering Guangdong Laboratory, Guangzhou, China

Xiaotong Zheng   Physical Oceanography Laboratory and Key Laboratory of Ocean-Atmosphere Interaction and Climate in Universities of Shandong, Ocean University of China, Qingdao, China; Key Laboratory of Physical Oceanography -Institute, for Advanced Ocean Studies, Ocean University of China, and Qingdao National Laboratory for Marine Science and Technology, Yushan Road, Qingdao, China

Foreword

India has two monsoons, summer monsoon during June to September and northeast or winter monsoon during October to December. The Indian summer monsoon (ISM) contributes to 70-90% of annual rainfall over the country. It exhibits variability in all timescales, right from diurnal to daily, subseasonal, seasonal, yearly, decadal, and even centennial time scales. Year to year variability of Indian summer monsoon is very crucial as it affects agricultural production, water resources and its management, and power production and consumption. More importantly, it affects India's GDP. For example, a deficient monsoon can bring down India's GDP by 1-2% in any particular year.

It has been observed that the summer monsoon does not always bring the same amount of rainfall; it changes one year to to the next. At the same time, year to year variability of Indian summer monsoon is well studied, right from the time of Sir Gilbert Walker, who discovered Walker Circulation and the teleconnections of Indian summer monsoon with the Southern Oscillation in early 1900s. Since then, there has been significant progress on understanding the remote forcing and its potential use in the prediction of ISM, with broad focus on El Niño–Southern Oscillation (ENSO). The availability of oceanic data, the teleconnection of ISM with El Niño sought the global attention from the early 1980s with hundreds of research papers published on various aspects of ENSO–ISM. For example, decadal variability and weakening of relationship, linkages to central equatorial Pacific El Niño (or El Niño Modoki), etc., have been examined in detail using both observations and climate models. The teleconnection with ENSO is also wisely used in long-range prediction models in India for predicting ISM rainfall.

However, over the recent years, several research papers explored the teleconnections of ISM beyond El Niño confirming the existence of drivers of Indian monsoon variability beyond ENSO. For example, the relationship with Eurasian snow cover/depth and midlatitude circulation, Arctic oscillation, sea surface temperature anomalies over the equatorial and subtropical Indian Ocean and Atlantic Oceans, etc., are some of the few new emerging areas of research on monsoon teleconnections.

Understanding the complex interaction among different modes that could influence the ISM rainfall is key to improving rainfall prediction. A book of this nature discussing the multifarious remote teleconnections to ISM rainfall and internal forcing is expected to fill this void. There are 23 chapters in this book addressing the emerging areas of research linking Indian monsoon variability.

This book is therefore an excellent compilation of latest research work related to all aspects of Indian monsoon variability and its observed teleconnection patterns covering both the historical aspects and future changes due to global warming. I am confident that this book will be an excellent reference material for students and scientists working on climate variability in general and Indian monsoon variability in particular. The editors have designed and executed this book by assembling various interesting articles/chapters that focused on monsoon variability, with deep insights and future prospects in mind.

I would like to congratulate all the contributors and the editors for bringing out this very valuable book on ISM variability.

(M Rajeevan)
Secretary,
Ministry of Earth Sciences,
Government of India

Preface

Monsoon, the seasonally reversing winds and associated rainfall, is the life line of any agriculture-based economy. Indian summer monsoon (ISM) or south Asian monsoon with its dynamic and complex features is one of the most fascinating phenomena in the field of climate science. One of the characteristic features of ISM is its temporal and spatial variability; the signatures include large-scale droughts and floods. The prediction of ISM, which is of paramount importance, depends on our understanding of the forcing mechanisms. It is well established that the remote forcing from the tropical Pacific plays an important role in the interannual variability of ISM. The east and central equatorial Pacific sea surface temperature (SST) is known to have strong effect on ISM, however, all the ISM droughts and floods can't be explained by the tropical Pacific SST associated with El Niño southern oscillation (ENSO). This in fact has opened up plethora of hypotheses and speculations on the possible teleconnection mechanisms of interannual ISM. In the recent decades, the changing patterns and strength of ENSO, such as canonical ENSO, central Pacific ENSO (ENSO Modoki), prolonged ENSO and lagged ENSO response, etc., further enhanced the complexity in our understanding of the ISM teleconnections. These different ENSO flavors could affect the ISM rainfall in different ways and could influence the predictability component of ISM. In addition to the forcing from Pacific, other modes of variability (unrelated to ENSO) in the tropical to subtropical Indian Ocean, Atlantic Ocean, and midlatitudes to extratropical regions could potentially alter the strength of ISM at various time scales. Hence, it is essential to compile our current understanding on the teleconnections of ISM rainfall variability from different climate modes and discuss the recent advancements and potential future research problems.

This book covers the important aspects of teleconnection pathways of ISM. A wide-spectrum of remote and local forcing mechanisms that potentially affect the ISM rainfall variability from interannual through decadal time scales and the climate change impacts are presented in this book. It emphasizes both ENSO and non-ENSO teleconnections to ISM variability. Unlike many other books on climate variability, this book highlights the potential research problems of relevance in different chapters, which may motivate young researchers and students. It also brings out the existing knowledge gap in the field and recommends the requirement of additional long-term observations. The current status of climate models is emphasized in many chapters to possibly help and motivate the modelers to address the challenging issues of climate science and ISM variability.

This book begins with Chapter 1, in which the authors ephemerally described the drivers of the monsoon variability on various temporal and spatial scales. In Chapter 2, particular focus is given on interannual variability of ISM rainfall and the dominant role of the Equatorial Indian Ocean Oscillation (EQUINOO). Part One of this book (Chapters 3–7) deals with the famous ENSO–ISM teleconnections and discusses the advancements on recent understanding mostly based on the observations. Chapter 3 discusses the importance of ENSO and non-ENSO induced rainfall patterns (deficit and excess rainfall) over India and associated circulation patterns. Chapter 4 provides a comprehensive review of ENSO Modoki teleconnections to ISM rainfall. To go beyond the concurrent relationship between El Niño–ISM, the antecedent El Niño influence on ISM rainfall is discussed in Chapter 5. Chapter 6 describes a nonlinear scale interactions perspective of El Niño–ISM relation in the energy exchange mechanism, providing a new dimension to our current understanding. Chapter 7 reviews the recent understanding of external and internal monsoon teleconnections and associated processes focusing mainly on the Holocene timescale.

Part Two (Chapters 8–11) is dedicated to understanding and exploring the influence of Indian and Atlantic Ocean climate modes on ISM rainfall. The role of the Indian Ocean Dipole (IOD) in altering the ENSO–ISM rainfall relationship is described in Chapter 8 with emphasis on both observations and current coupled models. It is well known that the ocean dynamics of south Tropical Indian Ocean (TIO) plays an important role in modulating the SST. Chapter 9 discusses the role of southern TIO dynamics on ISM variability in detail. Chapter 10 focuses on the impact of the Atlantic equatorial SST on ISM variability, discussed pathways connecting ISM through subtropical upper level Asian jet. Chapter 11 provides detailed analysis of teleconnections between tropical SST modes and ISM variability. This chapter is focused on assessing the current coupled models in capturing the teleconnections of the East Pacific (EP) type/canonical ENSO and the Central Pacific (CP) type/ENSO Modoki to ISM rainfall.

Part Three (Chapters 12–19) is devoted on subtropical and extratropical teleconnections to ISM variability. In this section, influence of non-ENSO teleconnections including Eurasian Snow cover, the Asian jet, western north Pacific climate, subtropical deserts, south Asian high, southern annular mode, the Atlantic multidecadal oscillation, and Pacific decadal oscillation, are discussed. Understanding the teleconnections to ISM variability unrelated to ENSO would advance our knowledge towards better prediction of monsoon. The relationship between winter-spring Eurasian snow and ISM rainfall is assessed in Chapter 12. Interaction among different monsoon systems provides useful guidelines to understand dynamics of monsoon system. Chapter 13 is focused on this issue and demonstrated the interaction of Indian, East Asian and Western North Pacific monsoon systems. The ISM variability associated with the Asian jet

(Chapter 14), subtropical deserts (Chapter 15) and South Asian High (Chapter 16) are described. The interdecadal variation of the ISM variability is strongly linked to subtropical climate modes. Chapter 17 discussed possible pathways that link southern annular mode and ISM variability. A detailed review on the relation between Atlantic multidecadal oscillation and ISM variability is documented in Chapter 18. The last chapter (19) of this part has demonstrated the influence of the Pacific decadal variability on ISM rainfall changes.

Climate change can potentially influence the monsoon rainfall characteristics; an area needing further understanding. Modulations in ISM variability associated with climate change are documented in Part Four (Chapters 20–23). ENSO-ISM teleconnection changes under the global warming (future) are outlined in Chapter 20, based on the latest state-of-the-art coupled models projections. Response of the ISM rainfall to changes in positive IOD to global warming (projections) is discussed in Chapter 21. Anthropogenic aerosol is also known to modulate the ISM variability and this issue is discussed in Chapter 22. Finally, response of the moisture recycling over the ISM region to global warming is demonstrated in Chapter 23.

First and foremost, we express our sincere appreciation to the contributing authors for their hard work and dedicated efforts. This book could not have been possible without their support. This book consists of 23 chapters, covering various aspects of ISM variability and teleconnections. It gives a comprehensive review of current literature and provides ideas for future research on the different themes connected to monsoon teleconnections. The discussions are not only limited to analysis of past observational data but also utilized the state-of-the-art coupled models to understand future changes in monsoon teleconnections.

We express our deep appreciation to the reviewers of various chapters (listed at the end of this book as Appendix) for valuable comments and support. We would like to thank the Indian Institute of Tropical Meteorology, Pune, Ministry of Earth Sciences for providing support over the years for carrying out research on Indian monsoon variability. This book can be served as a specialized reference for early career researchers and post-graduate and graduate students in Atmospheric Science and Meteorology.

Chapter 1

Drivers of the Indian summer monsoon climate variability

Jasti S. Chowdary[a], Shang-Ping Xie[b], Ravi S. Nanjundiah[a,c]
[a]*Indian Institute of Tropical Meteorology (IITM-MoES), Pune, India,* [b]*Scripps Institution of Oceanography, University of California San Diego, La Jolla, CA, United States,* [c]*Centre for Atmospheric and Oceanic Sciences, Indian Institute of Science, Bengaluru, India*

1.1   Indian Monsoon as a seasonal phenomena

Monsoons are associated with large seasonal movements of tropical convection (Charney, 1969; Sikka and Gadgil, 1980; Krishnamurti, 1985) driven by the seasonal cycle of the solar radiation (e.g., Turner and Annamalai, 2012). Rapid changes occur in the wind system, influenced by Earth's rotation (e.g., Hastenrath, 1995; Webster et al., 1998, Goswami et al., 2006). The Indian/South Asian monsoons feature dramatic wind reversals between winter and summer from Arabia through the South China Sea and are estimated to have occurred for 15–20 million years (e.g., Berkelhammer et al., 2012) in response to the rise of the Tibetan Plateau (Molnar et al., 2010). Most of the annual rainfall in India occurs from June to September (JJAS) referred to as the Indian summer monsoon (ISM; Fig. 1.1A–C). Larger seasonal variations in the circulation and rainfall over the monsoonal regions compared to the other tropical regions highlight the variability of the monsoon (e.g., Gadgil, 2007; Fig. 1.2A and B). Many sectors such as agriculture, food, energy, and water are influenced by changes in weather and climate associated with ISM. The moisture-laden southwesterly winds blowing from the Arabian Sea to the Indian subcontinent provide a large amount of rainfall (Figs. 1.1 and 1.2). These winds arrive at the southern tip of India by the last week of May to the first week of June, each year (e.g., Pisharoty, 1965). The rainfall maxima along the Western Ghats, near the foothills of the Himalayas and the Myanmar coast, represent orographic effects on rainfall distribution (e.g., Xie et al., 2006; Figs. 1.1B and 1.2B).

The Southwest/summer monsoon consists of some semipermanent features such as a heat low over Pakistan, Monsoon trough over central India, the low-level jet (LLJ) over the Arabian Sea, surface Mascarene High, the Tibetan anticyclone, and the easterly jet stream in the upper troposphere (e.g., Koteswaram,

Indian Summer Monsoon Variability: El Niño-teleconnections and beyond.
DOI: https://doi.org/10.1016/B978-0-12-822402-1.00020-X

1

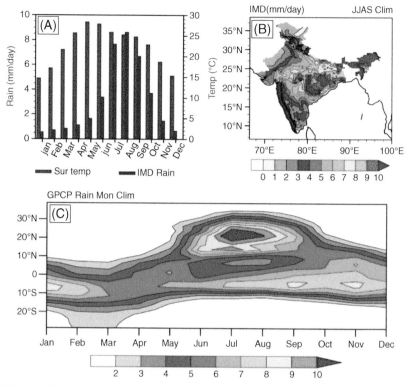

FIG. 1.1   (A) Annual cycle of surface air temperature (°C) and rainfall (IMD; mm/day) averaged over Indian land points, (B) JJAS mean rainfall (IMD; mm/day) and, (C) Latitude-time cross-section of mean precipitation (GPCP; mm/day) averaged from 70°E to 90°E. Gridded land rainfall data sets of 0.25° × 0.25° resolution for 1901–2019 from the India Meteorological Department (IMD, Pai et al. 2015), Global Precipitation Climatology Project v2.2, (GPCP, Adler et. al., 2003) for 1979–2019 and surface air temperature data from University of Delaware for 1901-2019 (Matsuura and Willmott, 2009) are used.

1958). The heat low is located over the central parts of Pakistan and neighborhood and is thought to be related to the monsoon activity. The monsoon trough, a major semipermanent feature of ISM extends from the heat low over Pakistan and adjoining region south-eastwards up to the Gangetic west Bengal region (e.g., Das, 1968). It is an elongated zone of low pressure extending along the Indo-Gangetic plains accompanied by cyclonic wind shear. In the vertical, the monsoon trough extends up to the mid-troposphere and tilts southward with height (Ramage, 1971). Moisture-laden winds are enticed toward the periphery of the trough. The maximum rainfall occurs to the south of the trough axis where tropical maritime air prevails up to a great depth. This monsoon trough occasionally shifts to the foothills of the Himalayas which causes heavy rainfall over extreme North India and break conditions over the central parts of India. Mascarene High is the high-pressure area at sea level south of the equator in the Indian Ocean near Mascarene Island, with its center located near 30°S; 50°E.

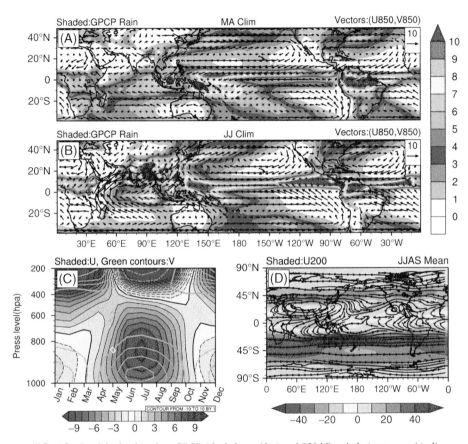

FIG. 1.2 Precipitation based on GPCP (shaded; mm/day) and 850 hPa winds (vectors; m/s) climatology over the global tropics for (A) March–April and (B) June–July. (C) Altitude-time cross-section of zonal (shaded and black contours; m/s) and meridional (green contours; m/s) wind climatology averaged over the ISM region (0°–25°N, 60°–100°E) and (D) JJAS streamlines of 200 hPa winds (m/s) and zonal wind (shaded; m/s). The European Centre for Medium-Range Weather Forecasts (*ECMWF*) Reanalysis Interim (ERA-I, Dee et al. 2011) winds from 1979 to 2019 are used. *ISM*, Indian summer monsoon.

The intensification of Mascarene High strengthens the cross-equatorial flow or LLJ (e.g., Joseph and Raman, 1966; Findlater, 1978). The low-level circulation over the ISM region is controlled by the cross-equatorial flow, the Somali jet, the cyclonic vortex over India (Fig. 1.2B). The high-pressure region is most prominent near 200 hPa centered over the Tibetan Plateau and extending eastward (Fig. 1.2D). The variation in the intensity and position of this upper level high and its orientation are closely related to the monsoon rainfall activity over South Asia (e.g., Raghavan, 1973; Koteswaram, 1958). Strong vertical wind shear during the summer monsoon period with dominant low-level westerlies and upper-level easterlies are apparent (Fig. 1.2C). The axis of the easterly jet generally extends from 5 to 20°N during the southwest monsoon season over the Indian subcontinent (Fig. 1.2D). The variability of ISM at different spatio-temporal scales is large and rainfall over this region can be altered through changes in above mentioned semipermanent systems. This Chapter discusses drivers of nonseasonal variability of the ISM from synoptic, interaseasonal to decadal time scale and climate change.

1.2 Synoptic variability and weather systems

Weather variations during ISM are associated with tropical lows, depressions, and semipermanent troughs. During the onset of the monsoon, increased cross-equatorial flow from the southern hemisphere reaches a broad region extending from the eastern Arabian Sea to the Bay of Bengal between the equator and 15°N (Fig. 1.3). The maximum zone of cloud bands with active convection triggers the onset of southwest monsoon rainfall over Kerala. The cross-equatorial LLJ is an important component that feeds convection over the southeast Arabian Sea (Fig. 1.3B and C). Over the south of Kerala and adjacent parts of the Arabian Sea, the vertical extent of southwesterlies strengthens during the onset period (e.g., Soman and Kumar, 1993; Joseph et al., 2006). The onset also involves the establishment of a low-pressure region that forms over the southeast Arabian Sea and moves in a northerly direction, which is known as monsoon onset vortex (e.g., Krishnamurti et al., 1981). Monsoon onset also involves the steady build-up of moisture and kinetic energy over the Arabian Sea (e.g., Krishnamurti,1985). The normal date of onset of the southwest monsoon over south Kerala is June 1. By June the low over Pakistan gets fully established and extends to the head Bay of Bengal across the Indo-Gangetic plains. Under vigorous and strong monsoon conditions the Somali LLJ extends across the Arabian Sea to the Indian peninsula (Fig. 1.2B). Once the southwest monsoon gets established, its strength fluctuates widely. After southwest monsoon current reaches Kerala, this branch advances northwards, reaching central India by June 15. On the other hand, the Bay of Bengal branch spreads over the entire Bay and eastern India by June 1. By the first week of July, the southwest monsoon is established over the entire subcontinent. Earlier studies noted that most of the late/early onset of ISM is influenced by El Niño-Southern Oscillation (ENSO). For example, Xavier et al. (2007) reported that most of the early monsoon onsets are associated with La Niña and late-onset with El Niño. They suggested that the changes in vertical and horizontal advection associated with the stationary waves forced by El Niño/La Niña over northern India and southern Eurasia influence the interannual variation of onset and could modulate the length of the rainy season.

Low-pressure systems (LPS) which include low, depression, deep depression, and cyclonic storms contribute largely to rainfall activity over the monsoon trough region (e.g., Mooley and Shukla, 1989). These systems are synoptic-scale disturbances that originate typically near the head of the Bay of Bengal and

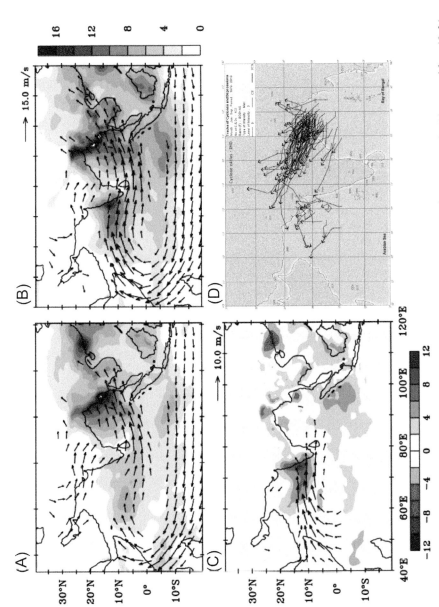

FIG. 1.3 Climatology of GPCP precipitation (shaded; mm/day) and 850 hPa winds (vectors from ERA-I; m/s) (A) averaged from 18–24 May, (B) averaged from 1–7 June, and (C) difference between (B) and (A) indicating the Onset of ISM, and (D) tracks of depressions in the north Indian Ocean (Data Source: http://www.rmcchennaieatlas.tn.nic.in) during summer (*JJAS*) for the period of 1979 to 2019. *ISM*, Indian summer monsoon.

in the Indian monsoon trough region (Fig. 1.3D; Asnani, 1973; Sikka, 1978). Synoptic LPS embedded in large-scale monsoon circulations and play a critical role in producing a large fraction of rainfall in ISM region (e.g., Goswami et al., 2003). The rainfall accompanying a low-pressure system covers a much larger area with heavy scattered showers. A majority of the LPSs normally form over the head of the Bay of Bengal, move northwestward and northward (Fig. 1.3D). These LPSs have a length scale of about 1000–2000 km and have a life cycle of 3–6 days (e.g., Mooley and Shukla, 1989; Mohapatra and Mohanty, 2004). The genesis of some of systems is seen over the South China Sea as a weak pressure wave traveling from the far east. A few systems also form over the Arabian Sea, move either northwestward or northeastward (Fig. 1.3D). The rainfall occurring on days with an LPS present gives that 60% of monsoon precipitation over the Indian subcontinent. (e.g., Praveen et al., 2015). The LPS tracks reach up to northwest India during flood years, whereas they are confined to central India during drought years (e.g., Krishnamurthy and Ajayamohan, 2010). They also noted that LPS in general contribute significantly to the seasonal monsoon rainfall over India. However, some studies reported a significant reduction in the frequency of higher intensity systems since the mid-1980s, despite the warming trend of north Indian Ocean Sea Surface Temperature (SST) (e.g., Rajeevan et al., 2000). Any change in frequency, intensity, and tracks of these systems have significant implications on floods and hydrology of various river basins (e.g., Prajeesh et al., 2013). Another type of quasi-stationary synoptic-scale disturbances that generally leads to heavy rainfall along the west coast of India near Gujarat are Midtropospheric cyclones (e.g., Miller and Keshavamurthy, 1968). It has been reported that dynamical instability mechanisms generate these systems (e.g., Mak, 1975). Further, Choudhury et al. (2018) recently found that slow northward propagation of summer monsoon rain belts on the subcontinent scale over North Arabian Sea could also generate midtropospheric cyclones.

1.3   Intraseasonal variability

The ISM rainfall exhibits prominent intraseasonal variability, manifesting as fluctuations between the active spell with good rainfall and the break spell (Fig. 1.4) with little rainfall over India (e.g., Goswami and Ajayamohan, 2001; Webster et al., 2002; Rajeevan et al., 2010). The active–break spells are largely caused by the northward-propagating, 30–60-day monsoon intraseasonal oscillations from the equatorial Indian Ocean (e.g., Sikka and Gadgil, 1980; Yasunari, 1981; Nanjundiah ct al., 1992). The subseasonal variability of the ISM exhibits two distinctive periodicities in 10–20 days and 30–60 days respectively (e.g., Krishnamurti and Bhalme,1976). These two periods, the 10–20 days and the 30–60 days have been related to the active and break cycles of the monsoon rainfall over the Indian subcontinent (e.g., Kulkarni et al., 2009; Pai et al., 2016). The 10–20 day oscillations are generally associated with westward

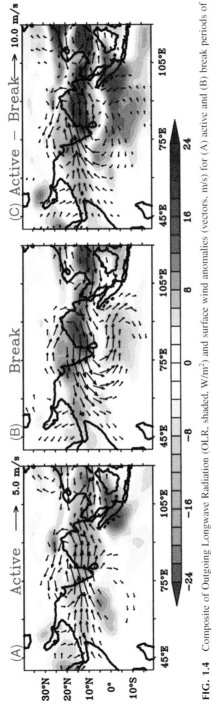

FIG. 1.4 Composite of Outgoing Longwave Radiation (OLR, shaded, W/m²) and surface wind anomalies (vectors, m/s) for (A) active and (B) break periods of ISM and (C) difference between active and break phase. Winds are based on ERA-interim data for 1979–2019. The OLR from the National Oceanic and Atmospheric Administration (NOAA; http://www.cdc.noaa.gov/) as a proxy for deep convection is utilized to identify active and break events. *ISM*, Indian summer monsoon.

propagating events entering the Indian land region from the Bay of Bengal and the 30–60 days oscillations with the northward propagation of cloudiness and rainfall from the equatorial Indian Ocean to the Indian subcontinent (e.g., Keshavamurty and Rao, 1992). The westward-moving 10–20-day oscillation between 10° and 15°N form over the South China Sea migrates toward India (e.g., Krishnamurti and Ardanuy, 1980) and influences the monsoon and interacts with the 30–40-day mode. Several studies highlighted the significance of northward propagating cloud bands, over the Indian subcontinent, on the timescale of 30–50 days (Sikka and Gadgil, 1980, Yasunari, 1981, Krishnamurti and Subrahmanyam, 1982). Recently using the multichannel singular spectrum analysis, Karmakar et al. (2017) identified the intra-seasonal oscillations (ISO) modes associated with ISM rainfall and investigated their behavior in space and time. They have reported the existence of a southeastward-propagating mode at a 10–20-day time scale, which propagates toward east-central India from the higher latitudes affects the canonical northwestward propagation.

These ISOs modulate large-scale atmospheric circulation and monsoon precipitation over the Indian subcontinent. The seasonality of ISOs has largely been attributed to the seasonal environment and ISO activities prefer those regions with low-level background westerlies and moisture convergence (e.g., Zhang and Dong, 2004). The physical mechanisms that give rise to the northward propagation of the 30–60-day oscillation have been attributed to convection–radiation–surface heat flux feedback (e.g., Goswami and Shukla, 1984), the air–sea interactions (e.g., Krishnan et al. 2000), the effects of easterly vertical shear on moist Rossby waves (Wang and Xie, 1997), and the moisture mode theory (e.g., Raymond and Fuchs, 2009). Conversely, the seasonal mean of intraseasonal anomalies explains about 50% of the interannual variability of the seasonal mean monsoon rainfall (e.g., Goswami and Xavier, 2005). Active and break episodes, characteristic of subseasonal variations of the ISM, are associated with enhanced (decreased) rainfall over central and western India and decreased (enhanced) rainfall over the southeastern peninsula and eastern India. The intraseasonal variations of rainfall (active-break cycles; Fig. 1.4) are strongly coupled to the intraseasonal variations of circulation (Webster et al., 1998).

The Madden–Julian Oscillation (MJO) is a large-scale, zonally oriented tropical convective disturbance that propagates east at about 5 m/s with a period of 30–70 days (e.g., Madden and Julian, 1972). In the tropics, the MJO is known to regulate wet and dry conditions (e.g., Zhang and Song, 2009), and influence the North and South American, African, Indian, and Asian-Australian monsoons (Wheeler and Hendon, 2004; Lorenz and Hartmann, 2006). The planetary-scale circulation anomalies associated with the MJO significantly affect monsoon development (e.g., DeMott et al. 2015). Many studies have noted that eastward-moving MJO plays important role in triggering the northward march of convection from the equatorial Indian Ocean (e.g., Madden and Julian, 1972). For example, Pai et al. (2011) suggested that the above normal convection over the equatorial Indian Ocean (EIO) during phases 1 and 2 of MJO causes large-scale anomalous subsidence over the monsoon trough region and helps to buildup break monsoon. They also reported that the phase and strength of the MJO could influence the duration and onset of break and active events over India during the summer season (e.g., Taraphdar et al., 2018). The intense rainfall events over the Indo-Gangetic plains during the ISM are linked with MJO phases 3–5 (e.g, Singh and Bhatla, 2019). In the intraseasonal timescale, convection over the eastern EIO is highly modulated by MJO (e.g., Karmakar and Krishnamurti, 2019). Recent studies demonstrated the success of moisture mode theory in explaining the maintenance and propagation mechanisms for the MJO (e.g., Raymond and Fuchs, 2009; Sobel and Maloney, 2013; Adames and Kim, 2016). In particular, Jiang et al. (2018) showed that the horizontal advection of the seasonal mean moisture distribution by the MJO wind anomalies plays an important role in the northward phase propagation during boreal summer. Thus the role of the MJO in modulating monsoon ISOs is a topic for further investigations.

1.4 Interannual variability

Slowly varying surface boundary conditions such as SST, land-surface temperature, snow cover, and soil moisture are believed to constitute a major forcing on the interannual variability of the monsoon rainfall (Charney and Shukla, 1981; Parthasarathy and Mooley, 1978). The extremes in the year-to-year variations of precipitation (Fig. 1.5A) manifest in the form of large-scale floods and droughts and cause devastating human and economic losses. The ENSO, equatorial Atlantic SSTs, and EIO climate anomalies (Fig. 1.5B and C) influence the interannual variability of the ISM, among the many other drivers (e.g., Shukla and Paolino, 1983). The ISM is known to have a strong association with ENSO events through ocean–atmosphere interactions (Walker, 1924, Kripalani and Kulkarni, 1997; Krishna Kumar, 1999; Annamalai et al., 2007; Ashok et al., 2019), also is the most significant and source of interannual variability (see Chapter 2–5). The high correlation between the interannual variability of ISM with that of Niño 3 SST (Fig. 1.5B) has been noted in many studies (Parthasarathy et al., 1994; Mehta and Lau, 1997; Krishnamurthy and Goswami, 2000). Apart from ENSO, the interaction of ISM with the Indian Ocean, Atlantic Ocean, Eurasian snow and the climate of other parts of the globe suggest that ISM is an integral part of the global climate system involving coupled atmosphere-land-ocean interactions (e.g., Webster et al., 1998; Sankarrao et al., 1996). Seasonal-mean rainfall during summer tends to be reduced (enhanced) over the Indian subcontinent during a developing El Niño (La Niña) event (Rasmusson and Carpenter, 1983). On the other hand, a dipole pattern tends to occur, with depressed rainfall over central and northeastern India and enhanced rainfall over the southern tip of India (Shukla, 1987), when the north Indian Ocean is anomalously warming during post-El Niño summers (e.g., Mishra et al., 2012; Chakravorty et al.,

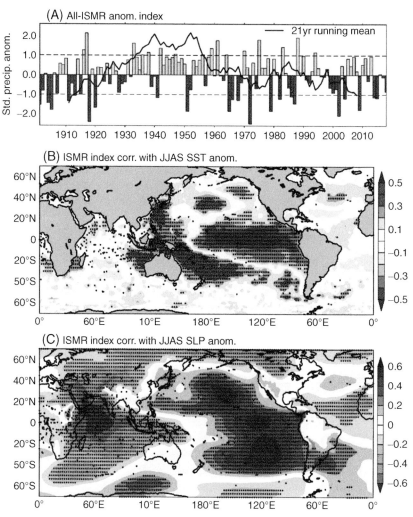

FIG. 1.5 (A) Standardized all India JJAS precipitation time series (bars) for the period of 1901–2019 and 21-year running mean (black line). (B) Correlation of all India JJAS rainfall time series with SST (HadISST1.1) and (C) same as in (B) but for SLP (NOAA-20CRv3) anomalies over the globe. Stippling indicates the relationship exceeds the 95% confidence level using a two-tailed t-test. Rainfall data from IMD, NOAA- twentieth-century reanalysis (20CRv3; Compo et al. 2011) of 2° × 2° resolution mean sea level pressure (SLP; 1901–2014), and Hadley center sea ice surface and sea surface temperature (*SST*) (HadISST1.1; Rayner et al., 2003) data, ranges from years 1901 to 2016 are utilized.

2016; Chowdary et al., 2017; Zhou et al., 2019). This suggests that the seasonal mean ISM rainfall is influenced by both concurrent and antecedent ENSO conditions. Mishra et al. (2012) have identified a prominent pattern of the year-to-year variability of ISM rainfall anomalies with a dipole structure between the Gangetic Plain and southern peninsular India. They showed that this dipole pattern in ISM rainfall is related to a well-defined pattern of SST anomalies over the Arabian Sea, Bay of Bengal, and South China Sea, reminiscent of those in the post-ENSO summers. Proxy data such as a regional tree-ring cellulose oxygen isotope ($\delta^{18}O$) record for the northern Indian subcontinent since 1820 exhibits significant interannual changes (*Chapter 7*), which are closely related to ENSO in the past (e.g., Xu et al., 2018; Chakraborty et al., 2012).

Central Pacific El Niño or El Niño Modoki events have a significant influence on ISM rainfall variability (e.g., Ashok et al., 2007, Krishna Kumar et al., 2006). Apart from ENSO, the Pacific Decadal Oscillation (PDO; *Chapter 19*) or Interdecadal Pacific Oscillation (IPO) has a significant impact on ISM rainfall variability at various time scales (e.g., Krishnan and Sugi, 2003; Krishnamurthy and Krishnamurthy, 2014; Joshi and Kucharski, 2017; Malik et al., 2017). Some studies suggested that the Atlantic Niño could influence the year-to-year variability of ISM rainfall (*Chapter 10*) through equatorial Kelvin waves (Kucharski et al., 2008) and by modulating the Asian jet (Yadav et al., 2018). Various effects of Indian Ocean SST variability on ISM have also been reported. Krishnan and Swapna (2009) suggest the presence of positive feedbacks between positive Indian Ocean Dipole (IOD) (Saji et al., 1999) events and ISM and Ashok et al. (2001, 2004) suggest that the influence of IOD on the ISM is opposite to the effect of ENSO (*Chapter 8*). Note that the relationship between ENSO, IOD, and the monsoon is not constant in time (e.g., Cherchi and Navarra, 2013). Several studies also showed that, in addition to ENSO, EIO oscillation (EQUINOO) also plays an important role in the interannual variation of ISM rainfall (e.g., Gadgil et al., 2004, Ihara et al., 2007) and the EQUINOO could be considered to be the atmospheric component of IOD though not every EQUINOO event corresponds to an IOD event. Gadgil et al. (2004) noted that ISM rainfall deficient and excess years are well separated in the phase-plane of EQUINOO and ENSO index for the last 6 decades (Chapter 2). Further, the El Niño induced TIO basin-wide warming directly affects the ISM rainfall (Yang et al., 2007; Park et al., 2010; Chowdary et al., 2015). Recent studies also highlighted the importance of Western North Pacific low-level circulation changes, such as the Pacific-Japan pattern in modulating the ISM rainfall on the interannual time scale (e.g., Srinivas et al., 2018). The Indo-Western Pacific Capacitor (Xie et al. 2016) which involves interbasin interaction between Western North Pacific and the north Indian Ocean also strongly influences the ISM interannual variability (e.g., Chowdary et al., 2019; Zhou et al., 2019; Gnanaseelan and Chowdary, 2020). Modulations in the Silk Road pattern, which is a teleconnection pattern along with the Asian jet in summer, could potentially influence the variation in monsoon rainfall (e.g., Wang et al., 2017; Kosaka et al., 2012; *Chapter 13-14*).

Several other modes or systems such as the Southern annular mode, changes in subtropical deserts, south Asian high, etc., (*Chapter 15 – 17*) also have some remote influence on ISM rainfall (e.g., Sooraj et al., 2019, Wei et al., 2019; Prabhu et al., 2017). However, most of these teleconnections to ISM rainfall are not stable in time (e.g., Krishna Kumar et al., 1999; Darshana et al., 2020). Thus, it is necessary to understand how these teleconnection patterns altered and modulating rainfall in multidecadal and climate change perspectives.

1.5 Decadal variability and climate change

1.5.1 Decadal variability

It is also known that the monsoon exhibits variability even on interdecadal time scales in association with other global climate variables. Variations in ISM rainfall, characterized by distinct epochs of above and below normal monsoon activity (Fig. 1.6A) typically lasting for about three decades (e.g., Kripalani

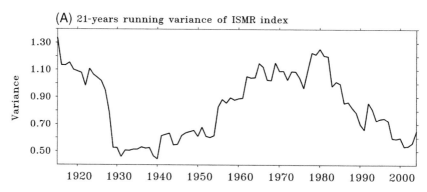

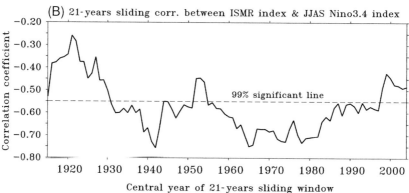

FIG. 1.6 (A) Twenty-one years running variance of all Indian summer monsoon (*JJAS*) rainfall time series (mm/day)^2 and (B) 21-year sliding correlation between IMD ISM rainfall and Niño 3.4 index (HadISST1.1) for the period of 1901–2016. The dashed line indicates 99% confidence level. *ISM*, Indian summer monsoon.

et al., 1997). It also noted that the variability in rainfall increases during the dry epochs and decreases during the wet epochs (Pant and Rupakumar, 1997), which is apparent in a 21-year moving variance of rainfall anomaly (Fig. 1.5A). Based on different proxies of 500 years of data, Goswami et al. (2015) suggested that that the Asian monsoon has a multidecadal oscillation with the period between 50 years and 80 years that change in time in an episodic manner. The PDO, Atlantic Multidecadal Oscillation (AMO), and external climate forcings, i.e., greenhouse gases (GHGs), volcanic eruptions, and total solar Irradiance are likely contributing to the decadal to multidecadal scale variability of the ISM rainfall (e.g., Malik et al., 2017). Krishnamurthy and Krishnamurthy (2014) suggested that the warm (cold) phase of the PDO is associated with the deficit (excess) summer rainfall over India and that the PDO modified the relationship between ISM and ENSO. During the boreal winter, the low pressure associated with PDO warm phase generates an SST anomaly in the subtropics and persists into the next boreal summer (Krishnamurthy and Krishnamurthy, 2014). They found that these anomalies in summer affect the equatorial winds which reinforce the equatorial Walker circulation with enhanced ascending motion in the east and central Pacific and subsidence over the Maritime Continent. Sankar et al. (2016) suggested that both ISM rainfall and North Atlantic SSTs display multidecadal variability with a period close to 60 years. They noted that the periods of warm (cold) North Atlantic SSTs are accompanied by periods of wetter (dryer) ISM rainfall and lower (higher) frequencies of dry years. Interdecadal/epochal modulation of ISM rainfall associated with ENSO is apparent in Fig. 1.6B. Note that the Niño 3.4 index is calculated as SST anomalies averaged in 5°S–5°N, 170°–120°W. Several earlier studies identified the interdecadal variability of the relationship between ENSO and ISM rainfall (e.g., Krishnamurthy and Goswami, 2000; Krishna Kumar et al., 1999). The presence of a see-saw between ENSO–ISM relationship is also connected to ENSO-West African Monsoon relationship (Srivastava et al., 2019). Regional rainfall patterns over India during summer are highly influenced by ENSO-ISM teleconnections modulations on a decadal time scale (e.g., Mahendra et al., 2020).

Studies identified that the positive AMO induces warm tropospheric temperature (TT) anomalies over Eurasia, strengthens the meridional gradient of TT, and influences ISM rainfall (Goswami et al., 2006; Joshi and Ha, 2019). Further, this TT warming over Europe and western Asia could excite a Rossby wave train from the North Atlantic to reach South Asia, which increases the meridional TT gradient between Eurasia and the Indian Ocean and enhances the ISM rainfall (e.g., Li et al. 2008) (see *Chapter 18*). Malik and Brönnimann (2018) argued that AMO is a potential source of the nonstationary relationship between ISM rainfall and ENSO. Large coherence between Asian monsoon multidecadal mode and AMO, ENSO, and PDO suggests that these modes all are an integral part of a global multidecadal mode with a periodicity of 50–80 years (e.g., Goswami et al., 2015). Niranjan Kumar et al. (2013) suggested there is a general increase in the intensity and percentage of areas affected by moderate droughts during the recent decades. This indicates that low-frequency modulation of the summer monsoon could largely influence rainfall over the Indian subcontinent.

1.5.2 Climate change

Understanding and reporting precipitation changes associated with climate change over the monsoon-dominated regions is one of the prime targets because of its direct impact on their livelihoods. Previous studies reported that the trends in the observed ISM rainfall over the Indian subcontinent, based on more than 100 years of data (1901–2015), showed a decreasing trend over the central-east and northern regions, and parts of the Western Ghats region (e.g., Roxy et al., 2015). Krishnan et al. (2016) showed that the regional land-use changes, anthropogenic-aerosol forcing, and the rapid warming signal of the EIO is mainly responsible for the observed monsoon weakening in recent decades. However, in some regions like Gujarat–Konkan coast, and extreme north India displayed significant increasing trend rainfall during the summer season (e.g., Kulkarni et al., 2020). These studies clearly suggest that the trends in precipitation over India exhibit large spatial variability. However, based on the coupled models' future projections, it has been reported that the mean ISM rainfall over India increases (e.g., Douville et al., 2000; Ueda et al., 2006; Rajendran et al., 2012; Kulkarni et al., 2020). This rainfall increase is attributed to enhanced atmospheric moisture content over the warmer Indian Ocean, and associated increased vertically integrated moisture fluxes toward India (e.g., May et al., 2002; Lee and Wang, 2014; Sharmila et al., 2015; Krishnan et al., 2016; Cherchi et al., 2011; Krishnan et al., 2013). Based on a set of 12 General Circulation Models (GCMs)–RCM combinations over the Coordinated Regional Climate Downscaling Experiment–South Asia (CORDEX-SA) domain, Maharana et al. (2020) suggested that the increasing land–sea temperature difference is strengthening the LLJ, which brings in moisture toward India with the rise of global temperature from 1.5 °C. Kitoh (2017) reported in future projections of Coupled Model Intercomparison Project (CMIP) version CMIP5 and CMIP3 models that increase in monsoon precipitation in South Asia, albeit with weakened circulation. Li et al. (2017) suggested that the land–sea thermal contrast intensifies around the south Asian Monsoon region in the future due to a reduction in wintertime snowfall over southwest Eurasia based on the analysis of CMIP5 models. They noted that the projected South Asia summer monsoon precipitation increase appears to be primarily driven by the thermodynamic response of water vapor flux; while the dynamic response largely determines the spatial pattern of precipitation change. However, the concept of land-ocean contrast in driving the monsoon is still debatable (Gadgil et al. 2021).

Studies have reported increased interannual variability of South Asian summer monsoon in the future owing to increased ENSO amplitude (e.g., Kitoh, 1997; Turner and Slingo, 2007; *Chapter 20*). The more frequent El Niño events in the future could trigger more droughts over India in response to greenhouse warming and maintain a stable inverse relationship between El Niño and ISM rainfall (e.g., Azad and Rajeevan, 2016; Kulkarni et al., 2020). The warmer SST, especially in the tropical Pacific, enhances changes in evaporation, which, in turn, intensifies the variability of ISM (e.g., Meehl and Arblaster, 2003). Based on a subset of CMIP5 models, Roy et al. (2019) reported that canonical and mixed canonical Modoki ENSO events teleconnections to ISM rainfall is stronger and spatially extended over most of India in the future and at the same time pure Modoki ENSO teleconnection to ISM rainfall have weakened/ shown no influence. Bhowmick et al. (2019) reported that over Indian land, the frequency of occurrence of daily precipitation extremes shows up to a three-fold increase under both RCP scenarios for global warming levels in the range of 1.5 °C–2.5 °C. Changes in frequency of ENSO and IOD and local changes in land use-land cover, soil moisture, and evapotranspiration (*Chapter 21 and 23*) would alter the extremes in ISM rainfall under global warming (Cai et al., 2014; Azad and Rajeevan, 2016).

Besides, to offset GHGs effects, aerosols are suggested to have competing influences at regional scales (e.g, Mascioli et al., 2016; Undorf et al., 2018). More recently, it was suggested that aerosol reductions will lead to a significant increase in precipitation extremes over Asia (e.g., Wang et al., 2019). By comparing CMIP5 historical experiments with those containing only a single varying forcing experiment (aerosol only and GHG only), Guo et al. (2015) attributed the decreasing trend of the ISM rainfall to aerosol, while, when only GHG forcing is used, rainfall increased. Ramanathan et al. (2005) suggested that aerosols had limited the surface warming over India related to GHG increases, reducing monsoon rainfall. Bollasina et al. (2011) suggested that the observed precipitation decrease can be attributed mainly to human-influenced aerosol emissions during the second half of the 20th century (*Chapter 22*). Polson et al. (2014) used a subset of CMIP5 historical single-forcing experiments to attribute the reduced Northern Hemisphere monsoon precipitation to increased aerosol emissions. Recently based on the IITM Earth System Model version2 (IITM-ESMv2) experiments Ayantika et al. (2021) revealed that over the Asian continent, aerosol-induced atmospheric absorption results in surface radiation reduction and warm above the boundary layer. This could stabilize the lower troposphere and weakens the monsoon south-westerly winds. Kim et al. (2016) noted that absorbing aerosol, particularly desert dust can strongly modulate ENSO influence, and possibly play important roles as a feedback agent in climate change in Asian monsoon regions. Pre-Monsoon aerosol loading could influence ISM rainfall during the El Niño years (e.g., Fadnavis et al., 2017).

Precipitation changes over the ISM region associated with global warming in future projections are considered based on a subset of the CMIP6 models (Eyring et al., 2016). Details of the models used in this Chapter are provided in Table 1.1. Only the first realization for each model is utilized. Monthly simulations of 15 GCMs data are used to estimate the ISM features in the historical

TABLE 1.1 Details of the CMIP6 models used in this chapter.

Modeling center (or group)	Model name	Grid size
Commonwealth Scientific and Industrial Research Organization (CSIRO) and Bureau of Meteorology (BOM), Australia	ACCESS-CM2	192 × 144
	ACCESS-ESM1-5	192 × 144
Beijing Climate Center, China Meteorological Administration, China	BCC-ESM2-MR	320 × 160
Chinese Academy of Meteorological Sciences, China	CAMS-CSM1-0	320 × 160
Canadian Centre for Climate Modeling and Analysis, Canada	CanESM5	128 × 64
National Center for Atmospheric Research, USA	CESM2-WACCM	288 × 192
Chinese Academy of Sciences, China	FGOALS-g3	180 × 80
The First Institute of Oceanography, China	FIO-ESM-2-0	288 × 192
Atmosphere and Ocean Research Institute (The University of Tokyo), National Institute for Environmental Studies, and Japan Agency for Marine-Earth Science and Technology, Japan	MIROC6	256 × 128
Max Planck Institute for Meteorology, Germany	MPI-ESM1-2-LR	192 × 96
	MPI-ESM1-2-HR	384 × 192
Meteorological Research Institute, Japan	MRI-ESM2-0	320 × 160
Norwegian Climate Centre, Norway	NorESM2-LM	144 × 96
	NorESM2-MM	288 × 192
Nanjing University of Information Science and Technology, China	NESM3	192 × 96

period of 1985–2014 and estimated the same for the far future period of 2071–2100 under the shared socioeconomic pathways (SSPs) of high-forcing emission scenario (SSP5-8.5). All the models used in the present study have been interpolated to a uniform grid resolution of $1° \times 1°$ using the bilinear interpolation method for making them consistent to compute the multimodel means. Future projections based on 15 CMIP6 models show a significant increase in ISM rainfall under SSP5-8.5 scenario (Fig. 1.7). This indicates the importance of global warming in modulating mean monsoon rainfall (see *Chapter 20*). Strong low-level easterlies over the EIO and southwesterlies over the Northern Arabian Sea are noted in future projections (Fig. 1.7E). This strong LLJ over the Arabian Sea generally favours enhanced rainfall over the Indian Subcontinent in summer. Further, low-level convergence over India as a result of an anomalous

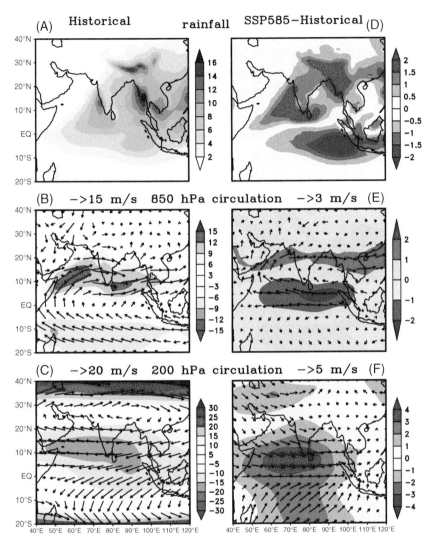

FIG. 1.7 (A) Historical mean precipitation (mm/day) JJAS climatology based on 15 CMIP6 models, (B) same as in (A) but for 850 hPa winds (vectors; m/s) and zonal wind (shaded; m/s), (C) same as in (A) but for 200 hPa winds (vectors; m/s) and zonal wind (shaded; m/s), (D) difference between SSP5-8.5 and historical mean precipitation during summer (JJAS), (E) same as in (D) but for 850 hPa winds (vectors; m/s) and zonal wind (shaded; m/s) and (F) same as in (D) but for 200 hPa winds (vectors; m/s) and zonal wind (shaded; m/s). The multi-model mean (*MMM*) is based on 15 models for historical period of 1985–2014 and the far future period of 2071–2100 under SSP5-8.5.

anticyclone over the Bay of Bengal and southwesterlies from the Arabian Sea clearly suggests the importance of low-level circulation changes in enhancing ISM rainfall in future projections. However, the strength of the easterly jet at 200 hPa is weaker under SSP5-8.5 as compared to the present. Apart from CMIP6 models, IITM-ESM also projects an increase in rainfall over the Indian region under both SSP5-8.5 and SSP2-4.5 scenarios (Krishnan et al. 2020). They reported that the far future precipitation increase over the Indian region is 0.51 mm day^{-1} under SSP5-8.5 and 0.03 mm/day under SSP2-4.5 as projected by IITM-ESM. Several other factors that influence the ISM rainfall in changing climate are discussed *in Chapter 20.*

1.6 Summary

Predictability of ISM originates from physical drivers, for example, IMD issues statistical forecast for all India Summer monsoon rainfall based on different parameters such as SST anomalies from the North Atlantic region, South East EIO and Niño 3.4 region, surface pressure anomalies from East Asia and North Atlantic and 850 hPa zonal wind anomaly from North Central Pacific region (Rajeevan et al. 2007). This collection of predictors clearly indicates the importance of remote forcing on ISM variability/predictability. In the recent decade, advances in the development of state of art coupled GCMs opened the way for dynamical prediction. With the ongoing development of dynamical atmosphere-ocean GCMs, expectations regarding seasonal forecasts of ISM are high. Despite the improvements in these models over the last two decades, the quality of their prediction of ISM, especially in operational mode, is still a matter of debate. The dynamical prediction skill originates mainly from the fact that monsoon strength is tightly linked to teleconnections. Current coupled models have limited skills in predicting teleconnections to ISM.

Understanding the spatio-temporal variations of the monsoon is an important, and challenging problem. High-frequency variability (within a season) is dominated by the active and break cycles involving Intraseasonal Oscillations, monsoon depressions, and MJOs. Low-frequency variations (on interannual or longer scales) of ISM are highly influenced by various drivers such as ENSO, Indian Ocean climate variability, Eurasian snow cover changes, the PDO, the AMO. On the other hand, the long-term variations are dominated by anthropogenic climate change influence.

This book highlights the recent advances in understanding various teleconnections that influence and drive ISM variability and change. In particular, physical mechanisms that link various global drivers to ISM variability are emphasized using a wide range of datasets, modern techniques, and models. Fig. 1.8 shows some of the possible global drivers with considerable impact on ISM prediction/variability; they include internal variability of the climate system and anthropogenic climate change. Understanding the actual physical

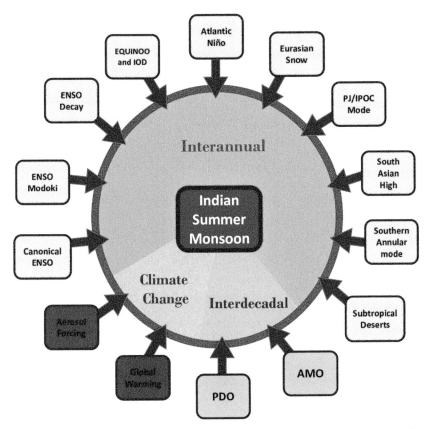

FIG. 1.8   Some possible climate drivers for ISM variability from interannual to decal time scale. *ISM*, Indian summer monsoon.

mechanisms that link these drivers to ISM is still a matter of serious research. Thus, an improved understanding of these drivers and their teleconnective effects on ISM variability is the way forward for further improvements in monsoon prediction.

Acknowledgments

J.S.C. wishes to thank Ministry of Earth Sciences, Government of India for support. The CMIP data used in this study are freely available through the Earth System Grid Federation (https://esgf.nci.org.au/) and GPCP data at https://www.esrl.noaa.gov/psd/data/gridded/data.gpcp.html. All CMIP6 simulations used are also freely available from the Earth System Grid Federation (https://esgf-node.llnl.gov/search/cmip6/). We thank Darshana P., Amol V., Dr. Srinivas G., and Dr. Koteswararao K. for helping in figures preparations.

References

Adames, Á.F., Kim, D., 2016. The MJO as a dispersive, convectively coupled moisture wave: theory and observations. J. Atmos. Sci. 73, 913–941. https://doi.org/10.1175/JAS-D-15-0170.1.

Adler, R.F., Huffman, G.J., Chang, A., Ferraro, R., Xie, P., Janowiak, J., Rudolf, B., Schneider, U., Curtis, S., Bolvin, D., Gruber, A., Susskind, J., Arkin, P., Nelkin, E., 2003. The version 2 Global Precipitation Climatology Project (GPCP) monthly precipitation analysis (1979-present). J. Hydrometeor. 4 (6), 1147–1167.

Annamalai, H., Hamilton, K., Sperber, K.R., 2007. The South Asian summer monsoon and its relationship with ENSO in the IPCC AR4 simulations. J. Clim. 20 (6), 1071–1092. https://doi.org/10.1175/JCLI4035.1.

Ashok, K., Chan, W-L., Motoi, T., Yamagata, T., 2004. Decadal variability of the Indian Ocean dipole. Geophys. Res. Lett. 31, L24207. https://doi.org/10.1029/2004GL021345.

Ashok, K., Feba, F., Tejavath, C.T., 2019. The Indian summer monsoon rainfall and ENSO. Mausam 70, 443–452.

Ashok, K., Guan, Z.Y., Yamagata, T., 2001. Impact of the Indian Ocean Dipole on the relationship between the Indian monsoon rainfall and ENSO. Geophys. Res. Lett. 28 (23), 4499–4502. doi:10.1029/2001gl013294.

Ashok, K., Saji, N.H., 2007. On the impacts of ENSO and Indian Ocean Dipole events on subregional Indian summer monsoon rainfall. Nat. Hazards. 42 (2), 273–285. https://doi.org/10.1007/s11069-006-9091-0.

Asnani, G.C., 1973. Origin of the Indian summer monsoon. Vayu Mandal 3 (105-106), 120.

Ayantika, D.C., Krishnan, R., Singh, M., et al., 2021. Understanding the combined effects of global warming and anthropogenic aerosol forcing on the South Asian monsoon. Clim Dyn. doi:10.1007/s00382-020-05551-5.

Azad, S., Rajeevan, M., 2016. Possible shift in the ENSO-Indian monsoon rainfall relationship under future global warming. Sci. Rep. 6, 20145.

Berkelhammer, M., Sinha, A., Stott, L., Cheng, H., Pausata, F.S.R., Yoshimura, K., 2012. An abrupt shift in the Indian monsoon 4000 Years ago. In: L., Giosan et al. (Eds.), Climates, Landscapes, and Civilizations: American Geophysical Monograph 198, pp. 75–87. American Geophysical Union Geophysical Monograph. https://doi.org/10.1029/2012GM001207.

Bhowmick, M., Sahany, S., Mishra, S.K., 2019. Projected precipitation changes over the South Asian region for every 0.5C increase in global warming. Environ. Res. Lett. 14 (5), 054005.

Bollasina, M., Ming, Y., Ramaswamy, V., 2011. Anthropogenic aerosols and the weakening of the South Asian summer monsoon. Science 334 (6055), 502–505.

Cai, W., Santoso, A., G.J., Wang, E., Weller, L., Wu, K., Ashok, Y., Masumoto, Yamagata, T., 2014. Increased frequency of extreme Indian Ocean Dipole events due to greenhouse warming. Nature 510, 254–258. doi:10.1038/nature13327.

Chakravorty, S., Gnanaseelan, C., Pillai, P.A., 2016. Combined influence of remote and local SST forcing on Indian Summer Monsoon Rainfall variability. Clim. Dyn. 47, 2817–2831.

Chakraborty, S., Goswami, B.N., Dutta, K., 2012. Pacific coral oxygen isotope and the tropospheric temperature gradient over Asian monsoon region: a tool to reconstruct past Indian summer monsoon rainfall. J. Quat. Sci. 27 (3), 269–278. doi:10.1002/jqs.1541.

Charney, J.G., 1969. The intertropical convergence zone and the hadley circulation of the atmosphere. Proc. WMO/IUCG Symp. Numer. Weather Predict. Jpn. Meteorol, 73–79 Agency III.

Charney, J.G., Shukla, J., 1981. Predictability of monsoons. In: Lighthill, S.J., Pearce, R.P. (Eds.), Monsoon Dynamics. Cambridge University Press, New York, NY, pp. 99–109.

Cherchi, A., Alessandri, A., Masina, S., Navarra, A., 2011. Effects of increased CO_2 levels on monsoons. Clim. Dyn. 37, 83–110.

Cherchi, A., Navarra, A., 2013. Influence of ENSO and of the Indian Ocean Dipole on the Indian summer monsoon variability. Clim. Dyn. 41 (1), 81–103.

Choudhury, A.D., Krishnan, R., Ramarao, M.V.S., et al., 2018. A phenomenological paradigm for midtropospheric cyclogenesis in the Indian Summer monsoon. J. Atmos. Sci. 75, 2931–2954. https://doi.org/10.1175/JAS-D-17-0356.1.

Chowdary, J.S., Bandgar, A., Gnanaseelan, C., Luo, J-J., 2015. Role of tropical Indian Ocean air–sea interactions in modulating Indian summer monsoon in a coupled model. Atmos. Sci. Lett. 16, 170–176. https://doi.org/10.1002/asl2.561.

Chowdary, J.S., Harsha, H.S., Gnanaseelan, C., Srinivas, G., Parekh, A., Pillai, P.A., Naidu, C.V., 2017. Indian summer monsoon rainfall variability in response to differences in the decay-phase of El Niño. Clim. Dyn. 48, 2707–2727. https://doi.org/10.1007/s00382-016-3233-1.

Chowdary, J.S., Patekar, D., Srinivas, G., Gnanaseelan Parekh, A., 2019. Impact of the Indo-Western Pacific Ocean capac-itor mode on South Asian summer monsoon rainfall. Clim. Dyn. 53, 2327–2338. https://doi.org/10.1007/s00382-019-04850-w.

Compo, G.P., et al., 2011. The twentieth century reanalysis project. Q. J. R. Meteor. Soc. 37 (654), 1–28.

Darshana, P., Chowdary, J.S., Gnanaseelan, C., Parekh, A., Srinivas, G., 2020. Interdecadal modula-tion of the Indo-western Pacific Ocean Capacitor mode and its influence on Indian summer monsoon rainfall. Clim. Dyn. 54 (3), 761–1777 10.1007/s00382-019-05085-5.

Das, P.K., 1968. The Monsoons. National Book Trust, New Delhi.

Dee, D.P., et al., 2011. The ERA-interim reanalysis: configuration and performance of the data as-similation system. Q. J. R. Meteorol. Soc. 137 (656), 553–597. https://doi.org/10.1002/qj.828.

DeMott, C.A., Klingaman, N.P., Woolnough, S.J., 2015. Atmosphere ocean coupled processes in the Madden Julian oscillation. Rev. Geophys. 53 (4), 1099–1154.

Douville, H., et al., 2000. Impact of CO_2 doubling on the Asian summer monsoon: Robust versus model-dependent responses. J. Meteorol. Soc. Jpn. 78, 421–439.

Eyring, V., Bony, S., Meehl, G.A., Senior, C.A., Stevens, B., Stouffer, R.J., Taylor, K.E., 2016. Overview of the Coupled Model Intercomparison Project Phase 6 (CMIP6) experimental design and organization. Geosci. Model Dev. 9, 1937–1958.

Fadnavis, S., Roy, C., Sabin, T.P., Ayantika, D.C., Ashok, K., 2017. Potential modulations of pre-monsoon aerosols during El Niño: impact on Indian summer monsoon. Clim Dynam 49 (7-8), 2279–2290.

Findlater, J., 1978. Observational aspects of the low-level cross-equatorial jet stream of the western Indian OceanMonsoon Dynamics. Birkhäuser, Basel, pp. 1251–1262.

Gadgil, S., 2007. The Indian Monsoon. Resonance 12 (5), 4–20.

Gadgil, S., Francis, P.A., Rajendran, K., Nanjundiah, R.S., Suryachandra, A.R., et al., 2021. Role of land-ocean contrast in the Indian summer monsoon rainfall. In: Chang, C.P. et al. (Eds.), The Multiscale Global Monsoon System. World Scientific Publishing Co, pp. 3–12. https://doi.org/10.1142/9789811216602_0001.

Gadgil, S., Vinayachandran, P.N., Francis, P.A., Gadgil, S., 2004. Extremes of the Indian summer monsoon rainfall, ENSO and equatorial Indian Ocean oscillation. Geophys. Res. Lett. 31 (12), L12213. https://doi.org/10.1029/2004GL019733.

Gnanaseelan, C., Chowdary, J.S., 2019. The Indo-Western Pacific climate variability and the impacts on India summer monsoon: two decades of advancement in India. Mausam 70, 731–752.

Goswami, B.N., Ajayamohan, RS, 2001. Intraseasonal Oscillations and Interannual Variabil-ity of the Indian Summer Monsoon. J. Clim. 14 (6), 1180–1198. doi:10.1175/1520-0442(2001)0142.0.co;2.

Goswami, B.N., Shukla, J., 1984. Quasi-periodic oscillations in a symmetric general circulation model. J. Atmos. Sci. 41, 20–37.

Goswami, B.N., Kripalani, R.H., Borgaonkar, H.P., Preethi, B., 2015. Multi-decadal variability in Indian summer monsoon rainfall using proxy data. In: Chih-Pei, C., Michael, G., Mojib, L., Wallace, J.M. (Eds.), Climate Change: Multidecadal and Beyond. World Scientific Publishing Co, Singapore. https://doi.org/10.1142/9789814579933_0021.

Goswami, B.N., Xavier, P.K., 2005. Dynamics of 'Internal' interannual variability of the Indian summer monsoon in a GCM. J. Geophys. Res. 110, D24 10.1029/2005jd006042.

Goswami, B.N., Ajayamohan, R.S., Xavier, P.K., Sengupta, D., 2003. Clustering of synoptic activity by Indian summer monsoon intraseasonal oscillations. Geophys. Res. Lett. 30 (8), 1431. https://doi.org/10.1029/2002GL016734.

Goswami, B.N., Madhusoodanan, M.S., Neema, C.P., 2006. A physical mechanism for North Atlantic SST influence on the Indian summer monsoon. Geophys. Res. Lett. 33, L02706.

Guo, L., Turner, A.G., Highwood, E.J., 2015. Impacts of 20th century aerosol emissions on the South Asian monsoon in the CMIP5 models. Atmos. Chem. Phys. 15, 6367–6378. https://doi.org/10.5194/acp-15-6367-2015.

Hastenrath, S., 1995. Climate Dynamics of the Tropics. Kluwer Academic Publishers. Dordrecht, Boston, London, second printing, p. 488.

Ihara, C., Kushnir, Y., Cane, M.A., De la Pena, V.H., 2007. Indian summer monsoon rainfall and its link with ENSO and Indian Ocean climate indices. Int. J. Clim. 27 (2), 179–187.

Jiang, X., Adames, F., Zhao, M., Waliser, D., Maloney, E., 2018. A unified moisture mode framework for seasonality of the Madden Julian oscillation. J. Clim. 31 (11), 4215–4224. https://doi.org/10.1175/JCLI-D-17-0671.1.

Joseph, P.V., Raman, P.L., 1966. Existence of low level westerly jet stream over peninsular India during July. Indian J. Meteorol. Geophys. 17 (1), 407–410.

Joseph, P.V., Sooraj, K.P., Rajan, C.K., 2006. The summer monsoon onset process over South Asia and an objective method for the date of monsoon onset over Kerala. Int. J. Clim. 4, 1549–1555.

Joshi, M.K., Ha, K., 2019. Fidelity of CMIP5-simulated teleconnection between Atlantic multidecadal oscillation and Indian summer monsoon rainfall. Clim Dynam 52, 4157–4176. https://doi.org/10.1007/s00382-018-4376-z.

Joshi, M.K., Kucharski, F., 2017. Impact of Interdecadal Pacific Oscillation on Indian summer monsoon rainfall: an assessment from CMIP5 climate models. Clim. Dynam. 48, 2375–2391.

Karmakar, N., Krishnamurti, T.N., 2019. Characteristics of northward propagating intraseasonal oscillation in the Indian summer monsoon. Clim. Dyn. 52, 1903–1916. https://doi.org/10.1007/s00382-018-4268-2.

Karmakar, N., Chakraborty, A., Nanjundiah, R.S., 2017. Space-time evolution of the low- and high-frequency intraseasonal modes of the Indian summer monsoon. Mon. Weather Rev. 145 (2), 413–435. https://doi.org/10.1175/MWR-D-16-0075.1.

Keshavamurty, R.N., Sankar Rao, M., 1992. The Physics of Monsoons. New Delhi: Allied Publishers.

Kim, M.K., Lau, W.K., Kim, K.M., Sang, J., Kim, Y.H, Lee, W.S., 2016. Amplification of ENSO effects on Indian summer monsoon by absorbing aerosols. Clim. Dyn. 46 (7-8), 2657–2671.

Kitoh, A., 2017. The Asian monsoon and its future change in climate models: a review. J. Meteorol. Soc. Jpn. Ser. II 95 (1), 7–33.

Kitoh, A., Yukimoto, S., Noda, A., Motoi, T., 1997. Simulated changes in the Asian summer monsoon at times of increased atmospheric CO2. J. Meteorol. Soc. Jpn 75, 1019–1031.

Kosaka, Y., Chowdary, J.S., Xie, S.P., Min, Y.M., Lee, J.Y., 2012. Limitations of seasonal predictability for summer climate over East Asia and the Northwestern Pacific. J. Clim. 25 (21), 7574–7589.

Koteswaram, P., 1958. The easterly jet stream in the tropics. Tellus 10 (1), 43–57 10.1111/j.2153-3490.1958.tb01984.x.

Kripalani, R.H., Kulkarni, A., 1997. Rainfall variability over South-East Asia—connections with Indian Monsoon and ENSO extremes: new perspectives. Int. J. Clim. 17 (11), 155–1168 10.1002/(sici)1097-0088(199709)17:113.0.co;2-b.

Kripalani, R.H., Kulkarni, A., Singh, S.V., 1997. Association of the Indian summer monsoon with the Northern Hemisphere mid-latitude circulation. Int. J. Clim. 17, 1055–1067. https://doi.org/10.1002/(SICI)1097-0088(199708)17:10%3c1055:AID-JOC180%3e3.0.CO;2-3.

Krishna Kumar, K., Rajagopalan, B., Cane, M.A., 1999. On the weakening relationship between the Indian monsoon and ENSO. Science 284 (5423), 2156–2159 10.1126/science.284.5423.2156.

Krishna Kumar, K., Rajagopalan, B., Hoerling, M., Bates, G., Cane, M., 2006. Unraveling the mystery of Indian monsoon failure during El Niño. Science 314, 115–119. https://doi.org/10.1126/science.1131152.

Krishnamurthy, L., Krishnamurthy, V.J.C.D., 2014. Influence of PDO on South Asian summer monsoon and monsoon–ENSO relation. Clim. Dyn. 42 (9-10), 397–2410.

Krishnamurthy, V., Ajayamohan, R.S., 2010. Composite structure of monsoon low pressure systems and its relation to Indian rainfall. J. Clim 23, 4285–4305.

Krishnamurthy, V., Goswami, B.N., 2000. Indian monsoon–ENSO relationship on interdecadal timescale. J. Clim. 13 (3), 579–595.

Krishnamurti, T.N., Ardanuy, P., 1980. The 10 to 20-day westward propagating mode and "breaks in the monsoons". Tellus 32, 15–26. https://doi.org/10.1111/j.2153-3490.1980.tb01717.x.

Krishnamurti, T.N., Bhalme, H.N., 1976. Oscillations of a monsoon system. Part I. Observational Aspects. J. Atmos. Sci. 33 (10), 1937–1954. doi:10.1175/15200469(1976)0332.0.co;2.

Krishnamurti, T.N., Subrahmanyam, D., 1982. The 30–50 Day Mode at 850 Mb During MONEX. J. Atmos. Sci. 39 (9), 2088–2095. doi:10.1175/1520-0469(1982)0392.0.co;2.

Krishnamurti, T.N, Ardanuy, P., Ramanathan, Y., Pasch, R., 1981. On the onset vortex of the summer monsoon. Mon. Weather Rev. 109, 344–363. doi:10.1175/1520-0493(1981)109<0344:OTO VOT>2.0.CO;2.

Krishnamurti, T.N., 1985. Numerical weather prediction in low latitudesAdvances in Geophysics, 28B. Elsevier, pp. 283–333.

Krishnan, R., Gnanaseelan, C., Sanjay, J., et al., 2020. Introduction to climate change over the Indian Region. In: Krishnan, R., Sanjay, J., Gnanaseelan, C. et al (Eds.), BT-assessment of climate change over the Indian region: a report of the Ministry of Earth Sciences (MoES), Government of India. Springer, Singapore, pp. 1–20.

Krishnan, R., Sabin, T.P, Vellore, R., Mujumdar, M., Sanjay, J., Goswami, B.N., Hourdin, F., Dufresne, J.L., Terray, P., 2016. Deciphering the desiccation trend of the South Asian monsoon hydroclimate in a warming world. Clim. Dyn. 47 (3–4), 1007–1027.

Krishnan, R., Sabin, T.P., Ayantika, D.C., Kitoh, A., Sugi, M., Murakami, H., Turner, A.G., Slingo, J.M., Rajendran, K., 2013. Will the South Asian monsoon overturning circulation stabilize any further? Clim. Dyn. 40, 187–211. https://doi.org/10.1007/s00382-012-1317-0.

Krishnan, R., Sabin, T.P., Vellore, R., Mujumdar, M., Sanjay, J., Goswami, B.N., Hourdin, F., Dufresne, J.L., Terray, P., 2016. Deciphering the desiccation trend of the South Asian monsoon hydroclimate in a warming world. Clim. Dyn. 47 (3-4), 1007–1027.

Krishnan, R., Sugi, M., 2003. Pacific decadal oscillation and variability of the Indian summer monsoon rainfall. Clim. Dyn. 21 (3-4), 233–242 10.1007/s00382-003-0330-8.

Krishnan, R., Swapna, P., 2009. Significance influence of the boreal summer monsoon flow on the Indian Ocean response during dipole events. J. Clim. 22, 5611–5634.

Krishnan, R., Zhang, C., Sugi, M., 2000. Dynamics of breaks in the Indian summer monsoon. J. Atmos. Sci. 57 (9), 1354–1372. https://doi.org/10.1175/1520-0469(2000)057%3c1354:DO BITI%3e2.0.CO;2.

Kucharski, F., Bracco, A., Yoo, J.H., Molteni, F., 2008. Atlantic forced component of the Indian monsoon interannual variability. Geophys. Res. Lett. 35. doi:10.1029/2007GL033037.

Kulkarni, A., et al., 2020. Precipitation changes in India. In: Krishnan, R., Sanjay, J., Gnanaseelan, C., Mujumdar, M., Kulkarni, A., Chakraborty, S. (Eds.), Assessment of Climate Change over the Indian Region. Springer, Singapore. https://doi.org/10.1007/978-981-15-4327-2_3.

Kulkarni, A., Sabade, S.S., Kripalani, R.H., 2009. Spatial variability of intra-seasonal oscillations during extreme Indian monsoons. Int. J. Clim. 29 (13), 1945–1955.

Lee, J.Y., Wang, B., 2014. Future change of global monsoon in the CMIP5. Clim. Dyn. 42, 101–119. https://doi.org/10.1007/s00382-012-1564-0.

Li, R., Lv, S., Han, B., Gao, Y., Meng, X., 2017. Projections of South Asian summer monsoon precipitation based on 12 CMIP5 models. Int. J. Clim. 37 (1), 94–108.

Li, S., Perlwitz, J., Quan, X., Hoerling, M.P., 2008. Modelling the influence of North Atlantic multidecadal warmth on the Indian summer rainfall. Geophys. Res. Lett. 35 (5), L05804. https://doi.org/10.1029/2007gl032901.

Lorenz, D.J., Hartmann, D.L., 2006. The effect of the MJO on the North American monsoon. J. Clim. 19, 333–343.

Madden, R., Julian, P., 1972. Description of global-scale circulation cells in the tropics with a 40–50 day period. J. Atmos. Sci. 29 (6), 1109–1123.

Maharana, P., Dimri, A.P., Choudhary, A., 2020. Future changes in Indian summer monsoon characteristics under 1.5 and 2°C specific warming levels. Clim. Dyn. 54 (1-2), 507–523.

Mahendra, N., Chowdary, J.S., Darshana, P., Sunitha, P., Parekh, A., Gnanaseelan, C., 2020. Interdecadal modulation of interannual ENSO-Indian Summer monsoon rainfall teleconnections in observations and CMIP6 models: regional patterns. Int. J. Clim. https://doi.org/10.1002/joc.6973.

Mak, M.-K., 1975. The monsoonal mid-tropospheric cyclogenesis. J. Atmos. Sci. 32, 2246–2253. https://doi.org/10.1175/1520-0469(1975)032<2246:TMMTC>2.0.CO;2.

Malik, A., Brönnimann, S., 2018. Factors affecting the inter-annual to centennial timescale variability of Indian summer monsoon rainfall. Clim. Dyn. 50 (11-12), 4347–4364.

Malik, A., Brönnimann, S., Stickler, A., Raible, C.C., Muthers, S., Anet, J., Rozanov, E., Schmutz, W., 2017. Decadal to multi-decadal scale variability of Indian summer monsoon rainfall in the coupled ocean-atmosphere-chemistry climate model SOCOL-MPIOM. Clim. Dyn. 49 (9-10), 3551–3572. https://doi.org/10.1007/s00382-017-3529-9.

Mascioli, N.R., Fiore, A.M., Previdi, M., Correa, G., 2016. Temperature and precipitation extremes in the United States: quantifying the responses to anthropogenic aerosols and greenhouse gases. J. Clim. 29, 2689–2701.

Matsuura, K., Willmott, C.J., 2009. Terrestrial air temperature: 1900-2008 gridded monthly time series (version 2.01). Center for Climatic Research, University of Delaware, digital Media. Available at: http://climate.geog.UDEL.edu/~climate/html_pages/ download.html#P2009.

May, W., 2002. Simulated changes of the Indian summer monsoon under enhanced greenhouse gas conditions in a global time-slice experiment. Geophys. Res. Lett. 29, 1118.

Meehl, G.A., Arblaster, J.M., 2003. Mechanisms for projected future changes in South Asian monsoon precipitation. Clim. Dynam. 21, 659–675. doi:10.1007/s00382-003-0343-3.

Mehta, V.M., Lau, K.M., 1997. Influence of solar irradiance on the Indian monsoon-ENSO relationship at decadal multidecadal time scales. Geophys. Res. Lett. 24 (2), 159–162.

Miller, F.R., Keshavamurthy, R.N., 1968. Structure of an Arabian Sea summer monsoon system. Meteorological Monographs1. East–West Center Press, Honolulu, p. 94.

Mishra, V., Smoliak, B.V., Lettenmaier, D.P., Wallace, J.M., 2012. A prominent pattern of year-to-year variability in Indian Summer Monsoon Rainfall. Proceedings of the National Academy of Sciences 109 (19), 7213–7217 10.1073/pnas.1119150109.

Mohapatra, M., Mohanty, U.C., 2004. Some characteristics of low pressure systems and summer monsoon rainfall over Orissa. Current Sci. 87 (9), 245–1255.

Molnar, P., Boos, R.W., Battisti, D.S., 2010. Orographic controls on climate and paleoclimate of Asia: thermal and mechanical roles for the Tibetan Plateau. Ann. Rev. Earth Planet. Sci. 38, 77–102. https://doi.org/10.1146/annurev-earth-040809-152456.

Mooley, D.A., Shukla, J., 1989. Main features of the westward-moving low-pressure systems which form over Indian region during the summer monsoon season and their relation to monsoon rainfall. Mausam 40, 137–152.

Nanjundiah, R.S., Srinivasan, J., Gadgil, S., 1992. Intraseasonal variation of the Indian summer monsoon. J. Meteor. Soc. Jpn. Ser. II. 70, 529–550.

Niranjan Kumar, K., Rajeevan, M., Pai, D.S., Srivastava, A.K., Preethi, B., 2013. On the observed variability of monsoon droughts over India. Weather Clim. Extremes 1, 42–50. https://doi.org/10.1016/j.wace.2013.07.006.

Pai, D.S., Bhate, J., Sreejith, O.P., Hatwar, H.R., 2011. Impact of MJO on the intraseasonal variation of summer monsoon rainfall over India. Clim. Dynam. 36 (1–2), 41–55.

Pai, D.S., Srhar, L., Badwaik, M.R., Rajeevan, M., 2015. Analysis of the daily rainfall events over india using a new long period (1901–2010) high resolution (0.25° × 0.25°) gridded rainfall data set. Clim. Dyn. 45 (3–4), 755–776. https://doi.org/10.1007/s00382-014-2307-1.

Pai, D.S., Sridhar, L., Kumar, M.R., 2016. Active and break events of Indian summer monsoon during 1901–2014. Clim. Dyn. 46 (11-12), 3921–3939.

Pant, G.B., Rupa Kumar, K., 1997. Climates of South Asia. John Wiley & Sons, Chichester, p. 320 ISBN 0-471-94948-5.

Park, H-S., Chiang, J.C.H., Lintner, B.R., Zhang, G.J., 2010. The delayed effect of major El Niño events on Indian monsoon rainfall. J. Clim. 23, 932–946.

Parthasarathy, B., Mooley, D.A., 1978. Some features of a long homogeneous series of Indian summer monsoon rainfall. Mon. Weather Rev. 106 (6), 771–781. doi:10.1175/1520-0493(1978)1062.0.co;2.

Parthasarathy, B., Munot, A.A., Kothawale, D.R., 1994. All-India monthly and seasonal rainfall series: 1871–1993. Theor. Appl. Climatol. 49, 217–224.

Pisharoty, P.R., 1965. Evaporation from the Arabian Sea and Indian Southwest monsoon, Proceedings of International Indian Ocean Expedition. Bombay, 22-26, 43–54.

Polson, D., Bollasina, M., Hegerl, G.C., Wilcox, L.J., 2014. Decreased monsoon precipitation in the Northern Hemisphere due to anthropogenic aerosols. Geophys. Res. Lett. 41 (16), 6023–6029. doi:10.1002/2014gl060811.

Prabhu, A., Kripalani, R., Oh, J., Preethi, B., 2017. Can the Southern annular mode influence the Korean summer monsoon rainfall? Asia-Pacific J. Atmos. Sci. 53 (2), 217–228.

Prajeesh, A.G., Ashok, K., Rao, D.B., 2013. Falling monsoon depression frequency: a Gray-Sikka conditions perspective. Sci. Rep. 3, 2989.

Praveen, V., Sandeep, S., Ajayamohan, R.S., 2015. On the relationship between mean monsoon precipitation and low pressure systems in climate model simulations. J. Clim. 28 (13), 5305–5324. doi:10.1175/jcli-d-14-00415.1.

Raghavan, K., 1973. Break-monsoon over India. Mon. Weather. Rev. 101 (1), 33–43. doi:10.1175/1520-0493(1973)1012.3.co;2.

Rajeevan, M., De, U.S., Prasad, R.K., 2000. Decadal variation of sea surface temperatures, cloudiness and monsoon depressions in the north Indian ocean. Current Sci. 79, 283–285.

Rajeevan, M., Gadgil, S., Bhate, J., 2010. Active and break spells of the Indian summer monsoon. J. earth syst. Sci. 119 (3), 229–247. doi:10.1007/s12040-010-0019-4.

Rajeevan, M., Pai, D.S., Kumar, R.A., Lal, B., 2007. New statistical models for long-range forecasting of southwest monsoon rainfall over India. Clim. Dyn. 28 (7–8), 813–828. https://doi.org/10.1007/s00382-006-0197-6.

Rajendran, K., Kitoh, A., Srinivasan, J., Mizuta, R., Krishnan, R., 2012. Monsoon circulation interaction with Western Ghats orography under changing climate. Theor. Appl. Climatol. 110 (4), 555–571. https://doi.org/10.1007/s00704-012-0690-2.

Ramage, C.S., Monsoon Meteorology. Academic Press, New York, NY and Lodon, 296, p. 1971.

Ramanathan, V., Chung, C., Kim, D., Bettge, T., Buja, L., Kiehl, J.T., Washington, W.M., Fu, Q., Sikka, D.R., Wild, M., 2005. Atmospheric brown clouds: Impacts on South Asian climate and hydrological cycle. Proc. Nat. Acad. Sci. 102 (15), 5326–5333. https://doi.org/10.1073/pnas.0500656102.

Rasmusson, E.M., Carpenter, T.H., 1983. The relationship between eastern equatorial pacific sea surface temperatures and rainfall over India and Sri Lanka. Mon. Weather Rev 111 (3), 517–528. doi:10.1175/1520-0493(1983)1112.0.co;2.

Raymond, D.J., Fuchs, Ž., 2009. Moisture modes and the Madden–Julian Oscillation. J. Clim. 22, 3031–3046. https://doi.org/10.1175/2008JCLI2739.1.

Rayner, N.A.A., Parker, D.E., Horton, E.B., Folland, C.K., Alexander, L.V., Rowell, D.P., Kent, E.C., Kaplan, A., 2003. Global analyses of sea surface temperature, sea ice, and night marine air temperature since the late nineteenth century. J. Geophys. Res. Atmos. 108 (D14). doi:1 0.1029/2002JD002670.

Roxy, M.K., Ritika, K., Terray, P., Murtugudde, R., Ashok, K., Goswami, B.N., 2015. Drying of Indian subcontinent by rapid Indian Ocean warming and a weakening land-sea thermal gradient. Nature Commun. 6 (1), 1–10.

Roy, I., Tedeschi, R.G., Collins, M., 2019. ENSO teleconnections to the Indian summer monsoon under changing climate. Int. J. Clim. 39 (6), 3031–3042.

Saji, N.H., Goswami, B.N., Vinayachandran, P.N., Yamagata, T., 1999. A dipole mode in the tropical Indian Ocean. Nature 401 (6751), 360–363.

Sankar, S., Svendsen, L., Gokulapalan, B., Joseph, P., Johannessen, O., 2016. The relationship between Indian summer monsoon rainfall and Atlantic multidecadal variability over the last 500 years. Tellus A 68, 1–14. https://doi.org/10.3402/tellusa.v68.31717.

Sankarrao, M., Lau, K.M., Yang, S., 1996. On the relationship between Eurasian snow cover and the Asian summer monsoon. Int. J. Clim. 16 (6), 605–616.

Sharmila, S., Joseph, S., Sahai, A.K., Abhilash, S., Chattopadhyay, R., 2015. Future projection of Indian summer monsoon variability under climate change scenario: an assessment from CMIP5 climate models. Glob. Planet. Change 124, 62–78.

Shukla, J., 1987 In J.S. Fein & P.L. Stephens (Eds.), Interannual Variability of Monsoons (pp. 399–464). New York, NY: Wiley.

Shukla, J., Paolino, D.A., 1983. The Southern oscillation and long-range forecasting of the summer monsoon rainfall over India. Mon. Weather Rev. 111 (9), 1830–1837. doi:10.1175/1520-0493(1983)1112.0.co;2.

Sikka, D.R., Gadgil, S., 1980. On the maximum cloud zone and the ITCZ over Indian, longitudes during the southwest monsoon. Mon. Weather. Rev. 108, 1840–1853.

Sikka, D.R., 1978. Some aspects of the life history, structure and movement of monsoon depressions. Monsoon Dyn. 1978, 1501–1529. doi:10.1007/978-3-0348-5759-8_21.

Singh, M., Bhatla, R., 2019. Intense rainfall conditions over Indo-Gangetic plains under the influence of Madden–Julian oscillation. Meteorol. Atmos. Phys. doi:10.1007/s00703-019-00703-7.

Sobel, A., Maloney, E., 2013. Moisture modes and the eastward propagation of the MJO. J. Atmos. Sci. 70, 187–192. https://doi.org/10.1175/JAS-D-12-0189.1.

Soman, M.K., Kumar, K.K., 1993. Space-time evolution of meteorological features associated with the onset of Indian summer monsoon. Mon. Weather. Rev. 121 (4), 1177–1194.

Sooraj, K.P., Terray, P., Masson, S., Crétat, J., 2019. Modulations of the Indian summer monsoon by the hot subtropical deserts: insights from coupled sensitivity experiments. Clim. Dyn. 52 (7-8), 4527–4555.

Srinivas, G., et al., 2018. Influence of the Pacific–Japan pattern on Indian summer monsoon rainfall. J. Clim. 31 (10), 3943–3958. doi:10.1175/jcli-d-17-0408.1.

Srivastava, G., Chakraborty, A., Nanjundiah, R.S., 2019. Multidecadal see–saw of the impact of ENSO on Indian and West African summer monsoon rainfall. Clim. Dyn. 52, 6633–6649.

Taraphdar, S., Zhang, F., Leung, L.R., Chen, X., Pauluis, O.M., 2018. MJO affects the monsoon onset timing over the Indian region. Geophys. Res. Lett. 45 (18), 10011–10018.

Turner, A.G., Annamalai, H., 2012. Climate change and the South Asian Monsoon. Nature Clim. Change 2, 587–595. doi:10.1038/nclimate1495.

Turner, A.G., Inness, P.A., Slingo, J.M., 2007. The effect of doubled CO2 and model basic state biases on the monsoon-ENSO system. I: mean response and interannual variability. Q. J. R. Meteorol. Soc. 133, 1143–1157.

Ueda, H., Iwai, A., Kuwako, K., Hori, M.E., 2006. Impact of anthropogenic forcing on the Asian summer monsoon as simulated by eight GCMs. Geophys. Res. Lett. 33, L06703. https://doi.org/10.1029/2005GL025336.

Undorf, S., Polson, D., Bollasina, M.A., Ming, Y., Schurer, A., Hegerl, G.C., 2018. Detectable impact of local and remote anthropogenic aerosols on the 20th century changes of West African and South Asian monsoon precipitation. J. Geophys. Res. 123 (10), 4871–4889. https://doi.org/10.1029/2017JD027711.

Walker, G.T., 1924. Correlation in seasonal variations of weather, IX. A further study of world weather. Memoirs of the India Meteorological Department 24 (9), 275–333.

Wang, B., Xie, X., 1997. A model for the Boreal summer intraseasonal oscillation. J. Atmos. Sci. 54 (1), 72–86. doi:10.1175/1520-0469(1997)0542.0.co;2.

Wang, L., Xu, P., Chen, W., Liu, Y., 2017. Interdecadal variations of the Silk Road pattern. J. Clim. 30, 9915–9932.

Wang, H., Xie, S.-P., Kosaka, Y., Liu, Q., Du, Y., 2019. Dynamics of Asian summer monsoon response to anthropogenic aerosol forcing. J. Clim. 32, 843–858. https://doi.org/10.1175/JCLI-D-18-0386.1.

Webster, P.J., Clark, C., Cherikova, G., Fasullo, J., Han, W., Loschnigg, J., et al., 2002. The monsoon as a self-regulating coupled ocean—atmosphere system. Int. Geophysics. 83, 198–219.

Webster, P.J., Magaña, V., Palmer, T.N., Shukla, J., Tomas, R.A., Yanai, M., Yasunari, T., 1998. Monsoons: processes, predictability and prospects for prediction. J. Geophys. Res. 103 (14), 451–510.

Wei, W., Wu, Y., Yang, S., Zhou, W., 2019. Role of the South Asian high in the onset process of the Asian summer monsoon during spring-to-summer transition. Atmosphere 10, 239.

Wheeler, M.C., Hendon, H.H., 2004. An all-season real-time multivariate MJO index: development of an index for monitoring and prediction. Mon. Weather Rev. 132 (8), 1917–1932. doi:10.1175/1520-0493(2004)1322.0.co;2.

Xavier, P.K., Marzin, C., Goswami, B., 2007. An objective definition of the Indian summer monsoon season and a new perspective on the ENSO–monsoon relationship. Quart. J. Roy. Meteor. Soc. 133, 749–764. https://doi.org/10.1002/qj.45.

Xie, S.-P., Kosaka, Y., Du, Y., Hu, K., Chowdary, J.S., Huang, G., 2016. Indo-western Pacific Ocean capacitor and coherent climate anomalies in post-ENSO summer: a review. Adv. Atmos. Sci. 33, 411–432. https://doi.org/10.1007/s00376-015-5192-6.

Xie, S.P., Xu, H., Saji, N.H., Wang, Y., Liu, W.T., 2006. Role of narrow mountains in large-scale organization of Asian monsoon convection. J. Clim. 19 (14), 3420–3429.

Xu, C., Sano, M., Dimri, A.P., Ramesh, R., Nakatsuka, T., Shi, F., Guo, Z., 2018. Decreasing Indian summer monsoon on the northern Indian sub-continent during the last 180 years: evidence from five tree-ring cellulose oxygen isotope chronologies. Climate of the Past 14 (5), 653–664. https://doi.org/10.5194/cp-14-653-2018.

Yadav, R.K., Srinivas, G., Chowdary, J.S., 2018. Atlantic Niño modulation of the Indian summer monsoon through Asian jet. npj Clim. Atmos. Sci. 1 (23). https://doi.org/10.1038/s41612-018-0029-5.

Yang, J., Liu, Q., Xie, S-P., Liu, Z., Wu, L., 2007. Impact of the Indian Ocean SST basin mode on the Asian summer monsoon. Geophys. Res. Lett. 34, L02708. doi:10.1029/2006GL028571.

Yasunari, T., 1981. Structure of an Indian summer monsoon system with around 40-Day Period. J. Meteorol. Soc. Jpn. Ser. II 59 (3), 336–354. doi:10.2151/jmsj1965.59.3_336.

Zhang, C., Min, D., 2004. Seasonality in the Madden–Julian oscillation. J. Clim. 17 (16), 3169–3180. doi:10.1175/1520-0442(2004)0172.0.co;2.

Zhang, G.J., Song, X., 2009. Interaction of deep and shallow convection is key to Madden Julian Oscillation simulation. Geophys. Res. Lett. 36. doi:10.1029/2009GL03740.

Zhou, Z.Q., Xie, S.P., Zhang, R.H., 2019. Variability and predictability of Indian rainfall during the monsoon onset month of June. Geophys. Res. Lett. doi:10.1029/2019gl085495.

Chapter 2

Interannual variation of the Indian summer monsoon, ENSO, IOD, and EQUINOO

Sulochana Gadgil[a], P.A. Francis[b], P.N. Vinayachandran[a], S. Sajani[c]
aCentre for Atmospheric and Oceanic Sciences, Indian Institute of Science, Bengaluru, India,
*bIndian National Centre for Ocean Information Services, Ministry of Earth Sciences, Hyderabad,
India, cCSIR-Fourth Paradigm Institute, Bengaluru, India*

2.1 Introduction

Most of the rainfall over the Indian region as a whole occurs during the summer monsoon season of June–September. During the summer monsoon, high rainfall occurs along the west coast of the peninsula closely associated with the orography parallel to the coast and over the northeastern regions. In addition, there is a broad zone around 20°N stretching northwestward from the head Bay of Bengal, called the monsoon zone (Sikka and Gadgil, 1980; Gadgil, 2003), which receives significant rainfall (Fig. 2.1). Monsoon is a manifestation of the seasonal migration of the intertropical convergence zone (ITCZ) onto the heated subcontinent in the summer (Gadgil, 2003,2018) and the large-scale summer monsoon rainfall is associated with the occurrence of the continental tropical convergence zone (CTCZ). At the culmination of the onset phase, toward the end of June, the CTCZ gets established over the monsoon zone. In the peak monsoon months of July and August, the CTCZ fluctuates primarily over this zone and starts retreating from the western part of this zone in September (Sikka and Gadgil, 1980). The variability of the monsoon rainfall on different time-scales is therefore related to the space-time variation of the CTCZ.

Here, we focus on the interannual variation of the Indian summer monsoon rainfall (ISMR), which is a scientifically challenging problem with great applied importance because of the large impact of the interannual variation on Indian agriculture and economy (Gadgil and Gadgil, 2006). For most of the studies on interannual variation of the ISMR, the time series derived by Parthasarathy et al. (1995) and updated by the Indian Institute of Tropical Meteorology is used. The all-India rainfall derived by India Meteorological Department (IMD), which is

Indian Summer Monsoon Variability: El Niño-teleconnections and beyond.
DOI: https://doi.org/10.1016/B978-0-12-822402-1.00014-4

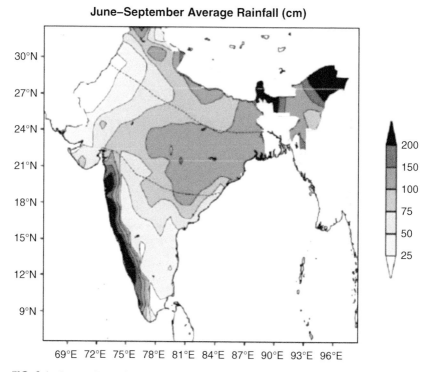

FIG. 2.1 June to September mean rainfall (cm) over the Indian region based on IMD gridded rainfall data (Rajeevan et al. 2006). The area between the *red lines* shows the core monsoon zone. "*IMD*, India Meteorological Department."

highly correlated with the Parthasarathy series, is used in developing forecasting methods and operational forecasts are issued for the IMD-all-India rainfall. Gadgil et al. (2019) have broughtout the appropriate region for deriving the rainfall associated with the monsoon during June–September using the gridded data that have recently become available from IMD (Rajeevan et al., 2006; Pai et al., 2014). The problem of understanding and prediction of the interannual variation of ISMR (Fig. 2.2) has been addressed for over a century and considerable progress has been made particularly since the 1980s. In this chapter, we review what is understood about the major players involved and the challenges that remain.

It is important to note that the monsoon rainfall over the Indian region arises primarily from the propagation of synoptic-scale systems over the surrounding oceans and of the tropical convergence zone (TCZ) over the equatorial Indian Ocean onto the region (Sikka and Gadgil, 1980; Gadgil, 2003). The CTCZ is sustained over the monsoon zone also partly by northward propagations of the TCZ over the equatorial Indian Ocean (Sikka and Gadgil, 1980; Gadgil, 2003). A substantial fraction of the large-scale rainfall over the Indian region occurs in association with the propagation of synoptic-scale systems generated over the

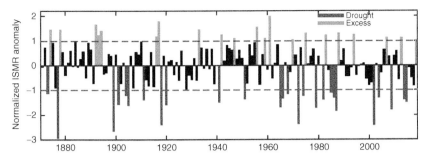

FIG. 2.2 Interannual variation of Indian summer monsoon rainfall. *Red (green) bars* represent drought (excess) seasons. *Horizontal dashed (red) lines* represent ±1 standard deviation. The figure is based on monthly all India rainfall from Parthasarathy et al. (1995) and update from the website of the Indian Institute of Tropical Meteorology (http://www.tropmet.res.in/).

Bay of Bengal onto the region. Hence, we expect the variation of the monsoon rainfall on subseasonal to interannual time-scales to be linked to that of organized deep convection over the surrounding oceans. Several of the synoptic systems over the Bay are believed to arise from the westward propagation of systems over the western Pacific (Krishnamurti and Ardanuy, 1980). Also, on several days, the CTCZ is a part of a planetary scale system stretching eastward to over the central Pacific and sometimes even beyond (Gadgil, 2003). Hence, the monsoon rainfall is also linked to convection over the Pacific. In fact, the major players in an interannual variation of ISMR involve phenomena over the equatorial Indian and Pacific Oceans. Here we focus on the links of the monsoon variability to the atmospheric events over the tropical Indian and Pacific Oceans (Fig. 2.3).

The dominant signal of interannual variation of the coupled atmosphere-ocean system over the Pacific is the El Niño Southern Oscillation (ENSO). ENSO involves oscillation between a warm phase, El Niño, characterized by an abnormal warming of surface ocean waters of the central and eastern equatorial Pacific and enhanced convection in the atmosphere above, and a cold phase, La Niña, characterized by abnormal cooling of these waters and suppressed convection in the atmosphere above (Fig. 2.4). Unraveling the physics of ENSO and development of models capable of predicting it is a major success story of the 1980s and 1990s. The development of a relatively simple model by Cane and Zebiak (1985) and Zebiak and Cane (1987) which incorporated the critical processes involved in ENSO identified by Bjerknes (1969), development of atmospheric general circulation models (AGCMs) which could simulate the atmospheric component of ENSO when forced with the observed sea surface temperature (SST) and Ocean GCMs which could simulate El Niño/La Niña when forced with the observed atmospheric conditions, paved the way for the development of coupled models capable of predicting ENSO.

Sikka (1980) and several subsequent studies (Pant and Parthasarathy, 1981; Rasmusson and Carpenter, 1983, etc.) have shown that there is an increased propensity of droughts during El Niño and of excess rainfall during La Niña.

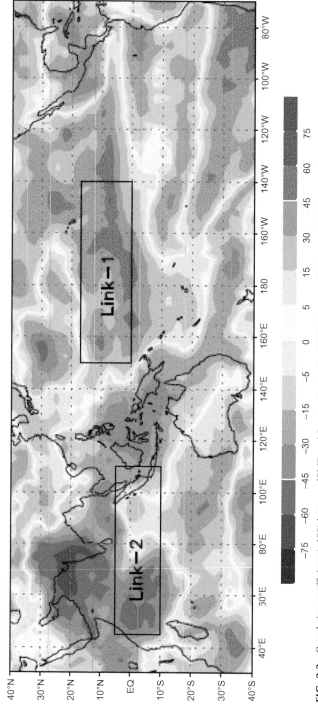

FIG. 2.3 Correlation coefficient (×100) between ISMR and June to September OLR. Important links are with (1) convection over equatorial central Pacific and (2) equatorial Indian Ocean. The outgoing longwave radiation (OLR) data are from Climate Diagnostic Center, United States (http://www.cdc.noaa.gov/). *"ISMR,* Indian summer monsoon rainfall."

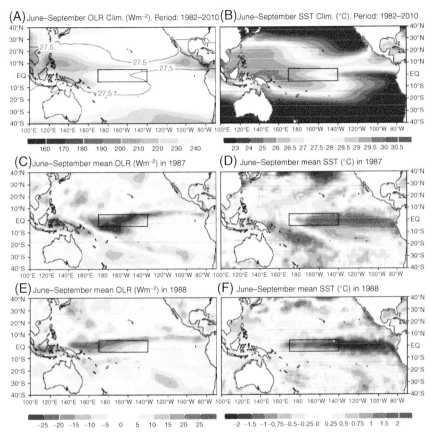

FIG. 2.4   June to September mean OLR/OLR anomaly (left) and mean SST/SST anomaly (right) patterns. Top: climatological mean for the period 1982–2010; middle: anomalies of 1987 and bottom: anomalies of 1988. Red contour in (A) encompasses the regions with SST above 27.5°C. SST data are based on HadISST (Rayner et al., 2003) data from http://badc.nerc.ac.uk. "*OLR*, outgoing longwave radiation; *SST*, sea surface temperature."

A mode of variability of the equatorial Indian Ocean, which to some extent, is independent of forcing by ENSO was discovered by Saji et al. (1999) and Webster et al. (1999). The distinguishing attribute of this mode is the opposite sign of the SST anomalies over the eastern equatorial Indian Ocean (EEIO) and the western equatorial Indian Ocean (WEIO) (Fig. 2.5C and D). Hence, it was called the Indian Ocean dipole (IOD) mode by Saji et.al. (1999) who defined the Dipole Mode Index (DMI) for the mode based on the difference of the SST anomalies of WEIO and EEIO. Climatologically, during June–September, the equatorial Indian Ocean is warmer in the east, and supports more convection in the atmosphere than in the west (Fig. 2.5A and B). Hence, a positive phase of IOD (i.e., with DMI positive) is characterized by weakening or reversal of the climatological zonal SST gradient. A strong positive IOD event (with high

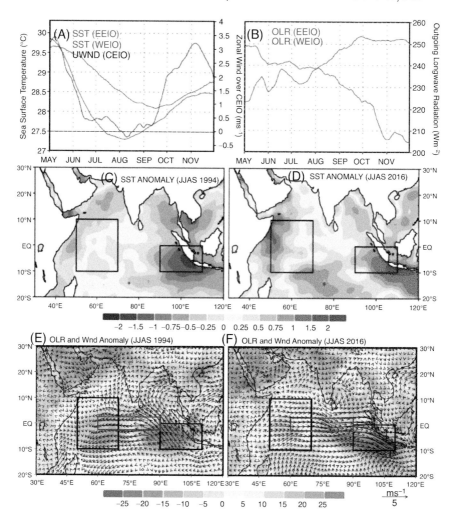

FIG. 2.5 Top: climatology of SST (left) and OLR (right) averaged over the WEIO (50–70°E, 10°S–10°N; *red lines*) and EEIO (90–110°E, 10°S EQ; *blue lines*) and zonal wind averaged over the CEIO (60–90°E, 2.5°S–2.5°N, *left black line*). Middle: June–September mean SST anomaly of 1994 (left) and 2016 (right). Bottom: June–September mean OLR and wind anomaly in 1994 (left) and 2016 (right). The surface wind data are from National Center for Environmental Prediction (Kalnay et al., 1996) (http://www.cdc.noaa.gov/). "*CEIO*, central equatorial Indian Ocean; *EEIO*, eastern equatorial Indian Ocean; *OLR*, outgoing longwave radiation; *SST*, sea surface temperature; *WEIO*, western equatorial Indian Ocean."

positive DMI, e.g., Fig. 2.5C) is associated with a positive outgoing longwave radiation (OLR) anomaly over EEIO and negative OLR anomaly over WEIO and anomalous easterly winds over the central equatorial Indian Ocean (Fig. 2.5E). The evolution of a positive IOD event involves coupled ocean-atmospheric processes in which convection, winds, SST, and thermocline take part actively.

Webster (1999) analyzed the 1997 positive IOD event in detail and concluded that it is an expression of an internal mode of the Indian Ocean and proposed a mechanism of ocean-atmosphere interaction governing the event.

Gadgil et al. (2004) showed that, in addition to ENSO, a mode called equatorial Indian Ocean oscillation (EQUINOO), plays an important role in determining the interannual variation of the monsoon. EQUINOO involves an oscillation between the positive and negative phases with the positive (negative) phase characterized by enhancement of convection over WEIO (EEIO) and suppression of convection over EEIO (WEIO) (Fig. 2.6A and B). Since the OLR anomaly pattern of the positive (negative) phase of EQUINOO is associated with negative (positive) OLR anomalies over the Indian region (Fig. 2.6A and B), positive (negative) phase of EQUINOO is favorable (unfavorable) for the monsoon rainfall. Also, since during the summer monsoon, climatological convection over EEIO is more than that over WEIO (Fig. 2.5B), the positive phase is associated with weakening or even reversal of the climatological zonal convection gradient, whereas the negative phase is associated with intensification of the climatological gradient in convection.

The positive phase of EQUINOO is associated with easterly anomalies of the zonal wind over the central equatorial Indian Ocean (CEIO), while the negative phase is characterized by westerly anomalies. Although the distinguishing attribute of EQUINOO is the see-saw in the convection over EEIO and WEIO, the index generally used for EQUINOO is based on the surface zonal wind over the CEIO, primarily because of the availability of wind data for a longer period compared to the satellite-derived OLR data. The wind index, EQWIN, used for EQUINOO, is defined as the *negative* anomaly of the surface zonal wind averaged over CEIO (Fig. 2.5C), normalized by its standard deviation (Gadgil et al., 2004) so that positive values represent the phase that is favorable for the monsoon.

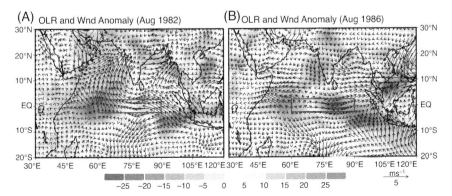

FIG. 2.6   Anomalous OLR and low-level winds in August 1982 (left) and August 1986 (right). CEIO, the wind anomaly over which is averaged to derive EQWIN, is also shown by the *green box*. "*CEIO*, central equatorial Indian Ocean; *OLR*, outgoing longwave radiation.".

2.2   Interaction of atmospheric convection with Pacific and Indian Ocean: ENSO, IOD, and EQUINOO

Climatologically during the summer monsoon season, the intense convection over the tropical Pacific extends from the equator to about 15°N (Fig. 2.4A), with the convection over the equatorial central Pacific being much less intense than that over the eastern and western parts. The convection anomalies associated with El Niño (e.g., 1987, Fig. 2.4C) differ markedly from those associated with La Niña (e.g. 1988, Fig. 2.4E) over the equatorial central Pacific, with the intense enhancement of convection during El Niño and suppression of convection during La Niña. ENSO involves convection anomalies over the equatorial west Pacific of opposite sign to those over equatorial central Pacific.

It is well known that there is a high propensity of convection over those parts of the tropical ocean characterized by SST above the threshold of about 27.5°C (Gadgil et al., 1984; Graham and Barnett, 1987) and that the mean convection increases rapidly with an increase in SST across the threshold. Climatologically, there is a region of large latitudinal extent over the west Pacific with SSTs well above the threshold and a region of smaller latitudinal extent over the east Pacific (Fig. 2.4B). Over the equatorial central Pacific there is a relatively narrow strip of warm water with mean SST a little above the threshold. The enhanced (suppressed) convection over the equatorial central Pacific characterizing El Niño (La Niña) is seen to be associated with warm (cold) SST anomalies over this region (Fig. 2.4D and E). This close link between anomalies of SST and atmospheric convection arises because the mean SST of the equatorial central Pacific during June–September is close to the threshold implying that with a relatively small positive anomaly, the equatorial central Pacific crosses the SST threshold, which leads to intensification of convection/precipitation during El Niño. The strong link between anomalies of SST and convection is a manifestation of strong ocean-atmosphere coupling which leads to a high correlation between the indices of El Niño such as Niño3, Niño 3.4 SST anomaly, and the southern oscillation index (with coefficients 0.85 and above). Hence the Niño SST anomaly indices are considered to be adequate to describe the phase of the coupled ENSO mode.

On the other hand, the mean SST of the equatorial Indian Ocean sector 60°–100°E, during June–September is well above the threshold for deep convection (Fig. 2.7). Thus while the variation of convection/precipitation over the equatorial central Pacific is rather sensitive to the SST (as first pointed out by Bjerknes, 1969), variability of convection over the central equatorial Indian Ocean is not. Hence, although EQUINOO is considered to be the atmospheric component of the coupled IOD mode, it is not as tightly coupled to the ocean component of IOD. Atmospheric dynamics, i.e., low-level convergence (Graham and Barnett, 1987), play a more important role in the variability of convection over the equatorial Indian Ocean.

It should be noted that the variability of convection over parts of the equatorial Indian Ocean is determined to a large extent by competition with convection over other favorable zones in the vicinity. The OLR over WEIO (EEIO) is

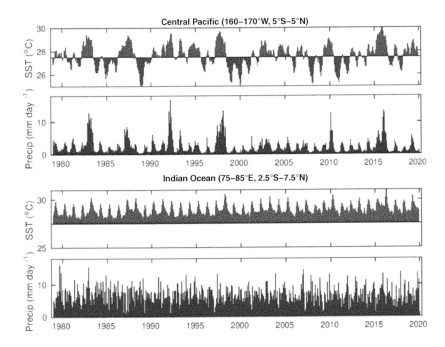

FIG. 2.7   First panel: SST averaged over equatorial central Pacific (160–170° W, 5°S–5°N), second panel: precipitation averaged over the equatorial central Pacific, third panel: SST averaged over equatorial Indian Ocean (75–85° E, 2.5°S–7.5°N), fourth panel: precipitation averaged over the equatorial Indian Ocean. "*SST*, sea surface temperature."

negatively correlated with OLR of EEIO (WEIO) on the seasonal scale (Fig. 2.8A and B), as well as monthly and smaller time-scales, because there is a competition between the convection over these regions, with the convection being enhanced (suppressed) over the WEIO (EEIO), when it is suppressed (enhanced) over EEIO (WEIO). This is the basis of the EQUINOO. It is interesting that a negative correlation between the precipitation over WEIO and that over EEIO is simulated by the latest versions of the coupled models. We analyzed Coupled Model Intercomparison Project Phase 6 (CMIP6) historical simulations of the four models which were amongst the best in simulating ISMR to study the simulation of the interannual variation and found that all of them simulate EQUINOO, i.e., negative correlation between the rainfall over WEIO and that over EEIO (Fig. 2.8E).

To analyze the daily variation of convection over relatively large regions such as the EEIO and WEIO of which convection may occur only on some grids on any day, we use a convection index (CI), which is a measure of the intensity and the horizontal extent of deep convection over a region. CI is derived from 2.5° resolution OLR data, using a threshold of 200 Wm^{-2} to identify the grids with deep convection. We take the difference of the OLR value from 200 Wm^{-2} to represent the intensity of deep convection at a grid and derive CI of any

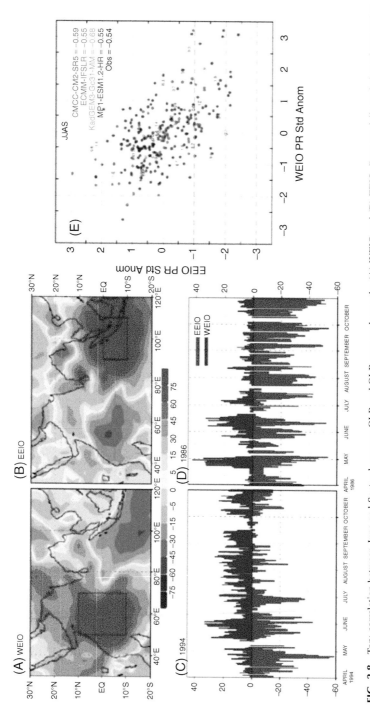

FIG. 2.8 Top: correlation between June and September mean OLR and OLR averaged over the (A) WEIO and (B) EEIO. Bottom: daily variation of convective index (CI) over WEIO and EEIO for (C) 1994 and (D) 1986. CI of EEIO is multiplied by −1 to represent in single graph. (E) Scatter diagram between rainfall over the EEIO and WEIO in the observation and selected coupled model simulations. "*EEIO*, eastern equatorial Indian Ocean; *OLR*, outgoing longwave radiation; *WEIO*, western equatorial Indian Ocean."

region as the sum of this intensity over all the convective grids in the region. The variation of CI over WEIO and EEIO during the summer monsoons of 1994 and 1986 is shown in Fig. 2.8C and D. It is seen that the positive phase (1994) is characterized by a high tendency of convection over WEIO. However, the convection over EEIO is not suppressed on all days.

2.3   Monsoon and ENSO

Consistent with the association of high propensity of deficit ISMR (excess rainfall) with warm events/El Niño (cold events/La Niña), deficit occurred in 21 out of 25 such warm events during 1875–1979, which included 9 of 11 seasons with largest rainfall anomalies (Rasmusson and Carpenter, 1983). Since the ISMR is negatively correlated with the commonly used Niño 3.4 index of ENSO which is the SST anomaly of the Niño 3.4 region, we define an ENSO index to be negative of the normalized Niño 3.4 SST anomaly, so that positive values of the ENSO index indicate a favorable phase of the ENSO for the monsoon. The relationship of ISMR with ENSO index for 1958–2010 is depicted in Fig. 2.9A. It

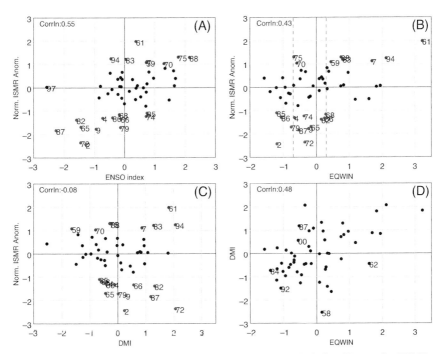

FIG. 2.9   Scatter plots of (A) normalized ISMR anomaly versus ENSO index, (B) normalized ISMR anomaly versus EQWIN, (C) normalized ISMR anomaly versus DMI, and (D) DMI versus EQWIN for the period 1958–2010. ISMR anomalies are normalized by the standard deviation. Niño 3.4 index, i.e., the SST anomaly over the Niño3.4 region (170° W–120° W, 5°S–5°N) is obtained from Climate Analysis Section, National Center for Atmospheric Research, United States (http://www.cgd.ucar.edu/). "*DMI*, Dipole Mode Index; *ISMR*, Indian summer monsoon rainfall; *SST*, sea surface temperature."

turns out that the maximum impact of ENSO is on the probability of categories with the highest and lowest monsoon rainfall, with the probability of the rainfall in the lowest (highest) category associated with warm (cold) events being 30% (25%) more than the climatological probability (Sajani et al., 2015)

It is seen in Fig. 2.9A that the El Niños of 1982 and 1987 were associated with droughts and the La Niña of 1988 with excess rainfall. However, during the strongest El Niño of the century in 1997, the ISMR was slightly above normal instead of the widely expected severe drought, and Kumar et al. (1999) suggested that the relationship between the Indian monsoon and ENSO had weakened in the last decades of the 20th century. This study has received considerable attention, leading to a widely held belief that the ENSO-monsoon relationship was unravelling, possibly due to global warming. However, it is seen from Fig. 2.9A that even in the earlier era, droughts occurred during the weak La Niña events of 1974 and 1985, and even after 1997, droughts occurred during the warm events of 2002, 2004, and 2009. Thus, while the monsoon rainfall anomaly cannot be attributed to ENSO in some years, ENSO has continued to be a major factor in determining the interannual variation of the monsoon even in the recent decades. Yet there are also seasons such as 1994 in which excess monsoon rainfall occurred during a warm event. In fact, 1994 was a positive IOD and positive EQUINOO event.

Systematic studies to improve the simulations of the monsoon-ENSO link by AGCMs were made in the 1990s beginning with the effort to simulate the drought of ISMR during the El Niño of 1987 and excess ISMR during La Niña of 1988 under the Monsoon Numerical Experimentation Group of the Tropical Ocean Global Atmosphere program (Palmer et al., 1992). The results of analysis of AMIP runs of 20 AGCMs (Gadgil and Sajani, 1998) and of the national atmospheric model intercomparison project SPIM (Gadgil and Srinivasan, 2011) showed that most of the models could simulate the extreme seasons associated with ENSO such as 1987 and 1988 but not those such as excess season of 1994 and drought of 1985 which were not associated with ENSO. Most of the state-of-the-art coupled models also successfully simulate the nature of the link between ENSO and the monsoon (Gadgil, 2012; Nanjundiah et al., 2013). While in most models the link is stronger than observed, we find that the latest versions of most of the models of the CMIP, i.e., CMIP6, simulate a realistic correlation coefficient between ENSO and the monsoon rainfall (see Chapter 20).

2.4 Monsoon and EQUINOO

The summer monsoon rainfall is well correlated with EQUINOO with the correlation coefficient between ISMR and EQWIN being 0.43 (Fig. 2.9B), which is somewhat smaller than that for the ENSO index. Since for June–September the correlation between ENSO index and EQWIN is poor, the two modes are independent. In fact, depiction of the extremes of ISMR (i.e., years with the magnitude of the ISMR anomaly normalized by the standard deviation greater

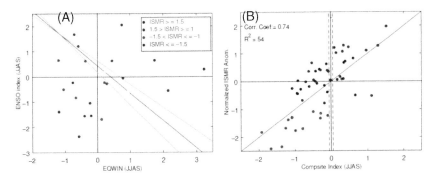

FIG. 2.10   (A) Extremes of ISMR are represented in the phase plane of June–September mean EQWIN and ENSO Index. The intensity of ISMR anomaly is represented as per the color given in the legend. (B) ISMR anomaly is plotted against the June–September mean composite index (0.58 × ENSO Index + 0.5 × EQWIN −0.27) for the period 1958–2010. *"ISMR,* Indian summer monsoon rainfall."

than 1) in the phase plane of these two indices (Fig. 2.10A), shows that there is a clear separation between droughts (normalized ISMR anomaly < −1) and the excess rainfall seasons (normalized ISMR anomaly > 1). The variation of the two indices and ISMR for the extreme seasons (Fig. 2.9A and B) clearly shows the dominant contribution to ISMR of the positive phase of EQUINOO in the excess monsoon seasons of 1994 and 1983 and negative phase of EQUINOO in the droughts of 1974 and 1985 while in the normal monsoon of 1997 the contribution of EQUINOO is comparable to that of ENSO. Since the two modes are independent, when a composite index is derived as a linear combination of the two indies, it explains much more of the variance of ISMR than that by either of the two modes (Sajani et al., 2015), and ISMR is highly correlated (coefficient 0.74) with the composite index (Fig. 2.10B).

At present, reasonably reliable predictions for ENSO are available before the commencement of the summer monsoon season. If it is predicted to be a warm event, we expect ISMR to be deficit/drought. In that case, it is important to be able to predict before the season, the chance of a positive phase of EQUINOO being triggered and being sustained during the season or at least a part of the season, which can lead to reduction of the adverse impact of ENSO on ISMR (or even overwhelming the adverse impact to result in excess monsoon as in 1994) by the favorable phase of EQUINOO. Even when a cold event is predicted, it is necessary to assess the chances of occurrence of a positive phase of EQUINOO and if that is small, try and assess the chance of a strong negative EQUINOO which can lead to a drought even in a La Niña year as in 1985

2.5   Monsoon and IOD

All strong positive EQUINOO events such as 1961,1994, and 2007 are also positive IOD events, some seasons with high positive DMI such as 1972 and

1987 are associated with negative EQUINOO events. It turns out that ISMR is rather poorly correlated with DMI (Fig. 2.9C) as expected from Saji et al. (1999) (Fig. 2.4). It is seen that while some years with large positive DMI such as 1961, 1994, and 2007 have excess monsoon rainfall, seasons with comparable values of DMI, such as 1972, 1982, and 1987 are droughts. The study of Ihara et al. (2007) on the relationship of the variation of the monsoon with ENSO, EQUINOO, and IOD, showed that the linear reconstruction of ISMR on the basis of a multiple regression from an ENSO index and EQUINOO index, better specifies the ISMR than the regression with only ENSO index, whereas no skill is added when DMI is considered along with ENSO. This is because DMI is well correlated with ENSO index (corr. Coeff of −0.44, Sajani et al. 2015), whereas the correlation of EQUINOO with ENSO index is poor. Hence although seasonal predictions of the IOD index are routinely made with climate models over the last two decades, they cannot contribute to the prediction of ISMR. For improvement of seasonal predictions of ISMR, better predictions of EQUINOO and its links with the ISMR are essential.

At this point, it does not appear to be possible to get such predictions from climate models because, although most models simulate the link between ENSO and the monsoon well, almost all the models simulate the link between EQUINOO and the monsoon to be opposite in sign to the observed. Nanjundiah et al. (2013) showed that while almost all models could simulate EQUINOO, i.e., negative correlation between the rainfall over WEIO and EEIO, none of the models participating in ENSEMBLES (except the ECMWF model), could simulate realistically the link between EQUINOO and rainfall over the Indian region during the summer monsoon. Vishnu et al. (2019) showed that CFSv2, the monsoon mission model, also could not simulate realistically the link between EQUINOO and rainfall over the Indian region during the summer monsoon.

2.6 Triggering of the favorable phase of EQUINOO/IOD and sustenance of positive EQUINOO during the season

The first step in the triggering of a positive IOD is the triggering of a positive EQUINOO (Vinayachandran et al., 2009). Francis et al. (2007) have shown that every positive IOD event in the period 1982–2004 (including the events of 2003 and 2008 which got aborted at the end of July) is preceded by a severe cyclone over the Bay of Bengal in April/May. This has turned out to be true for all the subsequent IOD events as well, including that of 2019. Francis et al. (2007) showed that the strong convection over the Bay of Bengal associated with the severe cyclones suppresses the convection over the EEIO by reducing the integrated water vapor content over EEIO. The suppression of convection over the EEIO leads to enhancement of convection over the WEIO and easterly anomaly. The cyclones are relatively short-lived and the suppression of convection over the EEIO associated with them does not last for more than a few days. However, with the establishment of convection over the WEIO and associated positive feedback, involving enhancement of the easterly anomalies and hence convergence into WEIO leading to further enhancement of convection over WEIO, convection over the EEIO can remain suppressed for several weeks, leading to a sustenance of the positive phase of the EQUINOO triggered by the cyclone. This illustrates how convection over parts of the north Indian Ocean can influence the convection over EEIO and hence EQUINOO. Suppressed convection over the EEIO associated with El Niño can also have an impact on EQUINOO.

Whether the positive EQUINOO phase triggered by the storm over the Bay in April/May is sustained till the commencement of the summer monsoon season depends on the convection occurring over the Bay of Bengal, tropical Pacific as well as interaction of the atmosphere with the equatorial Indian Ocean. The oceanic variables that are important in determining the occurrence and intensity of convection over the equatorial regions are the SST of the region as well as the zonal SST gradients. Even though the prediction of IOD index (DMI) does not contribute to prediction of ISMR, the studies of IOD have unraveled important mechanisms which determine the evolution of SST and the upper layer of the equatorial Indian Ocean. The SST variability, particularly of the WEIO, depends on the convection in the atmosphere above (Vinayachandran et al., 2002) with substantial cooling occurring when there is sustained intense convection. The outcome of this coupling between convection and SST is one of the factors determining the sustenance of convection over WEIO, i.e., the positive phase of EQUINOO. The ocean dynamics play a more important role in the SST variability of EEIO (Vinayachandran et al., 2002).

2.7 An experiment in prediction of the evolution of EQUINOO and its impact on the monsoon of 2019

The summer monsoon of 2019 began with a massive deficit in the all-India June rainfall of about 33% of the mean, which could be attributed to the El Niño (Gadgil et al., 2019). The phase of EQUINOO had been favorable from May onwards and led to the recovery of the monsoon in July (anomaly +4.2%) when the El Niño weakened (Fig. 2.11). It was expected that EQUINOO will play an important role in determining the rainfall in August and September and hence the seasonal rainfall in 2019. Gadgil et al. (2019) investigated the cases like 2003 and 2008 when the positive phase of EQUINOO lasted only during June and July and of 2007 when it lasted throughout the season (Fig. 2.12) and suggested that the unusually warm WEIO played an important role in the sustenance of the positive phase in 2020. Effy et al. (2020) with analysis of observations and simulations with a high-resolution ocean model had suggested that the anomalous warming of WEIO in 2007 was due to the combined effect of the wind-induced and reflected Rossby waves in the equatorial Indian Ocean. Anomalous westerly wind burst in the equatorial Indian Ocean in early April generated an eastward propagating downwelling Kelvin wave, which got reflected from the eastern boundary as a downwelling Rossby wave and

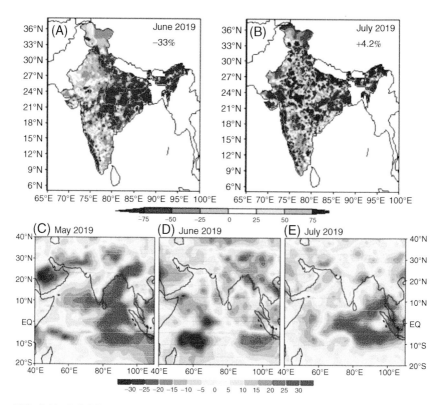

FIG. 2.11 Rainfall anomaly in mm for (A) June and (B) July 2019 (from http://imdpune.gov. in/ Seasons/). OLR anomaly for (C) May 2019, (D) June 2019, and (E) July 2019. The rainfall for 2019 is from the operational data in Climate Diagnostic Bulletin, India Meteorological Department. "*OLR*, outgoing longwave radiation."

propagated to the WEIO. In the second fortnight of April, easterlies appeared and generated another downwelling Rossby wave, which also propagated to the WEIO. The couple of downwelling Rossby waves, which propagated to the WEIO, deepened the thermocline and prohibited intense cooling in the WEIO during the early months of the summer monsoon.

In the case of 2019, unlike in 2003 and 2008, the SST of the WEIO started increasing in late July partly due to the mechanism which operated in 2007 and led to a positive SST anomaly of WEIO by August 10, while sustained cooling led to a negative SST anomaly over EEIO. Since conditions then appeared to be favorable for sustenance of a positive phase of EQUINOO, Gadgil et al. (2019) made an educated guess that the seasonal rainfall would be above normal. In fact, the warm anomaly over WEIO was sustained till the end of September 2019 (Fig. 2.13). With the positive phase of EQUINOO not only being sustained but intensifying in August and even more in September, the seasonal rainfall of 2019, turned out to be an excess (ISMR anomaly of +10%).

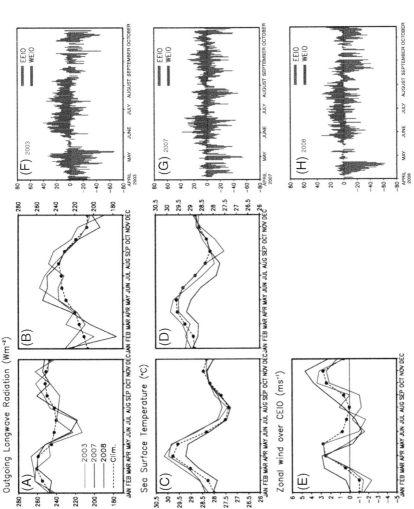

FIG. 2.12 Time series of monthly mean (A) OLR, and (C) SST averaged over the WEIO and (B) OLR and (D) SST averaged over the EEIO for 2003, 2007, and 2008. (E) Time series of monthly mean surface zonal wind averaged over the CEIO (60°–90°E, 2.5°S–2.5°N). Variation of daily convective index during April–October for the WEIO and EEIO (multiplied by −1) for 2003 (F), 2007 (G), and 1997 (H). "*CEIO*, central equatorial Indian Ocean; *EEIO*, eastern equatorial Indian Ocean; *OLR*, outgoing longwave radiation; *SST*, sea surface temperature; *WEIO*, western equatorial Indian Ocean."

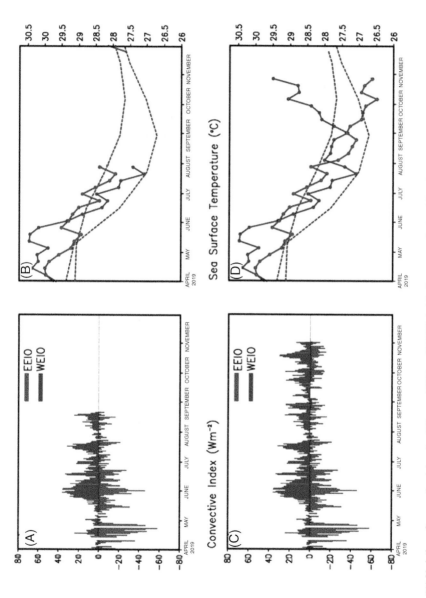

FIG. 2.13 (A) Variation of convective intensity (CI) over EEIO and WEIO (with sign reversed). (B) SST of WEIO (blue) and EEIO (red) for available data from April 1 until August 10, 2019, climatology shown as dashed lines. (C), (D), same as (A), (B), but for the period until October 31, 2019. Weekly mean optimum interpolation SST data from Climate Diagnostic Center, United States (Reynolds et al., 2002, http://www.cdc.noaa.gov/) are used. "*EEIO*, eastern equatorial Indian Ocean; *SST*, sea surface temperature. *WEIO*, western equatorial Indian Ocean."

2.8 Concluding remarks

Considerable progress has been made since the 1980s in understanding and predicting ENSO because of a combination of analysis of observations (including with Moored buoys over critical regions as well as satellites), development of simple models incorporating the critical physics and development of atmospheric GCMs, Ocean GCMs as well as coupled models capable of simulating and hence predicting different facets of ENSO. In our efforts to gain further insights into EQUINOO and its link with ISMR and developing models capable of predicting the impact on the monsoon, we can learn a lot from the success stories of ENSO predictions. We need to embark on concerted efforts with the active participation of scientists with different expertise to achieve success in simulating the impact of EQUINOO on ISMR and thus achieve improved predictions of ISMR.

References

Bjerknes, J., 1969. Atmospheric teleconnections from the equatorial pacific. J. Phys. Oceanogr. 97, 163–172.

Cane, M., Zebiak, S., 1985. A theory for El Niño and the southern oscillation. Science 228, 1085–1087. doi:10.1126/science.228.4703.1085.

Effy, B.J., Francis, P.A., Ramakrishna, S.S.V.S., Mukherjee, A., 2020. Anomalous warming of the western equatorial Indian Ocean in 2007: role of ocean dynamics. Ocean Model., 101542. doi:10.1016/j.ocemod.2019.101542.

Francis, P.A., Sulochana, G., Vinayachandran, P.N., 2007. Triggering of positive Indian Ocean Dipole events by severe cyclones over the Bay of Bengal. Tellus A 59, 461–475.

Gadgil, S., 2003. The Indian monsoon and its variability. Annu. Rev. Earth Planet Sci. **31**, 429–467.

Gadgil, S., 2012. Seasonal prediction of the Indian summer monsoon: science applications to Indian agriculture. ECMWF Seminar on Seasonal Prediction, 3-7 September, 105-130.

Gadgil, S., Gadgil, S., 2006. The Indian monsoon, GDP and agriculture. Econ. Pol. Wkly. **XLI**, 4887–4895.

Sulochana, G., Joseph, P.V., Joshi, N.V., 1984. Ocean-atmosphere coupling over monsoon regions. Nature 312, 141–143.

Gadgil, S., Sajani, S., 1998. Monsoon precipitation in the AMIP runs. Clim. Dyn. 14, 659–689. doi:10.1007/s003820050248.

Gadgil, S., Srinivasan, J., 2011. Seasonal prediction of the Indian monsoon. Curr. Sci. 100, 343–353 ISSN 0011-3891.

Gadgil, S., Vinayachandran, P.N., Francis, P.A., 2003. Droughts of the Indian summer monsoon: role of clouds over the Indian Ocean. Curr. Sci. **85**, 1713–1719.

Gadgil, S., Vinaychandran, P.N., Francis, P.A., Gadgil, S., 2004. Extremes of Indian summer monsoon rainfall, ENSO, equatorial Indian Ocean Oscillation. Geophys. Res. Lett., 31. doi:10.1029/2004GL019733.

Sulochana, G., Rajendran, K., Pai, D.S., 2019. A new rain-based index for the Indian summer monsoon rainfall. Mausam **70**, 485–500.

Graham, N.E., Barnett, T.P., 1987. Sea surface temperature, surface wind divergence, and convection over tropical oceans. Science 238, 657–659.

Gadgil, S., 2018. The monsoon system: land–sea breeze or the ITCZ? J. Earth Syst. Sci. 127, 1. doi:10.1007/s12040-017-0916-x.

Ihara, C., Kushnir, Y., Cane, M.A., De la Peña, V., 2007. Indian summer monsoon rainfall and its link with ENSO and the Indian Ocean climate indices. Int. J. Climatol. 27, 179–187.

Krishna murti, T.N., Ardanuy, P., 1980. The 10 to 20-day westward propagating mode and "breaks in the monsoons. Tellus 32, 15–26. doi:10.1111/j.2153-3490.1980.tb01717.x.

Kumar, K., Rajagopalan, B., Cane, M., 1999. On the weakening relationship between the Indian monsoon and ENSO. Science 284, 2156–2159.

Nanjundiah, R.S., Francis, P.A., Ved, M., Sulochana, G., 2013. Predicting the extremes of Indian summer monsoon rainfall with coupled ocean-atmosphere models. Curr Sci 104, 1380.

Pai, D.S., Latha, S., Rajeevan, M., Sreejith, O.P., Satbhai, N.S., Mukhopadhyay, B., 2014. Development of a new high spatial resolution (0.25 × 0.25) long period (1901-2010) daily gridded rainfall data set over India and its comparison with existing data sets over the region. Mausam **65**, 1–18.

Palmer, T.N., Brankovic C., Viterbo P., Miller M.J., 1992. Modeling interannual variations of summer monsoons. J. Climate 5, 399–417.

Pant, G.B., Parthasarathy, B., 1981. Some aspects of an association between the southern oscillation and Indian summer monsoon. Arch. Meteorol. Geophys. Bioklimatol. 1329, 245–252.

Parthasarathy, B., Munot, A.A. and Kothawale, D.R., 1995, "Monthly and seasonal rainfall series for all-India, homogenous regions and meteorological subdivisions: 1871–1994", IITM Pune Research *Rep. No. RR-065*, p 113, ISSN 0252-1075.

Rajeevan, M., Bhate, J., Kale, K.D., Lal, B., 2006. High resolution daily gridded rainfall data for the Indian region: analysis of break and active monsoon spells. Curr. Sci. **91**, 296–306.

Rasmusson, E.M., Carpenter, T.H., 1983. The relationship between eastern equatorial Pacific sea surface temperature and rainfall over India and Sri Lanka. Mon. Weather Rev. 111, 517–528.

Sajani, S., Gadgil, S., Francis, P.A., Rajeevan, M, 2015. Prediction of Indian rainfall during the summer monsoon season on the basis of links with equatorial Pacific and Indian Ocean climate indices. Environ. Res. Lett. 10 094004.

Saji, N.H., Goswami, B.N., Vinayachandran, P.N., Yamagata, T., 1999. A dipole mode in the tropical Indian Ocean. Nature 401, 360–363.

Sikka, D.R., Gadgil, S., 1980. On the maximum cloud zone and the ITCZ over India longitude during the southwest monsoon. Mon. Weather Rev. 108, 1122–1135.

Vinayachandran, P.N., Iizuka, S., Yamagata, T., 2002. Indian Ocean dipole mode events in an ocean general circulation model. Deep Sea Res. Part II 49, 1573–1596.

Vishnu, S., Francis, P.A., Ramakrishna, S.S.V.S., Shenoi, S.S.C, 2019. On the relationship between the Indian summer monsoon rainfall and the EQUINOO in the CFSv2. Clim. Dyn. 52, 1263–1281.

Webster, P.J., Moore, A.M., Loschnigg, J.P., Leben, R.R., 1999. Coupled ocean–atmosphere dynamics in the Indian Ocean during 1997–1998. Nature 401, 356–360.

Zebiak, S.E., Cane, M.A., 1987. A model El Niño-southern oscillation. Mon. Weather Rev. 97, 163–172.

Part I

ENSO-Indian Summer Monsoon teleconnections

Chapter 3

ENSO–Indian summer monsoon teleconnections

Ashwini Kulkarni[a], K. Koteswara Rao[b], Manish K. Joshi[a], Archana Rai[a], P. Darshana[a,c]

[a]Indian Institute of Tropical Meteorology, Ministry of Earth Sciences (IITM-MoES), Pune, Maharashtra, India, [b]Centre for Climate Change and Sustainability, Azim Premji University, Bengaluru, India, [c]Savitribai Phule Pune University, Pune, India

3.1 Introduction

The interannual variations of the Indian summer monsoon (ISM) rainfall (ISMR) are characterized by excess, deficit, and normal monsoons (Parthasarathy and Mooley, 1978; Shukla, 1987). The interannual variability of ISMR is governed by the slowly varying surface boundary conditions, such as sea surface temperatures (SSTs), soil moisture, snow cover, etc. (Charney and Shukla, 1981; Shukla, 1998). The most prominent and known driver of interannual variability of ISMR is the El Niño-Southern Oscillation (ENSO, Shukla and Paolino, 1983). An El Niño event is generally identified with the invasion of warm water from the western equatorial Pacific into the central and/or eastern equatorial Pacific Ocean, in conjunction with the cessation of upwelling of cold water along the equator. Toward the end of the 19th century, Sir Gilbert Walker initiated the research which was originally motivated not by ENSO but by occasional disastrous failures of the ISM. He was unaware of El Niño but knew the evidence that interannual pressure fluctuations over the Indian Ocean and the eastern tropical Pacific Ocean were out of phase. When pressure is high in the Pacific Ocean, it tends to be low in the Indian Ocean from Africa to Australia. Walker (1924, 1928) termed this phenomenon as Southern Oscillation (SO). Bjerknes (1969) reported the phase of El Niño to that of SO when the trade winds are weak and surface pressure is low over eastern and high over western tropical Pacific. Thus, ENSO is a coupled ocean-atmosphere process with SO as the atmospheric component of El Niño. The strength of SO is measured by the Southern Oscillation Index (SOI), which is the surface air pressure difference between Tahiti (in the Pacific) and Darwin, Australia (in the Indian Ocean). El Niño episodes have negative SOI, lower surface pressure over Tahiti, and higher over Darwin. The opposite phase of El Niño is anomalous cold temperatures

Indian Summer Monsoon Variability: El Niño-teleconnections and beyond.
DOI: https://doi.org/10.1016/B978-0-12-822402-1.00024-7

off the coast of Peru termed as La Niña. These La Niña episodes have positive SOI, higher surface pressure over Tahiti, and lower in Darwin. El Niño represents the oceanic component of climate variability while SO represents the atmospheric component hence the term ENSO is generally used to describe the combined ocean-atmosphere phenomenon. El Niño episodes are defined as sustained anomalous warming of the central and eastern tropical Pacific Ocean, thus resulting in a decrease in the strength of the Pacific trade winds, and a reduction in rainfall over Eastern and Northern Australia. La Niña episodes are defined as sustained anomalous cooling of the central and eastern tropical Pacific Ocean, thus increasing the strength of the Pacific trade winds, and the impacts are opposite in Australia compared to El Niño. Globally, El Niño events usually lead to a short-term rise in averaged temperatures while global-mean temperatures typically decrease during La Niña events (Cane et al., 1997).

The ENSO causes climatic fluctuations in the tropics and extratropics via atmospheric teleconnections. A developing El Niño shifts the Walker circulation to the east such that the torrential rain occurs over the eastern equatorial Pacific and anomalous subsidence occurs over the ISM region. A developing La Niña event has the opposite effects. The anomalous Pacific SST associated with ENSO modulates the Walker and Hadley circulations, causing profound impacts on rainfall and temperature over land and oceans. Major ENSO episodes lead to massive displacements of the rainfall regions of the tropics, bringing droughts to vast areas, and torrential rains to otherwise low rainfall regions. During El Niño events, anomalously dry conditions are generally observed in the Maritime Continent, Australia, Northern South America, Southern Asia, and Southern Africa, while anomalously wet conditions typically occur in Southwestern North America, Western Antarctica, and Eastern Africa (Taschetto et al., 2020). Though the global effects of La Niña are roughly opposite in sign, this is not true for all regions. This nonlinearity of ENSO atmospheric teleconnections is caused by variations in the location of the anomalous equatorial warming superimposed on the Pacific mean state and interactions of ENSO with off-equatorial regions and other ocean basins, as well as with the annual cycle and other modes of climate variability (Taschetto et al., 2020). Furthermore, the nonstationary behavior of ENSO teleconnections can occur either due to stochastic variability or deterministic low-frequency modulations.

The ENSO–monsoon relationship on an interannual timescale has been studied extensively (Sikka, 1980; Pant and Parthasarathy, 1981; Rasmusoon and Carpenter, 1983; Shukla and Paolina, 1983; Parthasarathy and Pant, 1985; Kripalani and Kulkarni, 1997; Goswami, 1998; Webster et al., 1998; Kumar et al., 1999; Lau and Nath, 2000, Ashok et al., 2007, Ashok et al., 2019). Though there is no one-to-one relationship between ENSO and ISMR, almost 50% of the Indian droughts are associated with El Niño however in the post-1980s the ENSO–monsoon relationship has weakened (Kripalani and Kulkarni, 1997; Kumar et al., 1999), inspite of increasing El Niño events. Some of the intense droughts in the recent period (e.g., 2004, 2014) are not associated with canonical/eastern Pacific El Niño.

In post-1990 more number of ENSO events with central Pacific warming is reported compared to canonical ENSO events, the former is termed as ENSO Modoki (pseudo-El Niño, Ashok et al., 2007). Another coupled-mode from the Indian Ocean is the Indian Ocean Dipole (IOD) (i.e., Saji et al., 1999), which is also known to modulate interannual variability of ISMR. In addition to IOD and ENSO, there is a strong link between ISMR and the equatorial Indian Ocean oscillation (EQUINOO; Gadgil et al., 2004; see Chapter-2). Eurasian snow cover also plays a major role in the year-to-year variability of ISMR (e.g., Blanford, 1884; Kripalani and Kulkarni, 1999; Bamzai and Shukla, 1999, also see Chapter 12). It has been observed that generally deficit monsoons over India which are not associated with El Niño have been associated with excessive snow depth over Eurasia (Kripalani and Kulkarni, 1999). Other than ENSO, the Atlantic Niño (e.g., Kucharski et al., 2008; Yadav et al., 2018), western North Pacific circulation changes (e.g., Chowdary et al., 2019; Srinivas et al., 2018), tropical Indian Ocean basin-wide warming (e.g., Yang et al., 2007; Chowdary et al., 2015; Chakravorty et al., 2016) also play a role in modulating monsoon interannual and decadal variability (e.g., Yadav, 2017).

In this chapter, we discuss the ENSO–ISM teleconnections in detail. Data used for this chapter are given in Section 3.2. ENSO evolution/cycle is discussed in Section 3.3. The ENSO–ISM teleconnections and the possible attribution of recent weakening relationships are discussed in Section 3.4. Section 3.5 presents the ISMR variability during ENSO and non-ENSO events. The summary of the chapter is provided in Section 3.6.

3.2   Data

The data used in the chapter are as follows.

(i)   The high resolution daily gridded ($0.25° \times 0.25°$ long/lat) rainfall data prepared by India Meteorological Department, for the period 1901–2018 has been used (Pai et al., 2014). The all-India summer monsoon rainfall time series for the period 1901–2018 has been prepared by averaging this high-resolution data over Indian landmass and is utilized to study the ISMR variability.

(ii)   The Extended Reconstructed Sea Surface Temperature (ERSST) dataset is a global monthly SST dataset derived from the International Comprehensive Ocean-Atmosphere Dataset. The newest version of Extended Reconstructed Sea Surface Temperature, version 5, uses new data sets from International Comprehensive Ocean-Atmosphere Dataset Release 3.0 SST. The monthly gridded ($2° \times 2°$ long/lat) data are available for 1854 to date on www.ncdc. noaa.gov. The data are used to define El Niño and La Niña events as well as to generate SST composites (Huang et al., 2017).

(iii)   The time series for Niño3.4 index ($5°$S–$5°$N, $170°$W–$120°$W) is prepared by averaging the SST anomaly (SSTA) over the Niño3.4 box for the period 1901–2018. This time series is normalized by the standard deviation of

SSTA in Niño3.4, El Niño (La Niña) year are identified if the normalized SSTA for the season JJAS (June through September) is more (less) than +1(−1).

(iv) To examine the circulation patterns the zonal and meridional winds at 850 hPa, 20th century reanalysis (20CR.V3) data from www.psl.noaa.gov (Slivinski et al., 2019) have been used.

3.3 El Niño-Southern Oscillation cycle evolution

ENSO is unique among all climate phenomena in its strength, predictability, global influence, and atmospheric teleconnections that affect patterns of weather and climate variability around the globe. It is the largest interannual climate signal and is the climate phenomenon that essentially involves coupled interactions of ocean and atmosphere (Philander, 1990). The ENSO cycle describes the fluctuations in temperature between the ocean and atmosphere in the east-central equatorial Pacific (approximately between the International Date Line and 120°W). El Niño and La Niña episodes typically last 9–12 months, but some prolonged events may last for years. El Niño and La Niña typically recur every 2–7 years and develop in association with fluctuations in atmospheric pressure patterns, the SO, in the tropical Indian and Pacific Oceans (Larkin and Harrison, 2002).

El Niño typically has four distinct phases in its life cycle: precursor, onset, growth, and decay. The precursor starts with an intensification of the prevailing weather patterns. The onset of El Niño occurs around December. The abnormally intense prevailing conditions suddenly change. SSTs drop in the Western Pacific and rise in the east. The growth phase is the continuation of the onset. SSTs of the coast of South America continue to rise, reaching a maximum in June. The flow of warm water from western to eastern Pacific raises sea level in the east and pushes the thermocline deeper. These conditions continue to intensify during the year. Winds continue to blow from west to east, rainfall decreases dramatically over Indonesia, and falls heavily over Central and Eastern Pacific and South American Pacific coast. The conditions reach a peak about a year after the onset and then westerly wind anomalies start weakening, heralding the start of the decay phase of ENSO. A year and a half after the onset the Pacific weather returns to normal. In general, El Niño events tend to only last for a single cycle (i.e., 1 year from boreal spring to spring), but it is not uncommon for multiyear La Niña events to occur (Yang et al., 2018; Wang, 2002; Cai et al., 2014). An especially strong Walker circulation causes La Niña, resulting in cooler ocean temperatures in the central and eastern tropical Pacific Ocean due to increased upwelling. A schematic for Walker circulation and SST changes over Pacific in normal, El Niño, and La Niña conditions during peak phase (boreal winter) season is shown in Fig. 3.1.

The ENSO-driven large-scale atmospheric teleconnections alter the near-surface air temperature, humidity, and wind as well as the distribution of clouds

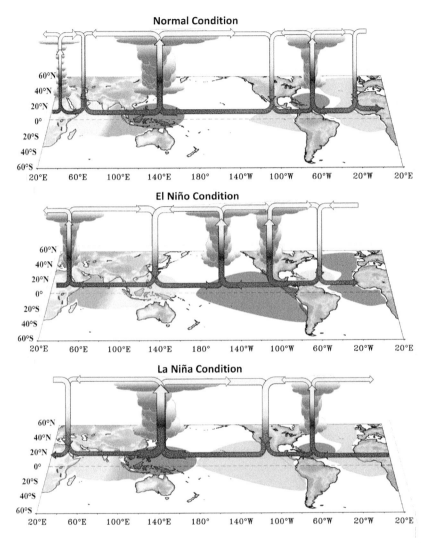

FIG. 3.1 Schematic showing changes in Walker circulation and SST anomalies during normal (top), canonical El Niño (middle), and canonical La Niña (bottom) conditions. *Blue* (*red*) shading represents cold (warm) SSTAs and *red* and *orange* (*blue*) arrows indicates low-level (upper-level) winds. Shading over the land region indicates Topography, green (brown) colour indicate low (high) topography. *SST*, sea surface temperature; *SSTAs*, sea surface temperature anomalies.

far from the equatorial Pacific. The resulting variations in the surface heat, momentum, and freshwater fluxes can induce changes in SST, salinity, mixed layer depth, and ocean currents. Thus, the atmosphere acts as a bridge spanning from the equatorial Pacific to the North Pacific and to the South Pacific, the Atlantic, and Indian Oceans (Alexander et al., 2002). The ENSO-related SST anomalies that develop over the world's oceans can also feedback on the original atmospheric response to ENSO. During El Niño event the regions of moist air convergence are displaced. This change is transmitted around the world through upper atmospheric processes leading to widespread shifts in the normal patterns of rain, temperature, and winds. During strong El Niño phases the displacement of warm waters in the Pacific causes anomalous convective activity over central-eastern Pacific, central western equatorial Indian Ocean, and off/near the Atlantic equatorial coast of Africa and northwestern South America (Allan et al., 1996).

3.4 ENSO–Indian summer monsoon teleconnections

Fig. 3.2A shows the time series of Niño3.4 normalized SSTA. It is observed that after the climate shift of 1975/76 the La Niña events have become less frequent. There are 11 El Niño events in pre-1975 and 6 events in post-1975. Hence there are 17 El Niño episodes (1902, 1905, 1914, 1919, 1930, 1941, 1951, 1957, 1963, 1965, 1972, 1982, 1987, 1997, 2002, 2009, 2015) and 21 La Niña episodes (1909,1910, 1916, 1924, 1933, 1938, 1942, 1945, 1950, 1954, 1955, 1956, 1964, 1970, 1971, 1973, 1975, 1988, 1998, 1999, 2010) during 1901–2018. The SOI index also shows similar variations as in Niño 3.4 but opposite in phase (Fig. 3.2B). The mean SST anomalies in preclimate shift period are 1.4°C while that in the postclimate shift period is 2.08°C. The difference is significant at 10% level. Fig. 3.2C shows Niño 3.4 SSTA correlation with SSTA, sea level pressure (SLP), and 850 hPa vector winds anomalies over the Indo-Pacific Ocean for the summer season (June through September). Niño 3.4 SSTAs display a strong negative correlation with SLP over the southeastern Pacific while the strong positive relationship with SLP over the Southern Indian Ocean. As Niño 3.4 SSTAs increase, the SLP over Southeast Pacific (south Indian Ocean) decreases (increase) which corresponds to negative SOI or weak Walker circulation. There are strong surface westerlies from the Southern Indian Ocean to the Western Pacific up to the east equatorial Pacific which confirms the eastward shift of Walker circulation during El Niño years. The southwesterlies over the Arabian Sea are weakened implying weak monsoon circulation and causes a reduction in rainfall over Indian region.

The composite of SST, SLP, and 850 hPa wind anomalies for El Niño and La Niña years for June through September are shown in Fig. 3.3. The shading shows SSTAs while contours depict the SLP anomalies. During El Niño the SSTA over the eastern equatorial Pacific are anomalously warm, the pressure is anomalously low, 6-10 hPa below normal, and high over the southern Indian Ocean which implies SO is weak and strong westerly winds are blowing from the western Pacific manifest shift of the Walker circulation to the east and hence there are torrential rains over eastern Pacific which is otherwise dry. Exactly opposite circulation is observed during La Niña episodes (Fig. 3.3B). The pressure departures are low over the Southern Indian Ocean, anomalously high over

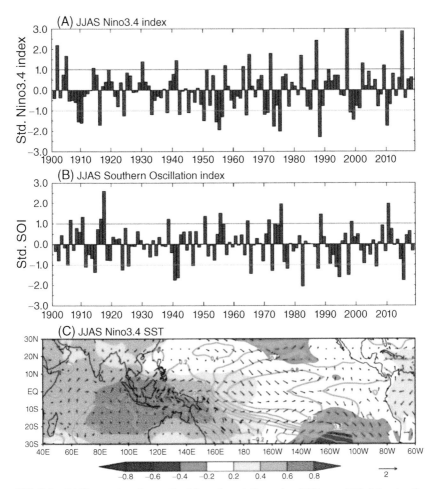

FIG. 3.2 (A) Time series of normalized June through September SSTA over Niño3.4 region for 1901–2018. Bars above (below) *red* (*green*) line are El Niño (La Niña) events. (B) Same as in (A) but for SOI and (C) Niño 3.4 SSTAs correlation with tropical Indo-Pacific SSTAs (contours) as well as with SLP (shading) and 850 hPa vector winds anomalies. *SLP*, sea level pressure; *SOI*, Southern Oscillation Index; *SSTA*, sea surface temperature anomaly.

the Eastern Pacific and Indian landmass. The SO is strong, there are strong surface easterlies from the Eastern Pacific which imply strong Walker circulation which is conducive for good monsoon activity over India, Indonesia, and Australia. Composite of El Niño events and correlation analysis show similar patterns (Fig. 3.2C, Fig. 3.3).

The basic mechanisms of the inverse ENSO–ISMR relationship involve Walker circulation, Hadley circulation, changes in the tropospheric temperature (TT) gradient, etc. A thermodynamic interpretation of teleconnection is given

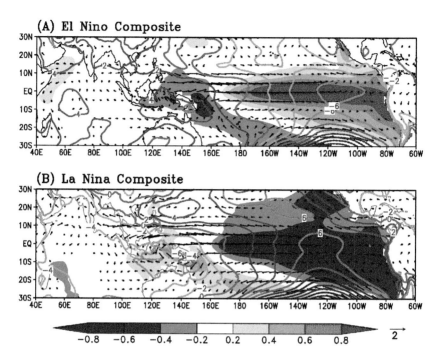

FIG. 3.3 Composite (JJAS) of SST anomalies (°C, shading) for the tropical Indo-Pacific region superimposed by SLP anomalies (hPa, contours) and 850 hPa vector wind anomalies (m/s) for El Niño (A) and La Niña (B) events. *SLP*, sea level pressure; *SST*, sea surface temperature.

as ENSO could impact ISMR by influencing TT gradient in the Asian monsoon region (Goswami and Xavier, 2005; Xavier et al., 2007). The heat source associated with El Niño or La Niña also results in stationary Rossby waves in the subtropics and leads to a persistent negative (positive) TT anomaly over Northern India and Southern Eurasia weakening (strengthening) the TT gradient over the ISM region and thereby the ISMR (Fig. 3.4A and B). TT is also responsible for the length of the rainy season. The rainy season is shorter in most of the El Niño years and longer during La Niña years (Goswami and Chakravorty, 2018). Further, the large-scale upper-level convergent winds (low-level divergence) representing strong subsidence over the ISM region associated with El Niño is apparent and vice-versa for La Niña events (Fig. 3.4C and D).

The long-term mean ISMR is 857.5 mm with a standard deviation of 77.7 mm based on the period 1901–2018. The ISMR time series has been normalized by subtracting the mean and normalized by the standard deviation. Fig. 3.5A shows the interannual variability of ISMR. Green bars show excess monsoons (normalized rainfall anomaly > 1), the red bars show deficit monsoons (normalized rainfall anomaly < −1) associated with El Niño and the hollow red bars, nondeficit or normal monsoons during El Niño years. The black

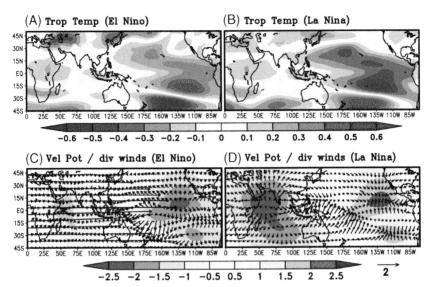

FIG. 3.4 Composite (JJAS) of tropospheric temperature anomalies (°C, shading) for (A) El Niño events (B) La Niña events. (C) and (D) are the same as in (A) and (B) but for 200 hPa velocity potential (10^{-6} m³/s) and divergent (/s) wind.

bars are normal monsoon years and hollow black bars are normal monsoons despite the occurrence of El Niño. It can be clearly seen that there are as many deficit monsoons associated with El Niño as those with non-El Niño cases; however, the most severe droughts are associated with El Niño events. During the period 1930–60 when ISMR was in the above normal epoch of its natural variability (Kripalani and Kulkarni, 1997), there were only three events when ISMR was deficit along with El Niño. As rightly mentioned by Kripalani and Kulkarni (1997), the impact of El Niño is severe on ISMR if El Niño happens in the below normal phase of natural variability of ISMR. Gadgil et al. (2003, 2004) have shown that the large deficits in ISMR are also associated with the EQUINOO, the difference in anomalous convection over the west and east equatorial Indian Ocean. ISMR is a large deficit if both EQUINOO and El Niño are unfavorable (see Chapter 2). The probability of occurrence of El Niño event remains the same in both the periods (14 %), before and after the climatic shift of 1975/76. The ENSO–ISMR relationship has weakened in post-1975 period (correlation = −0.49) as compared to pre-1975 period (correlation = −0.60).

The relationship between ENSO and ISMR can be best judged through scatter plot. Fig. 3.5B depicts the variation in normalized ISMR anomaly against Niño3.4 index. Green dots represent La Niña events, red represent El Niño events, while blue is normal/neutral conditions over the Pacific. Clearly, ENSO and ISMR hold the inverse relationship with a correlation −0.54 which is significant at 1% level. Hence El Niño explains 25% variability of ISMR. It is

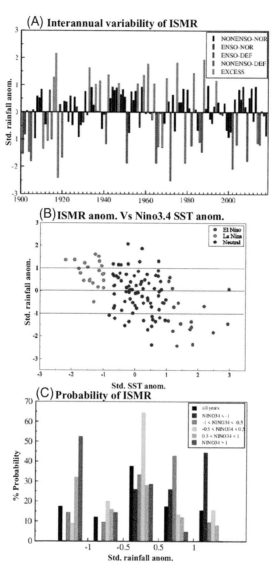

FIG 3.5 (A) Interannual variability of ISMR 1901–2018. The bars show excess (green), normal-non-ENSO (black), normal-ENSO (black hollow), deficit-El Niño (red), deficit-non El Niño (red hollow) monsoons, (B) scatter plot of normalized ISMR against normalized Niño3.4 SSTA. *Red (green)* dots represent El Niño (La Niña), *blue* dots are normal conditions over Pacific, and (C) the probability of ISMR in five categories, namely, deficit (normalized rainfall anomaly < −1), below normal (−1 < normalized rainfall anomaly ≤ −0.5), normal (−0.5 < normalized rainfall anomaly ≤ 0.5), above normal (+0.5 ≤ normalized rainfall anomaly < +1) and excess (normalized rainfall anomaly > +1) for June to September Niño3.4 anomaly categories namely cold (< −1), cool (−1 to −0.5), normal (−0.5 to +0.5), warm (+0.5 to 1), hot (> +1). *ISMR*, Indian summer monsoon rainfall; *SSTA*, sea surface temperature anomaly.

observed that not all El Niños are associated with deficit monsoons, also there are deficit monsoons when El Niño is absent. Similarly, when there is La Niña, ISMR is normal or excess. It is important to note that ISMR was not excess during El Niño events and it was not deficit when La Niña happens. The variations of normalized ISMR anomaly in various categories of ENSO index are shown in Fig. 3.5C. The deficit/below normal ISMR is mostly associated with high positive categories of SSTA in the Niño 3.4 region and vice versa. It is noted that 50% of the ISMR deficits are associated with El Niños and 70% of La Niña years witnessed above normal/excess ISMR.

3.4.1   Weakening of ENSO–monsoon relationship

The ENSO and ISMR relationship exhibits nonstationary behavior (Kumar et al., 1999; Kinter et al., 2002), and variations in both the amplitude and phase of these two oscillatory systems decide the evolution of this relationship (Maraun and Kurths, 2005). Many studies have documented that after the 1990s the ENSO–ISM relationship has weakened. Average 21-year sliding correlations between ISMR and Niño3.4 SSTs show changes in teleconnections (see Fig. 1.6 of Chapter-1). After the 1990s, the relationship seems to be weakening however in the recent decade it is strengthening again. The weakening can be attributed to many factors. The strengthening and poleward shift of the jet stream over the North Atlantic may disturb the conventional ENSO–ISMR relationship (Chang et al., 2001). Some suggest that changes in the Indo-Pacific circulation play a role in weakening of this relationship (Feba et al., 2019). Some studies attribute the weakening of ENSO–ISMR relationship to air–sea coupled interaction over the tropical Indian Ocean (Sreejith et al., 2015). The increasing frequency of different types of ENSO may be responsible for the weakening relationship between ISMR and canonical El Niño (Goswami and Chakravorty, 2018). Some studies suggest the southeastward shift of Walker circulation anomalies associated with ENSO events could lead to reduced subsidence over the Indian region which favors normal monsoon conditions (Kumar et al., 1999). Anomalous warming over Eurasia in winter and spring may favor the enhanced land–ocean thermal gradient conducive for a strong monsoon over India (Kripalani and Kulkarni, 1999; Kumar et al., 1999). The weakening ENSO–monsoon relationship is also attributed to recent period warming in Central Pacific flanked by cold SSTA on both sides along the equator, i.e., El Niño Modoki (Ashok et al., 2007; Kumar et al., 2006). In post-1990 period the weakening of equatorial easterlies associated with weakened zonal SST gradient which leads to more flattening of thermocline may cause more frequent Modokis (Ashok et al., 2007). Also, warm SSTA in central Pacific has more impact on ISMR which produces severe deficit monsoons as compared to canonical El Niño (Kumar et al., 2006) mainly over southern peninsular India. However, the contrary views are also there. Another study (Xavier et al., 2007) claims that the "real" ENSO–monsoon relationship has actually not decreased and the weakening correlation between ISMR and ENSO is a manifestation of increasing "climate noise" in ISMR owing to the increase in extreme rain events in recent years.

It is observed that post-1960 Indian Ocean has been warming rapidly and there are more frequent IOD events. When ENSO occurs during ISM season, IOD could influence the ENSO–ISMR relationship depending on the phase and amplitude of IOD and ENSO (Ashok et al., 2001). When ENSO–ISMR relationship is low, IOD–ISM relation is strong and vice versa. A coherence between IOD–ENSO plays a major role in determining the strength of ISMR (see Chapter 8). During the high coherence period of pre-1980s, ENSO drives both the IOD and regional rainfall, and the IOD's influence cannot manifest itself. During a weak coherence period/epoch in the post-1980s, a positive IOD leads to increased Indian rainfall, offsetting the impact from El Niño (Li et al., 2016). For El Niño events that co-occur with positive IOD events, Indian Ocean conditions act to counter El Niño's drought-inducing subsidence by enhancing moisture convergence over the Indian subcontinent, which results in ISMR being nondeficit. It is important to note that the ENSO is shown to have a strong adverse impact on ISMR if there is a phase-locking between ENSO and ISMR (Kripalani and Kulkarni, 1997). When ISMR is in the below (above) normal epoch of its natural variability and El Niño happens then it has a severe (weak) impact on ISMR.

3.5 ISMR variability: ENSO and non-ENSO

In this section, we discuss the characteristics of deficit and nondeficit monsoons during ENSO and non-ENSO years and their associated circulation patterns. Fig. 3.6A shows the mean rainfall pattern based on 1901–2018 and the contours represent the standard deviation in mm/day. The composite rainfall anomalies in excess and deficit ISMR are also shown in Fig. 3.6B and C, respectively. Strong positive rainfall anomalies are seen over the monsoon core zone in excess years and negative rainfall departures in deficit years. Thus the major difference in rainfall during excess and deficit monsoon years is seen over the monsoon core zone or Central India (20°–28°N; 72°–87°E). Now we discuss the mean rainfall patterns associated with (1) El Niño-deficit ISMR, (2) non-El Niño—deficit ISMR, and (3) El Niño-nondeficit ISMR. Fig. 3.6D–F shows the composite pattern of seasonal mean rainfall anomalies of deficit monsoons in El Niño (left; 1905, 1941, 1951, 1965, 1972, 1982, 1987, 2002, 2009, 2015), deficit monsoons in non-El Niño years (middle; 1901, 1904, 1911, 1913, 1915, 1918, 1920, 1966, 1968, 1979, 1986, 2004, 2014) and nondeficit monsoons during El Niño episodes (right; 1902, 1914, 1919, 1930, 1957, 1963, 1997). The deficit monsoons in El Niño and non-El Niño years show the major difference over central India and the west coast, which are the major seasonal rainfall regions over India. During El Niño episodes the monsoon core region shows negative anomalies with high magnitude, more than 3 mm/day. Whereas during non-El

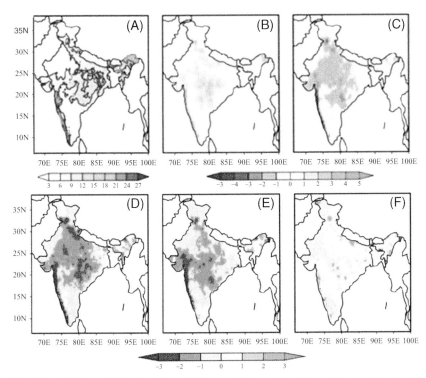

FIG. 3.6 (A) Mean ISMR (shaded; mm/day) over India and standard deviation (contours), (B) the composite mean rainfall in excess, and (C) deficit monsoons. (D) Composite mean summer monsoon rainfall anomalies (mm/day) for El Niño-deficit ISMR, (E) Non-El Niño-deficit ISMR (middle), and (F) El Niño-nondeficit ISMR. *ISMR*, Indian summer monsoon rainfall.

Niño years central India receives more rainfall as compared to El Niño years. Positive rainfall anomalies are seen in some places over central India. Thus, El Niño shows a strong inverse relationship mainly with the rainfall over the monsoon core zone.

Fig. 3.7 shows the composite SSTA during deficit monsoons associated with El Niño and non-El Niño conditions and nondeficit monsoons during El Niño episodes. During deficit ISM associated with El Niño, anomalous warming over the Pacific as well as basin-wide warming over the Indian Ocean are apparent (Fig. 3.7A). The Eastern Pacific is warmer than the Indian Ocean. This eastern equatorial Pacific warming is accompanied by strong westerly wind anomalies. Thus the Walker circulation display ascending branch over the Eastern Pacific and descending over the Indian Ocean which weakens the monsoon over India (Figs. 3.7 and 3.4). The weak westerlies over the equatorial Indian Ocean correspond to weak monsoon circulation. During Non-El Niño-deficit ISMR (Fig. 3.7B), SSTA are weak over the Pacific as well as over the Indian Ocean. Weak northeasterly wind anomalies over the ISM region during non-El Niño deficit

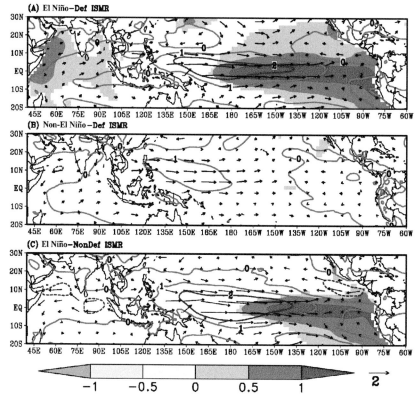

FIG. 3.7 Composite SST anomalies (shaded,°C), 850 hPa vector winds (m/s), and zonal wind (contours, m/s) for (A) El Niño-deficit ISMR, (B) Non-El Niño-deficit ISMR, and (C) El Niño-nondeficit ISMR. *ISMR*, Indian summer monsoon rainfall; *SST*, sea surface temperature.

years seem to be responsible for weak monsoon. However, the factors responsible for such deficit monsoons are yet to be explored. The deficit monsoons associated with non-El Niño episodes are not as severe as those associated with El Niño episodes.

The composite SST and 850 hPa wind anomalies of El Niño episodes but nondeficit monsoons are shown in Fig. 3.7C. Though El Niño conditions prevail over the Pacific, the strength of warm SSTs is much weaker than those in Fig 3.7A. Also, the anomalously warm SSTs do not extend beyond the dateline. The surface winds (Fig 3.7C) show easterlies over the equatorial Indian Ocean similar to the positive phase of EQUINOO and so favors ISMR, hence despite the presence of El Niño, ISMR is not deficit. It has been reported that when warming over the eastern equatorial Pacific starts from boreal winter and evolves early so that the West-Central Pacific and the Indian Ocean are warmer than normal during the summer monsoon season, and ISMR is not deficit in spite of presence of El Niño. In contrast when the eastern equatorial Pacific starts warming rapidly only about a season before the reference summer so that the Western-Central Pacific and Indian Oceans remain cold during the monsoon season, and the famous inverse relationship between ENSO–ISMR holds (Ihara et al., 2008).

3.6 Summary

ENSO is one of the major drivers of the earth's climate variability which impacts the climate all over the globe through teleconnections. The El Niño (warm phase) and La Niña (cold phase) are the two opposite phases of the ENSO cycle. It has a periodicity of 2–7 years. Generally, the anomalous warming over the eastern equatorial Pacific (El Niño) has adverse impacts on Indian summer monsoon rainfall (ISMR) but not all El Niños witness ISM deficit. The anomalous Pacific SST associated with ENSO modulates the Walker and Hadley circulations, causing profound impacts on rainfall and temperature over land and oceans. Major ENSO episodes lead to anomalously dry conditions over the Maritime Continent, Australia, Northern South America, Southern Asia, and Southern Africa, while anomalously wet conditions over Southwestern North America, Western Antarctica, and Eastern Africa. Though the global effects of La Niña are roughly of opposite sign, this is not true for all regions.

ENSO and ISMR hold an inverse relationship with a correlation of −0.54 which is significant at 1% level. Almost 50% of the Indian droughts are associated with El Niño. The basic mechanisms of the inverse ENSO–ISMR relationship involve Walker circulation, Hadley circulation, changes in TT gradient. Also, the rainy season is shorter in most of the El Niño years and longer during La Niña years (Goswami and Chakravorty, 2018). It is observed that not all monsoons are deficit when El Niño happens, also there are deficit monsoons when El Niño is absent. Similarly, when there is La Niña, ISMR is normal or excess. It is important to note that ISMR was not excess during El Niño events and was not deficit when La Niña events happened. The deficit monsoons in El Niño and non-El Niño years show the major difference over Central India and the west coast, which are important regions for seasonal rainfall over India. During El Niño episodes the monsoon core region shows negative anomalies with high magnitude, more than 3 mm/day. Whereas during non-El Niño ISMR deficit years central India receives more rainfall as compared to El Niño years. However, in the post-1980s the ENSO–monsoon relationship has been weakened (e.g., Kripalani and Kulkarni, 1997; Kumar et al., 1999), in spite of increasing El Niño events. Some of the intense droughts in the recent period (e.g., 2004, 2014) are not associated with eastern Pacific El Niño. After the 1990s the relationship shows consistent weakening however in the recent decade it seems to be strengthening again. Also, other types of El Niño became more frequent after the 1990s, such as El Niño Modoki (central Pacific warming) and basin-wide warming (see Chapter 4). Hence ENSO–monsoon teleconnections remain partly understood, their response under the background/future warming scenario (Chapter 20) needs future attention.

Acknowledgments

The authors thank Prof Ravi Nanjundiah, Director, Indian Institute of Tropical Meteorology, Ministry of Earth Sciences for providing all the facilities and encouragement.

References

Alexander, M.A., Bladé, I., Newman, M., Lazante, J.R., Lau, N.-C., Scott, J.D., 2002. The atmospheric bridge: the influence of ENSO teleconnections on air-sea interaction over the global oceans. J. Clim. 15, 2205–2231.

Allan, R., Lindesay, J., Parker, D., 1996. El Niño—Southern Oscillation and Climate Variability. CSIRO Publishing, Har/Cdr Edition, Australia, p. 416.

Ashok, K., Guan, Z., Yamagata, T., 2001. Impact of the Indian Ocean Dipole on the relationship between the Indian Monsoon rainfall and ENSO. Geophys. Res. Lett. 28, 4499–4502.

Ashok, K., Behera, S.K., Rao, S.A., Weng, H., Yamagata, T., 2007. El Niño Modoki and its possible teleconnection. J. Geophys. Res. 112, C11007. doi:10.1029/2006JC00379.

Ashok, K., Feba, F., Tejavath, C.T., 2019. The Indian summer monsoon rainfall and ENSO. Mausam 70 (3), 443–452.

Bamzai, A.S., Shukla, J., 1999. Relation between Eurasian snow cover, snow depth, and the Indian summer monsoon: an observational study. J. Clim. 12, 3117–3132.

Bjerknes, J., 1969. Atmospheric teleconnections from the equatorial Pacific. Mon. Weather Rev. 97, 163–172.

Blanford, H.F., 1884. On the connexion of Himalayan snowfall and seasons of drought in India. Proc. R. Soc. London 37, 3–22.

Chang, C.P., Harr, P., Ju, J., 2001. Possible roles of Atlantic circulation on the weakening Indian monsoon rainfall-ENSO relationship. J. Clim. 14 (11), 2376–2380.

Cai, W., Borlace, S., Lengaigne, M., Van Rensch, P., Collins, M., Vecchi, G., Timmermann, A., Santoso, A., McPhaden, M.J., Wu, L., et al., 2014. Increasing frequency of extreme El Niño events due to greenhouse warming. Nat. Clim. Chang. 4, 111–116.

Cane, M., et al., 1997. Twentieth-century sea surface temperature trends. Science 275, 957–960.

Charney, J.G., Shukla, J., 1981. Predictability of monsoons, in Monsoon Dynamics. Cambridge University Press, Cambridge, pp. 99–108 edited by J. Lighthill and R. P. Pearce.

Chakravorty, S., Gnaseelan, C., Pillai, P., 2016. Combined influence of remote and local SST forcing on Indian summer monsoon rainfall variability. Clim. Dyn. 47, 2817–2831. doi:10.1007/s00382-016-2999-5.

Chowdary, J.S., Bandgar, A., Gnanaseelan, C., Luo, Jing-Jia, 2015. Role of tropical Indian Ocean air–sea interactions in modulating Indian summer monsoon in a coupled mode. Atm. Sci. Let. 16, 170–176.

Chowdary, J.S., Patekar, D., Srinivas, G., Gnanaseelan, C., Parekh, A., 2019. Impact of the Indo-Western Pacific Ocean capacitor mode on South Asian summer monsoon rainfall. Clim. Dyn. 53, 2327.

Feba, F., Ashok, K., Ravichandran, M., 2019. Role of changed Indo-Pacific atmospheric circulation in the recent disconnect between the Indian summer monsoon and ENSO. Clim. Dyn. 52, 1461–1470.

Gadgil, S., Vinaychandran, P.N., Francis, P.A., 2003. Droughts of the Indian summer monsoon: role of clouds over the Indian Ocean. Curr. Sci. 85 (12), 1713–1719.

Gadgil, S., Vinayachandran, P.N., Francis, P.A., 2004. Extremes of the Indian summer monsoon rainfall, ENSO and equatorial Indian ocean oscillation. Geophys. Res. Lett. 31, L12213.

Goswami, B.N., 1998. Inter-annual variation of Indian summer monsoon in a GCM: external conditions versus internal feedbacks. J. Clim. 11, 501–522.

Goswami B.N. and Chakravorty S., 2018: Dynamics of the Indian summer monsoon climate. In Oxford Research Encyclopidea of Climate Science, DOI: 0.1093/acrefore/9780190228620.013.613.

Goswami, B.N., Xavier, P.K., 2005. Dynamics of "internal" inter-annual variability of Indian summer monsoon in a GCM. J. Geophys. Res. 110, D24104.

Huang, B., Peter, W., Thorne, et al., 2017. Extended Reconstructed Sea Surface Temperature version 5 (ERSSTv5), upgrades, validations, and intercomparisons. J. Clim. doi:10.1175/JCLI-D-16-0836.1.

Ihara, C., Kushnir, Y., Cane, M.A., Kaplan, A., 2008. Timing of El Niño–related warming and Indian summer monsoon rainfall. J. Clim. 21, 2711–2719.

Kinter, J.L., Miyakoda, K., Yang, S., 2002. Recent change in the connection from the Asian monsoon to ENSO. J. Clim. 15 (10), 1203–1215.

Kripalani, R.H., Kulkarni, A., 1997. Climatological impact of El Niño /La Niña on the Indian monsoon: a new perspective. Weather 52, 39–46.

Kripalani, R.H., Kulkarni, A., 1999. Climatology and variability of historical Soviet snow depth data:some new perspectives in snow—Indian monsoon teleconnections. Clim. Dyn. 15, 475–489.

Kucharski, F., Bracco, A., Yoo, J.H., Molteni, F., 2008. Atlantic forced component of the Indian monsoon interannual variability. Geophys. Res. Lett. 35, L04706.

Kumar, K.K., Rajagopalan, B., Cane, M.A., 1999. On the weakening relationship between the Indian Monsoon and ENSO. Science 284 (5423), 2156–2159.

Kumar, K.K., Rajagopalan, B., Hoerling, M., Bates, G., Cane, M., 2006. Unraveling the mystery of Indian monsoon failure during El Niño. Science 314, 115–119.

Larkin, N.K., Harrison, D.E., 2002. ENSO warm (El Niño) and cold (La Niña) event life cycles: ocean surface anomaly patterns, their symmetries, asymmetries and implications. J. Clim. 15, 1118–1140.

Lau, N.C., Nath, M.J., 2000. Impact of ENSO on the variability of the Asian-Australian monsoons as simulated in GCM experiments. J. Clim. 13, 4287–4309.

Li, Z., Cai, W., Lin, X., 2016. Dynamics of changing impacts of tropical Indo-Pacific variability on Indian and Australian rainfall. Sci. Rep. 6, 31767. https://doi.org/10.1038/srep31767.

Maraun, D., Kurths, J., 2005. Epochs of phase coherence between El Niño/Southern Oscillation and Indian monsoon. Geophys. Res. Lett. 32, L15709. https://doi.org/10.1029/2005 GL023225.

Pai, D.S., Shridhar, L., Rajeevan, M., Sreejith, O.P., Satbhai, N.S., Mukhopadhyay, B., 2014. Development of a new high spatial resolution (0.25° × 0.25°) long period (1901-2010) daily gridded rainfall data set over India and its comparison with existing data sets over the region. Mausam 65, 1–18.

Pant, G.B., Parthasarathy, B., 1981. Some aspects of an association between the southern oscillation and Indian summer monsoon. Arch. Meteor. Geophys. BioKlimatol. Ser. B. 29, 245–251.

Parthasarathy, B., Mooley, D.A., 1978. Some features of a long homogeneous series of Indian summer monsoon rainfall. Mon. Wea. Rev. 106, 771–781.

Parthasarathy, B., Pant, G.B., 1985. Seasonal relationship between Indian summer monsoon rainfall and the southern oscillation. Int. J. Climatol. 5, 369–378.

Philander, S.G., 1990. El Niño, La Niña, and the Southern Oscillation. Academic Press, San Diego, CA, p. 293.

Rasmusson, E.M., Carpenter, T.H., 1983. The relationship between the Eastern Pacific sea surface temperature and rainfall over India and Sri Lanka. Mon. Weather. Rev. 111, 517–528.

Saji, N.H., Goswami, B.N., Vinayachandran, P.N., Yamagata, T., 1999. A dipole mode in the tropical Indian Ocean. Nature 401, 360–363.

Shukla, J., 1987. Interannual variability of monsoons. In: Fein, J.S., Stephens, P.L. (Eds.), Monsoons. A Wiley Interscience Publication, USA, pp. 399–466.

Shukla, J., Paolino, D.A., 1983. The southern oscillation and long range forecasting of the summer monsoon rainfall over India. Mon. Weather Rev. 111, 1830–1837. doi:10.1175/1520-0493.

Shukla, J., 1998. Predictability in the midst of chaos: a scientific basis for climate forecasting. Science 282, 728–731.

Sikka, D.R., 1980. Some aspects of the large-scale fluctuations of summer monsoon rainfall over India in relation to fluctuations in planetary and regional scale circulation parameters. Proc. Indian Acad. Sci. Earth Planet Sci. 89, 179–195.

Slivinski, L.C., Compo, G.P., Whitaker, J.S., Sardeshmukh, P.D., Giese, B.S., McColl, C., Allan, R., Yin, X., Vose, R., Titchner, H., Kennedy, J., Spencer, L.J., Ashcroft, L., Brönnimann, S., Brunet, M., Camuffo, D., Cornes, R., Cram, T.A., Crouthamel, R., Domínguez-Castro, F., Freeman, J.E., Gergis, J., Hawkins, E., Jones, P.D., Jourdain, S., Kaplan, A., Kubota, H., Blancq, F.L., Lee, T., Lorrey, A., Luterbacher, J., Maugeri, M., Mock, C.J., Moore, G.K., Przybylak, R., Pudmenzky, C., Reason, C., Slonosky, V.C., Smith, C., Tinz, B., Trewin, B., Valente, M.A., Wang, X.L., Wilkinson, C., Wood, K., Wyszyński, P., 2019. Towards a more reliable historical reanalysis: improvements for version 3 of the Twentieth Century Reanalysis system. Q. J. Roy. Meteorol. Soc. 145, 2876–2908.

Sreejith, O.P., Panickal, S., Pai, S., Rajeevan, M., 2015. An Indian Ocean precursor for Indian summer monsoon rainfall variability. Geophys. Res. Lett. 42, 9345–9354.

Srinivas, G., Chowdary, J.S., Kosaka, Y., Gnanaseelan, C., Parekh, A., Prasad, K.V.S.R., 2018. The influence of the Pacific-Japan pattern on Indian summer monsoon rainfall. J. Clim. 31, 3943–3958.

Taschetto, A.S., Ummenhofer, C.C., Stuecker, M.F., Dommenget, D., Ashok, K., Rodrigues, R.R., SW, Y.e.h., 2020. Chapter14: ENSO atmospheric teleconnections. In: McPhaden, M.J., Santoso, A., Cai, W. (Eds.), El Niño Southern Oscillation in a Changing Climate, Geophysical Monograph1st ed. John Wiley and Sons, Inc, USA.

Walker, G.T., 1924. Correlation in seasonal variations of weather, IX: a further study of world weather. Memoirs India Meteorol. Depart. 24 (9), 275–333.

Walker, G.T., 1928. World weather. Q. J. R. Meteor. Soc. 54, 79–87.

Wang, C., 2002. Atmospheric circulation cells associated with the El Niño–Southern Oscillation. J. Clim. 15, 399–419.

Webster, P.J., Magana, V.O., Palmer, T.N., Shukla, J., Tomas, R.T., Yanai, M., Yasunari, T., 1998. Monsoons: processes, predictability and the prospects of prediction. Geophys. Res. 103 (C7), 14451–14510.

Xavier, P.K., Charline, M., Goswami, B.N., 2007. An objective definition of the Indian summer monsoon season and a new perspective on the ENSO–monsoon relationship. QJRMS 133, 749–764.

Yadav, R.K., 2017. On the relationship between east equatorial Atlantic SST and ISM through Eurasian wave. Clim. Dyn. 48 (1-2), 281–295.

Yadav, R.K., Srinivas, G., Chowdary, J.S., 2018. Atlantiv Nino modulation of Indian summer monsoon through Asian Jet. npj Clim. Atmos. Sci. 1, 1–23.

Yang, J., Liu, Q., Xie, S.P., Liu, Z., Wu, L., 2007. Impact of Indian Ocean SST basin mode on the Asian summer monsoon. Geophys. Res. Lett. 34, L02708. doi:10.1029/2006GL028571.

Yang, S., Li, Z., Yu, J.Y., Hu, X., Dong, W., He, S., 2018. El Niño–Southern Oscillation and its impact in the changing climate. Natl. Sci. Rev. 5, 840–857.

Chapter 4

ENSO Modoki teleconnections to Indian summer monsoon rainfall—A review

F. Feba, D. Govardhan, C.T. Tejavath, Karumuri Ashok
Centre for Earth, Ocean, and Atmospheric Sciences, University of Hyderabad, Hyderabad, India

4.1 Introduction to ENSO Modoki

The El Niño Southern Oscillation (ENSO) is probably the most studied tropical climate phenomenon. Gilbert Walker, with the discovery of the Southern Oscillation (Walker, 1923) and Jacob Bjerknes with Walker Circulation (Bjerknes, 1969), paved the way for most of the ensuing research on the ENSO. Meanwhile, the last few decades saw several ENSO-like warming events occurring in the tropical Pacific (e.g., 1986, 1990, 1991, 1994, 2002, 2004, etc.) These events were distinct from the known canonical El Niño events and named El-Niño Modoki[a] (Ashok, 2007; Ashok and Yamagata, 2009). The term "Modoki" is a Japanese word meaning "*similar yet different*" and hence El Niño Modoki (meaning "*Pseudo*-El Niño"). The signature of the El Niño Modoki in any particular season has also been referred to as the Dateline El Niño (Larkin and Harrison, 2005) or the Central Pacific El Niño (Kao and Yu, 2009; Yeh et al., 2009), remarkably, the definition of the Modoki events encompasses a persistence of the anomalous warming in the central tropical Pacific, as discussed later. Several studies have highlighted the increasing intensity and frequency of the El Niño Modoki since the late 1970s as compared to the canonical El Niños (Ashok et al., 2007; Freund et al., 2019; Kim and Yu, 2012; Lee and McPhaden, 2010; Yeh et al., 2009). Yeh et al. (2009) estimate the occurrence ratio of ENSO Modoki to ENSO and report an increase of Modoki as high as five times under the global warming scenario, though their subsequent simulations (Yeh et al., 2011) suggest that the changing Modoki frequency may be due to natural processes. Lee and McPhaden (2010) point out that the intensity of El Niño Modoki events has already doubled in the

---

[a] The term "Modoki" is a Japanese word meaning "similar yet different" and hence El Niño Modoki (meaning "Pseudo-El Niño").

Indian Summer Monsoon Variability: El Niño-teleconnections and beyond.
DOI: https://doi.org/10.1016/B978-0-12-822402-1.00003-X

69

recent decades, including the first decade of the new millennium. An important recent paper by Freund et al. (2019) shows that the increasing frequency of Modoki is observed only in the recent few decades compared to the past four centuries. All these lay emphasis on the compelling necessity to further study the ENSO Modoki phenomena, in a global warming scenario.

4.2   Data and methods

We used the Hadley Centre Sea Ice and Sea Surface Temperature dataset (HadISST; Rayner et al., 2003) for calculation of SST indices. We use the standard Niño3 and Niño4 indices (area-averaged SST anomaly over the regions 5°N–5°S, 150°W–90°W and 5°N–5°S, 160°E–150°W, respectively) to define the canonical ENSO events. For ENSO Modoki, we use the ENSO Modoki Index (EMI) as defined by Ashok et al. (2007) and Marathe et al. (2015).

$$\text{EMI} = [\text{SSTA}]_A - 0.5 \times [\text{SSTA}]_B - 0.5 \times [\text{SSTA}]_C$$

with SSTA area-averaged over the regions,
    A (165°E–140°W, 10°S–10°N), B (110°W–70°W, 15°S–5°N), and C (125°E–145°E, 10°S–20°N), respectively.

For our composite analysis, we use a threshold of the amplitude of the El Niño Modoki index (EMI) above one seasonal standard deviation through the summer is used to identify "strong" El Niño events. Based on these criteria, the El Niño Modoki years are 1967–1968, 1977–1978, 1991–1992, 1994–1995, 2002–2003, and 2004–2005. The six canonical El Niño years where the amplitude of the Niño3 index exceeds one standard deviation are 1957–1958, 1965–1966, 1972–1973, 1982–1983, 1987–1988, and 1997–1998. Furthermore, Ashok et al. (2017) identified the canonical La Niña and La Niña Modoki years, when the amplitude of the EMI is above its seasonal standard deviation during the summer and again in the consecutive winter. In this regard, the four La Niña Modoki events identified are 1975–1976, 1983–1984, 1998–1999, and 1999–2000 and the four canonical La Niña index which is below −1.0 of its standard deviation are 1970–1971, 1973–1974, 1988–1989, and 2007–2008. For the intraseasonal variability analysis, the statistical significance has been ascertained through a two-tailed Student's t test.

We used India Meteorological Department gridded rainfall data for Indian regional/subdivisional monthly rainfall (Rajeevan et al., 2006), and simple linear correlation analysis to show the impacts of ENSO and ENSO Modoki on the Indian summer monsoon (ISM). We also show the velocity potential fields, scaled at 10^{-6} (at 850 hPa and 200 hPa) derived from the NCEP wind data (Kalnay et al., 1996) for the period 1948–2019. The sign of the anomalous convergence/divergence signatures over the equatorial oceans at 200 hPa associated with the drivers such as ENSO, IOD, etc., in the equatorial Indo-Pacific will be more or less opposite to that at 850 hPa as can be envisaged from the Matsuno–Gill dynamics. Together, these large-scale fields represent the changes in the anomalous overturning circulations due to such climate drivers.

4.2.1 Criterion for break/active spells

In evaluating the potential impact of the ENSO Modoki on the active and break monsoon events, the rainfall anomalies are obtained by subtracting the daily rainfall climatology for the period 1951–2019 from the daily mean value. We follow the well-accepted Rajeevan et al. (2006) criterion to identify the active and break spells/phase of the ISM, which is based on daily rainfall data. According to the criterion, the active (break) episodes are identified during the peak monsoon months (July–August) as the period through which the standardized daily rainfall anomaly in the interested area exceeds (is less than) one standard deviation. If this continues for at least three consecutive days, then these are called Active (Break) spells. Here, we applied the criterion to identify the active and break episodes in the monsoon core zone limited to the region 18°N–28°N, 65°E–88°E.

4.3 ENSO Modoki and global impacts

4.3.1 Characteristics of ENSO Modoki

The El Niño Modoki (Ashok et al., 2007) is characterized by an anomalous tripole-like feature in the sea surface temperature anomalies (SSTA), with the maximum anomalous warming occurring in the tropical central Pacific with anomalous cooling on both the tropical Eastern and the Western Pacific (Fig. 4.1). The anomalous positive SSTA in the central tropical Pacific persists from boreal summer till the following spring. This is unlike the propagating trans-niño signal of canonical ENSO (Trenberth and Stepaniak, 2001) associated with the canonical ENSOs prior to the mid-1970s seen only for a season or so. The phase opposite to the El Niño Modoki, with persistent anomalous cooling in the central tropical Pacific, flanked by anomalous warming on both sides has been referred to as the La Niña Modoki. Fig. 4.1 shows that the distinction in La Niña flavors is relatively less significant as compared to that of El Niños. This is in agreement with Ren and Jin (2011) and Kug et al. (2011) who show that the distinction between the seasonal SST anomalies of the La Niña and those of La Niña Modoki are relatively less significant unlike that between the warm-phased ENSO types, though the distinction is clear in the subsurface (Ashok et al., 2017).

The anomalous central tropical Pacific warming associated with an El Niño Modoki event results in the anomalous formation of double cell structure with two Walker circulations, as shown in the schematic (Fig. 4.2; adopted from Ashok and Yamagata, 2009). An empirical orthogonal function (EOF) analysis by Ashok et al. (2007) for the post-1978 period shows that the first leading mode of EOF is the ENSO, whereas the second leading mode of the EOF shows the ENSO Modoki, explaining about 12% of the interannual variability in the

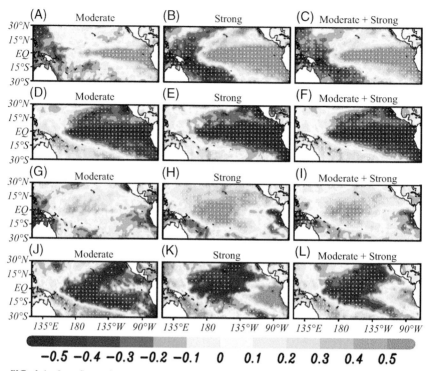

FIG. 4.1 June-September (JJAS) SST anomaly (°C) composites of ENSOs and ENSO Modokis for the period, 1949–2019 using HadISST. (A–C) Canonical El Niño, (D–F) Canonical La Niña, (G–H) El Niño Modoki and (J–L) La Niña Modoki. The stippled regions show significance at 90% confidence level using two-tailed Student's *t* test. "*ENSO*, El Niño Southern Oscillation; *SST*, sea surface temperature."

tropical Pacific. The anomalous zonal winds over the eastern tropical Pacific are associated with the canonical ENSO (hereafter ENSO), and show remarkable differences in seasonality with ENSO Modoki. As the zonal sea surface temperature (SST) gradients in the eastern equatorial Pacific are opposite between the two phenomena, the zonal wind direction is mainly opposite between El Niño Modoki and El Niño events. The low-level convergence area in El Niño Modoki remains west of that of an El Niño (Shinoda et al., 2011). During ENSO, the subsurface temperature anomalies originating in the Eastern Pacific propagate westward and change phase and return east forming a delayed oscillator (Suarez and Schopf, 1988). Whereas for the ENSO Modoki, the subsurface temperature anomalies originate, develop to a mature stage and decay, all in the central Pacific with no propagation. The ENSO Modoki also shows no phase-reversal behaving as an event or a series of events, rather than as a cycle. As a similarity, both ENSO and ENSO Modoki exhibit distinct recharge–discharge processes throughout their growth and phase transitions (Ren and Jin, 2013; Singh and Delcroix, 2013).

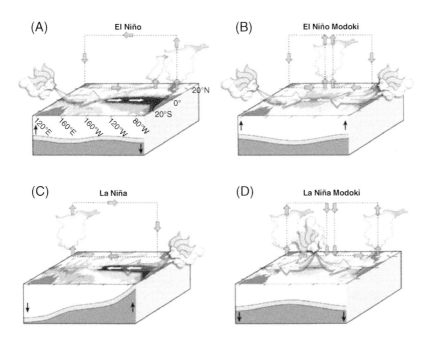

FIG. 4.2   A schematic representation of the Walker circulations during ENSO (single) and ENSO Modoki (double). A reproduction from Ashok and Yamagata (2009). "*ENSO*, El Niño Southern Oscillation."

4.3.2   Distinct phenomena, or diversity of ENSOs?

However, the discovery of the ENSO Modoki has led to a debate on whether the two types of ENSOs are indeed different phenomena. A study by Giese et al. (2011) describes both the ENSOs as a single continuum, while several others claimed that the distinction was an artifact due to the EOF analysis used (Lian and Chen., 2012; Marathe et al., 2015), through various linear methods such as the EOF, REOF methods, and composite analysis, and show that the distinction between two types of ENSOs is reasonable. They also suggest that the separation of these two phenomena is not just a result of classifying extreme El Niño events such as 1997 as a canonical El Niño event. A recent study by Freund et al. (2019), using a more extended proxy datasets from the last 400 years shows the increased frequency of ENSO in the recent few decades, thereby suggesting that the separation of two types of ENSOs may not be arising due to statistical sampling, confirming the arguments of Ashok et al. (2007) and also the potential role of global warming (Yeh et al., 2009). Further details on the discussion can be seen from (Behera and Yamagata, 2018; Capotondi et al., 2014; Marathe et al., 2015; (Marathe and Ashok, 2020)). We would like to conclude this section by quoting from Marathe et al. (2015) that "Further, even if the distribution (of

ENSOs) were unimodal, the difference between the evolution and teleconnections of the two types of El Niños stems out from the persistence, in central tropical Pacific for more than a few seasons during the occurrence of an El Niño Modoki, as compared to a much faster trans-Niño signal associated with a canonical El Niño." From a slightly different perspective, even if the two types of El Niños are part of a continuum based on the strength, these two types would be at the far ends of a diverse continuum (Capotondi et al., 2014), making the distinctions between these two types of the El Niños worth studying. Indeed, we also have observed basin-wide warming of tropical Pacific, with their distinct impacts, in 2009 and 2014 (Ashok et al., 2012; Jadhav et al., 2015).

4.3.3 Global impacts

The ENSO Modoki events have contrasting impacts, or similar impacts of differing strength, on the climates of various regions compared to those of the canonical ENSO events. The Western part of North America from Alaska to California, most of Japan, and the Yangtze River valley in China show higher temperatures during El Niño Modoki and normal (or even colder) temperatures during El Niño. The two types of ENSO show differences in associated rainfall patterns in most Pacific and Indian Ocean rim countries, including China, Japan, India, Taiwan, New Zealand, the United States, and as far as Africa (Ashok et al., 2007; Lin et al., 2015; Marathe and Ashok, 2020; Preethi et al., 2015; Ratnam et al., 2014; Roy et al., 2017; Weng et al., 2007). During an El Niño Modoki event, the Intertropical Convergence Zone (ITCZ) intensifies in the equatorial Western Pacific and shifts northward in the Eastern Pacific, and vice versa during a La Niña Modoki. The differences in the precipitation patterns and shifting of the ITCZ between ENSO Modoki and ENSO indicate a change in the midlatitude teleconnections as well. El Niño Modoki events are found to have a reverse effect on the middle-high latitudes stratosphere, compared to the effect of typical El Niño events. The northern polar vortex is stronger and colder; meanwhile, the southern polar vortex is weaker and warmer during El Niño Modoki events (Xie et al., 2012; Zubiaurre and Calvo, 2012). Xie et al. (2014) say that ENSO Modoki has had a significant effect on tropical ozone and could be a predictor for tropical ozone variations. Feng et al. (2019) show the delayed impacts of ENSO Modoki on the East Asian summer monsoon through an anomalous Western North Pacific anticyclone. The changes in SST and associated air-sea coupling induced by ENSO Modoki in the tropical Indian Ocean could play a significant role in governing Indo-Pacific teleconnections (Dogar et al., 2018).

The ENSO Modoki and ENSO have disparate impacts on high-frequency variability, such as tropical storm activity. The El Niño Modoki causes an anomalous blocking over Central-Eastern Australia, suppressing the storm track activity and associated rainfall in the southwest region of Australia (Ashok et al., 2009). The El Niño Modoki causes increased typhoon activity (Chen and Tam, 2010; Pradhan et al., 2011). This is due to the fact that the ENSO events modulate the background seasonal circulation. Given the relative uncertainties in SST, rainfall, and temperature observations, we should continue examining the distinct teleconnections with due caution (Yeh et al., 2018). We also have to be mindful of the non-linearity even in the impacts from the tropical Pacific (Taschetto et al., 2020).

4.4   Interannual variability of ENSO Modoki—ISM teleconnections

The ISM(June–September) accounts for about 75% of rainfall in the subcontinent driving the agriculture and economy (Gadgil and Gadgil, 2006; Parthasarathy et al., 1994). The interannual variability of ISM is majorly driven by the interannual variability of ENSO (Ashok et al., 2019; Bhalme and Mooley, 1980; Keshavamurthy, 1982; Sikka, 1980 and references therein) and Indian Ocean Dipole (Ashok et al., 2001; Ashok et al., 2004; Ashok and Saji, 2007; Behera et al., 1999). The El Niño phase is shown to weaken the ISM while a La Niña Phase strengthens the ISM. The IOD (Murtugudde et al., 2000; Saji et al., 1999; Webster et al., 1999) shows a relationship with the ISM opposite to that of ENSO; its positive phase strengthens the ISM, and the negative phase weakens it.

The failure of ISM in the year 2002 brought to the forefront the importance of tropical central Pacific for ISM prediction. Kumar et al. (2006) show that the El Niño Modoki affects the ISM more severely than a canonical El Niño. To an extent, the ENSO and ENSO Modoki have similar impacts on ISM circulation, but this conjecture stems out due to the definition of the area-averaged ISM rainfall, which is a large-scale spatial average. The ENSO Modoki has a significantly negative association with summer monsoon rainfall anomalies on peninsular India associated with a strong moisture divergence and resulting negative anomalies (Fig. 4.3). In comparison, the canonical ENSO shows a more negative correlation with central India, closer to the monsoon trough (Ashok et al., 2007, 2020). While the ENSO Modoki (similar to ENSO) strengthens or weakens the Hadley cell along with an induced variability in the Walker circulation influencing the ISM (Dogar et al., 2018), the impacts of the ENSO types on these large-scale circulations vary in terms of strength and localization (Amat and Ashok, 2018; Feba et al., 2018; Mujumdar et al., 2007). The ENSO and ENSO Modoki impacts on ISM affect the resulting rice production in several Indian states differently. Though, Haryana, United Andhra Pradesh, and Karnataka show similar impacts for both ENSOs, states like West Bengal, Bihar, Kerala, Punjab, and Uttar Pradesh show significantly different impacts on the rice production of the respective states (Amat and Ashok, 2018). The impact of ENSO Modoki over the Indian Ocean is also crucial in the understanding of ENSO Modoki–ISM relationship. Dandi et al. (2020) claim, using the CMIP5 datasets, that the El Niño Modoki induces changes in the North-West Pacific low-level circulation (850 hPa). This leads to a relatively weak moisture convergence over the monsoon trough at 850 hPa and a weak moisture divergence at 200 hPa (Fig. 4.4). This may result in a weak positive rainfall over the region. In reality, the El Niño

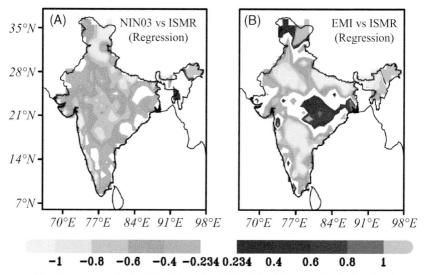

FIG. 4.3 Regression analysis of ENSO and ENSO Modoki with ISM rainfall. The significance has been calculated at 95% confidence level. The NINO3 index shows a negative relation over the entire subcontinent baring a few regions in the north-eastern part. In a contrast, the EMI index shows a positive influence on ISM rainfall over the central region. "*ENSO*, El Niño Southern Oscillation; *ISM*, Indian summer monsoon."

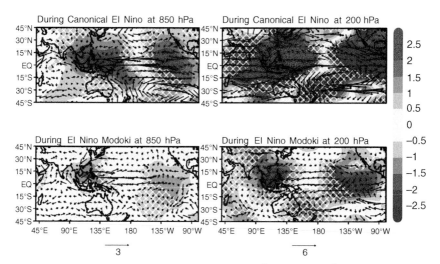

FIG. 4.4 The anomalous velocity potential fields (Units m²/s; scaled at 10⁻⁶) are calculated at 850 hPa and 200 hPa and overlaid with anomalous winds (m/s) at the respective heights for Canonical El Niño and El Niño Modoki years during 1948–2019. The cross-hatched region indicates 90% confidence of velocity potential fields using the two-tailed Student's *t* test.

Modoki events are associated with weak positive to near climatological signals over many places along the monsoon trough. This indicates that the CMIP5 models find it a challenge in distinguishing the impacts of various types of El Niños. However, there is another possibility that this strong negative rainfall signal during El Niño Modoki is due to an overestimation of the impact of Modoki during break monsoon conditions, as elaborated in the next section.

4.5   Intraseasonal variability of ENSO Modoki—ISM teleconnections

The intraseasonal variability of ISM rainfall is characterized by periods of wet (active) and dry (break) rainfall activity mainly prominent over the monsoon core region of central India. It is manifested as a 30-60-day oscillation (Gadgil and Joseph, 2003; Goswami and Mohan, 2001; Goswami and Xavier, 2003; Keshvamurthy and Sankar, Rao 1992; Krishnamurti and Bhalme, 1976; Krishnamurti and Subrahmanyam, 1982; Sikka and Gadgil, 1980; Yasunari, 1979). Goswami and Chakravorty (2017) elaborate on these aspects of active and break spells of monsoon in their recent review. Pai et al. (2016) showed that the composite rainfall patterns of active days during El Niños in the 1901–2014 period show only a slight difference from those during La Niñas, except for slightly stronger negative rainfall anomalies along the foothills of Himalayas during El Niños. However, in the case of break monsoon, the positive composite rainfall anomalies along the foothills of the Himalayas during La Niña years are stronger and extending more westwards (north-eastern states) compared to that of El Niño years.

Motivated by this, in our study, we carried out the composite analysis of intraseasonal summer monsoon rainfall anomalies for each type and phase of ENSOs during the 1951–2019 period by averaging daily rainfall anomalies during the break monsoon and active monsoon periods identified by Rajeevan et al. (2006; 2010) and are shown in Figs 4.5 and 4.6, respectively. Table 4.1 lists the aggregate number of break and active monsoon days over the Core Monsoon Region (CMR) in all the canonical El Niño years, those in El Niño Modoki years, and those in two types of La Niña years, respectively (Marathe et al., 2015), during the 1951–2007 period. The El Niño and El Niño Modoki are seen to be predominantly associated with higher number of break conditions over India (Fig. 4.5). Notably, unlike its seasonal signature (Fig. 4.5, Ashok et al., 2019), the El Niño Modoki events are associated with negative rainfall anomalies along the monsoon trough during the break periods (Fig. 4.5). However, the association between the type and phase of ENSO events with the active monsoon condition is not clear, unlike for the break events. Heuristically, we would expect that the number of active days is more during La Niñas. But we also see many active days during El Niños as well. Of course, the relatively high rainfall during El Niño Modokis over the core monsoon region may be due to the fact that the anomalously negative seasonal rainfall anomalies during these events are more concentrated in

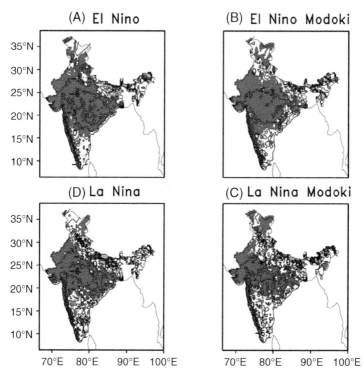

FIG. 4.5 (clockwise) Composite of rainfall anomalies during break monsoon days in the El Niño, El Niño Modoki, La Niña Modoki & La Niña years (in contours, at 3 mm/day intervals). The shaded region indicates 95% confidence using two-tailed Student's t test (orange—negative; sky-blue—positive).

peninsular India, with a positive association in the CMR (Fig. 4.4; also see Ashok et al., 2007,2019). There is an argument that propensity of the break and active cycles in any monsoon season decide whether the relevant seasonal monsoonal rainfall is below or above normal (Goswami and Chakravorty, 2017; Lau et al., 2012 and references therein).

We see from Fig. 4.7B that the highest number of break days in a year are seen in 2002, associated with a strong El Niño Modoki. The 1965 El Niño seems to be also associated with the third-highest number of break monsoon days. When the total number of break days are concerned (Table 4.1), the canonical El Niño events are more proficient. During years with a warm-phased event in the tropical Pacific, break spells of longer length are more common, such as the El Niño years during 1951–2007 showing longer break spells compared to the active spells and El Niño Modoki in 2002 and 2004, rather than during the La Niña years. All this suggests that the ENSOs affect the intraseasonal variability of the ISM through the modulation of background seasonal circulation. Moreover, there should be a significant role of internal variability, as evidenced

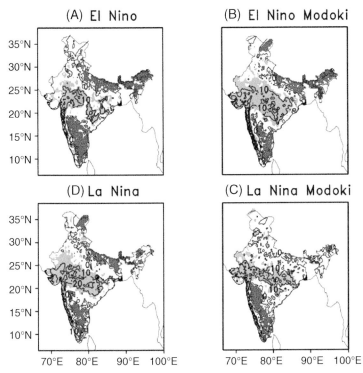

FIG. 4.6 (clockwise) Composite of rainfall anomalies during active monsoon days of Indian region in the El Niño, El Niño Modoki, La Niña Modoki & La Niña years (in contours, at 10 mm/day intervals). The shaded region indicates 95% confidence using two-tailed Student's *t* test (orange—negative; sky-blue—positive).

by that the second-highest number of breaks has occurred in 1966 (Table 4.1) when neither a strong warm ENSO nor a positive IOD has co-occurred (Figures not shown).

Importantly, we also see from Table 4.1 that the number of active days during either of these two El Niños is not negligible. The number of active days associated with La Niñas during 1951–2007 is higher than those associated with the La Niña Modokis. The highest number of active days during the summer monsoon occurred in 2006 when no discernible ENSO event was observed. These are likely associated with the co-occurring strong positive IOD event (Cai et al., 2009; Pillai and Chowdary, 2016; Kucharski et al., 2020). The above-normal active cycles during the strong canonical El Niño in 1997 and those during the strong El Niño Modoki during 1994 can be attributed to the opposing effect of the co-occurring strong positive IOD events in years, such as 1997 (Ashok et al., 2001). This of course needs to be verified through modeling experiments. In short, we observe that both El Niños and El Niño Modokis apparently facilitate higher than normal break days in ISM, the former particularly so. However,

TABLE 4.1 List of total number of active and break monsoon days (and spells) during El Niño, El Niño Modoki, La Niña Modoki, and La Niña years during the period of 1951–2007 over the Core Monsoon Region (CMR).

Event	Years	Total number of active rainfall days over CMR	Total average rainfall over CMR (mm)	Total number of break rainfall days over CMR	Total average rainfall over CMR (mm)
All	1951–2007	414	7.67	404	−4.06
El Niño Modoki	1967, 1977, 1991, 1994, 2002, 2004	36	8.84	52	−2.25
El Niño	1957, 1965, 1972, 1982, 1987, 1997	40	5.24	74	−4.29
La Niña Modoki	1975, 1983, 1998, 1999	19	7.37	32	−2.12
La Niña	1970, 1973, 1988, 2007	43	8.39	25	−3.05

an analogous unique impact of the La Niña's on active cycles is not that evident, which supports the argument of Pai et al. (2016). Interestingly, while the results from (Dwivedi et al., 2015) suggest an increasing frequency of breaks during El Niños, their results suggest an opposing influence from the La Niña events, a result that differs from that of Pai et al. (2016) as well as ours.

Fig. 4.5 depicts a composite analysis of daily rainfall anomalies during break conditions associated with the two types of El Niños, El Niño Modoki, and corresponding La Niñas. A similar analysis for the active cases is shown in Fig. 4.6. We see negative rainfall anomalies over most of India, except the anomalously positive rainfall anomalies along the Himalayan foothills. We do not see a big difference between the signatures of the two types of El Niños, except that the canonical events seem to introduce higher deficit in daily rainfall during the breaks. Notably, the impact of El Niño Modokis on the ISM rainfall along the monsoon trough (e.g., Fig. 4.5 of Ashok et al., 2019) is apparently opposite to that during the break monsoon days. In addition, while the El Niño Modokis and El Niños are associated with a positive rainfall anomaly in the break monsoon distribution and negative rainfall anomalies during active conditions over Northeast India, these indications are subject to the data quality in the region (Soraisam et al., 2018). The surplus rainfall footprints during active cases associated with all types ENSOs seem to be only confined to the monsoon trough

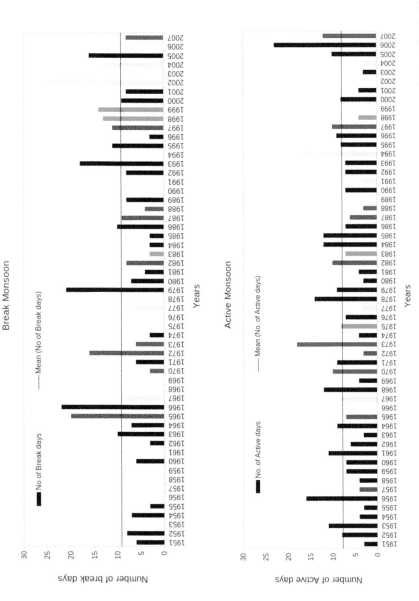

FIG. 4.7 (A—Top) Histogram of break days during 1951–2007 based on Rajeevan et al. (2008) criterion. The El Niño, El Niño Modoki, La Niña, La Niña Modoki are designated by the *Bars in red, yellow, blue,* and *green,* respectively. (B—Bottom) Same as Fig. 4.7A but for active days.

region (Fig. 4.6), unlike those during the corresponding break conditions. The La Niña Modokis do not seem to be as proficient as the La Niñas in causing surplus rainfall in the central Indian region during the active phase of ISM (Figs 4.5 and 4.6, lower panels).

Fig. 4.8A and B shows the composite of daily anomalous 850 hPa velocity potential and the divergent wind vectors over all the break and active cases during the 1951–2007 period. Notably, during the breaks, we see a relatively small zone anomalous convergence confined over the central tropical Pacific (Fig. 4.8A). On the other hand, during the active monsoon conditions, the zone of convergence extends well over the Western Pacific region (Fig. 4.8B); this looks like an enhanced seasonal mean Walker circulation, with a strong

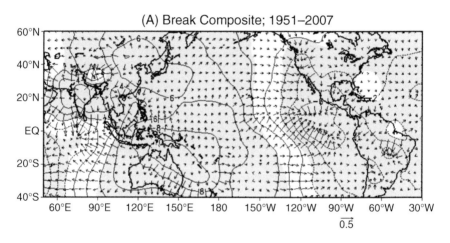

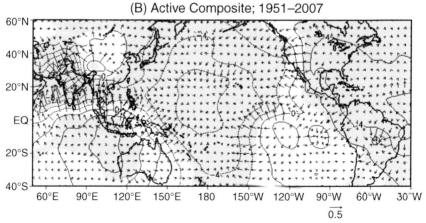

FIG. 4.8 850 hPa velocity potential anomalies (in contours, at 2 m²/s intervals) and divergent (convergent) wind vectors for all the (A) break and (B) active cases during the 1951–2007 period. The shaded region indicates 95% confidence using two-tailed Student's *t* test.

signal over the tropical Western Pacific. In conformation with the active conditions, we find a convergence over the Indian subcontinent too, along with convergence from the equatorial Indian Ocean (Fig. 4.8B). This is more prominent over the eastern portion, importantly, over the equatorial Pacific region at the level of 850 hPa, the magnitude of the divergent vectors is strong during the active phase of the ISM, with an opposite signature during the breaks. (Fig. 4.8B).

Fig. 4.9 illustrates the composites of the daily anomalous 850 hPa velocity potential and the divergent wind vectors for summer monsoon breaks during all phases of co-occurring ENSOs and ENSO Modokis. Fig. 4.9A shows that during canonical El Niños, we find a strong zone of anomalous divergence over the Western Indian subcontinent. An anomalous zone of convergence over the northeast Indian region is also seen. These anomalous signatures explain the dynamics behind the rainfall anomalies during breaks associated with the canonical El Niños (Fig. 4.5). During the El Niño Modokis as well, we still see anomalous divergence signature, statistically significant at 95% confidence level. The composites of the active cases during all phases of ENSO and ENSO Modoki indicate anomalous convergence mainly over the Northern Head Bay of Bengal (Fig. 4.10), except in the case of La Niña Modoki where the anomalous convergence is more prominent over the Arabian Sea. We also observe a strong convergence zone (Fig. 4.10A) in the tropical Eastern Pacific case of El Niño, and in northern tropical central Pacific region in the case of El Niño Modokis with a slight weakened convergence over the western equatorial Pacific region (Fig. 4.10B) and vice versa in the cases of La Niña (Fig. 4.10C) and La Niña Modoki (Fig. 4.10D), respectively, in these locations. The anomalous convergence signal over the Indian regions falls slightly short of statistical significance during active cases associated with canonical La Niñas.

The above composite analysis, essentially a preliminary effort, when combined with Table 4.1 suggests that changes in seasonal circulation may play a potential role in the active-break cycles. There seems to be a relatively stronger association between the El Niños—be it canonical or Modokis—and breaks. However, a similar relatively stronger relation between active conditions with La Niñas, which we expect because of the propensity of La Niñas to provide a favorable background condition for the active conditions, is not that conspicuous. These subtle conjectures need to be supported by further observations using multiple observations—particularly station level observations, and a more exhaustive dynamical analysis. We would also note from Figs 4.9 and 4.10 that the large-scale circulation changes associated with monsoon break and active conditions due to ENSOs have their signatures as far as Australia, indicating the complex dynamics that may be involved, and need further attention. The analysis is also subject to the fact that we have not factored the impacts from other climate drivers such as the IOD, Atlantic Zonal Mode, and Madden–Julian Oscillation into our analysis.

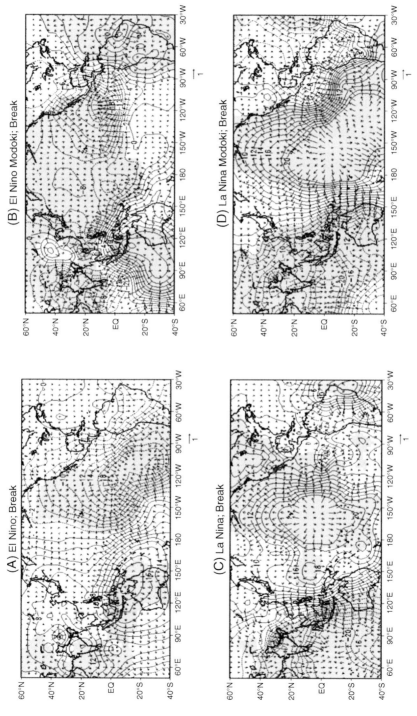

FIG. 4.9 850 hPa velocity potential anomalies (in contours, at 2 m²/s intervals) and divergent (convergent) wind vectors for the break cases in (A) El Niño, (B) El Niño Modoki, (C) La Niña, and (D) La Niña Modoki years. The shaded region indicates 95% confidence using two-tailed Student's *t* test.

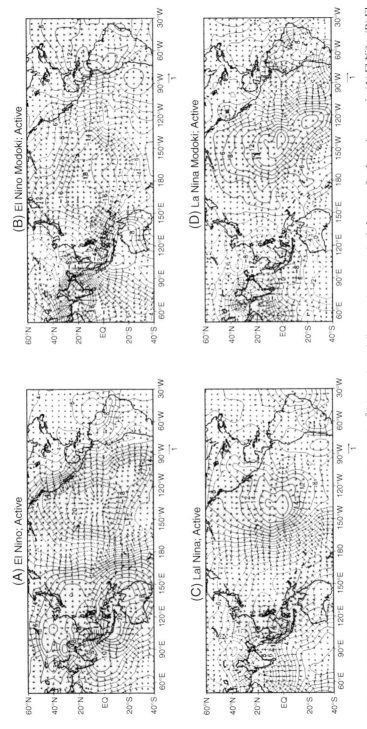

FIG. 4.10 850 hPa velocity potential anomalies (in contours, at 2 m²/s intervals) and divergent (convergent) wind vectors for the active cases in (A) El Niño, (B) El Niño Modoki, (C) La Niña, and (D) La Niña Modoki years. The shaded region indicates 95% confidence using two-tailed Student's t test.

References

Amat, H.B., Ashok, K., 2018. Relevance of Indian summer monsoon and its tropical Indo-Pacific climate drivers for the kharif crop production. Pure Appl. Geophys. 175 (3), 1221.

Ashok, K., Feba, F., Tejavath, C.T, 2019. The Indian summer monsoon rainfall and ENSO. Mausam 70, 443–452.

Ashok, K., Guan, Z., Yamagata, T., 2001. Impact of the Indian Ocean Dipole on the relationship between the Indian monsoon rainfall and ENSO. Geophys. Res. Lett. 28, 4499–4502.

Ashok, K., Guan, Z., Saji, N.H., Yamagata, T., 2004. Individual and combined influences of the ENSO and Indian Ocean Dipole on the Indian summer monsoon. J. Climate 17, 3141–3155.

Ashok, K, Sabin, TP, Swapna, P, Murtugudde, RG, 2012. Is a global warming signature emerging in the tropical Pacific? Geophys. Res. Lett. 39, L02701. doi:10.1029/2011GL050232.

Ashok, K., Saji, N.H., 2007. Impacts of ENSO and Indian Ocean Dipole events on the sub-regional Indian summer monsoon rainfall. J. Nat. Hazards 42, 273–285.

Ashok, K., Shamal, M., Sahai, A.K., Swapna, P., 2017. Nonlinearities in the evolutional distinctions between El Niño and La Niña types. J. Geophys. Res.: Oceans 122, 9649–9662. doi:10.100 2/2017JC013129.

Ashok, K., Tam, C.-Y., Lee, W.-J., 2009. ENSO Modoki impact on the southern hemisphere storm track activity during extended austral winter. Geophys. Res. Lett. doi:10.1029/2009 GL038847.

Ashok, H., Yamagata, T., 2009. Climate change: the El Niño with a difference. Nature 461, 481–484. doi:10.1038/461481a.

Behera, S.K., Krishnan, R., Yamagata, T., 1999. Unusual ocean-atmosphere conditions in the tropical Indian Ocean during 1994. Geophys. Res. Lett. 26, 3001–3004.

Behera, S., Yamagata, T., 2018. Climate dynamics of ENSO Modoki phenomenon. Oxford Research Encyclopaedia of Climate Science. https://oxfordre.com/climatescience/view/10.1093/acre-fore/9780190228620.001.0001/acrefore-9780190228620-e-612.

Bhalme, H.N., Mooley, D.A., 1980. Large-scale drought/floods and monsoon circulation. Mon. Weather Rev. 108, 1197–1211.

Bjerknes, J., 1969. Atmospheric teleconnections from the equatorial Pacific. J. Phys. Oceanogr. 97 (3), 163–172.

Cai, W. and Power, S. (2009). El Niño and climate change. International ENSO Workshop held at Greenhouse 2009, Perth, 25-27 March 2009. CSIRO Marine an Atmospheric Research, As-pendale.

Capotondi, A., Wittenberg, A.T., Newman, M., Di Lorenzo, E., Yu, J.-Y., Braconnot, P., et al., 2014. Understanding ENSO diversity. Bull. Am. Meteorol. Soc. 96, 921–938. doi:10.1175/BAMS-D-13-00117.1.

Chen, G., Tam, C.-Y., 2010. Different impact of two kinds of Pacific Ocean warming on tropical cyclone frequency over western North Pacific. Geophys. Res. Lett. 37, L01803. doi:10.102 9/2009GL041708.

Dandi, R, Chowdary, J.S., Pillai, P.A., Sidhan, N.S., Kundeti, K., Ramakrishna, S.S.V.S., 2020. Impact of El Niño Modoki on Indian summer monsoon rainfall: Role of western north Pacific circulation in observations and CMIP5 models. Int. J. Climatol 40, 2117–2133. doi:10.1002/ joc.6322.

Dogar, M.M., Kucharski, F., Sato, T., Mehmood, S., Ali, S., Gong, Z., et al., 2019. Towards understanding the global and regional climatic impacts of Modoki magnitude. Global Planet. Change 172, 223–241. doi:10.1016/j.gloplacha.2018.10.004.

Dwivedi, S., Goswami, B.N., Kucharski, F., 2015. Unraveling the missing link of ENSO control over the Indian monsoon rainfall. Geophys. Res. Lett. 42, 8201–8207. doi: https://doi.org/1 0.1002/2015GL065909.

Feba, F., Ashok, K., Ravichandran, M., 2018. Role of changed Indo-Pacific atmospheric circulation in the recent disconnect between the Indian summer monsoon and ENSO. Clim. Dyn. 52 (3-4), 1461–1470. doi:10.1007/s00382-018-4207-2.

Feng, J., Chen, W., Gong, H., et al., 2019. An investigation of CMIP5 model biases in simulating the impacts of central Pacific El Niño on the East Asian summer monsoon. Clim Dyn **52**, 2631–2646. doi:10.1007/s00382-018-4284-2.

Freund, M.B., Henley, B.J., Karoly, D.J., et al., 2019. Higher frequency of Central Pacific El Niño events in recent decades relative to past centuries. Nat. Geosci. **12**, 450–455. doi:10.1038/ s41561-019-0353-3.

Gadgil, S., Gadgil, S., 2006. The Indian monsoon, GDP and agriculture. Econ. Polit. Wkly XLI, 4887–4895.

Gadgil, S., Joseph, P.V., 2003. On breaks of the Indian monsoon. Indian Acad. Sci. (Earth Planet. Sci.) 112, 529–558.

Giese, B.S., Ray, S., 2011. El Niño variability in simple ocean data assimilation (SODA), 1871–2008. J Geophys Res 116, C02024. doi:10.1029/2010JC006695.

Goswami, B., & Chakravorty, S. (2017). Dynamics of the Indian summer monsoon climate. Oxford Research Encyclopaedia of Climate Science. https://oxfordre.com/climatescience/ view/10.1093/acrefore/9780190228620.001.0001/acrefore-9780190228620-e-613 (Accessed 29 November 2020).

Goswami, B.N., Mohan, R.S.A, 2001. Intraseasonal oscillations and interannual variability of the Indian summer monsoon. J. Clim. 14, 1180–1198.

Goswami, B.N., Xavier, P.K., 2003. Potential predictability and extended range prediction of Indian summer monsoon breaks. Geophys. Res. Lett. 30 (18). doi:10.1029/2003SGL017810 1966.

Jadhav, J., Panickal, S., Marathe, S., et al., 2015. On the possible cause of distinct El Niño types in the recent decades. Sci. Rep. 5, 17009. doi:10.1038/srep17009.

Kalnay, E., Kanamitsu, M., Kistler, R., Collins, W., Deaven, D., Gandin, L., et al., 1996. The NCEP/ NCAR 40-year reanalysis project. B. Am. Meteorol. Soc. 77, 437–471. doi:10.1175/1520-0477(1996)077<0437:TNYRP>2.0.CO;2.

Kao, H.-Y., Yu, J.-Y., 2009. Contrasting eastern-Pacific and central-Pacific types of ENSO. J. Clim. 22, 615–632.

Keshavamurty, R.N., 1982. Response of the atmosphere to sea surface temperature anomalies over the equatorial Pacific and the teleconnections of the southern oscillation. J. Atmos. Sci. 39, 1241–1259. doi:10.1175/1520 469(1982)039<1241:ROTATS>2.0.CO;2.

Keshavamurty, R.N., Sankar Rao, M., 1992. The Physics of Monsoons. Allied Publishers Ltd, New Delhi.

Kim, S.T., Yu, J.-Y., 2012. The two types of ENSO in CMIP5 models. Geophys. Res. Lett. 39, L11704. doi:10.1029/2012GL052006.

Krishnamurti, T.N., Bhalme, H.N., 1976. Oscillations of a monsoon system, part I: observational aspects. J. Atmos. Sci. 33, 1937–1954.

Krishnamurti, T.N., Subrahmanyam, D., 1982. The 30–50 Day Mode at 850 mb During MONEX. J. Atmos. Sci. 39, 2088–2095. doi:10.1175/1520-0469(1982)039<2088:TDMAMD >2.0.CO;2.

Kucharski, F., Biastoch, A., Ashok, K., Yuan, D., 2020. Mechoso, C. (Ed.), Indian Ocean variability and interactions. Interacting Climates of Ocean Basins: Observations, Mechanisms, Predictability, and Impacts. Cambridge University Press, Cambridge, pp. 153–185, ISBN-13: 978-1108492706.

Kug, J.S., Ham, Y.G., 2011. Are there two types of la Nina? Geophys. Res. Lett. 38, 2–7. doi:10.1029/2011GL048237.

Kumar, K.K., Rajagopalan, B., Hoerling, M., Bates, G., Cane, M., 2006. Unravelling the mystery of Indian monsoon failure during El Niño. Science 314, 115–119.

Lau, W.K.M., Waliser, D.E., Goswami, B.N, 2012. South Asian monsoonIntraseasonal Variability in the Atmosphere-Ocean Climate System. Springer, Berlin. doi:10.1007/978-3-642-13914-7_2.

Larkin, N.K., Harrison, D.E., 2005. On the definition of El Niño and associated seasonal average U.S. weather anomalies. Geophys. Res. Lett., 32.

Lee, T., McPhaden, M.J., 2010. Increasing intensity of El Niño in the central-equatorial Pacific. Geophys. Res. Lett. 37, L14603. doi:10.1029/2010GL044007.

Lian, T., Chen, D., 2012. An evaluation of rotated EOF analysis and its application to tropical pacific SST variability. J Clim. 25, 5361–5373.

Lin, C.-C., Liou, Y.-J., Huang, S.-J., 2015. Impacts of two-type ENSO on rainfall over Taiwan. Adv. Meteor. 2015, 658347. doi:10.1155/2015/658347.

Marathe, S., and K. Ashok, (2020) Tropical and Extra-Tropical Air-Sea Interactions, Elsevier, Amsterdam, ISBN 10 : 9780128181577.

Marathe, S., Ashok, K., Swapna, P., Sabin, T.P., 2015. Revisiting El Niño Modokis. Clim. Dyn. 45, 3527–3545. doi:10.1007/s00382-015-2555-8.

Mujumdar, M., Kumar, V., Krishnan, R., 2007. The Indian summer monsoon drought of 2002 and its linkage with tropical convective activity over northwest Pacific. Clim. Dyn. 28, 743–758. doi:10.1007/s00382-006-0208-7.

Murtugudde, R., McCreary, J.P., Busalacchi, A.J., 2000. Oceanic processes associated with anomalous events in the Indian Ocean with relevance to 1997-1998. J. Geophys. Res.: Oceans 105 (C2), 3295–3306.

Pai, D.S., Sridhar, L., Kumar, M.R., 2016. Active and break events of Indian summer monsoon during 1901–2014. Clim. Dyn. 46 (11–12), 3921–3939.

Parthasarathy, B., Munot, A.A., Kothawale, D.R., 1994. All-India monthly and seasonal rainfall series: 1871–1993. Theor. Appl. Climatol. 49, 217–224.

Pillai, P.A, Chowdary, J.S., 2016. Indian summer monsoon intra-seasonal oscillation associated with the developing and decaying phase of El Niño. Int. J. Climatol. 36, 1846–1862.

Pradhan, P.K., Preethi, B., Ashok, K., Krishnan, R., Sahai, A.K., 2011. Modoki, Indian Ocean Dipole, and western North Pacific typhoons: possible implications for extreme events. J. Geophys. Res. 116, D18108. doi:10.1029/2011JD015666.

Preethi, B., Sabin, T.P., Adedoyin, J.A., Ashok, K., 2015. Impacts of the ENSO Modoki and other tropical Indo-Pacific climate-drivers on African rainfall. Sci. Rep. 5, 16653. doi:10.1038/srep16653.

Rajeevan, M., Bhate, J., Kale, J.D., Lal, B., 2006. High resolution daily gridded rainfall data for the Indian region: analysis of break and active monsoon spells. Curr. Sci. 91 (3), 296–306.

Rajeevan, M., Gadgil, S., Bhate, J., 2010. Active and break spells of the Indian summer monsoon. J. Earth Syst. Sci. **119**, 229–247. doi:10.1007/s12040-010-0019-4.

Ratnam, J.V., Behera, S.K., Masumoto, Y., Yamagata, T., 2014. Remote effects of El Niño and Modoki events on the austral summer precipitation of Southern Africa. J. Clim. 27, 3802–3815.

Rayner, N.A., Parker, D.E., Horton, E.B., Folland, C.K., Alexander, L.V., Rowell, D.P., et al., 2003. Global analyses of sea surface temperature, sea ice and night marine air temperature since the late nineteenth century. J. Geophys. Res. 108 (D14), 4407.

Ren, H.-L., Jin, F.-F., 2011. Niño indices for two types of ENSO. Geophys. Res. Lett. 38, L04704. doi:10.1029/2010GL046031.35.

Ren, H.L., Jin, F.F., 2013. Recharge oscillator mechanisms in two types of ENSO. J. Clim. 26, 6506–6523. doi:10.1175/JCLI-D-12-00601.1.

Roy, I., 2017. Indian summer monsoon and El Niño–Southern Oscillation in CMIP5 models: a few areas of agreement and disagreement. Atmosphere 8 (8), 154. doi:10.3390/atmos8080154.

Saji, N.H., Goswami, B.N., Vinayachandran, P.N., Yamagata, T., 1999. A dipole mode in the tropical Indian Ocean. Nature 401, 360–363.

Shinoda, T., Hurlburt, H.E., Metzger, E.J., 2011. Anomalous tropical ocean circulation associated with La Niña Modoki. J. Geophys. Res. 116, C12001. doi:10.1029/2011JC007304.

Sikka, D.R., 1980. Some aspects of the large-scale fluctuations of summer monsoon rainfall over India in relation to fluctuations in planetary and regional scale circulation parameters. Proc. Indian Acad. Sci. Earth Planet Sci. 89, 179–195.

Sikka, D.R., Gadgil, S., 1980. On the maximum cloud zone and the ITCZ over Indian longitudes during southwest monsoon. Mon. Weather Rev. 108, 1840–1853.

Singh, A., Delcroix, T., 2013. Eastern and Central Pacific ENSO and their relationships to the recharge/discharge oscillator paradigm. Deep Sea Res. part I: Oceanogr. Res. Papers 82, 32–43.

Soraisam, B., Ashok, K., Pai, D.S., 2018. Uncertainties in observations and climate projections for the North East India. Global Planet. Change 160, 96–108. doi:10.1016/j.gloplacha.2017.11.010 ISSN 0921-8181.

Suarez, M.J., Schopf, P.S., 1988. A delayed action oscillator for ENSO. J. Atmos. Sci. 45, 3283–3287. doi:10.1175/1520-0469(1988)045<3283:ADAOFE>2.0.CO;2.

Taschetto, A.S., Ummenhofer, C.C., Stuecker, M.F., Dommenget, D., Ashok, K., Rodrigues, R.R., et al., 2020. ENSO atmospheric teleconnections. In: McPhaden, M.J., Santoso, A., Cai, W. (Eds.), El Niño Southern Oscillation in a Changing Climate. doi:10.1002/9781119548164. ch14.

Trenberth, K.E., Stepaniak, D.P., 2001. Indices of El Niño evolution. J. Clim. 14, 1697–1701.

Walker, G.T., 1923. Correlation in seasonal variations of weather, VIII: a preliminary study of world weather. Mem. India Meteorol. Dep. 24, 75–131.

Webster, P.J., Moore, A., Loschnig, J., Leban, M., 1999. Coupled ocean-atmosphere dynamics in the Indian Ocean during 1997–98. Nature 401 (23), 356–360.

Weng, H., Ashok, K., Behera, S., Rao, S.A., Yamagata, T., 2007. Impacts of recent El Niño Modoki on dry/wet conditions in the Pacific rim during boreal summer. Clim. Dyn. 29, 113–129. doi:10.1007/s00382-007-0234-0.

Xie, F., Li, J., Tian, W., Feng, J., Huo, Y., 2012. The Signals of El Niño Modoki in the tropical tropopause layer and stratosphere. Atmos. Chem. Phys 12 (11), 125259–125273.

Xie, F., Li, J.P., Tian, W.S., Zhang, J.K., 2014. The relative impacts of El Niño Modoki, Canonical El Niño, and QBO on tropical ozone changes since the 1980s. Environ. Res. Lett. **9**, 064020.

Yasunari, T., 1979. Cloudiness fluctuations associated with the northern hemisphere summer monsoon. J. Meteorol. Soc. Japan 57, 227–242.

Yeh, S.-W., Cai, W., Min, S.-K., McPhaden, M.J., Dommenget, D., Dewitte, B., et al., 2018. ENSO atmospheric teleconnections and their response to greenhouse gas forcing. Rev. Geophys. 56, 185–206. doi:10.1002/2017RG000568.

Yeh, S.-W., Kirtman, B.P., Kug, J.-S., Park, W., Latif, M., 2011. Natural variability of the central Pacific El Niño event on multi-centennial timescales. Geophys. Res. Lett. 38, L02704.

Yeh, S.-W., Kug, J., Dewitte, B., Kwon, M.-H., Kirtman, P., Jin, F.F., 2009. El Niño in a changing climate. Nature 461, 511–514.

Zubiaurre, I., Calvo, N., 2012. The El Niño–Southern Oscillation (ENSO) Modoki signal in the stratosphere. J. Geophys. Res. 117 (D4).

Chapter 5

The decaying phase of El Niño and Indian summer monsoon rainfall

Hyo-Seok Park[a], Won-il Lim[b], Kyong-Hwan Seo[b]
[a]*Department of Ocean Science and Technology, Hanyang University, Ansan, South Korea,*
[b]*Department of Atmospheric Sciences, Pusan National University, Busan, South Korea*

5.1 Introduction

Intense research has been devoted to elucidate the mechanisms associated with the year-to-year fluctuations of South Asian summer monsoon, specifically rainfall over India. While Indian summer monsoon (ISM) rainfall is suppressed during the development of El Niño conditions in the tropical Pacific, rainfall over India increases during the summer following the winter peak of El Niño (Chowdary et al., 2017; Lau and Nath, 2012; Park et al., 2010; Shukla, 1995; Tao et al., 2016; Webster et al., 1998; Yang et al., 2007), which is referred to as the "delayed effect" (Park et al., 2010). One potential cause of this delayed effect is the transition to La Niña conditions after strong El Niño events (Chowdary et al., 2017; Shukla, 1995), but La Niña does not always follow El Niño events. On the other hand, the basin-wide tropical Indian Ocean warming that develops during, and persists after, El Niño events may contribute to enhanced ISM rainfall (Park et al., 2010; Terray et al., 2003).

Attribution of ISM rainfall variability to tropical Indian Ocean warming remains elusive. Indeed, tropical Indian Ocean warming after the mature phase of El Niño events has been regarded as a factor for suppressing the monsoon onset (Joseph et al., 1994; Lau and Nath, 2012), as persistent El Niño–induced off-equatorial South Indian Ocean warming may delay ISM onset by a week (Annamalai et al., 2005). Boreal spring warming of the southwest tropical Indian Ocean could generate North Indian Ocean (NIO) warming by inducing anomalous northerlies, which in turn suppress surface latent heat flux and thus establish a positive tendency in the ocean surface heat budget (Lau and Nath, 2012; Park et al., 2010). As the NIO warms, the thermodynamic regulation of surface latent heat flux (Park et al., 2010) begins to dominate over windspeed,

Indian Summer Monsoon Variability: El Niño-teleconnections and beyond.
DOI: https://doi.org/10.1016/B978-0-12-822402-1.00004-1

increasing latent heat flux, and ultimately ISM rainfall (Park et al., 2010; Terray et al., 2003).

In the summer of 1983, following the 1982–1983 major winter El Niño, a notable NIO warming occurred, whereas the anomalously warm sea surface temperatures (SSTs) over the equatorial Indian Ocean somewhat dissipated. Although such an SST pattern may favor increased ISM rainfall (Chung and Ramanathan, 2006), this pattern does not occur with every El Niño event. In 2016, for example, anomalously high SSTs in the equatorial and the off-equatorial South Indian Ocean persisted throughout the summer (Chowdary et al., 2019b); the anomalous SST distribution in 2016 may have contributed to suppressing ISM rainfall. Furthermore, the atmospheric circulation pattern over the South Asian monsoon region during the post El Niño summers is not only affected by the Indian Ocean SST pattern but also by subtropical Northwest Pacific SST anomalies (e.g., Chowdary et al., 2017; Kosaka et al., 2013; Park et al., 2010; Xie et al., 2009).

In other words, the predictability of the peak-to-decaying phase El Niño subsequent summer-time ISM rainfall appears to be highly dependent on the details of the spatial pattern and temporal evolution of the Indo-Western Pacific SST anomalies. Likewise, the delayed effect is weakly supported by the statistical correlation on interannual time scales. The interannual correlation coefficient between the preceding winter Niño 3.4 index and ISM rainfall averaged from June to August is ~0.4, which is statistically weak (Park et al., 2010). Although it has been suggested that the relationship between decay phase El Niño conditions and ISM rainfall has strengthened recently (Chakravorty et al., 2016), the statistical relationship is still marginally significant.

In this chapter, we revisit the seasonal evolution of Indian Ocean SSTs and the accompanying impact on the ISM rainfall after three major El Niño events, 1982–1983 (82/83), 1997–1998 (97/98), and 2015–2016 (15/16). It has been previously reported that the El Niños of 1982–1983 and 1997–1998 caused off-equatorial South Indian Ocean warming in the spring and the subsequent northward progress of the warm anomalies that eventually led to the anomalously strong rainfall over the Indian subcontinent in August–September (Park et al., 2010). However, the northward progression of SST anomalies and the associated rainfall increase does not appear clearly in the spring and summer of 2016 (Chowdary et al., 2019b). Here, we revisit the delayed effect by adding the recent 2015–2016 El Niño event, which is one of the strongest events recorded in the last 145 years, comparable in magnitude to the 1997–1998 El Niño of the century (e.g., Huang et al., 2016; Jacox et al., 2016; Paek et al., 2017). We examine the Indian Ocean SST anomalies in the spring and summer of 2016 and the associated response of the ISM rainfall and compare with previous strong events, such as 1982–1983 and 1997–1998. Section 5.3.1 describes the spatial and temporal evolution of SST and low-level wind anomalies in the Indian Ocean following the three major El Niño events. In section 5.3.2, the ISM rainfall responses to the three major El Niño events are investigated in detail.

5.2   Methods and data used

To understand the dynamical processes associated with ISM precipitation during the decaying phase of El Niño, used monthly mean atmospheric reanalysis data. The atmospheric variables, including 10 m zonal and meridional winds, vertically integrated moisture flux, and moisture flux convergence from the European Center for Medium-Range Weather Forecasts version 5 (ERA5) reanalysis hourly gridded dataset, have a horizontal resolution of $0.25° \times 0.25°$ and cover January 1979 through December 2018 (Hersbach et al., 2020). To study the seasonal variability of SST, the Extended Reconstructed Sea Surface Temperature version 5 (ERSSTv5) dataset with a horizontal grid of $0.25° \times 0.25°$ is employed for the period from January 1979 to December 2018 (Huang et al., 2017). To examine the Indo-Western Pacific precipitation anomalies, we used the Global Precipitation Climatology Project (GPCP) v2.3 monthly data, spanning January 1979 to December 2018. The GPCP v2.3 combines rain gauge observations and satellite precipitation and has a $2.5° \times 2.5°$ horizontal resolution (Adler et al., 2018).

To identify the statistical relationship between El Niño and ISM rainfall, we analyzed the monthly El Niño and Indian summer monsoon rainfall indices. In addition to GPCP, we have used the Homogeneous Indian Monthly Rainfall Data Sets, which comprises monthly mean rainfall data spanning from 1987 to 2016 provided by the Indian Institute of Tropical Meteorology. We used Niño 3.4 index, which is calculated by averaging SST anomalies over the eastern Pacific from 5°S–5°N, 170°–120°W in the analysis. To identify the time-varying relationship between the preceding winter El Niño and the Indian summer monsoon rainfall index, 15-year running correlation between the preceding winter (Niño 3.4 index averaged from November to January) and All-ISM rainfall index (from June to September) is calculated, which is presented in Fig. 5.1A. For example, the correlation coefficient in the year 2000 is calculated using the time period 1993–2007.

The climatological means are calculated using 40-year (1979–2018) monthly long-term means of the individual variable, including surface latent heat flux, 10 m winds, vertically integrated moisture flux, and moisture flux convergence, SST, and precipitation. Because of the obvious increasing trend of SST, the seasonal-mean linear trends of SST at each grid are removed prior to calculating the anomalies.

5.3   Indian summer monsoon rainfall during El Niño decay

All ISM rainfall index over the entire monsoon season (June–September, hereafter JJAS) from 1979 to 2018 shows a large interannual variability (Fig. 5.1). Fig. 5.1A shows the 15-year moving window correlation between the preceding winter (November–January) El Niño, specifically the Niño 3.4 index averaged from November to January, and the JJAS ISM rainfall index. The correlation coefficient ranges from 0.3 to 0.5, which is statistically weak. The 39-year correlation

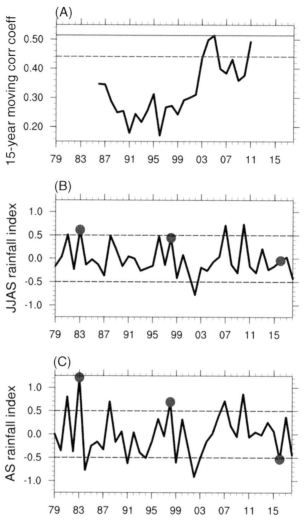

FIG. 5.1 (A) The 15-year running-window correlation between the preceding winter Niño 3.4 index and the standardized JJAS mean All-India summer monsoon rainfall index. Correlation coefficients above the red-solid line (*red-dotted line*) are statistically significant at 95% (90%) confidence interval. The 39-year correlation coefficient is 0.38. (B) Standardized June–September mean All-Indian summer monsoon rainfall index and the years with the three major winter El Niños before the summer monsoon season (*red dots*). (C) same as (A), except for late season (August–September; AS) Indian summer monsoon rainfall. In (B, C) dotted lines indicate ±0.5 standard deviations of monsoon rainfall.

from 1979/1980 to 2017/2018 between the preceding winter Niño and the ISM rainfall is 0.38, which explains approximately 15% of the interannual variance. As noted by Park et al. (2010), the ISM rainfalls were higher than normal following the two major winter El Niño events, 1982–1983 and 1997–1998 (Fig. 5.1B) and these above-normal rainfalls were particularly notable during the late monsoon season, August–September (Fig. 5.1C). However, the recent major El Niño event of 2015–2016 was not followed by above-normal monsoon rainfall in the summer of 2016 (Fig. 5.1B).

Although the delayed impact of the preceding winter El Niño on ISM rainfall is statistically weak, the NIO warming in the summer monsoon season is far more robust (Fig. 5.2). The lagged correlation between the preceding winter (November–January) Niño 3.4 index and the summer (JJAS average) SSTs from 1979 to 2018 shows a clear NIO warming (Fig. 5.2A), which is consistent with previous studies (Park et al., 2010). A majority of grid cells over the NIO shows statistically significant ($P < .05$) correlation coefficients, generally higher than 0.4, verifying the relatively robust impact of the preceding winter El Niño on the summer NIO SSTs. Unlike the NIO warming, precipitation responses are less robust. While the lagged correlations between the preceding winter Niño3.4 index and the summer precipitation are mostly positive over the NIO and neighboring Indian subcontinent, only a few grid cells exhibit statistically significant values (Fig. 5.2B). It is worth noting that the lower tropospheric wind anomalies are westward, implying weaker monsoonal westerlies (vectors in Fig. 5.2B), which suppress surface latent heat flux and thereby contributing to the summer NIO SST warming (Park et al., 2010; Lau and Nath., 2012).

While the lagged correlation maps present canonical patterns of summer SST and precipitation following the winter El Niño, the individual event shows a different SST pattern and a different precipitation response. Examining the SST responses to the individual El Niño event is necessary toward a better understanding of the delayed impact on ISM rainfall. In the next sections, we examine the seasonal SST and precipitation anomalies for the major winter El Niño events, 1982–1983, 1998–1999, and 2015–2016, respectively.

FIG. 5.2 Lagged correlations between the preceding winter (November–January) Niño 3.4 index and June–September mean (A) SST anomalies and (B) precipitation (shadings) with 850-hPa wind (vectors) anomalies. Statistically significant values ($P < .05$) are stippled. As for 850-hPa winds, only statistically significant values ($P < .05$) are presented. "*SST*, sea surface temperature."

5.3.1   Seasonal SST anomalies

The major winter El Niño events are robustly followed by basin-wide tropical Indian Ocean warming (Fig. 5.3A–C) and this El Niño induced remote tropical SST warming has been extensively investigated. Specifically, El Niño events warm the tropical Indian Ocean (20°S–20°N) by suppressing surface heat fluxes (Chiang and Lintner, 2005; Wu et al., 2008; Wu and Yeh, 2010) and by ocean dynamic processes (Du et al., 2009; Liu and Alexander, 2007; Xie et al., 2002; Chowdary and Gnanaseelan, 2007; Kakatkar et al. 2020). While all the three major El Niño events are followed by basin-wide tropical Indian Ocean warming, the detailed spatial structure and the seasonal evolution of SST anomalies for the three events contain some substantial differences.

In the following spring of all three major winter El Niño events, 1983, 1998, and 2016, warm SST anomalies develop in the equatorial and off-equatorial South Indian Ocean (Fig. 5.3A–C), which can be interpreted as a typical response to El Niño (Chen et al., 2019; Chowdary and Gnanaseelan, 2007; Chowdary et al., 2009; Lau and Nath, 2003; Xie et al., 2002). However, their seasonal progressions over the Indian Ocean contain similarities and differences. In the summer of 1983 and 1998, the equatorial and south equatorial warm Indian Ocean SST anomalies dissipated, whereas the warm NIO signal slightly strengthened (Fig. 5.3D and E). This seasonal progression of SSTs over the Indian Ocean, specifically the NIO warming in the summer being preceded by the equatorial Indian Ocean warming, can be clearly seen in Hovmöller plot (Fig. 5.4A and B). Here, the longitudinal range is averaged from 40°E to 90°E that covers the entire Western Indian Ocean including the Arabian Sea and the Bay of Bengal. As noted by Park et al. (2010), the SST anomalies averaged over

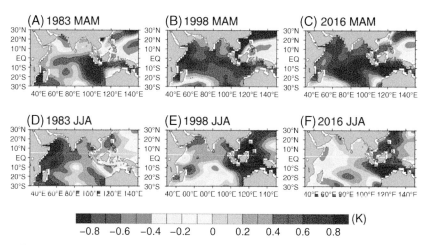

FIG. 5.3   Seasonal transitions of anomalous SSTs (K) during the (A–C) spring (March–May) and the (D–F) summer (June–August) after the three major winter El Niño events: (A, D) 1983, (B, E) 1998, and (C, F) 2016. "*SST*, sea surface temperature."

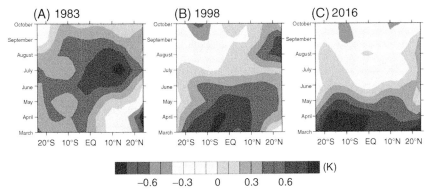

FIG. 5.4   Hovmöller diagram (40°–90°E longitudinal mean) of anomalous SST (K) following (A) 1982–1983, (B) 1997–1998 and (C) 2015–2016 major winter El Niño events. "*SST*, sea surface temperature."

the longitudinal sector of the Indian Ocean exhibit a northward progression of warm SST anomalies both in 1983 and 1998, although the zonal (east-west) SST anomalies were remarkably different each other: the maximum warm SST anomalies exist over the Western Pacific and South China Sea in the summer of 1998, whereas 1983 event has its maximum warm anomalies over the Arabian Sea (Fig. 5.3D and E).

In the summer of 2016, northward progression of warm SST anomalies over the Indian Ocean does not appear. The equatorial and the off-equatorial South Indian Ocean warming, which is a typical response to El Niño, appears (Fig. 5.3C) and the basin-wide Indian Ocean warming is stronger than those of 1983 and 1998. Unlike 1998, however, the warm SST anomalies over the NIO in the spring rapidly dissipated in the summer of 2016 (Fig. 5.3C and F). Therefore, the meridional gradient of SST anomalies over the Indian Ocean in the summer of 2016 is reversed, i.e., more warming in the south, which can weaken the ISM rainfall (Chung and Ramanathan, 2006). The Hovmöller plot averaged over the Indian Ocean clearly shows that the anomalously warm NIO relative to the equatorial Indian Ocean in the summer of 1983 and in the late summer (August–September) of 1998 (Fig. 5.4A and B) does not appear in 2016 (Fig. 5.4C). Because the warm SST anomalies over the off-equatorial South Indian Ocean persist into the summer of 2016 (Figs 5.3F and 5.4C), the meridional SST gradient effect on the ISM rainfall rather weakens. Furthermore, it is worth noting that the warm Indian Ocean SST anomalies in the spring of 2016 are wider and stronger than those of 1983 and 1998. More pronounced tropical Indian Ocean warming is observed in the spring and these SST anomalies are spatially uniform, including the Arabian Sea SST warming, which is up to ~0.7 K (Fig. 5.3C). The long-term background warming in 2015–2016 El Niño, which is associated with unusually warm conditions in 2014 (Santoso et al., 2017) is likely to be a factor for the stronger Indian Ocean warming than previous major El Niño events. In summary, the Indian Ocean warming in the spring of 2016 is generally stronger than that of previous major El Niño events, but this strong warming is not followed by the NIO warming in the summer.

5.3.2 Monsoonal winds and moisture transport

The surface wind weakening, specifically the weakening of westerlies over NIO and the associated surface latent heat suppression in the spring and summer, has been suggested as a principal driver of the NIO warming (Lau and Nath, 2012; Park et al., 2010). In the summer of 1983 and 1998, the surface wind anomalies over NIO are easterlies, both in the Bay of Bengal and Arabian Sea (Fig. 5.5A and B). These easterly anomalies are not limited to NIO but extend to the subtropical Western North Pacific (10°–25°N), where anomalous anticyclonic winds develop (Xie et al., 2009). The internal feedback between the anomalously warm NIO and the anticyclonic winds over the subtropical North Pacific, so-called the Philippine Sea anticyclones, has been well established (Annamalai et al., 2005; Kosaka et al., 2013; Stuecker et al., 2015; Watanabe and Jin, 2003; Xie et al., 2009). This interbasin feedback is referred to as the Indo-Western Pacific ocean capacitor mode (Xie et al., 2016). The warmer NIO generates eastward propagating atmospheric Kelvin waves, which would maintain the Philippine Sea anticyclones (Xie et al., 2009). Conversely, the weakening of convection over the subtropical western North Pacific (10°–25°N) generates a cold Rossby wave that can suppress precipitation westward to the Bay of Bengal (Kosaka et al., 2013; Srinivas et al. 2018). In the summer of 1983 and 1998, the development of surface anticyclones over the subtropical Western North Pacific and the associated decrease in precipitation over South China Sea and the Bay of Bengal appear (Fig. 5.5A and B).

Despite the weakening of monsoonal westerlies over NIO, precipitation is above normal over the Indian subcontinent in the summer of 1983 and 1998 (right panel figures in Fig. 5.5A and B). The paradox of the increased monsoon rainfall associated with the weaker monsoon circulation has been previously addressed by a few general circulation model (GCM) studies under global warming scenarios (Kitoh et al., 1997; Stowasser et al., 2009) and by a more realistic El Niño case with atmospheric GCM coupled to a slab ocean model (Park et al., 2010). As noted in Park et al. (2010), the monsoonal low-level wind strength does not necessarily represent the monsoonal convective strength, but moist processes, such as moisture flux convergence associated with the warmer NIO can compensate the wind effect. Indeed, Fig. 5.6 shows that the vertically integrated moisture flux anomalies are eastward over the subcontinent of India, Bay of Bengal, and the South China Sea (vectors in Fig. 5.6), which are generally consistent with the lower-tropospheric wind anomalies (vectors in Fig 5.5). However, the weakening of monsoonal winds and the associated moisture fluxes are accompanied by strengthened moisture flux convergence over subcontinent

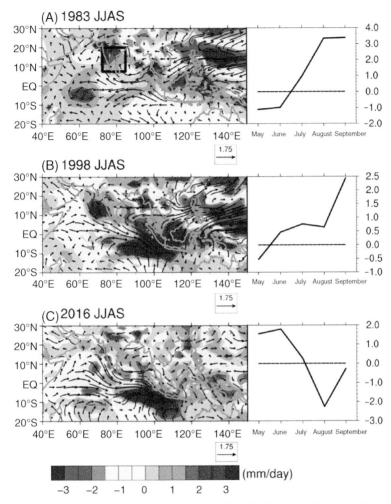

FIG. 5.5 June–September mean precipitation anomalies (shadings: mm/day) and surface wind anomalies (vectors: m/s) in (A) 1983, (B) 1998, and (C) 2016. Right-hand side panels are monthly precipitation anomalies from May to September averaged over the black square in (A) (8°–20°N, 70°–85°E).

of India (shadings in Fig. 5.6), which may explain the paradox of increased monsoon rainfall with the weaker monsoon circulation. Recent studies suggest that northwestward propagating Rossby waves induced by anomalous heating over the Maritime continent can enhance the rainfall over southern peninsular India during decay phase of El Niño years (Chowdary et al., 2019a; Srinivas et al., 2018), which can partly explain the paradox.

The weakening of NIO-region surface wind speeds during the early phase of the monsoon followed by a later rebound toward climatological values

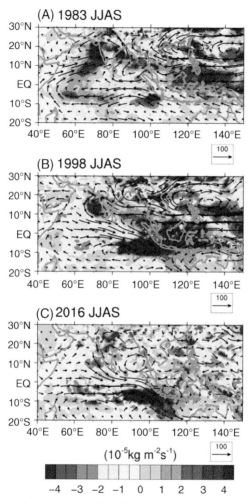

FIG. 5.6 June–September mean composites of the anomalous moisture flux (vectors: (kg/m/s) and moisture flux convergence (shadings: kg/m²/s) in (A) 1983, (B) 1998, and (C) 2016.

during 1983 and 1998, especially in September (Park et al., 2010), contributes to the notable increase in monsoon rainfall (right panel figures in Fig. 5.5A and B). In the summer of 2016, however, the basin-wide weakening of South Asian monsoon circulation that is also dynamically tied to the anomalous Philippine Sea anticyclone does not appear (Fig. 5.5C). The absence of the systematic weakening of NIO-region surface wind speed might explain the relatively weak NIO warming in the summer of 2016 (Figs 5.3F and 5.4C). Because the NIO warming signal is weak, whereas the anomalously high SSTs persist over the off-equatorial South Indian Ocean, all ISM rainfall index is close to normal in 2016 (Fig. 5.1). The subseasonal precipitation anomalies in 2016 are almost opposite to those of previous El Niño events: while precipitation is above-normal in the early monsoon season (May–June), the late monsoon season (August–September) precipitation is lower than average (right panel figure in Fig. 5.5C). The absence of the anomalous Philippine Sea anticyclone in the summer of 2016 suggests that the development of the Philippine Sea anticyclone, which contributes to weakening the low-level monsoonal winds, which in turn warms the NIO, is probably an important factor for increasing the moisture flux convergence and the associated ISM rainfall after major winter El Niño events. This analogy is supported by recent GCM experiments (Chowdary et al., 2019a; Srinivas et al., 2018), which verified the close connection between the anomalous Philippine Sea anticyclone and the increased ISM rainfall after major winter El Niño events.

5.4   Summary and discussion

In this review article, we have revisited the so-called delayed effect of major winter El Niño events on subsequent Indian Ocean SST warming and the ISM rainfall increases by including the 2015–16 El Niño event, an event of similar magnitude to prior events (1983 and 1998) for which a delayed effect is inferred. In the earlier events, the influence of preceding winter El Niño event persists into the ISM season via an anomalously warm NIO, with the latter in turn driving increasing rainfall over the NIO and neighboring Indian subcontinent. During both 1983 and 1998, the occurrence of weakening of low-level monsoonal winds and the suppression of surface latent heat flux during the late spring and summer is reported. Moreover, the presence of an anomalous Philippine Sea anticyclone is consistent with the weakening of low-level winds over NIO. Relaxation of the winds, through latent heat flux suppression, is in turn consistent with NIO warming. In 2016, however, the basin-wide NIO warming did not follow the basin-wide spring tropical Indian Ocean warming, possibly because a strong Philippine sea surface anticyclone failed to develop (Chowdary et al., 2019b). As suggested by GCM experiments (Chowdary et al., 2019a; Srinivas et al., 2018), the feedback between the NIO and the Philippine Sea surface anticyclones is a key to the NIO warming and increasing the ISM rainfall. In 1983 and 1998, the warm El Niño conditions in the eastern Pacific slowly decayed and persisted into the spring season, which has been suggested as a key driver of the NIO warming in the summer (Park et al., 2010).

It is noted that the Niño 3 index representing the eastern equatorial Pacific SST anomaly was more positive in the spring (April–May) of 1998 than that of 2016, although a transition into La Niña condition was suppressed in the summer of 2016 (Kakatkar et al., 2018). It is likely that the early decay of the anomalously warm SSTs in the Eastern Pacific in the spring of 2016 led to the weakening of the Philippine Sea surface anticyclones. Predicting the summer basin-wide tropical Indian Ocean/NIO warming patterns during El Niño decay would be useful for ISM rainfall prediction. Further numerical modeling studies on the decaying phase of El Niño, such as the persistence of the warm eastern Pacific SST in the spring, are needed to better understand the delayed impact of winter El Niño on the monsoon circulation strength and ISM rainfall.

Acknowledgments

We would like to thank Benjamin Lintner for helpful comments and for proofing the manuscript. H.-S.P. is supported by the National Research Foundation of Korea (NRF) no. 2020R1A2C2010025.

References

Adler, R., Sapiano, M., Huffman, G., Wang, J.-J., Gu, G., Bolvin, D., et al., 2018. The Global Precipitation Climatology Project (GPCP) monthly analysis (New Version 2.3) and a review of 2017 global precipitation. Atmosphere (Basel) 9, 138. doi:10.3390/atmos9040138.

Annamalai, H., Liu, P., Xie, S.P., 2005. Southwest Indian Ocean SST variability: its local effect and remote influence on Asian monsoons. J. Clim. 18, 4150–4167. doi:10.1175/JCLI3533.1.

Chakravorty, S., Gnanaseelan, C., Pillai, P.A., 2016. Combined influence of remote and local SST forcing on Indian summer monsoon rainfall variability. Clim. Dyn. 47, 2817–2831. doi:10.1007/s00382-016-2999-5.

Chen, Z., Du, Y., Wen, Z., Wu, R., Xie, S.P., 2019. Evolution of south tropical Indian ocean warming and the climatic impacts following strong el niño events. J. Clim. 32, 7329–7347. doi:10.1175/JCLI-D-18-0704.1.

Chiang, J.C.H., Lintner, B.R., 2005. Mechanisms of remote tropical surface warming during El Niño. J. Clim. 18, 4130–4149. doi:10.1175/JCLI3529.1.

Chowdary, J.S., Gnanaseelan, C., 2007. Basin-wide warming of the Indian Ocean during El Niño and Indian Ocean dipole years. Int. J. Climatol. 27, 1421–1438. doi:10.1002/joc.1482.

Chowdary, J.S., Gnanaseelan, C., Xie, S.P., 2009. Westward propagation of barrier layer formation in the 2006–07 Rossby wave event over the tropical southwest Indian Ocean. Geophys. Res. Lett. 36 (4). doi:10.1029/2008GL036642.

Chowdary, J.S., Harsha, H.S., Gnanaseelan, C., Srinivas, G., Parekh, A., Pillai, P., et al., 2017. Indian summer monsoon rainfall variability in response to differences in the decay phase of El Niño. Clim. Dyn. 48, 2707–2727. doi:10.1007/s00382-016-3233-1.

Chowdary, J.S., Patekar, D., Srinivas, G., Gnanaseelan, C., Parekh, A., 2019a. Impact of the Indo-Western Pacific Ocean Capacitor mode on South Asian summer monsoon rainfall. Clim. Dyn. 53, 2327–2338. doi:10.1007/s00382-019-04850-w.

Chowdary, J.S., Srinivas, G., Du, Y., Gopinath, K., Gnanaseelan, C., Parekh, A., et al., 2019b. Month-to-month variability of Indian summer monsoon rainfall in 2016: role of the Indo-Pacific climatic conditions. Clim. Dyn. 52, 1157–1171. doi:10.1007/s00382-018-4185-4.

Chung, C.E., Ramanathan, V., 2006. Weakening of north Indian SST gradients and the monsoon rainfall in India and the Sahel. J. Clim. 19, 2036–2045. doi:10.1175/JCLI3820.1.

Du, Y., Xie, S.P., Huang, G., Hu, K., 2009. Role of air-sea interaction in the long persistence of El Niño-induced north Indian Ocean warming. J. Clim. 22, 2023–2038. doi:10.1175/2008JCLI2590.1.

Hersbach, H., Bell, B., Berrisford, P., Hirahara, S., Horányi, A., Muñoz-Sabater, J., et al., 2020. The ERA5 global reanalysis. Q. J. R. Meteorol. Soc. 146, 1999–2049. doi:10.1002/qj.3803.

Huang, B., L'Heureux, M., Hu, Z.Z., Zhang, H.M., 2016. Ranking the strongest ENSO events while incorporating SST uncertainty. Geophys. Res. Lett. 43, 9165–9172. doi:10.1002/2016GL070888.

Huang, B., Thorne, P.W., Banzon, V.F., Boyer, T., Chepurin, G., Lawrimore, J.H., et al., 2017. Extended reconstructed Sea surface temperature, Version 5 (ERSSTv5): upgrades, validations, and intercomparisons. J. Clim. 30, 8179–8205. doi:10.1175/JCLI-D-16-0836.1.

Jacox, M.G., Hazen, E.L., Zaba, K.D., Rudnick, D.L., Edwards, C.A., Moore, A.M., Bograd, S.J., 2016. Impacts of the 2015–2016 El Niño on the California Current System: Early assessment and comparison to past events. Geophys. Res. Lett. 43, 7072–7080. doi:10.1002/2016 GL069716.

Joseph, P.V., Eischeid, J.K., Pyle, R.J., 1994. Interannual variability of the onset of the Indian summer monsoon and its association with atmospheric features, El Nino, and sea surface temperature anomalies. J. Clim. 7, 81–105. doi:10.1175/1520-0442 (1994)007<0081:IVOTOO >2.0.CO;2.

Kakatkar, R., Gnanaseelan, C., Chowdary, J.S., 2020. Asymmetry in the tropical Indian Ocean subsurface temperature variability. Dyn. of Atmos. Ocean. 90, 101142. doi:10.1016/j.dynatmoce.2020.101142.

Kakatkar, R., Gnanaseelan, C., Deepa, J.S., Chowdary, J.S., Parekh, A., 2018. Role of ocean-atmosphere interactions in modulating the 2016 La Niña like pattern over the tropical Pacific. Dyn. Atmos. Ocean. 83, 100–110. doi:10.1016/j.dynatmoce.2018.07.003.

Kitoh, A., Yukimoto, S., Noda, A., Motoi, T., 1997. Simulated changes in the Asian summer monsoon at times of increased atmospheric CO_2. J. Meteorol. Soc. Japan. Ser. II 75, 1019–1031. doi:10.2151/jmsj1965.75.6_1019.

Kosaka, Y., Xie, S.P., Lau, N.C., Vecchi, G.A., 2013. Origin of seasonal predictability for summer climate over the Northwestern Pacific. Proc. Natl. Acad. Sci. U. S. A. 110, 7574–7579. doi:10.1073/pnas.1215582110.

Lau, N.C., Nath, M.J., 2012. A Model study of the air-sea interaction associated with the climatological aspects and interannual variability of the South Asian summer monsoon development. J. Clim. 25, 839–857. doi:10.1175/JCLI-D-11-00035.1.

Lau, N.C., Nath, M.J., 2003. Atmosphere-ocean variations in the Indo-Pacific sector during ENSO episodes. J. Clim. 16, 3–20. doi:10.1175/1520-0442 (2003)016<0003:AOVITI>2.0.CO;2.

Liu, Z., Alexander, M., 2007. Atmospheric bridge, oceanic tunnel, and global climatic teleconnections. Rev. Geophys. doi:10.1029/2005RG000172.

Paek, H., Yu, J.Y., Qian, C., 2017. Why were the 2015/2016 and 1997/1998 extreme El Niños different? Geophys. Res. Lett. 44, 1848–1856. doi:10.1002/2016GL071515.

Park, H.S., Chiang, J.C.H., Lintner, B.R., Zhang, G.J., 2010. The delayed effect of major El Niño events on Indian monsoon rainfall. J. Clim. 23, 932–946. doi:10.1175/2009JCLI2916.1.

Santoso, A., Mcphaden, M.J., Cai, W., 2017. The defining characteristics of ENSO extremes and the strong 2015/2016 El Niño. Rev. Geophys. doi:10.1002/2017RG000560.

Shukla, J., 1995. Predictability of the tropical atmosphere, the tropical oceans and TOGA, Proc. Int. Sci. Conf. Trop. Ocean Glob. Atmos. Program. WCRP-91, WMO/TD 717, 2. Melbourne, pp. 725–730.

Srinivas, G., Chowdary, J.S., Kosaka, Y., Gnanaseelan, C., Parekh, A., Prasad, K.V.S.R., 2018. Influence of the Pacific-Japan pattern on Indian summer monsoon rainfall. J. Clim. 31, 3943–3958. doi:10.1175/JCLI-D-17-0408.1.

Stowasser, M., Annamalai, H., Hafner, J., 2009. Response of the South Asian summer monsoon to global warming: mean and synoptic systems. J. Clim. 22, 1014–1036. doi:10.1175/2008JC LI2218.1.

Stuecker, M.F., Jin, F.F., Timmermann, A., Mcgregor, S., 2015. Combination mode dynamics of the anomalous northwest pacific anticyclone. J. Clim. 28, 1093–1111. doi:10.1175/JCLI-D-14-00225.1.

Tao, W., Huang, G., Hu, K., Gong, H., Wen, G., Liu, L., 2016. A study of biases in simulation of the Indian Ocean basin mode and its capacitor effect in CMIP3/CMIP5 models. Clim. Dyn. 46, 205–226. doi:10.1007/s00382-015-2579-0.

Terray, P., Delecluse, P., Labattu, S., Terray, L., 2003. Sea surface temperature associations with the late Indian summer monsoon. Clim. Dyn. 21, 593–618. doi:10.1007/s00382-003-0354-0.

Watanabe, M., Jin, F.F., 2003. A moist linear baroclinic model: coupled dynamical-convective response to El Niño. J. Clim. 16, 1121–1139. doi:10.1175/1520-0442 (2003)16<1121:AMLBMC>2.0.CO;2.

Webster, P.J., Magaña, V.O., Palmer, T.N., Shukla, J., Tomas, R.A., Yanai, M., et al., 1998. Monsoons: processes, predictability, and the prospects for prediction. J. Geophys. Res. Ocean. 103, 14451–14510. doi:10.1029/97jc02719.

Wu, R., Kirtman, B.P., Krishnamurthy, V., 2008. An asymmetric mode of tropical Indian Ocean rainfall variability in boreal spring. J. Geophys. Res. Atmos. 113. doi:10.1029/2007JD009316 n/a-n/a.

Wu, R., Yeh, S.-W., 2010. A further study of the tropical Indian Ocean asymmetric mode in boreal spring. J. Geophys. Res. 115, D08101. doi:10.1029/2009JD012999.

Xie, S.P., Annamalai, H., Schott, F.A., McCreary, J.P., 2002. Structure and mechanisms of South Indian Ocean climate variability. J. Clim. 15, 864–878. doi:10.1175/1520-0442 (2002)015<0864:SAMOSI>2.0.CO;2.

Xie, S.P., Hu, K., Hafner, J., Tokinaga, H., Du, Y., Huang, G., et al., 2009. Indian Ocean capacitor effect on Indo-Western pacific climate during the summer following El Niño. J. Clim. 22, 730–747. doi:10.1175/2008JCLI2544.1.

Xie, S.P., Kosaka, Y., Du, Y., Hu, K., Chowdary, J.S., Huang, G., 2016. Indo-western Pacific ocean capacitor and coherent climate anomalies in post-ENSO summer: a review. Adv. Atmos. Sci. 33, 411–432. doi:10.1007/s00376-015-5192-6.

Yang, J., Liu, Q., Xie, S.-P., Liu, Z., Wu, L., 2007. Impact of the Indian Ocean SST basin mode on the Asian summer monsoon. Geophys. Res. Lett. 34, L02708. doi:10.1029/2006GL028571.

Chapter 6

El Niño–Indian summer monsoon relation—a nonlinear scale interactive energy exchange perspective

S. De, N.K. Agarwal

Indian Institute of Tropical Meteorology, Pune, Maharashtra, India

6.1 Introduction

El Niño is the largest air–sea interaction phenomena in the tropics that affects the global monsoon, in particular the Indian summer monsoon (hereafter referred to as ISM). One of the major manifestations of air–sea interactions is the energy exchange between the ocean and atmosphere. Hence, a comprehensive energetics study is inevitable to focus on the air–sea coupling processes for its dynamical and physical understanding through which El Niño evolves. A historical quest for the physical and dynamical relationship between El Niño and ISM was started to establish after the invention of Walker circulation in 1923 (Walker 1924). The Walker circulation determines the large-scale out-of-phase relation in surface pressure over the South Pacific/South American region with an Indian Ocean/Indonesian pressure, known as Southern Oscillations (SO). The SO redistributes the heat source/sink over the equatorial Pacific, first found by Bjerknes (1969), resulting in thermally driven east-west Walker circulation over the Indo-Pacific region (Sikka, 1980; Pant and Parthasarathy, 1981). Krishnamurti (1971) documented Walker circulation as the part of global circulation with equivalent strength of regional Hadley cell over the Indian region during the summer monsoon period. Rasmusson and Carpenter (1983) established a strong correlation (−0.62) between ISM rainfall (ISMR) and sea surface temperature (SST) anomaly over Date Line and South American coast following December–January period with the inverse relation between SST anomaly and ISMR. Shukla and Paolino (1983) also documented the relations among Asian summer monsoon, El Niño and SO from 81 years of observations during 1901–81. They illustrated that these three phenomenon either influence or were influenced by each other

Indian Summer Monsoon Variability: El Niño-teleconnections and beyond.
DOI: https://doi.org/10.1016/B978-0-12-822402-1.00025-9

at the different stages of their life cycles. Another landmark work was done by Webster and Yang (1992) where the authors proposed a monsoon index involving both the observed upper and lower tropospheric winds and showed the selective interactions between south Asian monsoon, El Niño and SO (ENSO) through Walker circulation and documented the dynamical aspects of ENSO–monsoon nonlinearity. Limited studies have been so far documented revealing the dynamics of this nonlinearity (Klinter III et al., 2002; Singh et al., 2020, etc.). How the large volcanic eruption affects ENSO–ISM nonlinearity and the potential improvement of ISM forecast are discussed recently by Singh et al. (2020).

The first modeling approach to recognize the relationship of the evolution of monsoon with ENSO was documented by Ju and Slingo (1995) using the atmospheric general circulation model (AGCM) over a period of ten years with the revelation that large scale monsoon circulation is more robustly associated with the SST variability than with the regional aspect of monsoon-like local rainfall. Another modeling initiative was taken by Soman and Slingo (1997) where the ENSO monsoon relationship was seen in terms of tropical convective maxima (TCM) in the equatorial region applied to the year 1983 (El Niño) and 1984 (La Niña). ISM strength was evaluated from monsoon onset date derived from low level (850 hPa) kinetic energy (KE) averaged over a region covering India and Bay of Bengal region. The study inferred that the warm (cold) SST anomaly in central/east Pacific modulates the Walker circulation by shifting it toward the east (west) that causes the eastward (westward) transfer of TCM attributing to the delayed (early) monsoon onset during the El Niño year 1983 (La Niña year 1984). Krishnan and Kasture (1996) investigated how the positive SST anomaly in equatorial central and Eastern Pacific during the El Niño event modulated the low-frequency oscillations (LFOs) over the Indian subcontinent by inducing the anomalous changes in large scale monsoon basic flow. The main conclusion of the study was that El Niño elongated the period of intraseasonal cloudiness fluctuations over the Indian subcontinent and the period was found to be 60-day, whereas the non-El Niño years showed 30–40 days intraseasonal oscillations (ISOs). The summer time SST anomaly has a strong influence in detection and prediction of intrinsic nature of low-frequency intraseasonal variability of ISM. Recent severe Indian monsoon rainfall deficit during 2009 was examined by VenkataRatnam et al. (2010) in terms of perturbation experiment by atmospheric GCM using observed and climatological SST. The study reveals that the strong El Niño in 2009 is due to the interaction between strong low-level westerlies observed near west Pacific and suppressed convection resulting regional Walker circulation with updraft near central Pacific and downdraft near west Pacific and Indian region. From the above literature survey, it has been inferred that the large-scale circulations are the robust metrics rather than the regional scale rainfall to study the modulation of Indian monsoon by El Niño. As the ISM variability comprises of low and high frequency oscillations which regulate the monsoon rainfall, it may be hypothesized that the strong interactions may prevail among El Niño, low and high frequency oscillations. Unraveling these interactions may add a new dimension as a deeper insight in revisiting El Niño–monsoon relation.

Although wind is the potential parameter to study the relation between El Niño and ISM, very few research works have so far been focused on energetics study. Pioneering work by Goddard and Philander (2000) documented that atmospheric KE is fed into the ocean available potential energy (APE) through the coupling between the wind stress and ocean current as a result of dynamical air–sea interactions. The authors concluded that the ultimate energy sources come from the work done by the wind in KE equation. Oceans gain (loose) energy in transition from El Niño to La Niña (La Niña to El Niño) (Oort et al. 1989). Due to this, the mean atmospheric potential energy (P_M) and mean KE (K_M) are observed to be enhanced by 2% and 4%, respectively, during El Niño event, whereas La Niña episode has been witnessed to 2%–3% decrease for both P_M and K_M revealed from the study Li et al. (2011). Moreover, Li et al. (2011) showed strong correlations of Niño 3.4 index with P_M and K_M with a time lag of 2–4 months. The same index is also correlated significantly with eddy KE (K_E) with 7-month time lag. The authors speculated from the above correlations that the variation of energy conversion terms may affect the atmospheric energies in El Niño and La Niña events. From the above energetics study it was not well understood how the energy exchanges were taking place between ocean and atmosphere in terms of the SST anomaly over the tropical Pacific and for that, recently, Dong et al. (2017) derived a parameter perturbation potential energy (PPE) from total potential energy. The authors explored the variability of the intensity of Hadley circulations and its relation with atmospheric PE and its conversion to KE to complete the ocean-atmosphere cycle through air–sea interface during El Niño, La Niña, and their transition phase. As far as energy conversions are concerned, Dong et al. (2017) restricted their work to the conversion from PPE to KE, only to the tropical Pacific and not extended up to the ISM region. Hence, from that particular study, it is not clear what role the KE is playing to regulate the ISM by El Niño? How does El Niño influence the low and high frequency oscillations pertinent to ISM through the exchange of KE among El Niño, low and high frequency oscillations? This entails the objective of this work to explore the intrinsic dynamical relation between El Niño and ISM.

The variability of any atmospheric system is due to the compound effect of different scales associated with the system and which can be evaluated through a scale interaction mechanism. The scale interaction is essentially a complex nonlinear tool where one scale can be dominated at the cost of the other scale by exchanging the KE between the two participating scales. Saltzman (1957) invented this technique which was later modified by Murakami and Tomatsu (1964) and Hayashi (1980). This technique was extensively used by Kanamitsu et al. (1972), Krishnamurti et al. (2003), Krishnamurti and Chakraborty (2005), Sheng and Hayashi (1990), etc., for tropical energetics and to delineate the nonlinear aspects of the intraseasonal oscillations in wavenumber and frequency domain. De et al. (2019) utilized the technique to unfold the cloud-circulation interactions in both domains. The first time the El Niño–ISM relations have been explored in the light of scale interactions through the process of nonlinear KE exchanges among El Niño, low and high frequency oscillations to get into the in-depth science behind the relations. The chapter has been organized as follows. Methodology and formulations are described in Section 6.2. Section 6.3 delineates results, whereas Section 6.4 summarizes the work with conclusions and discussions.

6.2 Methodology and formulations

The scale interactions are basically computed in two spectral domains—one is in wavenumber domain, derived from discrete spatial Fourier transform, whereas the other is in the frequency domain, obtained from discrete temporal Fourier transform. This work has been restricted to frequency domain study as it deals with the waves of different time periods. Moreover, the frequency domain has an advantage to study for a particular region (here for El Niño–monsoon study it is 45°E–90°W; 25°S–25°N) by virtue of its methodology, whereas for wavenumber domain, the region of study requires either whole globe or global tropical /subtropical belt which is not applicable for this work. Initial scale interaction formulation by Saltzman (1957) was modified by Murakami and Tomatsu (1964) and Tomatsu (1979) as triad interaction technique. It is a specific as well as unique mathematical tool in the sense that this tool is used only in the Fourier method and may be applied to a term of triple product variables. This mathematical instrument invokes the interscale information in the form of the nonlinear energy exchange into individual triad interactions that contribute to the total exchange associated with any particular scale whose triad interaction is to be computed. Thus the effect of the one scale to other scale may also be known. When the energy budgets are evaluated in the spectral domain, one may ask question on the growth of energy at a certain frequency from its interaction with other temporal scales. This entails the computation of the Fourier series of nonlinearities in wave–wave energy exchanges associated with triple products. Here coherence of Fourier spectra in the space of nonlinearities constitutes the main computations.

A temporal Fourier analysis is used to separate the El Niño, low-frequency, and high-frequency modes from the wind fields. The nonlinear KE transfer into frequency "n" after interactions with different frequencies (excluding time mean) denoted by $<Km.\ Kn>$ is the sum of energy transfers by all possible individual triad interactions $L(n,\ r,\ s)$ among three frequencies $n,\ r,\ s$, where $n,\ r,$ and s are related following some trigonometric selection rules as $n = r + s$ and $n = |\ r - s\ |$ (Saltzman 1957). Therefore, $<Km\ .\ Kn> = \Sigma L(n,\ r,\ s) = <L(n)>$. The nonlinear KE transfer spectra among different frequencies i.e. wave-wave interaction $<Km.\ Kn>$ has been computed following Eq. (1) shown in Appendix. The energy transfer processes are evaluated in barotropic atmosphere and hence, baroclinic terms are not considered here.

6.2.1 Scale selection of El Niño and data

El Niño is generally a recursive phenomenon of 2–7 years having large spatial scale of influence expanding from ISM region to the west coast of South America but it has no definite time period. The time span of El Niño depends

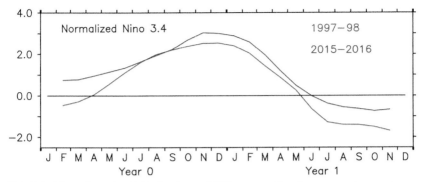

FIG. 6.1 Normalized Niño 3.4 index for two El Niño events, Year 0 represents the current year and year 1 stands for the following year.

upon its length of its decay phase (e.g., Chowdary et al., 2017). There are El Niño events of early decay whose decay period ended before the spring time having the time period of around 12 months. Another type of El Niño is having decay period that lasts up to mid-summer and then, the phenomenon turns into La Niña for which the time span of El Niño may go up to 18 months. For the third type of El Niño the decay phase may go beyond the boreal summer period. Fig. 6.1 delineates the normalized Niño 3.4 SST index for 1997–98 and 2015–16 El Niño events starting from current year, i.e., 1997 (2015) denoted by Year 0 and ending to the following year 1998 (2016) represented by Year 1. The detailed justifications for selecting the above two El Niño events are given in the following paragraph. The figure demonstrates that 1997–98 El Niño initiated from the month of April 1997 and ended in May 1998 spanning around 14 months. Similarly, 2015–16 El Niño period may be taken as 18 months starting from January 2015 and ending in June 2016. Thus the time period of El Niño may be selected approximately 12–18 months. Neelin et al. (2000) and Krishnamurti and Chakraborty (2005) considered El Niño as LFOs and exhibited explosive growth of interactions between the waves of two different time scales under the phase-locking condition. Therefore, in view of that, El Niño may be treated broadly as the LFOs of time period 12–18 months.

Two extreme/super El Niño events are chosen—one is 1997–98 El Niño and other is 2015–16 El Niño having a great similarity of maximum SST anomaly of about 3.5°C initiated from South American coast at their developing stage but at the same time showing large dissimilarity at their decay phase (Paek et al. 2017; also see Chapter 5). The positive SST anomaly moved westward from South American coast for both the cases but 1997–98 El Niño retreated back to the coast showing eastern Pacific (EP) El Niño. On the contrary, the peak value of the SST anomaly was observed around International Date Line in the case of 2015–16 event and it lasted during the decay phase of El Niño exhibiting central Pacific El Niño (Paek et al. 2017). Another significant dissimilarity between the two El Niño(s) is their relationship with associated ISM. The years 1997–98 reported 103% of long-period average ISMR (Srinivasan and Nanjundiah, 2002) with 102% and 105%, during 1997 and 1998, respectively (India Meteorological Department, 1998; India Meteorological Department, 1999), although there is an inverse relation between El Niño and ISMR. Unlike to 1997–98, below normal rainfall is observed in 2015 and 2016 showing 86% (India Meteorological Department, 2016) and 97% long period average (India Meteorological Department, 2017) over India, respectively. So, two similar types of El Niño(s) are related oppositely with their associated ISM and to unfold the reasons behind this selective El Niño–ISM relation (Webster and Yang, 1992), above two events have been chosen consciously. The European reanalysis wind (ERA) of $1°$ resolution at 850 hPa and 200 hPa, representative of lower and upper atmosphere, respectively, are used in this study (Dee et al., 2011). The periods of two datasets are from January 1, 1997 to December 31, 1999 and from January 1, 2015 to December 31, 2017, encompassing the two El Niño(s) 1997–98 and 2015–16, respectively. The maximum period of El Niño is 18 months representing half of the cycle (shown in Fig. 6.1) which is Nyquist frequency and so, for capturing the full cycle to avoid the nonlinear aliasing error (Hayashi 1980) 3 years continuous daily data are used in each dataset.

6.3   Nonlinear scale interactions between El Niño and Indian summer monsoon

The interannual variability of ISM is mostly regulated by the high and LFOs. Most dominating high frequency wave is 3–5 days oscillations, which represent the synoptic scale disturbances in the form of lows, depressions, deep depressions, cyclonic disturbances, etc., designated as low pressure systems that are the main rain bearing disturbances during ISM (Mooley and Shukla, 1987; Sikka, 1980). The prominent LFOs during monsoon are 10–20 day and 30–60 day modes. The 10–20 day oscillations represent the active-break spell of ISM by influencing the westward propagation of low-pressure systems (Goswami et al. 2003) and 30–60 day mode of ISOs regulates the latitudinal position of monsoon trough by controlling the northward propagation of trough (Sikka and Gadgil, 1980; Krishnamurti and Subrahmanyam, 1982; see Chapter 1). As the above high and low frequency waves are the inseparable parts of ISM which is strongly connected with the large scale El Niño, it may be postulated that there must be a strong interactions among El Niño, 30–60 day, 10–20 day, and 3–5 day waves through the exchange of KE. Already there was evidence of sensitizing the 30–50 day intraseasonal monsoon mode during the El Niño episode (Krishnan and Kasture, 1996), but the quantification of interactions between El Niño and ISO is absent in that study. Moreover, the interconnections between El Niño and other low and high frequency monsoon modes have not been studied so far in previous works except Krishnamurti et al. (2017), which documented the scale interactions during extreme rainfall event over southeast India during

North-East Indian monsoon. Keeping in view of above, assuming El Niño a large scale low frequency mode having the period of 12–18 months, the interactions among El Niño and dominant low and high frequency monsoon modes are addressed and quantified in terms of triad interactions through the process of nonlinear KE exchanges among different scales following Eq. (1).

The global mean atmospheric energy cycle following Lorenz (1955) was successfully examined by Li et al. (2007) for two reanalysis datasets: NCEP and ERA. The global P_M, K_M, P_E (eddy APE), and K_E are resembled with the global distribution of those parameter evaluated by Lorenz (1955), showing the similar pattern including the conversion terms—$P_M \rightarrow K_M$; $P_E \rightarrow K_E$; and $K_M \rightarrow K_E$. This study was extended further by Li et al. (2011) considering the El Niño and La Niña years which showed that eddy KE is highly correlated with Niño 3.4 index with a time lag of around 7 months (which is the response time of large scale atmospheric process to SST) as observed in both NCEP and ERA datasets. In this view, as an initial study in this direction, only ERA data are considered and for each El Niño episode 3 years continuous daily data have been utilized to include the response time of atmospheric processes to SST. Fig. 6.2 delineates the total nonlinear KE associated with El Niño, 30–60 day, 10–20 day, and 3–5 day oscillations during two El Niño events. The total nonlinear KE associated with any particular wave may be evaluated as the sum of the KE exchanges related to that wave while interacting with all possible waves. The following are revealed from the figure.

1. Total KE associated with an oscillation obeys the hierarchy of the scale of the wave which implies that El Niño possesses maximum KE following 30–60 day, 10–20 day, and 3–5 day for both El Niño(s) shown at lower and upper atmosphere.

2. All the modes except the 3–5 day synoptic-scale possesses higher amount of energy during 1997–98 El Niño event compared to those in 2015–16 El Niño at 850 hPa. Unlike to lower troposphere, upper troposphere shows almost equal amount of energy associated with all modes for the two El Niño events with larger magnitude of KE compared to that of all waves in lower troposphere.

It is quite understandable from Lorenz (1969) that the amount of KE associated with all waves must be proportional to the scale (here time period) of that wave implying the waves of higher time scale possess larger amount of energy compared to the waves of smaller time period. Due to the stronger upper-tropospheric winds KE is appeared to be larger in magnitude at 200 hPa (Fig. 6.2B) compared to the KE at 850 hPa (Fig. 6.2A). Hence, Fig. 6.2 is following the basic energy rules given by Lorenz (1969). From the total KE it is very difficult to evaluate the variability in energetics between the two El Niño(s) unless the energy interaction terms are computed. Therefore, the nonlinear scale interactions among El Niño, low and high frequency waves in the form of triad interactive KE exchanges are described in the following sub-section.

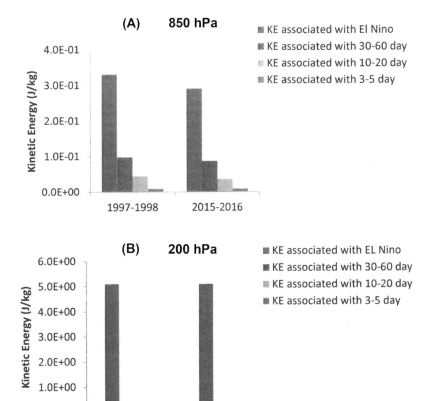

FIG. 6.2   Total nonlinear kinetic energy associated El Niño, 30–60 day, 10–20 day, and 3–5 day oscillations for two El Niño episodes (1997–98 and 2015–16) at (A) lower (850 hPa) and (B) upper (200 hPa) atmosphere.

6.3.1   El Niño 1997–98

The positive (negative) value in the geographical distribution of interaction terms associated with El Niño implies KE transfer to (from) El Niño.

Fig. 6.3 represents the lower tropospheric KE exchanges between El Niño and (A) 3–5 day synoptic-scale systems as high-frequency waves, (B) 10–20 day, and (C) 30–60 day LFOs. The following aspects may be revealed from the figure.

1. There are three predominant zones of interactions shown in three subplots. These are (1) ISM region (50°E–90°E, 5°N–25°N); (2) west north Pacific (WNP) region (100°E–135°E, 10°N–25°N); (3) Maritime continent (MC) region (105°E–160°E, 5°S–15°S). Moreover, another significant area is observed at EP region around 135°W–100°W and 10°N–15°N in Fig. 6.3A

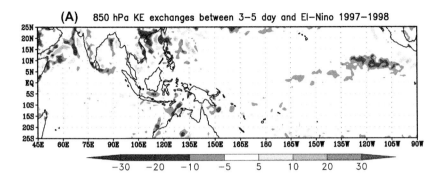

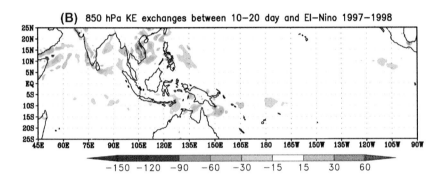

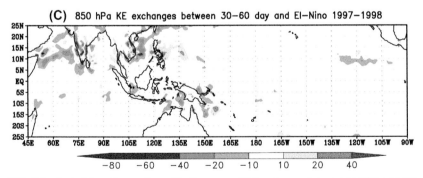

FIG. 6.3 Scale interactive nonlinear kinetic energy (*KE*) exchanges between El Niño and (A) 3–5 day, (B) 10–20 day, and (C) 30–60 day oscillations in tropical Pacific and Indian region during 1997–98 El Niño event at lower (850 hPa) troposphere. Positive (negative) values represent the KE transfer to (from) El Niño (unit: 10^{-6} watts/kg).

and a small region is also observed in Fig. 6.3C at 120°W in the same latitudinal belt. As far as ISM, WNP, and MC regions are concerned, more negative values are observed than the of positive value implying the net negative value that means 3–5 days synoptic scales are drawing energy from El Niño scale. Considerable amount of KE is also transferred to 3–5 days from El

Niño at EP region, as shown in Fig. 6.3A. Most dominant energy transfer is occurred at WNP region compared to other regions.

2. Fig. 6.3B exhibits net negative value leading to the nonlinear KE gain by 10–20 day LFOs from El Niño at WNP, ISM, and MC region. Arabian Sea, west coast of India, Central Bay of Bengal, and equatorial Indian Ocean have shown dominant transfer of energy to 10–20 day mode in ISM region. WNP region shows strongest interactions like Fig. 6.3A. A net gain of KE by 30–60 day ISO mode from El Niño has revealed from Fig. 6.3C, observed in ISM and MC regions. Almost equal amount of positive and negative value of interactions are observed in WNP region resulting net weak interactions which imply the small gain of KE by 30–60 day mode, unlike to ISM and MC regions.

The upper tropospheric KE exchanges are basically confined to midlatitudes as revealed from the interactions of El Niño with 3–5 day (Fig. 6.4A), 10–20 day (Fig. 6.4B), and 30–60 day (Fig. 6.4C) oscillations. The upper tropospheric energy exchange is found to be stronger than the energy exchange in lower troposphere. The dominant interactions are observed in midlatitude regions of ISM, WNP, and the East of the central Pacific confined to the north of the equator. No significant interactions are occurred at MC region. The following aspects are depicted from Fig. 6.4.

1. Synoptic scale waves are receiving KE from El Niño at WNP region and the northern part of India and Arabian Sea area of ISM region (Fig. 6.4A). But, at the north of Central Pacific around 20°N–25°N of 165°W there is a strong positive interaction where energy is transferred to El Niño from 3 to 5 days waves. A noticeable amount of energy was gained by high-frequency modes at East of the central Pacific region.

2. The most significant observation in Fig. 6.4B is that large amount of KE has been transferred to 10–20 day LFO from El Niño over the northern part of India stretching from Arabian Sea to Bay of Bengal, whereas the same LFO is losing KE to El Niño at WNP region. Another upper tropospheric LFO of 30–60 day mode is showing the significant transfer of KE to El Niño in Central and North East India, however, a small amount of KE is also received by 30–60 day mode north-western part of India resulting net gain of KE to El Niño from 30 to 60 day oscillations (Fig. 6.4C).

The interaction between high and LFOs at 850 hPa during 1997–98 El Niño reveal the dominant interactions at WNP region where the synoptic scale is gaining KE from 10–20 day and 30–60 day LFOs, as shown in Figs. 6.5A and B, respectively, due to the more negative value than positive. The equivalent amount of positive and negative interactions are appeared in WNP and south-west Pacific region leading to the net weak interactions between 10–20 day and 30–60 day modes at 850 hPa, as revealed from Fig. 6.5C. The upper tropospheric KE exchanges between synoptic scale and LFOs exhibit its presence

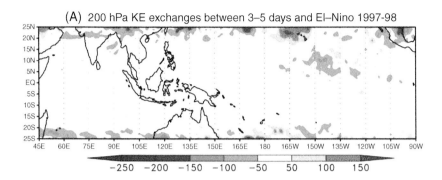

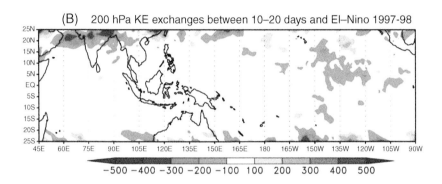

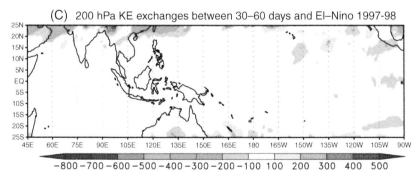

FIG. 6.4 Scale interactive nonlinear kinetic energy (KE) exchanges between El Niño and (A) 3–5 day, (B) 10–20 day, and (C) 30–60 day oscillations in tropical Pacific and Indian region during 1997–98 El Niño event at upper (200 hPa) troposphere. Positive (negative) values represent the KE transfer to (from) El Niño (unit: 10^{-6} watts/kg).

in midlatitude but with almost equal magnitude of positive and negative value resulting weak interactions between these two modes, revealed from Fig. 6.6.

One of the major reasons for non-occurrence of below normal ISMR during 1997–98 is that the 3–5 day synoptic scale and 10–20 day LFO are strengthened

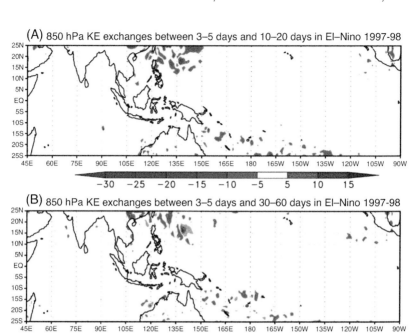

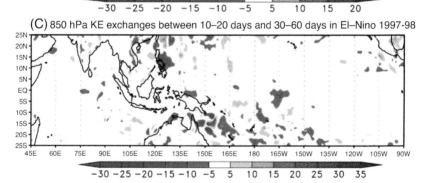

FIG. 6.5 Scale interactive nonlinear kinetic energy (*KE*) exchanges between (A) 3–5 day and 10–20 day, (B) 3–5 day and 30–60 day, and (C) 10–20 day and 30–60 day oscillations in tropical Pacific and Indian region during 1997–98 El Niño event at lower (850 hPa) troposphere. Positive (negative) values represent the KE transfer from (to) 3–5 day in (A); (B) and 10–20 day in (C) (unit: 10^{-6} watts/kg).

due to the feeding of KE to these scales from El Niño 1997–98 at both lower and upper troposphere. For a good ISM 3–5 day wave and 10–20 day oscillations should be pronounced as the synoptic scales are responsible for wide spread ISMR and the 10–20 days LFOs control the active-break phase of ISM (Sikka 1980; Mooley and Shukla, 1987; Goswami et al., 2003). On the other hand, 30–60 day mode, observed to be weak (strong) in normal (deficient) ISMR (Kripalani et al., 2004; Kulkarni et al., 2011), is losing more KE to El

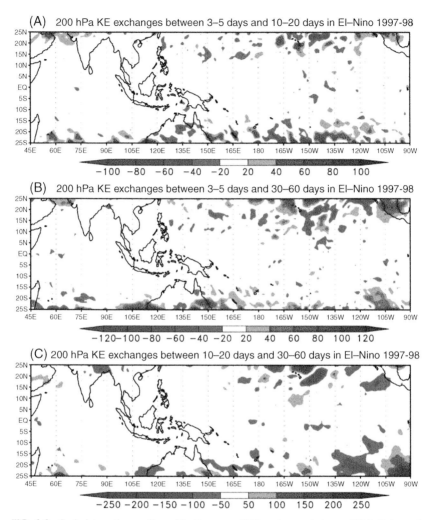

FIG. 6.6 Scale interactive nonlinear kinetic energy (*KE*) exchanges between (A) 3–5 day and 10–20 day, (B) 3–5 day and 30–60 day, and (C) 10–20 day and 30–60 day oscillations in tropical Pacific and Indian region during 1997–98 El Niño event at upper (200 hPa) troposphere. Positive (negative) values represent the KE transfer from (to) 3–5 days in (A); (B) and 10–20 days in (C) (unit: 10^{-6} watts/kg).

Niño at 200 hPa than receiving the same at 850 hPa leading to the weak 30–60 day mode during 1997–98 El Niño. The midlatitude dominance of upper tropospheric energy interactions described in Figs. 6.4 and 6.6 are strongly resembled with the study documented by Li et al. (2007, 2011) where K_E and the conversion between K_M and K_E were confined to the midlatitude using ERA40 data. The 200hPa interaction terms are predominant at the places where K_E in Li et al.'s study (2007) showed strong presence.

6.3.2   El Niño 2015–16

El Niño 2015-2016 is the strongest El Niño in the 21st century as of now (Paek et al. 2017) in which ISM showed below normal rainfall, unlike to 1997–98 El Niño as per end of seasons reports (1997, 1998, 2015, 2016) of India Meteorological Department.

The spatial distribution of nonlinear KE exchanges among El Niño, synoptic scale and LFOs are elucidated in Figs. 6.7 and 6.8 for lower and upper troposphere, respectively. The following observations are delineated from the figures.

1. Strong interactions are found in WNP region with more negative value than the positive leaving the net negative value after summation which implies large transfer of energy from El Niño to synoptic waves, whereas opposite feature is observed in ISM region. Positive and negative interactions of small magnitude coexist over the Arabian Sea and Bay of Bengal. Indian landmass is void of any significant interactions. This signifies the net weak interactions leading to the reduced KE exchanges between 3–5 day and El Niño over ISM region (Fig. 6.7A). The synoptic scale has gained comparatively less amount of energy from El Niño 2015–16 in respect of 1997–98 El Niño (Fig. 6.3A) over EP region. The MC region exhibits weak interactions during 2015–16 indicating low energy transfer to 3–5 day mode. The 10–20 day LFO has received KE from El Niño over ISM, WNP and MC regions as revealed in Fig. 6.7B. As far as the 30–60 day LFOs and El Niño interactions are concerned over ISM region (Fig. 6.7C), Arabian Sea and west equatorial Indian Ocean show larger negative value interactions compared to the interactions of positive value signifying gain of energy by LFO from El Niño, whereas net weak interactions are found in favor of LFO at southern part of India. Due to presence of small positive values over the Bay of Bengal small KE has been transferred to El Niño from 30–60 day. Considerable amount of opposite interactions is occurred in MC region leading to small energy gain by LFO. Overall a weak interaction has been prevailed in favor of synoptic and 30–60 day modes at lower troposphere during El Niño 2015–16 over the ISM region.
2. Similar to Fig. 6.4, upper tropospheric interactions are confined to midlatitudes, observed from Fig. 6.8. The Bay of Bengal and north-eastern part of the Indian landmass are the regions where small gain of KE by 3–5 day from El Niño is observed due to the presence of weak negative interactions. In north Pacific, opposite transfer of KE between synoptic scale and El Niño are observed at the east and west of Date Line (Fig. 6.8A). Most significant interaction between 10–20 day LFO and El Niño are found over the Indian region (Fig. 6.8B) where two equally strong positive and negative values are present implying the net exchange of upper tropospheric KE to LFO from El Niño 2015–16 is small. Unlike to Fig. 6.8B, 6.8C shows strong energy exchange over ISM region where the area covered by negative value is more than the area of positive value. Moreover, the

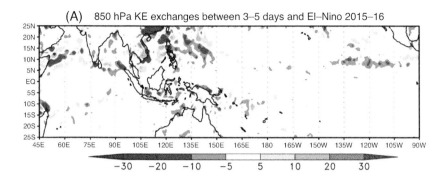

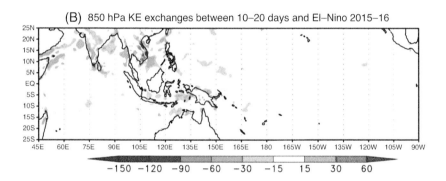

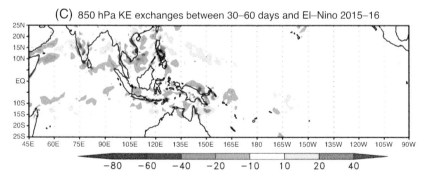

FIG. 6.7 Scale interactive nonlinear kinetic energy (*KE*) exchanges between El Niño and (A) 3–5 day, (B) 10–20 day, and (C) 30–60 day oscillations in tropical Pacific and Indian region during 2015–16 El Niño event at lower (850 hPa) troposphere. Positive (negative) values represent the KE transfer to (from) El Niño (unit: 10^{-6} watts/kg).

highest magnitude of negative interactions is one and half times more than that of positive interactions which ultimately results large negative interactions and thus, 30–60 day LFO has received significant amount of KE from El Niño 2015–16.

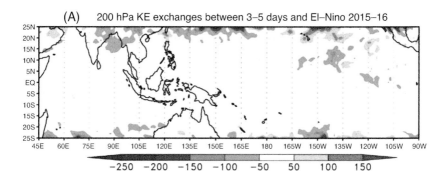

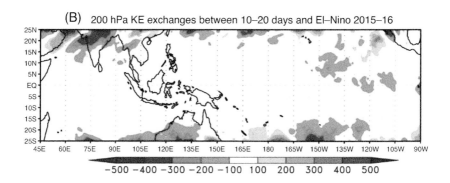

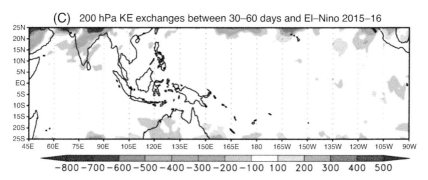

FIG. 6.8 Scale interactive nonlinear kinetic energy (*KE*) exchanges between El Niño and (A) 3–5 day, (B) 10–20 day, and (C) 30–60 day oscillations in tropical Pacific and Indian region during 2015–16 El Niño event at upper (200 hPa) troposphere. Positive (negative) values represent the KE transfer to (from) El Niño (unit: 10^{-6} watts/kg).

No significant interactions are observed except in WNP region while high frequency is interacting with low frequencies during El Niño episode at 850 hPa (Fig. 6.9A and B). Small exchange of KE is occurred between synoptic scale and LFOs as both the positive and negative values coexist in that area.

Fig. 6.9C shows small KE is transferred from 30–60 days to 10–20 days in equatorial West Pacific at 850 hPa due to the presence of opposite transfers of KE between two LFOs. Except the midlatitude region, prominent 200 hPa energy exchanges are observed in ISM region where synoptic scale is losing KE to LFOs in the Indian landmass (Fig. 6.10A and B). Weak scale interaction between two LFOs is observed as because almost equal positive and negative

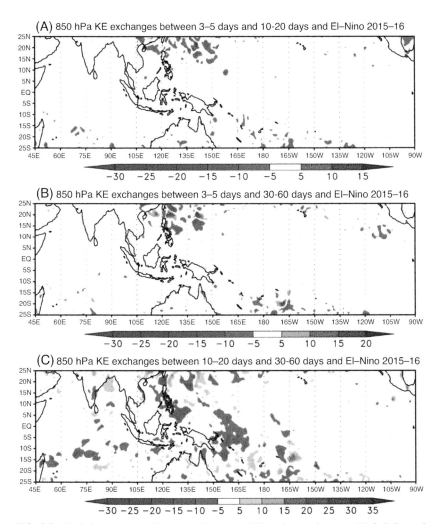

FIG. 6.9 Scale interactive nonlinear kinetic energy (*KE*) exchanges between (A) 3–5 day and 10–20 day, (B) 3–5 day and 30–60 day, and (C) 10–20 day and 30–60 day oscillations in tropical Pacific and Indian region during 2015–16 El Niño event at lower (850 hPa) troposphere. Positive (negative) values represent the KE transfer from (to) 3–5 day in (A); (B); and 10–20 day in (C) (unit: 10^{-6} watts/kg).

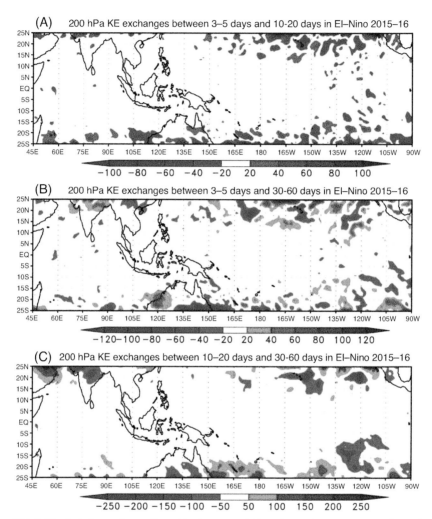

FIG. 6.10 Scale interactive nonlinear kinetic energy (*KE*) exchanges between (A) 3–5 day and 10–20 day, (B) 3–5 day and 30–60 day, and (C) 10–20 day and 30–60 day oscillations in tropical Pacific and Indian region during 2015–16 El Niño event at upper (200 hPa) troposphere. Positive (negative) values represent the KE transfer from (to) 3–5 day in (A); (B) and 10–20 day in (C) (unit: 10^{-6} watts/kg).

interactions are present in ISM region leading to small net interactions by nullifying opposite interactions (Fig. 6.10C).

A deficient ISMR is generally characterized by less intensified rain bearing synoptic scale transients with their restricted northwestward propagation (Agarwal et al. 2016), weak 10–20 day mode regulating active-break phase of ISM (Goswami, 2003) and a strong 30–60 day oscillations influencing northward propagation of monsoon trough (Sikka, 1980; Kripalani et al., 2004).

From the energetics perspective, 3–5 day synoptic scale waves and 10–20 day LFO are showing deficiency in KE causing weak transients (Agarwal et al., 2016) and long (short) spell of break (active) phase during the below normal ISMR and El Niño event (Dwivedi et al. 2015), whereas the strong 30–60 day mode in deficient ISM requires KE for its poleward propagation from Indian Ocean to foothills of Himalaya through the mechanism of maximum moisture convergence at the north of the convective maxima (Goswami, 2005; references therein). Now, the question arises how does the El Niño play a role to weaken the synoptic scale and 10–20 day LFO and strengthen 30–60 day LFO in terms of energy exchanges across the scales? It may be inferred from the scale interaction study among El Niño 2015–16 (associated with below-normal ISMR), high and LFOs that synoptic disturbances are losing KE to El Niño and LFOs at both upper and lower troposphere leading to feeble synoptic-scale wave that may cause subdued rainfall during 2015–16. Although 10–20 day mode is receiving energy from El Niño 2015–16 but the net transfer of energy to 10–20 day LFO is smaller than that during 1997–98 El Niño (comparing Figs. 6.3 and 6.4–6.7 and 6.8, respectively). Stronger 30–60 day waves due to the extracting KE from El Niño 2015–16 particularly at 200 hPa is observed compared to the same mode during 1997–98 El Niño and that was substantiated by Krishnan and Kasture (1996) study where authors documented that the energy transfer to 30–50 LFO in upper troposphere favors the oscillation to make it stronger and regular during El Niño event. So, El Niño has an immense potential for the variability and maintenance of high and low frequency monsoon modes by interacting nonlinearly with these modes through scale-dependent KE exchanges. Therefore, it may be inferred from the study that the reasons for two strong and similar type of El Niño(s) 1997–98 and 2015–16 of recent past behaving oppositely with ISM may be revealed from the variability in nonlinear scale interactive energy exchanges between El Niño and ISM attributing to the El Niño–monsoon selective relation.

6.4   Conclusions and discussions

One of the major large-scale drivers of ISM variability is El Niño, which is nonlinearly related to ISM. Very few studies addressed the nonlinear aspect of El Niño–monsoon connectivity. The mechanistic exploration of the relation is imperative to quantify the nonlinearity between El Niño and ISM. The first time, the El Niño–monsoon nonlinearity has been unveiled in terms of interscale transfer of KE among El Niño, high and low frequency monsoon modes through the mechanism of nonlinear triad interactions in frequency domain. For this purpose, El Niño has been treated as a LFOs having a time period of 12–18 months. As the inverse relation between ISMR and El Niño does not always hold well, two strongest El Niño events 1997–98 and 2015–16 are chosen purposefully where the opposite relationships between El Niño and ISM are established. How El Niño, being a trans-oceanic phenomena dictate monsoon, has been unraveled by computing KE exchanges among the prominent high, low frequency monsoon modes, and El Niño through the scale interaction mechanism. The following inferences may be drawn from the whole study.

- Total KE possessed by any oscillation follows the hierarchy of the scale, i.e., a wave of longer time period contains more energy than the wave of shorter time scale and that obeys the basic energy rules followed by Lorenz (1969).
- As good (bad) monsoon has been characterized by strong (weak) synoptic scale waves of 3–5 day, and 10–20 day LFO and weak (strong) 30–60 day LFO, one of the major reasons of above average ISMR during 1997–98 El Niño is that both 3–5 day and 10–20 day modes are intensified by the supply of KE to these scales from El Niño observed in both lower and upper atmosphere while El Niño is interacting nonlinearly with these waves. Opposite feature is seen for 30–60 day oscillation which is weakened due to the loss of more KE at 200 hPa than its gain from lower level at the time of interactions with El Niño.
- Below normal ISM rainfall during 2015–16 El Niño is attributed to weak synoptic scale wave (3–5 day) due to the energy transfer from this scale to El Niño, 10–20 day and 30–60 day LFOs at lower and upper troposphere. Moreover, this deficient ISMR is also due to the smaller KE transfer to 10–20 day from El Niño 2015–16 compared to the energy transfer during 1997–98 El Niño and strong KE transfer to the 30–60 day mode from 2015–2016 El Niño as revealed in upper troposphere.
- The dominance of upper tropospheric KE interactions at midlatitudes is mainly because of availability of K_E due to the conversion of K_M to K_E in abundance at those regions, revealed from previous literatures.

Therefore, it is noted that the variability in strength of high and low frequency dominant monsoon modes may be governed by the interaction among El Niño and these modes through the nonlinear interscale KE exchanges which may be one of the causes of the interannual variability of rainfall over ISM region, revealed from two contrasting El Niño-monsoon relation and thus, the mechanism and quantification of nonlinearity between these two phenomena have been unraveled. Moreover, El Niño–monsoon selective relation (Webster and Yang, 1992) can be explained by the way, the scale interactions happen between ISM and two similar types of El Niño(s).

Other than ISM area, there are dominant interactions over WNP and MC regions. Separate study may be executed taking into consideration of individual area. Authors have restricted this study using ERA data only as an initial approach in this direction. Another popular dataset like NCEP can be used for the robustness of this study. However, the correlation between K_E and Niño 3.4 index was equivalent for both the ERA and NCEP datasets and the meridional distribution of K_E at different vertical levels were exhibited similar pattern for both datasets revealed from previous studies. The work may be extended further by considering the cold phase of ENSO episode, i.e., La Niña and the Indian

Ocean dipole events taken into account individually as well as simultaneously to explore the variability in nonlinearities between ENSO and monsoon.

Acknowledgments

Thanks to the Director, Indian Institute of Tropical Meteorology (IITM), Pune, and Earth System Science Organisation, Ministry of Earth Sciences, Government of India for encouragement to execute the work. Authors are indebted to Dr. Jasti S. Chowdary and Ms. Darshana Patekar, IITM who has given special effort for providing data and figure. Authors are grateful to ECMWF for making the data available freely. Thanks are due to Brain Doty, COLA for GrADS software.

Appendix

Formulation of KE equation in terms of nonlinear scale interactive energy exchanges in the Frequency Domain

The equation for nonlinear exchanges of KE among different frequencies n, r, and s following Chakraborty and Agarwal (1996) can be written as:

$$\langle L(n)\rangle = -\frac{1}{2}\left[\begin{array}{c}+\sum_{r+s=n}\\+\sum_{r-s=n}\\+\sum_{r-s=-n}\end{array}\right]\left[\begin{array}{c}UOCn.UTCr.\left(2.\frac{\partial}{\partial x}UTCs+\frac{\partial}{\partial y}VTCs+\frac{\partial}{\partial p}WTCs-\frac{\tan\varphi}{a}.VTCs\right)\\+UOCn.\left(\frac{\partial}{\partial y}UTCr.VTCs+\frac{\partial}{\partial p}UTCr.WTCs\right)\\+VOCn.UTCr\left(\frac{\partial}{\partial x}VTCs+\frac{\tan\varphi}{a}.UTCs\right)\\+VOCn.VTCr\left(2.\frac{\partial}{\partial y}VTCs+\frac{\partial}{\partial p}WTCs\right)\\+VOCn\left(\frac{\partial}{\partial x}UTCr.VTCs+\frac{\partial}{\partial p}VTCr.WTCs\right)\end{array}\right]$$

$$+\frac{1}{2}\left[\begin{array}{c}-\sum_{r+s=n}\\+\sum_{r-s=n}\\+\sum_{r-s=-n}\end{array}\right]\left[\begin{array}{c}UOCn.UTSr.\left(2.\frac{\partial}{\partial x}UTSs+\frac{\partial}{\partial y}VTSs+\frac{\partial}{\partial p}WTSs-\frac{\tan\varphi}{a}.VTSs\right)\\+UOCn.\left(\frac{\partial}{\partial y}UTSr.VTSs+\frac{\partial}{\partial p}UTSr.WTSs\right)\\+VOCn.UTSr\left(\frac{\partial}{\partial x}VTSs+\frac{\tan\varphi}{a}.UTSs\right)\\+VOCn.VTSr\left(2\frac{\partial}{\partial y}VTSs+\frac{\partial}{\partial p}WTSs\right)\\+VOCn.\left(\frac{\partial}{\partial x}UTSr.VTSs+\frac{\partial}{\partial p}VTSr.WTSs\right)\end{array}\right]$$

$$+\frac{1}{2}\left[\begin{array}{l}+\sum_{r+s=n}\\-\sum_{r-s=n}\\+\sum_{r-s=-n}\end{array}\right]\left[\begin{array}{l}UOSn.UTCr\left(2.\frac{\partial}{\partial x}UTSs+\frac{\partial}{\partial y}VTSs+\frac{\partial}{\partial p}WTSs-\frac{\tan\varphi}{a}.VTSs\right)\\+UOSn.\left(\frac{\partial}{\partial y}UTCr.VTSs+\frac{\partial}{\partial p}UTCr.WTSs\right)\\+VOSn.UTCr\left(\frac{\partial}{\partial x}VTSs+\frac{\tan\varphi}{a}.UTSs\right)\\+VOSn.VTCr\left(2.\frac{\partial}{\partial y}VTSs+\frac{\partial}{\partial p}WTSs\right)\\+VOSn.\left(\frac{\partial}{\partial x}UTCr.VTSs+\frac{\partial}{\partial p}VTCr.WTSs\right)\end{array}\right.$$

$$+\frac{1}{2}\left[\begin{array}{l}+\sum_{r+s=n}\\+\sum_{r-s=n}\\-\sum_{r-s=n}\end{array}\right]\left.\begin{array}{l}UOSn.UTSr\left(2.\frac{\partial}{\partial x}UTCs+\frac{\partial}{\partial y}VTCs+\frac{\partial}{\partial p}WTCs-\frac{\tan\varphi}{a}.VTCs\right)\\+UOSn.\left(\frac{\partial}{\partial y}UTSr.VTCs+\frac{\partial}{\partial p}UTSr.WTCs\right)\\+VOSn.UTSr\left(\frac{\partial}{\partial x}VTCs+\frac{\tan\varphi}{a}.UTCs\right)\\+VOSn.VTSr\left(2.\frac{\partial}{\partial y}VTCs+\frac{\partial}{\partial p}WTCs\right)\\+VOSn.\left(\frac{\partial}{\partial x}UTSr.VTCs+\frac{\partial}{\partial p}VTSr.WTCs\right)\end{array}\right] \quad (1)$$

where "a" is the earth's radius, "ϕ" is the latitude, and n, r, and s are the frequency indices. The wind fields U, V, and W represent the zonal, meridional, and vertical wind, respectively. Here (UOC, UOS), (VOC, VOS) are the temporal Fourier cosine and sine coefficients of the observed U, V fields, respectively, associated with frequency n, whereas (UTC, UTS), (VTC, VTS), and (WTC, WTS) are the same except for transient U, V, and W fields, respectively, associated with frequencies r and s. Most of the terms for nonlinear interactions in Eq. (1) involve triple products.

References

Agarwal, N.K., Naik, S.S., De, S., Sahai, A.K., 2016. Why are the Indian monsoon transients short-lived and less intensified during droughts vis-à-vis good monsoon years?—An inspection through scale interactive energy exchanges in frequency domain. Int. J. Climatol. 36, 2958–2978. doi:10.1002/joc.4531.

Bjerknes, J., 1969. Atmospheric teleconnections from the equatorial Pacific. Mon. Wea. Rev. 97, 163–172.

Chakraborty, D.R., Agarwal, N.K., 1996. Role of triad kinetic energy interactions for maintenance of upper tropospheric low frequency waves during summer monsoon. Adv. Atmos. Sci. 13, 91–102.

Chowdary, J.S., Harsha, H.S., Gnanaseelan, C., Srinivas, G., Parekh, Anant, Pillai, Prasanth, C.V., Naidu, C.V., 2017. Indian summer monsoon rainfall variability in response to differences in the decay phase of El Niño. Clim. Dyn. 48, 2707–2727. https://doi.org/10.1007/s00382-016-3233-1.

De, S., Agarwal, N.K., Hazra, A., Chaudhari, H.S., Sahai, A.K, 2019. On unravelling mechanism of interplay between cloud and large scale circulation: a grey area in climate science. Clim. Dyn. 52, 1547–1568. https://doi.org/10.1007/s00382-018-4211-6.

Dee, D.P., Uppala, S.M., Simmons, A.J., et al., 2011. The ERA-interim reanalysis: configurations and performance of the data assimilation system. Qart. J. R. Meteorol. Soc. 137 (656), 553–597. https://doi.org/10.1002/qj.828.

Dwivedi, S., Goswami, B.N., Krucharski, F., 2015. Unravelling the missing link of ENSO control over the Indian monsoon rainfall. Geophys. Res. Lett. 42, 8201–8207. doi:10.1002/2015GL065909.

Dong, D., Li, J., Huyan, L., Xue, J., 2017. Atmospheric energetics over the tropical Pacific during the ENSO Cycle. J. Clim. 30, 3635–3654. https://doi.org/10.1175/JCLI-D-16-0480.1.

Goddard, L., Philander, S.G., 2000. The energetics of El Niño and La Niña. J. Clim. 13, 1496–1516.

Goswami, B.N., Ajayamohan, R.S., Xavier, P.K., Sengupta, D., 2003. Clustering of low pressure systems during the Indian summer monsoon by intra-seasonal oscillations. Geophys. Res. Lett. 30 (8). https://doi.org/10.1029/2002GL016734.

Goswami, B.N., 2005. Chapter 2: The South Asian monsoon. In: Lau, W.K.M., Waliser, D.E. (Eds.), In Intraseasonal Variability of the Atmosphere-Ocean Climate System. 2nd ed. Springer Praxis, Berlin, pp. 19–65.

Hayashi, Y., 1980. Estimation of non-linear energy transfer spectra by the cross spectral method. J. Atmos. Sci. 37, 299–307.

India Meteorological Department, 1998. Weather in India; Monsoon season (June–September 1997). Mausam 49, 405–432.

India Meteorological Department, 1999. Weather in India; Monsoon season (June–September 1998). Mausam 50, 311–333.

India Meteorological Department, 2016. Weather in India; Monsoon season (June–September 2015). Mausam 67, 723–744.

India Meteorological Department, 2017. Weather in India; Monsoon season (June–September 2016). Mausam 68, 405–432.

Ju, J., Slingo, J., 1995. The Asian summer monsoon and ENSO. Q. J. R. Meteorol. Soc. 121, 1133–1168.

Kanamitsu, M., Krishnamurti, T.N., Depradine, C., 1972. On scale interactions in the Tropics during northern summer. J. Atmos. Sci. 29, 698–706.

Klinter III, J.L., Miyakoda, K., Yang, S, 2002. Recent change in the connection from the Asian monsoon to ENSO. J. Clim. 15, 1203–1215.

Kripalani, R.H., Kulkarni, A., Sabade, S.S., Revadekar, J.V., Patwardhan, S.K., Kulkarni, J.R., 2004. Intra-seasonal oscillations during monsoon 2002 and 2003. Curr. Sci. 87, 325–331.

Krishnamurti, T.N., 1971. Tropical East-West circulations during the northern summer. J. Atmos. Sci. 1971 (28), 1342–1347.

Krishnamurti, T.N., Subrahmanyam, D., 1982. 30-50 day mode at 850mb during Monex. J. Atmos. Sci. 39, 2088–2095.

Krishnamurti, T.N., Chakraborty, D.R., Cubukcu, N., Stefanova, L., Vijaya Kumar, T.S.V., 2003. A mechanism of the Madden–Julian oscillation based on interactions in the frequency domain. Quart. J. Roy. Meteor. Soc. 129, 2559–2590. https://doi.org/10.1256/qj.02.151.

Krishnamurti, T.N., Chakraborty, D.R., 2005. The dynamics of phase locking. J. Atmos. Sci. 62, 2952–2964. https://doi.org/10.1175/JAS3507.1.

Krishnamurti, T.N., Dubey, S., Kumar, V., Deepa, R., Bharadwaj, A., 2017. Scale interaction during an extreme rain event over southeast India. Ind. Q. J. R. Meteorol. Soc. 143, 1442–1458. https://doi.org/10.1002/qj.3016.

Kulkarni, A., Kripalani, R.H., Sabade, S.S., Rajeevan, M., 2011. Role of intra-seasonal oscillations in modulating Indian summer monsoon rainfall. Clim. Dyn. 36, 1005–1021.

Krishnan, R., Kasture, S.V., 1996. Modulation of low frequency intraseasonal oscillation of northern summer monsoon by El Niño and Southern Oscillation (ENSO). Meteorol. Atmos. Phys. 60, 237–257.

Li, L., Ingersoll, A., Jiang, P., Xun, F., Daniel, Y., Yuk, L., 2007. Lorenz energy cycle of the global atmosphere based on reanalysis datasets. Geophys. Res. Lett. 34, L16813. https://doi.org/10.1029/2007GL029985.

Li, L., Chahine, M.T., Jiang, X., Yung, Y.L., 2011. The mechanical energies of the global atmosphere in El Niño and La Niña years. J. Atmos. Sci. 68, 3072–3078. doi:10.1175/JAS-D-11-072.1.

Lorenz, E.N., 1955. Available potential energy and the maintenance of the general circulation. Tellus 7, 157–167.

Lorenz, E.N., 1969. The predictability of a flow which possesses many scales of motion. Tellus 21, 289–308.

Mooley, D.A., Shukla, J., 1987. Characteristics of the westward moving summer monsoon low pressure systems over the Indian region and its relationship with the monsoon rainfall. COLA Technical Report. University of Maryland, Maryland.

Murakami, T., Tomatsu, K., 1964. The spectrum analysis of energy interaction terms in the atmosphere. J. Meteor. Soc. Japan 42, 14–25.

Neelin, J.D., Jin, F.-F., Sue, H.-H., 2000. Variations in ENSO phase-locking. J. Clim. 13, 2570–2590.

Oort, A.H., Ascher, S.C., Levitus, S., Piexoto, J.P., 1989. New estimates of the available potential energy in the world ocean. J. Geophys. Res. 94, 3187–3200.

Pant, G.B., Parthasarathy, B., 1981. Some aspects of an association between southern oscillation and Indian summer monsoon. Arch. Met. Geoph. Biokl., Ser. B 29, 245–252.

Paek, H., Yu, J.-Y., Qian, C., 2017. Why were the 2015/2016 and 1997/1998 extreme El Niños different? Geophys. Res. Lett. 10.1002/2016GL071515.

Rasmusson, E.M., Carpenter, T.H., 1983. The relationship between Eastern Equatorial Pacific Sea Surface Temperatures and Rainfall over India and Sri Lanka. Mon. Wea. Rev. 111, 517–528.

Saltzman, B., 1957. Equations governing the energetics of the large scales of atmospheric turbulence in the domain of wave number. J. Meteor. 14, 513–523.

Sheng, J., Hayashi, Y., 1990. Estimation of atmospheric energetics in frequency domain during the FGGE year. J. Atmos. Sci. 47, 1255–1268.

Shukla, J., Paolino Daniel, A., 1983. The southern oscillation and long-range forecasting of the summer monsoon rainfall over India. Mon. Wea. Rev. 111, 1830–1837.

Sikka, D.R., 1980. Some aspects of large-scale fluctuations of summer monsoon rainfall over India in relation to fluctuations in the planetary and regional scale circulation parameters. Proc. Ind. Acad. Sci. (Earth & Planet. Sci.) 89, 179–195.

Sikka, D.R., Gadgil, S., 1980. On the maximum cloud zone and the ITCZ over Indian longitudes during the southwest monsoon. Mon. Wea. Rev. 108, 1840–1853.

Singh, M., Krishnan, R., Goswami, B., Choudhury, A.D., Swapna, P., Vellore, R., Prajeesh, A.G., Sandeep, N., Venkataraman, C., Donner, R.V., Marwan, N., Kurths, J., 2020. Fingerprint of volcanic forcing on the ENSO–Indian monsoon coupling. Sci. Adv. 6, eaba8164. https://doi.org/10.1126/sciadv.aba8164.

Soman, M.K., Slingo, J., 1997. Sensitivity of the Asian summer monsoon to aspects of sea-surface-temperature anomalies in the tropical Pacific ocean. Q. J. R. Meteorol. Soc. 123, 309–336.

Srinivasan, J., Nanjundiah, R.S., 2002. The evolution of Indian summer monsoon in 1997 and 1983. Meteorol. Atmos. Phy. 79, 243–257.

Tomatsu, K., 1979. Spectral energetics of the troposphere and lower stratosphere. In: Saltzman, B (Ed.). Advances in Geophysics, 21. Academy Press, New York, NY, pp. 289–401.

VenkataRatnam, A., Behera, S.K., Matsumoto, Y., Takahashi, K., Yamagata, T., 2010. Pacific Ocean origin for the 2009 Indian summer monsoon failure. Geophys. Res. Lett. 37, L07807. https://doi.org/10.1029/2010GL042798.

Walker, G.T., 1924. Correlations in seasonal variations of weather, VIII and IX. Mem. India Met. Dept. 24, 75–133 .

Webster, P.J., Yang, S., 1992. Monsoon and ENSO: selective interactive systems. Q. J. R. Meteorol. Soc. 118, 877–926.

Chapter 7

Teleconnections between the Indian summer monsoon and climate variability: a proxy perspective

S. Chakraborty[a], Aasif M. Lone[b], Anant Parekh[a], P.M. Mohan[c]
[a]Indian Institute of Tropical Meteorology, Ministry of Earth Sciences, Pune, Maharashtra, India, [b]Indian Institute of Science Education and Research, Bhopal, India, [c]Pondicherry University, Port Blair, India

7.1 Introduction

7.1.1 Monsoon variability

Monsoon is a natural process composed of multiple complex interactions among atmosphere, land, ocean, vegetation, ice sheet, etc., over a broad spectrum of temporal and spatial scales (e.g., Gadgil 2003). Its variability is influenced by several global-scale events, such as the El Niño-Southern Oscillation (ENSO) (Tudhope et al., 2001; Moy et al., 2002;), Indian Ocean Dipole (IOD) (Ashok et al., 2001; DiNezio et al., 2020), Pacific Decadal Oscillation (PDO) (Newman et al., 2016; Zhang and Delworth, 2016), Atlantic Multi Decadal Oscillation (AMO) (Knudsen et al., 2011) shifts in the moisture source (Polanski et al., 2014; Jasechko, 2015) volcanic forcing (Singh et al., 2020) and changes in earth's orbital configurations (Jalihal et al., 2020).

One of the reliable means of studying the interactions among atmosphere, land, ocean, vegetation, and ice over a long period is proxy-based paleoclimatic investigations. The palaeoclimate reconstruction methods are widely used to study past monsoon variability and their causative mechanism(s) (Bolton et al., 2013; Dahl and Oppo, 2006; Tiwari et al., 2015). The use of paleoclimate reconstruction methods has significantly advanced in the past few decades involving changes in the chemical, physical, and biological processes of the Earth system in coherence with temperature and rainfall variations (Mann et al., 2008; PAGES2k Consortium, 2013). Among these methods, one of the means is to study the geophysical, geochemical, and biological characteristics of natural archives known to preserve records of these interaction processes reliably. The physical processes that govern these interactions and, in turn, determine the

Indian Summer Monsoon Variability: El Niño-teleconnections and beyond.
DOI: https://doi.org/10.1016/B978-0-12-822402-1.00070-3

climatic phenomena are evaporation, precipitation, atmospheric and oceanic circulation, transfer of heat energy, etc. (Doose-Rolinski et al., 2001).

In recent times, several studies have been compiled to draw inferences on Indian summer monsoon variability using single-multi-archived proxy records (Banerji et al., 2020 and reference therein). These climate archives include cave deposits, i.e., speleothems, ocean and lake sediments, tree rings, corals, etc. They are known to respond to changes in the atmospheric conditions associated with the monsoon through physical, chemical, or biological processes. Hence, variability in the Indian summer monsoon over decadal, centennial, and millennial timescales is apparent in these climate archives.

The physio-chemical changes could be recorded by various means, one of them being variations in the heavy to the light isotopic distribution pattern of certain low mass elements, such as hydrogen, carbon, oxygen, nitrogen, etc. The changes in heavy to light isotopic distribution, the so-called *isotope fractionation* process, both in biogenic and abiogenic materials, are governed by temperature and precipitation changes, pH variability, and isotope composition of the ambient environment (de Villiers et al., 1995). The fractionation processes have been extensively studied in biogenic and abiogenic carbonates and employed to reconstruct past atmospheric conditions and biomineralization processes (McConnaughey, 1989). The isotopic ratio is denoted in δ notation; see Box 7.1 for definition. Some of the natural archives that rely on the isotopic distribution characteristics are discussed below.

7.1.2 Speleothem

The speleothems are inorganic carbonate deposits formed in caves when the environmental conditions are conducive. The basic premise is that rainwater

---

Box 7.1 Isotopic Proxies

Various kinds of proxies are used to study the past climate depending on their sensitivity to environmental conditions. Isotopic ratios of a few low mass elements (H, C, N, O, S) are widely used for this purpose. Since the relative difference in isotope ratios can be measured more precisely than the absolute isotopic ratios, the concept of δ notation was introduced to report the isotopic data. For example, in the case of the oxygen isotopic ratio, the formulation is shown below:

$$\delta^{18}O = \left[\frac{\left({}^{18}O/{}^{16}O \right)_{sample} - \left({}^{18}O/{}^{16}O \right)_{reference}}{\left({}^{18}O/{}^{16}O \right)_{reference}} \right] \times 1000\%o$$

where the numerator on the right-hand side is the difference between the $({}^{18}O/{}^{16}O)$ ratio of a sample and that of reference material; for oxygen or hydrogen, typically the VSMOW (Vienna Standard Mean Ocean Water) is used as a reference material (Criss, 1999).

percolating through the bedrock carries the isotopic signature of rainfall and is preserved in the speleothem calcium carbonate matrix. They record changes in the heavy to light isotopes of H and O in the rain owing to differences in precipitation amount and moisture sources (Fairchild et al., 2006; Lachinet, 2009). The rainwater, which percolates through the bedrock, integrates several years (up to 10 years) of climate information (Fairchild et al., 2006; Berkelhammer et al., 2014). As a result, the speleothem records act as a low-pass filter yielding low-frequency information better than the high-frequency variability (Berkelhammer et al., 2014). However, a near-continuous flow of groundwater in cave systems, in principle, enables us to retrieve climatic information nearly on an annual scale (Fairchild and Baker, 2012). This is aided by advancement in analytical techniques that offer high-resolution speleothem sampling and hence improved climatic information (Affolter et al., 2014). In the Indian context, however, contrasting patterns of subdecadal to decadal-scale Indian summer monsoon (ISM) variability (strong versus weak monsoon) have been better-studied (Sinha et al., 2007; Kathayat et al., 2017) than the near-annual scale variability (Sinha et al., 2011; Sinha et al., 2018).

7.1.3 Tree rings

In the continental environment, analysis of tree ring width and their isotopic ratios ($^{18}O/^{16}O$, $^{2}H/^{1}H$, $^{15}N/^{14}N$) offers a reliable means for studying the high-resolution records of past precipitation variability (Roden et al., 2000; Zhang et al., 2020). The tree ring width is usually susceptible to ambient environmental conditions mostly linked to temperature and precipitation changes (Yadav et al., 2011; Borgaonkar et al., 2018). As such, normal tree-ring growth may be impeded under climate-stressed conditions, such as insufficient water availability due to declining rainfall under drought-like conditions. Such kinds of environmental changes are reliably recorded in tree rings, which manifest in the form of varying tree ring widths. The annual changes in the tree ring width dominantly controlled by ambient environmental conditions have been reliably used to reconstruct past Indian monsoonal shifts (Yadav, 2013) and El Niño induced drought conditions (Cook et al., 2010; Borgaonkar et al., 2010; Zaw et al., 2020). As trees source their water from the root zone, the isotopic signature of soil water is believed to be recorded by the distribution of oxygen or hydrogen isotopes in the tree ring structure (Evans et al., 2006; Bose et al., 2014). Other isotopes, such as ^{13}C, ^{14}C, ^{15}N, that are sensitive to changes in CO_2 concentrations in the atmosphere have also been used (Chakraborty et al., 2008).

7.1.4 Marine sediments

The paleo-monsoonal shifts are also apparent from the biological response to the monsoonal activity preserved in the form of increased abundance of the planktic foraminifera in the marine environments. For example, certain species, such as the *Globigerina bulloides,* reliably provided records of multidecadal to century-scale changes in the strength of the Indian summer monsoon (Gupta et al., 2003, 2005). During shell formation in foraminifera or corals, a relatively greater number of heavier isotopes are incorporated in the calcium carbonate matrix during cooler sea surface conditions. This process provides an effective means to quantify changes in the thermal and physical behavior of the surface ocean conditions (Cobb et al., 2003; Abram et al., 2008; Chakraborty, 2006, 2020), including surface and deep-ocean circulation (Chakraborty et al., 1994; Goldstein et al., 2001). Moreover, coralline oxygen isotopes are highly sensitive to thermal conditions; hence, changes in Sea Surface Temperature (SST) are reliably recorded in their oxygen isotopic distribution. Hence, the coralline archives are useful to study the ISM teleconnections with ENSO and IOD (Charles et al., 1997; Chakraborty et al., 2012; Chakraborty, 2015). Box 7.2 shows a schematic diagram illustrating how the tropical Pacific temperature anomalies set up a stationary wave response in the subtropics over Eurasia as well as zonal and meridional circulation in the Indian and Pacific region through the atmosphere-ocean interaction (also see Chapters 2–4). This perturbation, in turn, influences the north–south tropospheric temperature (TT) gradient and the circulation and convective activity associated with the monsoon, thereby controlling the spatial and temporal evolution of ISM rainfall (e.g., Kumar et al., 1999; Xavier et al., 2007).

7.1.5   Proxy's response to forcing mechanism

The climatic information retrieved from proxy records depends primarily on the sensitivity of the archives toward the changes in environmental conditions. This is because each climate archive and proxy has its own response for the same climatic perturbation. Paleo-climatologists have studied the above-mentioned proxy records and proposed several mechanisms that highlight the role of short-(multidecadal) and long-term (centennial to millennial) ocean-atmospheric forcing factors in modulating the monsoon variability (Agnihotri et al., 2002; Dutt et al., 2015). One significant aspect concerns the temporal relationship of the forcing mechanism between the multiproxy records. Interestingly, most of these proxy records highlight similar phase relationships and synchronous periodicities, suggesting a similar forcing mechanism (Menzel et al., 2014; Dutt et al., 2015; Gupta et al., 2019). Each of the proxy records suggests the role of independent and/or coupled forcing mechanism over land and ocean, allowing us to make robust inferences on coupled ocean-atmosphere-land dynamics. Because of these characteristics, the proxy-based paleoclimate reconstruction offers a reliable means to study the past monsoon variability modulated by external and internal factors occurring over multiple timescales. One such method involves reconstructing past monsoonal shifts in the Asian monsoon system and changes in the ISM (Gupta et al., 2005; Zhisheng et al., 2011; 2015; Saraswat et al.,

Box 7.2 How do the proxies record the teleconnections process?

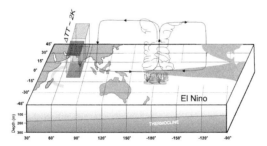

Tropical Pacific climatology during a normal year: the trade winds blow across the Pacific, and the air rises in the convergence zone over the warm SSTs in the west. At the upper level, it diverges and subsequently subsides over the Indian Ocean (IO) and the Pacific Ocean. The thick blue arrow shows the low-level southwest monsoon circulation

During an El Niño year, the equatorial Pacific warms considerably (as shown by reddish shading in the equatorial region). The Walker cell is shifted, as a result, anomalous convection takes place in the central equatorial Pacific. The heating anomaly is then transferred to the atmosphere, which in turn sets up a stationary wave response in the subtropics. This perturbation influences the TT anomalies over Eurasia, which subsequently modulates the north–south gradient of TT (denoted as ΔTT) over the Indian monsoon region and hence the Indian summer monsoon rainfall. The N-S gradient or ΔTT over the Indian region are calculated between the two vertical columns, as shown by two vertical boxes, red box (5°N–35°N, 40°E–100°E) and the blue box (15°S-5°N, 40°E-100°E; see details in Xavier et al., 2007). During a normal year, ΔTT is relatively high, approximately 4K (shown in the upper panel). But in an El Niño year, the north–south temperature gradient is decreased, and ΔTT is low (~2K). Corals growing in the equatorial Pacific can accurately record the SST anomaly in their oxygen isotopic composition (Cobb et al., 2003; Cole 2005). Since the anomaly in the SST is translated to a heating anomaly in the atmosphere through moist convection that drives the stationary wave and influences the ΔTT, the coralline oxygen isotopic ratios show a strong association with the TT gradient. This characteristic feature was used to quantify the past ISM rainfall variability (Chakraborty et al., 2012).

2011; Gill et al., 2017). More discussion in this aspect is provided in the later sections.

7.1.6   Proxy studies of Indian summer monsoon: scope of the work

The ISM undergoes a range of variability from intraseasonal to multimillennium timescales (e.g., Goswami et al., 2014). To study the natural modes of variability in the Indian summer monsoon system, various terrestrial and marine proxy records are used by paleo-climatologists (Ramesh et al., 2010).

In this chapter, we discuss how the proxy-based paleo-monsoon records have been used to study the nature and pattern of the *teleconnections mechanism* aimed at understanding the past ISM variabilities, approximately for the Holocene. Though we will rely on the proxy-based ISM reconstructions, a detailed review of the literature, however, is beyond the scope of this work. Recent studies have already presented the proxy-based reconstruction of the Indian summer monsoon system on millennium (Dixit and Tandon, 2016) and multimillennium times scales (Staubwasser, 2006; Banerji et al., 2020). Fig. 7.1 provides a schematic

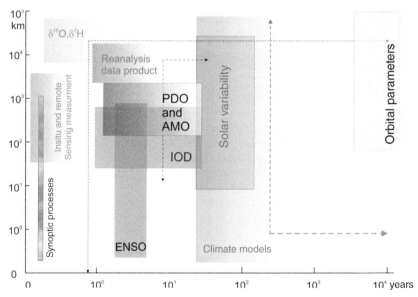

FIG. 7.1   A schematic representation of the processes controls the Indian summer monsoon variability at various temporal and spatial scales. The processes are shown as vertical boxes with black labels; their effect on monsoon covering the temporal and spatial scales is approximately shown by x- and y-axes. The intertropical convergence zone (*ITCZ*), which strongly modulates the monsoon dynamics both in seasonal and millennium timescales, is not shown here. Similarly, the greenhouse gas and volcanic forcing are not represented. The methods and tools of studying the monsoon processes are shown as boxes with varying shades and labeled in orange color (processes are indicative only).

representation of the processes that control monsoon variability across spatial and temporal scales. The analytical and observational methods used to study the monsoon system are shown in the shaded boxes (orange-colored labels).

Paleo-climatic information is largely retrieved by studying the isotopic composition of carbonate systems. These carbonate systems inherit their isotopic signature either through rainfall (i.e., speleothems) or through the ambient water environment (i.e., foraminifera, corals, gastropods, etc.). The isotopic characteristics of precipitation reliably record the atmospheric convective systems, which are transferred through the climate proxies (i.e., speleothems). Hence, it is important to understand how the 'atmospheric information' is passed on to the climate archives.

To provide a comprehensive review of the short and long-term forcing factors on the Indian monsoonal variability, we have divided this chapter into two sections. Firstly, we present a summary of the water isotope and an associated case study to explain how the rainfall information is recorded by land-based speleothem archives. The second part will discuss the proxy-based reconstruction of ISM variability on decadal to multimillennial time scales and its possible global teleconnections; this part is based on the available literature.

7.2 Precipitation isotopes: a proxy for large-scale moisture source signature

On a daily and monthly time scales, the stable oxygen isotope variations in precipitation are widely used to study the regional hydrology and short-term monsoonal variations (Araguas and Araguas et al., 1998; Rahul et al., 2016 Chakraborty et al., 2016, 2018; Midhun et al., 2018; Sinha et al., 2019; Lone et al., 2020; Sengupta et al., 2020). During this process, the evaporation from the sea surface preferentially removes the lighter isotopes (^{16}O, ^{1}H). As a result, the proportion of heavier isotopes (^{18}O, ^{2}H) is reduced in the vapor phase causing isotopic fractionation. Moreover, the initial composition of water vapor formed from evaporation over the ocean and land surface is subsequently modified by rainout and cloud microphysical processes, moisture recycling, advective mixing, raindrop evaporation, rain-vapor interaction, etc. (Dar and Ghosh, 2017; Rahul et al., 2018; Sinha and Chakraborty, 2020).

One pertinent question is how the isotopic signature of rainfall is transferred to individual geological deposits (i.e., speleothem) that enables us to estimate the past monsoon variability? A consequential matter is that how does an individual natural archive (considered as a point source) provide rainfall information over a vast region? To discuss this issue, one should understand that there is a fundamental difference in the dynamical properties of precipitation and its isotopic values. For example, in principle, rainfall is an episodic phenomenon having a limited spatial and temporal domain. However, its isotopic composition reflects the overall status of the water vapor reservoir at the time of precipitation formation (Rozanski 2005). Further, this means the isotopic values of precipitation carry the signature of the water vapor, which is characterized by a much larger spatial domain than the individual rainfall events. To demonstrate this behavior, precipitation isotopic values on a point scale may be compared with rainfall data at different spatial scales. We present the isotopic records of precipitation from the Andaman Island, Bay of Bengal. The reason for choosing this region is that the island contains a variety of speleothems, which are known to provide past monsoonal information (Laskar et al., 2013). Secondly, being an island, the moisture comes predominantly from the ocean; hence the terrestrial controls on the precipitation isotopic variability are negligible. Fig. 7.2 shows oxygen isotope values of rainfall at Port Blair, situated in the southern part of this island. Isotopic time profiles (black line) of four years (2012, 2014, 2016, and 2018) are plotted on an intraseasonal time scale (May–Oct.). The rain gauge data representing point-scale rainfall variability are shown as red bars.

On the other hand, Tropical Rainfall Measuring Mission satellite-derived rainfall data averaged over a larger area ($10°N–12°N$, $90°E–92°E$) has been examined to infer the impact of large-scale rainfall variability on the isotopic values. Rainfall anomaly was calculated by subtracting rainfall climatology of 20 years (1998–2018) over the above-mentioned grid box. The plot clearly shows that the depleted (enriched) isotopic values are mostly associated with high positive (negative) rainfall anomalies. Only for a few cases, heavy rainfall

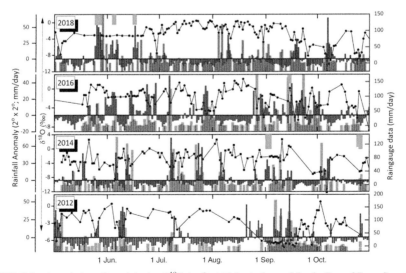

FIG. 7.2 Association of precipitation $\delta^{18}O$ (at Port Blair, Andaman Islands, Bay of Bengal) and the rainfall variability on two different spatial scales. The isotopic variability (black line) has been plotted on an intraseasonal timescale. The rain gauge data are shown as red bars. The satellite-derived rainfall anomaly (averaged over a $2° \times 2°$ grid box) are shown as blue bars. Large isotopic depletions are mostly associated with high positive rainfall anomaly, except in a few cases (indicated as *grey bars*) when they deviate from the expected behavior.

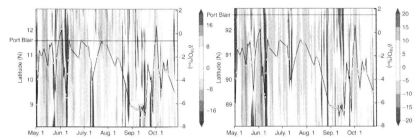

FIG. 7.3 Precipitation isotopes response to large scale rainfall anomaly: (left panel) time latitude Hovmoller diagram showing the rainfall anomaly averaged over a large area (88°E –93°E,
8°N–13°N) over the Bay of Bengal. Superimposed is the precipitation $\delta^{18}O$ record observed at Port
Blair for the year 2012. (right panel) same as before but the time longitude Hovemoller diagram
and the precipitation isotopes covariability are shown. Large isotopic depletions almost always
coincided with the high positive rainfall anomaly represented by blue shading in both cases. The
horizontal black lines show the position of Port Blair.

events did not produce large isotopic depletions (shown as grey bars). The
graphical presentation of the result indeed demonstrates that isotopic values of
rainfall, even on a point scale, could capture the signature of rainfall variability
over a large spatial scale. We have also examined the behavior of the precipitation isotope records yet on a larger spatial scale, covering approximately an area
of 250,000 km^2 over the Bay of Bengal (8°N –13°N, 88°E –93°E). Fig. 7.3 left
panel (right panel) shows the time-latitude (time-longitude) Hovmoller diagram
of rainfall anomaly calculated over the above-mentioned grid box. The precipitation isotope record of Port Blair-2012 has been superposed onto this diagram
to observe the covariability. The data reveal a strong association of depleted
isotopic values and the positive rainfall anomalies, both in zonal and meridional dimensions. It is apparent that the rainfall-isotope antiphase relationship
improves when studied on a larger spatial scale. The records demonstrate that
the isotopic signature of rainfall over a large region could be captured even by a
climate archive having a short spatial extent. Intraseasonal variability of rainfall
information may also be retrieved from a fast-growing biological proxy (Ghosh
et al., 2017).

7.3   Proxy evidence of multiple-scale oscillation in Indian summer monsoon

7.3.1   Marine records

Paleoclimatic investigations highlight the occurrence of persistent variability
in the monsoonal rainfall throughout the Holocene that was closely related to
millennial-scale climate events suggesting a mechanistic link (Overpeck et al.,
1996; Neff et al., 2001; Gupta et al., 2003, 2005; Fleitmann et al., 2007). The
records retrieved from various geo-archives have suggested that monsoonal

variability is very sensitive to relatively small changes in the natural forcing (0.25% change in solar output, or a 2°C change in SST; Shindell et al., 2001). Gupta et al. (2003, 2005) postulated a solar-monsoon link, which suggests that ISM winds were stronger (high solar activity) during the early Holocene that steadily weakened over time. Similarly, the Medieval Warm Period (MWP; CE 800-1200; Kamae et al., 2017) also witnessed higher solar activity, but the Little Ice Age (LIA; CE 1400-1800) experienced low solar activity resulting in enhanced and weakened monsoonal intensity, respectively. In this context, the role of Tibetan Plateau (TP) was also discussed. According to Overpeck et al. (1996), reduced glacial boundary conditions triggered strong warming of the TP and generated a stronger pressure gradient to produce an intensi-fied monsoon during the early Holocene period. These findings have prompted climate scientists to investigate the role of heating of the Eurasian continent and, more specifically, the role of TP (an indicator of surface heating) in modu-lating the Asian monsoon system (Liu and Yanai, 2002). An improved record of understanding the role of TP in linking ISM variations has confirmed that during the last two millennia, epochs of cool (warm) temperatures in the TP coincided with weak (strong) monsoon conditions (Feng and Hu, 2005). This study further confirms that the monotonic rise of the TP surface temperature had a dominant influence on the ISM variations in contrast to the Eurasian continent during the last ~350 years. However, some investigators (Neff et al., 2001; Zorzi et al., 2015) have put forward alternative hypotheses and opine that solar output might have played only an indirect role in modulating the monsoon intensity on a millennium timescale. Zorzi et al. (2015) suggested that direct insolation changes only control the long-term trends in the Indian summer monsoon precipitation. Whereas the ocean-atmospheric-land teleconn-nections dominantly modulate its millennial-scale intensity and variability with the partial role of solar activity. These studies suggest that minor variations over smaller timescales are unlikely to have directly caused significant differences in the sensible heating of the TP. Hence, it is argued that solar output may not have a direct impact on the ISM rainfall variability; instead, it leads to oceanic circulation changes (Neff et al., 2001).

7.3.2 Speleothem records

Previous studies using marine proxy records from the Arabian Sea have reliably reconstructed variations in the ISM and highlighted significant changes in the wind intensity, river discharge, and extent of upwelling (Anderson et al., 2002; Gupta et al., 2003). One of the fundamental limitations of these studies is that they do not directly reflect the precipitation changes, which is integral for assess-ing the actual monsoon precipitation behavior (Berkelhammer et al., 2010). Hence, appropriate proxy records from reliable archives such as speleothem, corals, tree rings, etc., are required to quantify the monsoon rainfall variability.

Speleothems (cave carbonate deposits) occurring in wide geographic regions such as the monsoon core zone (MCZ; approximately a box encompassing an area of 18°N–28°N, and 65°N–88°E in central India, defined in Rajeevan et al., 2010) and the Himalayan region, form an ideal climate archive to reconstruct past changes in the Indian monsoon (Yadava and Ramesh, 2005; Sinha et al., 2007; 2011; Dutt et al., 2015; Kathayat et al., 2017; Kaushal et al., 2018; Kotlia et al., 2015; Gautam et al., 2019). These studies highlighted that variations in speleothem oxygen isotopic composition are primarily governed by changes in the precipitation/moisture sources and hence provide direct evidence of past monsoonal variability. Speleothem proxy records from Northeast India (Berkelhammer et al., 2012; Dutt et al., 2015) reveal significant changes in the ISM during the late Pleistocene and a prominent weakening during the mid-Holocene (the 4.2 ka event). Apart from inferring the monsoonal precipitation variations, these proxy records suggest a coupled relation of the ISM shifts to ocean-atmospheric processes (Gautam et al., 2019); Northern Hemisphere temperature changes (Kathayat et al., 2017), and changing moisture transport pathways (Kotlia et al., 2015). Moreover, speleothem oxygen isotope records of Central and Northeast India reveal that the frequency distribution of active-break periods in ISM varies on centennial timescales and suggests a significant role of internal dynamics in governing the ISM response to slowly evolving changes in the external boundary conditions (Sinha et al., 2011). Additionally, Berkelhammer et al. (2010) studied annually resolved and absolute dated speleothem oxygen isotopic records from the MCZ, the Dandak cave, (19°N and 82°E at 400 m). In their study, the robustness of the extracted climatic information was ascertained by comparing its isotopic profile with a Chinese speleothem record approximately 3000 km away (Wanxiang cave; Zhang et al., 2008) and observed three contemporaneous maxima occurring around AD 900, AD 1050, and AD 1375. Moreover, a high correlation coefficient (0.8) was obtained between these records during AD 880–1420, asserting a common forcing mechanism for the Asian monsoon system.

7.3.3   Pacific teleconnections with Indian summer monsoon

Several regional proxy records have revealed that periodic changes in natural forcing factors such as solar activity influence the monsoonal conditions on centennial to millennial time scales. For example, spectral analysis of the $\delta^{18}O$ record shows a single statistically significant peak centered around 90 years/cycle, commonly referred to as the Gleissberg solar cycle (Berkelhammer et al., 2010). However, this periodicity was restricted only during the MWP, which prompted these authors to suggest either varying ENSO dynamics and/or changing frequency, including the amplitude of IOD events intermittently amplify or neutralize the ENSO impact on monsoon circulation. Hence, the authors hypothesize that changes in the ocean-atmospheric system, specifically in the ENSO domain, may have operated to disrupt or enhance the solar output–monsoon relationship. This implies that the dynamical mechanism linking ISM variability and solar production is modulated by external boundary conditions, in which ENSO plays a critical role. According to this hypothesis, sustained change in ENSO dynamics during the MWP could have provided the conditions necessary for a consistent solar flux variability–monsoon relationship.

To investigate further the coupling of the ISM with the ENSO and the North Atlantic climate variability, Berkelhammer et al. (2014) analyzed additional speleothem records from the MCZ of India. A composite time series of speleothem oxygen isotopes (~1400 years) was created using Jhumar cave (19°N, 82°E; AD 1075–2007) and Dandak cave (AD 625–1562) isotopic records. Additionally, coral and tree ring-based ENSO indices were also used to study the dynamical behavior of ENSO and monsoon.

Using a suite of proxy records, Berkelhammer et al. (2014) focused on two windows of the time spectrum. One is a high-frequency band that can capture all the power in the 5–15 years window, which provides insight into the long-term stability of the known dynamical connection between the ISM and ENSO during the instrumental period (Kumar et al., 1999). The other one is a low-frequency band that can extract the power in the 30–90 years window, aimed at providing information on the effect of the multidecadal climate variability on the ISM. This study further investigated the ENSO–ISM teleconnections mechanism by examining the coherency and phase angle between the two systems (ENSO and ISM) using multiple combinations of the proxy records (e.g., coral ENSO vs speleothem ISM, tree ring ENSO vs tree ring ISM, speleothem ISM vs composite ENSO). This study highlighted that in the high-frequency domain (5–15 years), periods of positive ENSO amplitude were associated with reduced ISM rainfall. However, no statistically significant phase relationship was found between ENSO amplitude and the speleothem-based ISM reconstruction in the low-frequency (30–90 years) domain. The result may appear inconsistent with the conventional wisdom as it shows a more active ENSO state coinciding with an intensified monsoon rainfall regime. Therefore, the authors suggest two hypotheses to explain this apparent inconsistency: (1) changes in ENSO variance are being driven by an increase in La Niña events, (2) in the low-frequency domain; the two systems are phase-locked through a common forcing mechanism, which leads to both an increase (decrease) in ENSO variance and a strengthened (weakened) ISM. In this context, it may be noted that Tejavath et al. (2019) performed a multimodel simulation of ISM for the last millennium and highlighted that majority of the models consistently simulated more strong El Niño events during the MWP as compared to the number of intense El Niño events during the LIA. On the other hand, most of the models consistently simulated more robust La Niña events during the LIA than the number of strong La Niña events during the MWP. Considering this, the second hypothesis proposed by Berkelhammer et al. (2014) seems to offer a more realistic mechanism within the last millennium timeframe. Accordingly, shared multidecadal power between these systems comes through periodic variability in the North Atlantic SSTs. In this case, positive (negative) SST anomaly in the North Atlantic would lead to both an increase (decrease) in ISM rainfall and ENSO variance. Berkelhammer et al. (2014) further argued for a dominant role of dynamics internal to the Indian Ocean (IO) in producing multidecadal hydroclimate variability across the IO basin. This implies the presence of a relationship between IO dynamics and the ENSO system. If basin-wide shifts in the IO are influencing changes in ENSO variance, then the results here suggest that the IO's direct influence on the monsoon is more substantial than from the teleconnections that affect the ISM through ENSO. The results suggest that a continuous trend of increased variance in the ENSO system during the late 20th century will lead to a weakened monsoon. If this coupling between the systems remains stable in the future, the opposite trend, i.e., an increase in ISM rainfall, would be more likely. However, there is one caveat in this ISM projection by Berkelhammer et al. (2014). The antiphasing nature of ENSO–monsoon relation on the high-frequency domain, derived from the proxy records is characterized by high confidence. But the observed similar phasing on the multi-decadal time scale (30–90 years) is characterized by low confidence.

7.3.4 Atlantic Teleconnections with Indian summer monsoon

Recent studies have also suggested a significant influence of North Atlantic SST variations and AMO on the ISM conditions at various timescales (Gupta et al., 2003; Goswami et al., 2006; also see Chapter 18). More recently, multiple studies were proposed to establish the mechanism of this link. The records reveal that warm (cold) phases of AMO correspond to abundant (deficit) ISM rainfall conditions (Saraswat et al., 2011). The AMO is thought to influence the ISM rainfall by altering the temperature and heat source in the TP and thereby the Meridional Temperature Gradient (MTG) between TP and the tropical IO before the onset of the ISM season. The rising surface at TP causes increased MTG and strengthens the south-westerly monsoonal wind circulation, and contributes to extended wet ISM season. To further substantiate this hypothesis, Feng and Hu (2005) suggested that periods of weak TP heat sources concurred with phases of weak monsoon, and to confirm whether these phases also correspond to cooler North Atlantic SST, these authors examined the link between North Atlantic SST, TP, and MTG variations. A comparison of these variations with that of TP surface temperature and the proxy records of the ISM suggests a persistent impact of AMO variability on the ISM intensity. One of the strongest links between the North Atlantic SST and ISM was documented by Dixit et al. (2014a). Around 8200 years ago, the North Atlantic SST fell sharply (ca $3.3\pm1.1^{\circ}C$) due to glacial outbursts resulting in a massive amount of freshwater influx (the 8.2 ka event). The resulting decrease in ocean heat transport forced the Intertropical Convergence Zone to move southward and, in turn, caused a weakening of the ISM intensity (Haug et al., 2001; Mukherjee et al., 2016). Feng and Hu (2005) consider that the impact of AMO on ISM is less likely through the North Atlantic Oscillation related atmospheric circulation changes and views TP heat source as an essential mechanism that transitions the AMO–ISM coupling. However, a comparison of the ISM rainfall records with instrumental records shows that AMO–ISMR link has weakened during the last decade, suggesting that the teleconnections between these two may not be stable (Sankar et al., 2016).

Another notable monsoonal shift was observed around 4200 years BP. The so-called "4.2 ka BP event" has been detected in several paleoclimate archives across the globe, and a concurrent decrease in ISM strength was reported in the Indian context (Berkelhammer et al., 2012; Dixit et al., 2014b). Different mechanisms were proposed to explain the nature and cause of this event. For example, Staubwasser et al. (2003) attributed solar variability, while Berkelhammer et al. (2012) invoked that changes in the mid-latitude circulation pattern led to this event. A weakened mid-latitude westerly system provided dry extratropical air into the ISM system, which suppressed the ISM. Dixit et al. (2014b) speculated that a coupled mechanism of negative IOD and an enhanced ENSO caused such a reduction in ISM at 4.2 ka BP. However, most of these descriptions are not supported by concrete evidence, and a consistent explanation of teleconnections mechanism is still lacking.

Proxy-based evidence suggests continuous drying of the ISM during the mid to late Holocene epoch (Gupta et al., 2003; Staubwasser et al., 2006; Misra et al., 2019). Although there have been various factors that caused this aridification, the predominantly cited records suggest that decreased solar insolation during the middle and later part of Holocene caused reduced land-ocean temperature gradient and reduced monsoonal wetness. According to these studies, the enhanced land surface temperature would contribute to more significant monsoonal precipitation, specifically over the Indian region. However, instrumental and simulation records generated over the Indian region have suggested a dramatic rise in the surface temperature corresponding to below-average rainfall (Krishnan et al., 2020). This apparent inconsistency in the proxy-based and instrumental records signifies that other processes besides solar insolation might be affecting the ISM dynamics.

The biological proxy-based work of past monsoonal circulation reconstruction (Gupta et al., 2003) and its characteristic variation over the Holocene was further strengthened by several other investigators. For example, Gill et al. (2017) analyzed a large pool of SST-based proxy records across the equatorial Pacific. Using a multiproxy reduced dimension methodology, these authors demonstrated that the Arabian Sea summer wind stress curl was higher by 30%–50%, and ISM rainfall was up to 60% higher in certain parts of India at 10 ka compared to recent times. The precipitation pattern across most Indian landmass, excluding the Western Ghats and Northeast India, experienced a declining trend throughout the Holocene. However, unlike Gupta et al. (2003), solar-monsoon linking mechanism and the land-ocean temperature gradient hypothesis, Gill et al. (2017) emphasized the role of ENSO in modulating the rainfall pattern. According to them, ENSO teleconnections during an early to mid-Holocene, La Niña state contributed an intensified rainfall in north-western parts of India. This hypothesis agrees with the observation of Moy et al. (2002), who demonstrated a weak state of El Niño during the early phase of Holocene (see Fig. 7.4A). However, the weakening of the ISM intensity since the mid-Holocene was attributed to the orbital parameters (Cretat et al. 2020). Simulation of the monsoon seasonality during the mid-Holocene was found to be stronger than the modern times by these authors, which led them to invoke the ISM response to the orbital parameters. Interestingly, fossil coral records demonstrated that the mid-Holocene (ca. 6500 years BP) was characterized by low interannual variance in ENSO compared to modern conditions (Cole, 2005). Several studies reported that ENSO variance was significantly reduced during the mid-Holocene compared to today (see a review by Emile-Geay et al., 2021).

Fig. 7.4 provides an overview of the ISM variability and its association with the global climate variability for the past 10,000 years. A speleothem oxygen isotopic record (red line with a 51-year smoothing) from the Oman coast is considered to be a proxy for wind strength over the Arabian Sea, and ISM precipitation over India indicates a progressively declining trend (Fleitmann et al. 2007). A concurrent southward movement of the Intertropical Convergence Zone is illustrated by titanium concentration records from Cariaco Basin in Venezuela (Haug et al., 2001). Moy et al. (2002) documented the frequency of El Niño events through the Holocene, shown in this figure as grey shading. It clearly shows that the early Holocene experienced weak El Niño events but progressively intensified through the later part of the Holocene. The middle panel shows the reduced ENSO variance during the mid-Holocene (Cole, 2005), as evident from a fossil coral record (purple color) compared to a modern coral record (red color). The lower panel shows a north Indian cave (Sahiya cave) isotopic record (Kathyat et al., 2017) representing the ISM variability for the past two millennia. Since intense rainfall reduces the salinity and hence the isotopic composition of seawater, reconstruction of sea surface $\delta^{18}O$ is also used as a proxy for ISM variability. One such record from the western coastal region of the Bay of Bengal (Naidu et al., 2020) has also been shown. It is noteworthy that these two records show a good match as far as the long-term trend in ISM is concerned.

The records mentioned above highlight that variations in the ISM conditions are influenced by a set of variable factors acting either in association or independently. These factors include solar insolation, atmospheric circulation, elevation, altitude, SST, land-sea temperature gradient, ice/snow area, etc. Changes in these factors yielded substantial variations in the regional intensity and amount of ISM rainfall over orbital and millennial/decadal timescales. However, whether the variations in these factors influence the monsoonal precipitation directly (external forcing only) or more indirectly (internal processes playing a significant role) is still debated. On longer time (orbital) scales, ISM variations are known to follow the solar insolation forcing, whereas, on centennial to decadal timescales, solar output changes perhaps played a partial role in

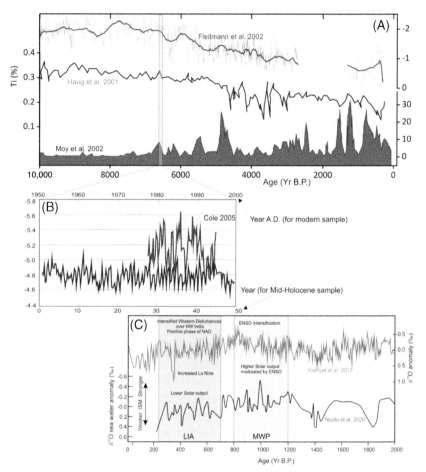

FIG. 7.4 Proxy records of the ISM variability and its association with global climate variables for the last 10,000 years. (A) Speleothem $\delta^{18}O$ record from Oman coast representing the circulation over the Arabian Sea (Fleitmann et al. 2007); *blue line* is the variation of titanium in the Carriaco Basin representing the movement of the ITCZ (Haug et al. 2001); the *grey shading* depicts the frequency of El Niño events during this timeframe (Moy et al. 2002). The middle panel (B) shows the coral oxygen isotopic records on a paleo (approximately 6500 years BP) scale (*purple color*) and a modern scale (*red color*; site: Houn Peninsula, Papua Guinea; Tudhope et al., 2001; Cole, 2005). Lower panel (C): the orange line shows the oxygen isotopic record of the Sahiya Cave speleothem from North India (Kathyat et al. 2017). The *blue line* depicts the seawater oxygen isotopic variability of the western coastal region of the Bay of Bengal (Naidu et al., 2020). *ISM*, Indian summer monsoon; *ITCZ*, intertropical convergence zone.

modulating the ISM intensity. This is because small changes (0.09%) in the total solar irradiance are not adequate to directly modulate the ISM oscillations via the radiation effects alone.

Apparent inconsistencies are found in explaining these mechanisms at different timescales. For example, the early Holocene is believed to have experienced more (little) La Niña (El Niño) type episodes but intensified ISM (Gill et al., 2017). On the other hand, the LIA also experienced increased La Niña type events but was characterized by significantly reduced ISM (Tejavath et al., 2019). Furthermore, there seems an apparent inconsistency between the proxy and instrumental records generated over the Indian region to reconstruct past and modern fluctuations in the monsoonal precipitation and the controlling factors. Although, the proxy records show that much of the variations in the ISM rainfall are attributed to climate teleconnections processes, but the exact control is demonstrated by the type of process that dominantly controls the monsoon-teleconnections mechanism. Therefore, it is necessary to identify and quantify the role of predominant teleconnections processes that characterize the ISM variations at different timescales. A unified mechanism needs to be developed that can satisfactorily address the ISM variability at all timescales. Furthermore, most of these records suggest an ocean-atmosphere-land teleconnections link that characterizes changes in the ISM rainfall over different timescales. Yet, few of the studies consider that solar output drives the ISM oscillation directly. Hence, the role of solar insolation in modulating the ISM circulation directly or indirectly remains a crucial question and needs to be thoroughly studied.

7.4 Summary and recommendations

The Indian summer monsoon has undergone significant variability in its intensity and pattern over the Holocene epoch. Several proxy records revealed a significant rise in the monsoon precipitation during the early Holocene; simultaneously, it was characterized by substantial spatial heterogeneity. Subsequently, the ISM intensity weakened, believed to be caused by a decrease in the solar insolation. However, proxy records also indicated a rise in La Niña events during this time, resulting in relatively enhanced rainfall than the late Holocene. An abrupt reduction in ISM intensity was observed at ~8.2 ka, believed to be caused by a sharp cooling in the North Atlantic, indicating a strong connection between ISM and North Atlantic climate. If global warming forces the Arctic ice sheets to melt at an accelerated pace, the resulting freshening of the North Atlantic may adversely affect the ISM intensity in the foreseeable future.

The mid-Holocene experienced ISM intensification, which is believed to have been modulated by the tropical Pacific SST variability. Afterward, a progressive weakening of the ISM intensity was documented by various proxy records and climate model simulations. At the beginning of the last millennium, a moderate intensification of ISM activity was witnessed owing to increased solar activity. However, during the latter part of this millennium, proxy records indicated a significant reduction in ISM intensity owing to increased ENSO episodes.

Though the proxy records have made a significant contribution in enhancing our understanding of the ISM teleconnections to the global climate variables, they, however, suffer from some limitations. The proxies are not uniformly distributed, and the data are too sparse to capture the broad spectrum of the climate variability. An accurate chronology of the records is imperative to decipher the lead-lag characteristics of the climatic events. However, analytical uncertainties exert a limit in this endeavor. Although several proxy records have been generated with some success to reconstruct past shifts in the Indian summer monsoon; however, climate records established from individual proxy materials are inadequate and represent a smoothed signal owing to poor sample resolution and inability to decipher short-term variations (Staubwasser, 2006). Individual proxies are unlikely to capture the diverse aspects of a natural event. Emile-Geay et al. (2021) opine that single sites provide only local-scale information; in order to capture the diversity of ENSO's spatial footprint, multiple records must be analyzed to understand its full range of variability. Another issue is that any given site may not be adequately *teleconnected* to a particular event. In the Indian context, collection of speleothem samples are preferred from the monsoon core zone with a perception that this area would be most sensitive to external processes such as ENSO. But an objective analysis of the ISM and ENSO teleconnections (Mahendra et al., 2021) shows that the northern parts of central India (approximately an area covering 74°N – 84°N, 26°E – 32°E) is more sensitive to ENSO processes. Hence, careful selection of sample collection sites is a prerequisite to capture signals of the teleconnected processes. Similarly, they will help in identifying and quantifying the dominant processes controlling ISM variability, which will guide future paleo-monsoon research. High-resolution multiproxy records with well-constrained chronology integrated over the immense ocean and continental provinces coupled with global climate model simulations would likely provide a better understanding in this endeavor.

Acknowledgments

IITM is fully funded by the Ministry of Earth Sciences, Government of India. We thank Amey Datye and Charuta Murkute for preparing the graphics. Comments and suggestions provided by two anonymous reviewers helped improve the presentation. This work is part of a project work of the International Atomic Energy Agency, CRP F31006.

References

Abram, N.J., Gagan, M.K., Cole, J.E., et al., 2008. Recent intensification of tropical climate variability in the Indian Ocean. Nature Geosci. 1, 849–853.

Agnihotri, R., Dutta, K., Bhushan, R., Somayajulu, B,L,K., 2002. Evidence for solar forcing on the Indian monsoon during the last millennium. Earth Planet. Sci. Lett. 198, 521–527.

Affolter, S., Fleitmann, D., Leuenberger, M., 2014. New-on-line method for water isotope analysis of speleothem fluid inclusions using laser absorption spectroscopy (WS-CRDS). Clim. Past 10, 1291–1304.

Anderson, D.M., Overpeck, J.T., Gupta, A.K., 2002. Increase in the Asian southwest monsoon during the past four centuries. Science 297, 596–599.

Araguas-Araguas, L., Froehlich, K, 1998. Stable isotope composition of precipitation over southeast Asia. J. Geophys. Res. 103, 28721–28742.

Ashok, K., Guan, Z., et al., 2001. Impact of the Indian Ocean Dipole in the relationship between the Indian monsoon rainfall and ENSO. Geophys. Res. Lett. 28, 4499–4502.

Banerji, U., Arulbalaji, P., Padmala, D., 2020. Holocene climate variability and Indian Summer monsoon: an overview. Holocene. doi:10.1177/0959683619895577.

Berkelhammer, M., Sinha, A., et al., 2010. Persistent multi-decadal power of the Indian summer monsoon. Earth Planet. Sci. Lett. 290, 166–172.

Berkelhammer, M., Sinha, A., Stott, L., Cheng, H., Pausata, F.S.R., Yoshimura, K, 2012. An Abrupt Shift in the Indian monsoon 4000 years ago. Geophys. Monogr. Ser. 198. doi:10.1029/2012 GM001207.

Berkelhammer, M., Sinha, A., et al., 2014. On the low frequency component of the ENSO-Indian monsoon relationship: a paired proxy perspective. Clim. Past. 10, 733–744.

Bolton, C.T., Chang, L., Clemens, S.C., Kodama, K., Ikehara, M., MedinaElizalde, M., Yamamoto, Y., 2013. A 500,000 year record of Indian summer monsoon dynamics recorded by eastern equatorial Indian Ocean upper water-column structure. Quat. Sci. Rev. 77, 167–180.

Borgaonkar, H., Sikder, A., et al., 2010. El Niño and related monsoon drought signals in 523 year long ring width records of teak (Tectona grandis) trees from South India. Paleogr. Paleoclimtol. Paleoecol. 285, 74–84.

Borgaonkar, H.P., Gandhi, Naveen, Ram, Somaru, Krishnan, R., 2018. Tree-ring reconstruction of late summer temperatures in northern Sikkim (eastern Himalayas). Palaeogeogr., Palaeoclimatol., Palaeoecol. 504, 125–135.

Bose, T., Chakraborty, S., Borgaonkar, H.P., Sengupta, S., Ramesh, R., 2014. Estimation of past atmospheric carbon dioxide levels using tree-ring cellulose δ^{13}C. Curr. Sci. 107 (6), 971–982.

Chakraborty, S., Ramesh, R., Krishnaswami, S., 1994. Air-sea exchange of CO2 in the Gulf of Kutch based on bomb carbon in corals and tree rings. Procd. Ind. Acad. Sci. (Earth Palnet. Sci.) 103, 329–340.

Chakraborty, S., 2006. Coral records from the Northern Indian Ocean: understanding the monsoon variability. J. Geological. Soc. India 68, 395–405.

Chakraborty, S., Dutta, K., Bhattacharyya, A., Nigam, M., Schuur, E.A.G., Shah, S., 2008. Atmospheric ^{14}C variability recorded in tree rings from Peninsular India: implications for fossil fuel CO_2 emission and atmospheric transport. Radiocarbon 50 (3), 321–330.

Chakraborty, S., Goswami, B.N., Dutta, K., 2012. Pacific coral oxygen isotope and the tropospheric temperature gradient over Asian monsoon region: a tool to reconstruct past Indian summer monsoon rainfall. J. Quat. Sci. 27 (3), 269–278. doi:10.1002/jqs.1541.

Chakraborty, S., 2015. Interannual climate variabilities of the tropical Indian Ocean and coral isotopic records. Gond. Geol. Mag. 29 (1 & 2), 61–66.

Chakraborty, S., Sinha, N., Chattopadhyay, R., Sengupta, S., Mohan, P.M., Datye, A., 2016. Atmospheric controls on the precipitation isotopes over the Andaman Islands, Bay Bengal. Sci. Rep. 6, 19555. https://doi.org/10.1038/srep19555.

Chakraborty, S., Belekar, A.R., Datye, A., Sinha, N., 2018. Isotopic study of intra-seasonal variations of plant transpiration: an alternative means to characterise the dry phases of monsoon. Sci. Rep. 8, 8647. https://doi.org/10.1038/s41598-018-26965-6.

Chakraborty, S., 2020. Potential of reef building corals to study the past Indian monsoon rainfall variability. Curr. Sci. 119 (2), 273–281. doi:10.18520/cs/v119/i2/273-281.

Charles, C.D., Hunter, D.E., Fairbanks, R.G., 1997. Interaction between the ENSO and the Asian monsoon in a coral record of tropical climate. Science 277, 925–928.

Cobb, K., Charles, C.D., Cheng, H., Edwards, L., 2003. El Niño/Southern Oscillation and tropical Pacific climate during the last millennium. Nature 424, 271–276.

Consortium, PAGES 2k, 2013. Continental scale temperature variability during the past two millennia. Nat. Geosci. doi:10.1038/NGEO1797.

Cook, E.R., Anchukaitis, K.J., Buckley, B.M., D'Arrigo, R.D., Jacoby, G.C., Wright, W.E., 2010. Asian monsoon failure and megadrought during the last millennium. Science 328, 486–489.

Cole, J.E., et al., 2005. Holocene coral records: windows on tropical climate variability. In: Mackay, A. et al (Ed.), Global Change in the Holocene. Hodder Arnold, New York, NY, p. 528.

Criss, R.E., 1999. Principles of Stable Isotope Distributions. Oxford University Press, New York, NY, p. 254.

Crétat, J., Braconnot, P., Terray, P., et al., 2020. Mid-Holocene to present-day evolution of the Indian monsoon in transient global simulations. Clim. Dyn. 55, 2761–2784. https://doi.org/10.1007/s00382-020-05418-9.

Dahl, K.A., Oppo, D.W., 2006. Sea surface temperature pattern reconstructions in the Arabian Sea. Paleoceanography 21 (1).

Dar, S.S., Ghosh, P., 2017. Estimates of land and sea moisture contributions to the monsoonal rain over Kolkata, deduced based on isotopic analysis of rainwater. https://doi.org/10.5194/esd-8-313-2017.

deVilliers, S., Nelson, B.K., Chivas, A.R., 1995. Biological controls on coral Sr/Ca and δ18O reconstructions of sea surface temperatures. Science 269, 1247–1249.

DiNezio, P.N., Puy, M., Thirumalai, K., Jin, F., Tierney, J.R., 2020. Emergence of an equatorial mode of climate variability in the Indian Ocean. Sci. Adv. 6, eaay7684.

Dixit, Y., Tandon, S.K., 2016. Hydroclimate variability on the Indian subcontinent in the past millennium: review and assessment. Earth Planet. Sci. Lett. 161, 1–15.

Dixit, Y., Hodell, D.A., Sinha, R., et al., 2014a. Abrupt weakening of the Indian summer monsoon at 8.2 kyr B.P. Earth. Planet. Lett. 391, 16–23.

Dixit, Y., Hodell, D.A., Petrie, C.A., 2014b. Abrupt weakening of the summer monsoon in northwest India~ 4100 yr ago. Geology 42 (4), 339–342.

Doose-Rolinski, H., Rogalla, U., Scheeder, G., Lückge, A., von Rad, U., 2001. High-resolution temperature and evaporation changes during the Late Holocene in the northeastern Arabian Sea. Paleoceanography 16, 358–367.

Dutt, S., Gupta, A.K., Clemens, S.C., Cheng, H., Singh, R.K., Kathayat, G., Edwards, R.L., 2015. Abrupt changes in Indian summer monsoon strength during 33,800 to 5500 years BP. Geophys. Res. Lett. 42 (13), 5526–5532.

Emile-Geay, J., Cobb, K.M., Cole, J.C., Elliot, M., Zhu, F., 2021. Past ENSO variability: reconstructions, models, and implications, in: El Nino southern oscillation in a changing climate. In: McPhaden, M.J., Santoso, A., Cai, W. (Eds.). Geophysical Monograph, 253. AGU, Wiley, p. 506.

Evans, M.N., Reichert, B.K., Kaplan, A., Anchukaitis, K.J., Vaganov, E.A., Hughes, M.K., Cane, M.A., 2006. A forward modeling approach to paleoclimatic interpretation of tree-ring data. J. Geophys. Res. 111, G03008. doi:10.1029/2006JG000166.

Fairchild, I.J., Smith, C.L., Baker, A., Fuller, L., Spotl, C., Mattey, D., McDermott, F., 2006. Modification and preservation of environmental signals in speleothems. Earth-Sci. Rev. 75, 105–153.

Fairchild, I.J., Baker, A., 2012. Speleothem Science: From Process to Past Environments. John Wiley & Sons.

Feng, S., Qi, Hu, 2005. Regulation of Tibetan Plateau heating on variation of Indian summer monsoon in the last two millennia. Geophys. Res. Lett. 32. doi:10.1092/2004GL021246.

Fleitmann, D., Burns, S.J., Mangini, A., Mudelsee, M., Kramers, J., Villa, I., Neff, U., Al-Subbary, A.A., Buettner, A., Hippler, D., Matter, A., 2007. Holocene ITCZ and Indian monsoon dynamics recorded in stalagmites from Oman and Yemen (Socotra). Quat. Sci. Rev. 26, 170–188.

Gadgil, S., 2003. The Indian monsoon and its variability. Ann. Rev. Earth. Planet. Sci. 31, 429–467.

Gautam, P.K., Narayana, A.C., Band, S.T., Yadava, M.G., Ramesh, R., Wu, C.C., Shen, C.C., 2019. High-resolution reconstruction of Indian summer monsoon during the Bølling-Allerød from a central Indian stalagmite. Palaeogeogr. Palaeoclimatol. Palaeoecol. 514, 567–576.

Ghosh, P., Rangarajan, R., Thirumalai, K., Naggs, F., 2017. Extreme monsoon rainfall signatures preserved in the invasive terrestrial gastropod Lissachatina fulica. Geochem. Geophys. Geosyst. 18, 3758–3770. https://doi.org/10.1002/.

Gill, E.C., et al., 2017. Reconstruction of Indian summer monsoon winds and precipitation over the past 10,000 years using equatorial pacific SST proxy records. Paleoceanogr. Paleoclimatol. 32, 195–216.

Goldstein, S., Lea, D., et al., 2001. Uranium series and radiocarbon geochronology of deep-sea corals: implications for Southern Ocean ventillation rates and the oceanic carbon cycle. Earth Planet. Sci. Lett. 193, 167–182.

Goswami, B.N., Kripalani, R.H., Borgaonkar, H.P., Preethi, B., 2014. Multi-decadal variability in Indian summer monsoon rainfall using proxy data. In: Chang, C.P. (Ed.), Climate Change: Multi Decadal and Beyond. World Scientific, p. 376.

Goswami, B.N., Venugopal, V., et al., 2006. Increasing trend of extreme rain events over India in a warming environment. Science 314, 1442–1446.

Gupta, A.K., Dutt, S., Cheng, H., Singh, R.K., 2019. Abrupt changes in Indian summer monsoon strength during the last~ 900 years and their linkages to socio-economic conditions in the Indian subcontinent. Palaeogeogr. Palaeoclimatol. Palaeoecol. 536, 109347.

Gupta, A.K., Anderson, D.M., Overpeck, J., 2003. Abrupt changes in the Asian southwest monsoon during the Holocene and their links to the North Atlantic Ocean. Nature 421, 354–357.

Gupta, A.K., Das, M., Anderson, D.M., 2005. Solar influence on the Indian summer monsoon during the Holocene. Geophys. Res. Lett. 32, L17703. doi:10.1029/2005GL022685.

Haug, G.H., Hughen, K.A., et al., 2001. Southward migration of the Intertropical Convergence Zone through the Holocene. Science 293, 1304–1308.

Jalihal, C., Srinivasan, J., Chakraborty, A., 2020. Different precipitation response over land and ocean to orbital and greenhouse gas forcing. Sci. Rep. 10, 11891. https://doi.org/10.1038/s41598-020-68346-y.

Jasechko, S., et al., 2015. Late-glacial to late-Holocene shifts in global precipitation $\delta^{18}O$. Climate Past 11 (11), 1375–1393.

Kamae, Y., Kawana, T., Oshiro, M., Ueda, H., 2017. Seasonal modulation of the Asian summer monsoon between the Medieval Warm Period and Little Ice Age: a multi model study. Progr. Earth Planet. Sci. 4 (1), 1–13.

Kathayat, G., Cheng, H., et al., 2017. The Indian monsoon variability and civilization changes in the Indian subcontinent. Sci. Adv. 3, e17001296.

Kaushal, N., Breitenbach, Sebastian F.M., Lechleitner, F.A., Sinha, A., Tewari, V.C., Ahmad, S.M., Berkelhammer, M., et al., 2018. The Indian summer monsoon from a Speleothem $\delta^{18}O$ perspective—a review. Quaternary 1 (3), 29.

Knudsen, M., Seidenkrantz, M., Jacobsen, B., et al., 2011. Tracking the Atlantic multidecadal oscillation through the last 8,000 years. Nat. Commun. 2, 178.

Kotlia, B.S., Singh, A.K., Joshi, L.M., Dhaila, B.S., 2015. Precipitation variability in the Indian Central Himalaya during last ca. 4,000 years inferred from a speleothem record: Impact of Indian Summer Monsoon (ISM) and Westerlies. Quat. Int. 371, 244–253.

Krishnan, R., Sanjay, J., Gnanaseelan, C., Mujumdar, M., Kulkarni, A., Chakraborty, S., 2020. Assessment of Climate Change over the Indian Region: A Report of the Ministry of Earth Sciences (MoES), Government of India. Springer, Singapore.

Kumar, K., Rajagopalan, B., et al., 1999. On the weakening relationship between the Indian monsoon and ENSO. Science 284, 2156–2160.

Lachinet, 2009. Climatic and environmental controls on speleothem oxygen-isotope values. Quat. Sci.Rev. 28, 412–432.

Laskar, A., Yadava, M., Ramesh, R., et al., 2013. A 4 kyr stalagmite oxygen isotopic record of the past Indian summer monsoon in the Andaman Islands. Geochem., Geophys., Geosyst 14 (9). doi:10.1002/ggge.20203.

Liu, X., Yanai, M., 2002. Influence of Eurasian spring snow cover on Asian summer rainfall. Int. J. Climatol. 22 (9), 1075–1089.

Lone, A.M., Achyuthan, H., Chakraborty, S., et al., 2020. Controls on the isotopic composition of daily precipitation characterized by dual moisture transport pathways at the monsoonal margin region of North-Western India. J. Hydrol. https://doi.org/10.1016/j.jhydrol.2020.125106.

Mahendra, N., Chowdary, J.S., Darshana, P., Sunitha, P., Parekh, A, Gnanaseelan, C., 2021. Inter-decadal modulation of interannual ENSO-Indian summer monsoon rainfall teleconnections and CMIP6 models: Regional patterns. Int. J. Climatol. doi:10.1002/joc.6973.

McConnaughey, T.A., 1989. ^{13}C and ^{18}O isotopic disequilibrium in biological carbonates: I patterns. Geochim. Cosmochim. Acta 53, 151–162.

Mann, M.E, et al., 2008. Proxy-based reconstructions of hemispheric and global surface temperature variations over the past two millennia. Prod. Natl. Acad. Sci. 105, 13252–13257.

Menzel, P., Gaye, B., Mishra, P.K., Anoop, A., Basavaiah, N., Marwan, N., Plessen, B., et al., 2014. Linking Holocene drying trends from Lonar Lake in monsoonal central India to North Atlantic cooling events. Palaeogeogr., Palaeoclimatol. Palaeoecol. 410, 164–178.

Midhun, M., Lekshmy, P.R., Ramesh, et al., 2018. The effect of monsoon circulation on the stable isotopic composition of rainfall. J. Geophys. Res. Atmos. 123, 5205–5221. https://doi.org/1 0.1029/2017JD027427.

Misra, P., Tandon, S.K., Sinha, R., 2019. Holocene climate records from lake sediments in India: assessment of coherence across climate zones. Earth-Sci. Rev. 190, 370–397.

Moy, C.M., Seltzer, G.O., Rodbell, D.T., Anderson, D.M., 2002. Variability of El Nino/Southern oscillation activity at millennial timescales during the Holocene epoch. Nature 420, 162–165.

Mukherjee, P., Sinha, N., Chakraborty, S., 2016. Investigating the dynamical behavior of the Inter-tropical Convergence Zone since the last glacial maximum based on terrestrial and marine sedimentary records. Quat. Int. doi:10.1016/j.quaint.2016.08.030.

Naidu, P., Ganeshram, R., et al., 2020. Coherent response of the Indian monsoon rainfall to Atlantic multi-decadal variability over the last 2000 year. Sci. Rep. doi:10.1038/s41598-020-58265-3.

Neff, U., Burns, J., et al., 2001. Strong coherence between solar variability and the monsoon in Oman between 9 and 6 kyr ago. Nature 411, 290–293.

Newman, M., et al., 2016. The Pacific decadal oscillation, revisited. J. Climate 29, 4399–4427. https://doi.org/10.1175/JCLI-D-15-0508.1.

Overpeck, J., Anderson, D., Trumbore, S., Prell, W., 1996. The southwest Indian monsoon over the last 18,000 years. Clim. Dyn. 12, 213–225.

Polanski, S., Fallah, B., Befort, D.J., Prasad, S., Cubasch, U., 2014. Regional moisture change over India during the past Millennium: a comparison of multi-proxy reconstructions and climate model simulations. Global Planet. Change 122, 176–185.

Rahul, P., Ghosh, P., Bhattacharya, S.K, 2016. Rainouts over the Arabian Sea and Western Ghats during moisture advection and recycling explain the isotopic composition of Bangalore summer rains. J. Geophys. Res. (Atmos.) 121, 6148–6163.

Rahul, P., Prasanna, K., Ghosh, P., Anilkumar, N., Yoshimura, K, 2018. Stable isotopes in water vapor and rainwater over Indian sector of Southern Ocean and estimation of fraction of recycled moisture. Sci. Rep. 8, 7552. https://doi.org/10.1038/s41598-018-25522-5.

Rajeevan, M., Gadgil, S., Bhate, J., 2010. Active and break spells of the Indian summer monsoon. J Earth Syst. Sci. 119 (3), 229–247.

Ramesh, R., Tiwari, M., Chakraborty, S., Managave, S.R., Yadava, M.G., Sinha, D.K., 2010. Retrieval of South Asian monsoon variation during the Holocene from climate natural archives. Curr. Sci. 99, 1770–1786.

Roden, S.L., James, G., Ehleringer, R., 2000. A mechanistic model for interpretation of hydrogen and oxygen isotope ratios in tree-ring cellulose. Geochimica et Cosmochimica Acta 64 (1), 21–35.

Rozanski, K., 2005. Isotopes in atmospheric moisture. In: Aggarwal, P.K., Gat, J.R., Froehlich, K.F. (Eds.), Isotopes in the Water Cycle. Springer, Dordrecht. https://doi.org/10.1007/1-4020-3023-1_18.

Saraswat, R., Lea, D., Nigam, R, et al., 2011. Deglaciation in the tropical Indian Ocean driven by interplay between the regional monsoon and global teleconnection. Earth Planet. Sci. Lett. 375, 166–175.

Sengupta, S., Bhattacharya, S.K., Parekh, A., et al., 2020. Signatures of monsoon intra-seasonal oscillation and stratiform process in rain isotope variability in northern Bay of Bengal and their simulation by isotope enabled general circulation model. Clim. Dyn. doi:10.1007/s00382-020-05344-w.

Shindell, D.T., Schmidt, G.A., Mann, M.E., Rind, D., Waple, A., 2001. Solar forcing of regional climate change during the Maunder Minimum. science 294 (5549), 2149–2152.

Sinha, N., Chakraborty, S., 2020. Isotopic interaction and source moisture control on the isotopic composition of rainfall over the Bay of Bengal. Atmos. Res. 35, 104760. doi:10.1016/j.atmosres.2019.104760.

Sinha, N., Chakraborty, S., Chattopadhyay, R., et al., 2019. Isotopic investigation of the moisture transport processes over the Bay of Bengal. J. Hydrol. X. doi:10.1016/j.hydroa.2019.100021.

Sinha, A., Berkelhammer, M., Stott, L., Mudelsee, M., Cheng, H., Biswas, J., 2011. The leading mode of Indian summer monsoon precipitation variability during the last millennium. Geophys. Res. Lett. 38 (15). doi:10.1029/2011GL047713.

Sinha, A., Cannariato, K.G., Stott, L.D., Cheng, H., Lawrence Edwards, R., Yadava, M.G., Ramesh, R., Singh, I.B., 2007. A 900-year (600 to 1500 AD) record of the Indian summer monsoon precipitation from the core monsoon zone of India. Geophys. Res. Lett. 34 (16).

Sinha, N., Gandhi, N., Chakraborty, S., Krishnan, R., Yadava, M.G., Ramesh, R., 2018. Abrupt climate change at ~ 2800 Yr BP evidenced by high-resolution oxygen isotopic record of a Stalagmite from Peninsular India. The Holocene 28 (11), 1720–1730. https://doi.org/10.1177/0959683618788647.

Singh, M., Krishnan, R., Goswami, B., Choudhury, A.D., Swapna, P., Vellore, R., Prajeesh, A.G., Sandeep, N., Venkataraman, C., Donner, R.V., Marwan, N., Kurths, J., 2020. Fingerprint of volcanic forcing on the ENSO–Indian monsoon coupling. Sci. Adv. 6, eaba8164. doi:10.1126/sciadv.aba8164.

Staubwasser, M, 2006. An overview of Holocene South Asian monsoon records–monsoon domains and regional contrasts. J. Geol. Soc. India. 68, 433–446.

Staubwasser, M., Sirocko, F., Grootes, P.M., Segl, M., 2003. Climate change at the 4.2 ka BP termination of the Indus valley civilization and Holocene south Asian monsoon variability. Geophys. Res. Lett. 30 (8).

Sankar S., L. Svendsen, et al. (2016) The relationship between Indian summer monsoon rainfall and Atlantic multidecadal variability over the last 500 years, Tellus A: Dyn. Meteorol. Oceanogr., 68:1, 31717, DOI: 10.3402/tellusa.v68.31717.

Tejavath, C., Ashok, K., Chakraborty, S., Ramesh, R., 2019. A PMIP3 narrative of modulation of ENSO teleconnections to the Indian summer monsoon by background changes in the Last Millennium. Clim. Dyn. doi:10.1007/s00382-019-04718-z.

Tiwari, M., Nagoji, S.S., Ganeshram, R.S., 2015. Multi-centennial scale SST and Indian summer monsoon precipitation variability since the mid-Holocene and its nonlinear response to solar activity. Holocene. doi:10.1177/0959683615585840.

Tudhope, A.W., et al., 2001. Variability in the El Nino: southern oscillation through a glacial–interglacial cycle. Science 291, 1511–1517.

Xavier, P.K., Marzina, C., Goswami, B.N., 2007. An objective definition of the Indian summer monsoon season and a new perspective on the ENSO–monsoon relationship. Q. J. R. Meteorol. Soc. 133, 749–764. doi:10.1002/qj.45.

Yadav, R.R., Braeuning, A., Singh, J., 2011. Tree ring inferred summer temperature variations over the last millennium in western Himalaya, India. Clim. Dyn. 36 (7-8), 1545–1554.

Yadav, R., 2013. Tree ring–based seven-century drought records for the Western Himalaya, India. J. Geophys. Res. (Atmos.) 118 (10). doi:10.1002/jgrd.50265.

Yadava, M.G., Ramesh, R., 2005. Monsoon reconstruction from radiocarbon dated tropical Indian speleothems. Holocene 15 (1), 48–59.

Zaw, Z., et al., 2020. Drought Reconstruction Over the Past Two Centuries in Southern Myanmar Using Teak Tree-Rings: Linkages to the Pacific and Indian Oceans. Geophys. Res. Lett. 47. https://doi.org/10.1029/2020GL087627.

Zhang, L., Delworth, T.L., 2016. Simulated response of the Pacific Decadal oscillation to climate change. J. Clim. 29, 5999–6018. https://doi.org/10.1175/JCLI-D-15-0690.1.

Zhang, P., Cheng, H., Edwards, L., et al., 2008. A test of climate, sun, and culture relationships from an 1810-year Chinese cave record. Science 322, 940–942.

Zhang, P., Jee-Hoon, Jeong, Yoon, Jin-Ho, et al., 2020. Abrupt shift to hotter and drier climate over inner East Asia beyond the tipping point. Science 370, 1095–1099.

Zhisheng, A., Clemens, S.C., Shen, J., et al., 2011. Glacial-Interglacial Indian summer monsoon dynamics. Science 333, 719–723.

Zhisheng, A., Clemens, S.C., Shen, J., et al., 2015. Indian monsoon variations during three contrasting climatic periods: The Holocene, Heinrich Stadial 2 and the last interglacial-glacial transition. Quat. Sci. Rev. doi:10.1016/j.quascirev.2015.06.009.

Zorzi, C., Krishnamurthy, A, et al., 2015. Indian monsoon variations during three contrasting climatic periods: The Holocene, Heinrich Stadial 2 and the last interglacial-glacial transition. Quat. Sci. Rev. doi:10.1016/j.quascirev.2015.06.009.

Part II

Indian and Atlantic Ocean – Indian Summer Monsoon teleconnections

Chapter 8

Indian Ocean Dipole influence on Indian summer monsoon and ENSO: A review

Annalisa Cherchi[a], Pascal Terray[b], Satyaban B. Ratna[c], Syam Sankar[d], K P Sooraj[e], Swadhin Behera[f]

[a]National Research Council of Italy, Institute of Atmospheric Sciences and Climate (CNR-ISAC), Bologna, Italy, Istituto Nazionale di Geofisica e Vulcanologia, Bologna, Italy; [b]LOCEAN/IPSL, Sorbonne Universités (UPMC, Univ Paris 06)-CNRS-IRD-MNHN, Paris, France; [c]Climatic Research Unit, School of Environmental Sciences, University of East Anglia, Norwich, United Kingdom, Climate Research and Services, India Meteorological Department, Pune, India; [d]Advanced Centre for Atmospheric Radar Research (ACARR), Cochin University of Science and Technology, Kochi, India; [e]Centre for Climate Change Research, Indian Institute of Tropical Meteorology, Ministry of Earth Sciences (IITM-MoES), Pune, India; [f]Application Laboratory, Research Institute for Value-Added-Information Generation, Japan Agency for Marine-Earth Science and Technology, Yokohama, Japan

8.1 Introduction

The Indian summer monsoon (ISM) is one of the main components of the South Asian summer monsoon, representing the largest source of moisture and precipitation over the tropical sector (Webster et al., 1998). The ISM is highly variable and its variability is partly modulated by external factors, the El Niño Southern Oscillation (ENSO) being one of the most important. The remote connection between ENSO and ISM is known since the beginning of the 19th century and it has been largely investigated in the past (Walker, 1924; Sikka, 1980; Rasmusson and Carpenter, 1983; Kirtman and Shukla, 2000, etc.). Schematically, during warm ENSO episodes, the rising limb of the Walker circulation over West Pacific shifts eastward in response to a warming of the eastern Pacific, causing descent of air to the west of it and aiding decreased monsoon rainfall over India (Goswami, 1998; Lau and Wang, 2006). It has been natural also to explore the possible influence of the neighboring Indian Ocean on the ISM variability, with many studies pointing out significant connections (Rao and Goswami, 1988; Ashok et al., 2001, 2004; Gadgil et al., 2004, 2005, 2007; Krishnan et al., 2003; Terray et al., 2005, 2007; Cherchi et al., 2007; Izumo et al., 2008; Boschat et al., 2011, 2012; Cherchi and Navarra, 2013; Shukla and Huang, 2016, etc.).

Indian Summer Monsoon Variability: El Niño-teleconnections and beyond.
DOI: https://doi.org/10.1016/B978-0-12-822402-1.00011-9

The Indian Ocean Dipole (IOD) was discovered at the end of the 1990s (Saji et al., 1999; Webster et al., 1999) and it is recognized as one of the dominant modes of variability of the tropical Indian Ocean. Toward the end of the 20th century, weakening in the strength of the ENSO-monsoon relationship have been documented (Kumar et al., 1999; Kinter et al., 2002) and, since its discovery, the IOD has been identified as one potential trigger of that connection (Ashok et al., 2001; Li et al., 2003). Contrasting literature about its active or passive role has been produced since then and the debate is still open (Ashok et al., 2001; Li et al., 2003; Meehl et al., 2003; Ashok et al., 2004; Wu and Kirtman, 2004; Cherchi et al., 2007; Krishnan et al., 2011; Cherchi and Navarra, 2013; Krishnaswamy et al., 2015; Chowdary et al., 2016; Srivastava et al., 2019, to mention a few).

This chapter intends to provide an updated review on the current understanding of the influence of the IOD on the ISM and its teleconnection with ENSO. In particular, the chapter is organized as follows: Section 8.2 is dedicated to the description of the IOD, whereas Section 8.3 is focused on the processes at work in IOD influencing the monsoon and its relationship with ENSO, also from a modeling point of view. Section 8.4 reviews the literature about past evidences, present case studies, and future projections about the topic, and Section 8.5 is dedicated to the discussion of the results reviewed, highlighting some of the associated challenges, and related future perspectives. Finally, Section 8.6 collects the main conclusion derived from the review.

8.2 Some salient features of the Indian Ocean Dipole

The IOD is characterized by a zonal dipole in the tropical Indian Ocean with positive Sea Surface Temperature (SST) anomalies in the western equatorial Indian Ocean (50°–70°E, 10°S–10°N) and negative SST anomalies in the southeastern equatorail Indian Ocean (90°–110°E, 10°S-EQ) in its positive phase (Fig. 8.1; Saji et al., 1999). The formation of the IOD relies on the Bjerknes feedback, requiring background surface easterlies and thermocline shallowing in the eastern part along the Equator (Fig. 8.1; Schott et al., 2009). These conditions set in boreal spring and persist until autumn, explaining the IOD development in boreal summer, its peak toward autumn, and its rapid termination before winter, because of the monsoon wind swing (Schott et al., 2009).

Before the discovery of the IOD as one of the dominant modes of variability of the tropical Indian Ocean, earliest suggestions of inherent coupled dynamics in the basin identified periods of anomalous easterlies in the central Indian Ocean concurrent with anomalous cold (warm) SST in the eastern (western) part, that occurred during boreal fall in the absence of ENSO (Reverdin et al., 1986). Significant feedback mechanisms at play between zonal SST and pressure gradient, equatorial easterly wind anomalies, and precipitation anomalies in the western equatorial Indian Ocean were later identified (Hastenrath et al., 1993). An east-west seesaw in sea level anomaly in the tropical Indian Ocean was noticed and correlated with thermocline depth changes (Murtugudde et al., 1995). This east-west sea level dipole has

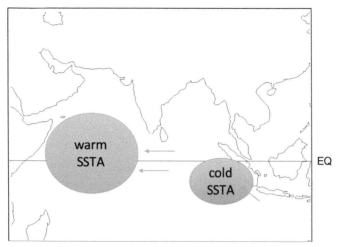

FIG. 8.1 Schematic of the Indian Ocean Dipole in its positive phase with warm SST anomalies on the west and cold SST anomalies toward the coast of Sumatra. The *green arrows* indicate the direction of the prevailing corresponding surface winds. SST, sea surface temperature.

been associated with the east-west SST gradient and strongly correlated with the ISM (Murtugudde et al., 1998). The SST gradient was shown to be in phase with equatorial zonal wind anomalies (Murtugudde and Busalacchi, 1999).

Since its discovery, there has been strong debate and controversy whether the IOD is an intrinsic mode of Indian Ocean coupled variability, or whether and how it is driven by external forcing, like ENSO (e.g., Hastenrath, 2002; Dommenget and Latif, 2002; Yamagata et al., 2003; Murtugudde et al., 2003; Behera et al., 2003). Many studies claim that most of the IOD variability is driven by ENSO (Allan et al., 2001; Baquero-Bernal et al., 2002; Huang and Kinter, 2002; Dommenget, 2011; Zhao and Nigam, 2015; Zhao et al., 2019), while others suggest that IOD is a self-sustained mode of oscillation (Ashok et al., 2003; Yamagata et al., 2004; Behera et al., 2006). About 20% of the IOD events seem to co-occur with ENSO (Fig. 8.2; Saji, 2018). IOD events that could be categorized as with or without the influence of ENSO have systematic differences in their temporal evolution and spatial distribution, including periodicity, strength, and formation processes (Behera et al., 2006; Hong et al., 2008). Modeling studies confirm that IOD events are often triggered by ENSO, but they also demonstrate that IOD events can exist without ENSO by means of dedicated sensitivity experiments in which ENSO is removed by different nudging techniques (Lau and Nath, 2004; Fischer et al., 2005; Behera et al., 2006; Wang et al., 2016; 2019; Cretat et al., 2017, 2018). In the absence of ENSO, the interannual IOD variability is mostly biennial (Behera et al., 2006; Cretat et al., 2018), while in years of co-occurrences ENSO affects the periodicity, strength, and formation processes of IODs (Cretat et al., 2018).

Through changes in the atmospheric circulation, the IOD exerts its influence on, among others, the Southern Oscillation (Behera and Yamagata, 2003), the

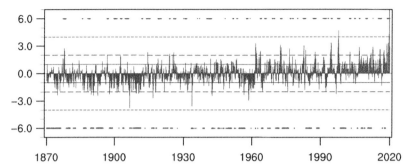

FIG. 8.2 Normalized monthly IOD index (std) defined as anomalous SST gradient between the western equatorial Indian Ocean (50°E–70°E and 10°S–10°N) and the south eastern equatorial Indian Ocean (90°E–110°E and 10°S–0°N). Anomalies (with respect to 1981–2010 mean) have been downloaded from https://psl.noaa.gov/gcos_wgsp/Timeseries/DMI/, but 2019 values have been integrated from JAMSTEC repository (http://www.jamstec.go.jp/virtualearth/general/en/index.html). Red and blue markers along 6 and −6 std correspond to Niño3.4 anomalies (1981–2010 mean removed) larger than 0.5°C. Niño3.4 anomalies have been downloaded from https://psl.noaa.gov/gcos_wgsp/Timeseries/Nino34/. As indicated in the respective websites, both IOD and Niño3.4 values are computed from the HadISST1 dataset (Rayner et al., 2003). "*IOD*, Indian Ocean Dipole."

summer climate condition in Europe (Behera et al., 2012), East Asia (Guan and Yamagata, 2003; Guan et al., 2003; Chen et al., 2019), and streamflows in the western part of Indonesia (Sahu et al., 2012), as well as on rainfall over Africa (Black et al., 2003; Manatsa and Behera, 2013; Endris et al., 2019), Sri Lanka (Zubair et al., 2003), Australia (Ashok et al., 2003; Ummenhofer et al., 2013; Dey et al., 2019; Hossain et al., 2020), and Brazil (Chan et al., 2008; Taschetto and Ambrizzi, 2012; Bazo et al., 2013). In the following, we focus on the IOD influence on the summer monsoon rainfall over India and linkages with ENSO, which are both subject of important controversies in the literature.

8.3 IOD and the ENSO-monsoon teleconnections: processes at work

To influence the ISM and its ENSO teleconnection, IOD events should develop from boreal summer (June to August). As described by Schott et al. (2009), this is in fact the case for many IOD events. However, some IODs peak in the monsoon core season, though a few others develop later (Du et al., 2013). According to that, strong interactions between IOD and ISM can be expected during boreal summer and the withdraw phase of the monsoon.

Some authors suggested a direct influence of the IOD on the ISM rainfall (ISMR) through moisture transport over the western Indian Ocean or modifications of the local Hadley cell, with enhanced ascendance and a northward shift of its uplift branch during positive IOD events, both enhancing ISMR (Fig. 8.3A; Ashok et al., 2001, 2004; Gadgil et al., 2004; Behera et al., 2005; Ashok

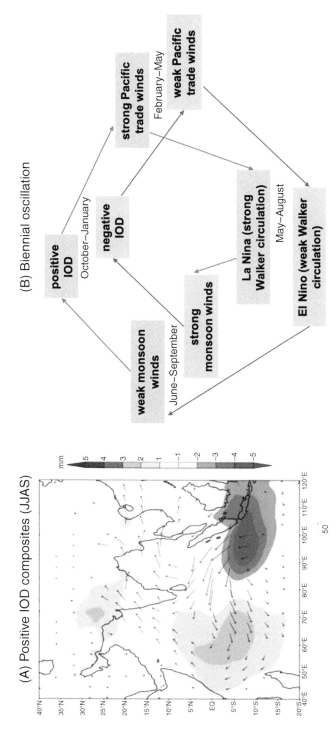

(B) Biennial oscillation

(A) Positive IOD composites (JJAS)

FIG. 8.3   (A) Positive IOD composite of specific humidity (mm, shaded) and moisture flux (integrated up to 300 hPa, kg/m/s, vectors) anomalies averaged in summer (JJAS). The figure is taken from Behera and Ratnam (2018); (B) schematic of the IOD influence on ISM rainfall embedded in the TBO (adapted from the scheme in Webster and Hoyos (2010)). "*IOD*, Indian Ocean Dipole; *ISM*, Indian summer monsoon)."

and Saji, 2007; Behera and Ratman, 2018). Recent investigations also reveal that early IODs (i.e., peaking in July) or prolonged IODs (i.e., lasting longer, typically more than 8 months) have an excess of evaporation from the Arabian Sea and stronger cross-equatorial flow, leading to enhanced monsoon activity with decreased numbers of break spells (Anil et al., 2016). Others suggest that positive IOD events during boreal fall normally follow weak ISMs, and vice versa, as ISM circulation during boreal summer can also induce an equatorial anomalous SST gradient in the Indian Ocean during the following boreal fall (Loschnigg et al., 2003, Meehl et al., 2003; Terray et al., 2005, 2007). In this framework, ENSO, ISM, and IOD appear as strongly inter-related components of the tropospheric biennial oscillation (TBO) in the tropics (Fig. 8.3B; Meehl and Arblaster, 2002; Meehl et al., 2003; Li et al., 2006; Drbohlav et al., 2007; Webster and Hoyos, 2010). The development of the IOD during boreal summer and autumn can lead to SST warming in the tropical southwest Indian Ocean via ocean dynamics during the next boreal winter and spring (Xie et al., 2002; Du et al., 2009). This warming can further influence the ISM onset in the following year, especially for IOD events co-occurring with El Niño in the Pacific Ocean and followed by a basin-wide Indian Ocean warming (Annamalai et al., 2005; Yang et al., 2007; Hong et al., 2010).

The ISMR response is not necessarily spatially coherent to the IOD phases (Behera and Ratman, 2018). The anomalous moisture transports to India associated with a positive IOD strengthen the monsoon trough and rainfall through an intensified monsoon-Hadley circulation (Behera et al., 1999; Ashok et al., 2001; Anil et al., 2016), with below-normal rainfall to the south and to the north of the trough. During positive IODs, the north–south precipitation (heating) gradient over the eastern Indian Ocean dominates over the one in the equatorial Indian Ocean, resulting in a regional meridional circulation with uplift over the monsoon trough and sinking in the eastern lobe (Annamalai et al., 2003). On the other hand, in a negative IOD event, a regional Walker circulation and the moisture distribution favor moisture divergence (convergence) in the eastern (western) part of India. This gives rise to a zonal dipole in the rainfall anomalies with abundant rainfall on the western part and scanty rainfall on the east. The resulted regional asymmetry is a unique feature associated with the ISMR response to IOD but it is not well simulated by coupled GCMs, though regional model experiments with different physical parameterization schemes may provide few combinations able to realistically reproduce the asymmetric response to the two phases of the IOD (Behera and Ratman, 2018).

IOD can influence the ENSO–ISM relationship indirectly (Ashok et al., 2001; Behera and Yamagata, 2003; Cai et al., 2011; Weller and Cai, 2013), by modulating ENSO events in the tropical Pacific itself (Luo et al., 2010; Izumo et al., 2010; Cai et al., 2019). Considering IOD events as triggered by ENSO, they may counteract its simultaneous influence on ISM (Ashok et al., 2001, 2004; Ashok and Saji, 2007; Ummenhofer et al., 2011; Lau and Nath, 2003, 2012; Krishnaswamy et al., 2015). More recently IOD has been suggested as potential trigger of ENSO, adding more complexity to the emerging picture (Luo et al., 2010; Izumo et al., 2010; Zhou et al., 2015; Jourdain et al., 2016; Wieners et al., 2017a,b; Wang et al., 2019; Cai et al., 2019). In fact, the Indian Ocean (either IOD and Indian Ocean Basin-wide Warming) has a highly significant impact on both the variability and predictability of ENSO, as evidenced by decoupled experiments using different coupled climate models (Luo et al., 2010; Izumo et al., 2010; Santoso et al., 2012; Terray et al., 2016; Kajtar et al., 2016; Wang et al., 2019). For example, IOD events co-occurring with ENSO may fasten its phase transition (Kug and Kang, 2006; Luo et al., 2010; Izumo et al., 2010; Kug and Ham, 2012).

In the absence of ENSO, IOD still exists (usually known as "pure IOD") and its variability is mainly driven by the eastern Indian Ocean in a suite of coupled climate model simulations nudging the tropical Pacific SSTs toward an SST climatology estimated from observations or a control simulation (Cretat et al., 2017, 2018), as consistently seen in other coupled model studies (Gualdi et al., 2003; Fischer et al., 2005; Behera et al., 2005, 2006; Luo et al., 2010; Izumo et al., 2010; Wang et al., 2016, 2019). In the nudged experiments, the strong diabatic heating associated with enhanced rainfall over the eastern IOD lobe modulates the local Hadley circulation and induces a negative (positive) rainfall anomaly in the northern Indian Ocean during boreal summer during negative (positive) IOD events, as suggested from observational studies (Fig. 8.4; Cretat et al., 2017). Such changes in the local Hadley circulation are attenuated in the presence of ENSO because ENSO-induced changes in the (zonal) Walker circulation dominate (Cretat et al., 2017). It has been also found that rainfall anomalies over India associated with these pure IODs are modest and not statistically significant, especially at the beginning of the monsoon, although the simulated SST variability in the eastern Indian Ocean is overestimated (Fischer et al., 2005; Terray et al., 2012; Cretat et al., 2017, 2018). However, pure IODs promote a quadrupole rainfall pattern linking the tropical Indian Ocean and the Western North Pacific, and induce important zonal shifts of the Walker circulation in the absence of ENSO (Fig. 8.4), in agreement with earlier findings (Li et al., 2006; Chowdary et al., 2011). The circulation patterns with and without ENSO largely differ, confirming potential opposite effects between IOD and ENSO (e.g., Ashok et al., 2001; Lau and Nath, 2012; Pepler et al., 2014).

Pure IOD events may also help to sustain the TBO: a stronger-than-observed biennial spectrum of the IOD is found after removing ENSO's impacts (Behera et al., 2006; Cretat et al., 2018). Moreover, coupled ocean-atmosphere interactions in the Indian Ocean can sustain its own TBO without ENSO (Cretat et al., 2018). First, subsurface ocean dynamics play a key role in the biennial anomalies during boreal winter (Rao et al., 2002, 2009; Schott et al., 2009; McPhaden and Nagura, 2014; Delman et al., 2016) with a sudden reversal of thermocline anomalies in the eastern equatorial IO forced by intraseasonal disturbances reminiscent of the Madden–Julian Oscillation (Rao and Yamagata, 2004; Han et al., 2006). Second, tropical-extra-tropical interactions within the Indian

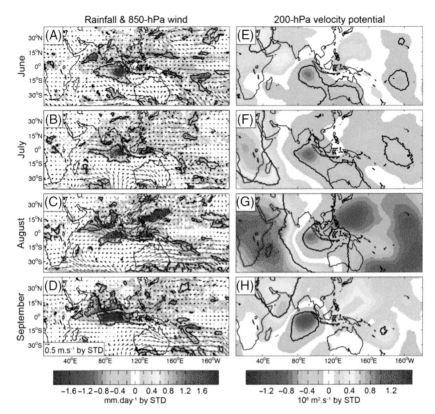

FIG. 8.4 June to September (A–D) rainfall (shading), 850-hPa wind (vectors), and (E–H) 200-hPa velocity potential anomalies regressed onto normalized boreal fall (i.e., SON) SST anomalies over the eastern IOD pole (domain: 90°–110°E, 10°S-Eq) when ENSO is removed. Significant anomalies at the 90% level are shown with black contours for rainfall and 200-hPa velocity potential anomalies, and with purple vectors for 850-hPa wind anomalies. Positive 200-hPa velocity potential anomalies correspond to abnormal upper-level mass flux convergence. "*ENSO*, El Niño Southern Oscillation; *IOD*, Indian Ocean Dipole."

Ocean appear to be the main trigger of IODs in the absence of ENSO (Cretat et al., 2018). In nudged experiments, both the power spectra of the ISMR and IOD indices during boreal summer shift toward increased biennial variability compared to the control simulation, which may be more consistent with a possible coupling of the IOD with ISM in the absence of ENSO, but in a regional TBO framework (Cretat et al., 2018). However, this TBO framework is again mainly based on the strong influence of the ISM circulation on the Indian Ocean SSTs despite the absence of ENSO in the nudged experiments.

In synthesis, these recent modeling studies do suggest that IOD exists without ENSO, but the exact relationships between IOD and ISM remain elusive, even in the absence of ENSO.

8.4   Past, present, and future IOD influence on the ENSO-monsoon teleconnection

At long time scales, the influence of IOD on ISM seems opposite to the effect of ENSO, and the IOD–ISM relationship seems to vary complementarily to that between ENSO and ISM (Ashok et al., 2001; Krishnaswamy et al., 2015). In fact, the IOD–ISM relationship has strengthened in the recent decades (Ashok et al., 2001, 2004; Ashok and Saji, 2007; Izumo et al., 2010; Ummenhofer et al., 2011) due to nonuniform warming of the Indian Ocean (Ihara et al., 2007; Cai et al., 2009), whereas the ENSO–ISM relationship has weakened (Kumar et al., 1999; Ashrit et al., 2001; Ihara et al., 2007).

On longer IOD records, changes in frequency and teleconnections have been identified (e.g., Abram et al., 2008; Kayanne et al., 2006; Abram et al., 2020). Coral proxy records from the Lake Victoria in Kenya suggest that the influence of ENSO has decreased over the Western Indian Ocean in recent decades (Nakamura et al., 2009). A mode shift in IOD variability related to the warming trend in the western Indian Ocean has raised the mean SST to a threshold value that encourages tropical convections (Nakamura et al., 2009). A recent reconstruction of the last millennium indicates clustering of positive IOD events with extreme IOD variability and a persistent tropical Indo-Pacific climate coupling (Abram et al., 2020). The frequency and strength of IOD events exceptionally increased during the 20th century associated with enhanced upwelling in the eastern pole of the IOD, likely making more direct the influence of the IOD on the Asian monsoon (Abram et al., 2008). These processes and associated changes in the Walker circulation, linked to global warming, may precondition the mean state to trigger frequent positive IOD events, together with intense short rains in East Africa.

In the 20th century, the ENSO–IOD correlation was strongly positive and significant since the mid-1960s (Cherchi and Navarra, 2013), with ENSO and IOD almost independent before 1970 (Yuan and Li, 2008). A recent weakening of the coupling has been identified during 1999–2014 compared to the previous two decades (i.e., 1979–1998), associated with different spatial patterns in ENSO evolution during boreal spring and summer (Ham et al., 2017). The stronger/weaker correlation may correspond with either strong or weak ENSO-monsoon relationship and with strong or weak IOD-monsoon relationship, with differences arising from the relationship between ISM and SST in other ocean basins rather than the Indo-Pacific sector alone (Cherchi and Navarra, 2013).

The IOD–ENSO–ISM relationship appears to work differently with different seasons (Agrawal et al 2017). The connection between ISM and IOD is mostly confined in the summer and autumn, while that with ENSO is stronger and extends more in time (Cherchi and Navarra, 2013). In fact, the evolution of the correlation between ISMR and monthly Niño3.4 is maximum in August–November, remaining strong and stable until March of the following year (Gershunov et al., 2001).

The 1997 El Niño was one of the strongest events that occurred in the 20th century, but ISMR was slightly above normal (Srinivasan and Nanjundiah, 2002), and this has been attributed to the influence from the Indian Ocean (Slingo and Annamalai, 2000; Sreejith et al., 2015). Similarly, the influence of IOD helped nullifying the effect of ENSO on the monsoon during 1999 (Saji et al., 1999; Webster et al., 1999). Positive IOD events, as those that occurred during 2007–2008, coexisted with La Niña episodes (e.g., Ashok et al., 2003; Cai et al., 2009). A very strong positive IOD event occurred in the summer of 1994, as a clear coupled ocean-atmosphere phenomenon of the Indian Ocean (Behera et al., 1999). The 1994 event lasted more than 8 months from March to October and positively influenced the ISMR that recorded 265 mm/month, a value 19% above the climatological mean (Guan and Yamagata, 2003).

In 2019 the monsoon onset was delayed by about 7 days over India with the June rainfall recording a deficit of about 33% with respect to the mean climatology (Gadgil et al., 2019; http://www.imd.gov.in). According to the India Meteorological Department, subsequent to this monsoon onset, further northward progression of the monsoon remained slow due to the formation of a very severe cyclone over central eastern Arabian Sea (i.e., the cyclone VAYU that formed during June 10–17, 2019, https://mausam.imd.gov.in/imd_latest/contents/season_report.php). El Niño weakened in July while a strong positive IOD started to develop (Fig. 8.5B) and rainfall picked up strength during the latter stages of the monsoon season (late July to September; Fig. 8.5B–D), remaining above normal. For that year, the seasonal rainfall recorded for India has been quantified at 110% percent of the long period average as defined by India Meteorological Department, with the September rainfall at 152%. The positive IOD that occurred during late summer in 2019 was one of the strongest in the recent times (Fig. 8.2), with its predictability linked with the existence of the El Niño Modoki in the Equatorial Pacific (Doi et al., 2020) and to a strong pressure dipole between the Australian High and the South China Sea/ Philippine Sea region (Lu and Ren, 2020). The exceptional intensity of the event remains even after the IOD index is detrended (not shown). This very recent case illustrates how the interactions between ISM, IOD, and ENSO are subtle and complex, and highly influenced by internal dynamical processes.

In CMIP5 models, a correct representation of the coupled processes (i.e., Bjerknes feedback) in the equatorial Indian Ocean is a necessary condition for realistic monsoon simulations (Annamalai et al., 2017). At the same time, a too weak simulated southwest summer monsoon over the Arabian Sea may generate a warm SST bias over the western equatorial Indian Ocean that in the following fall may amplify the error toward an IOD-like SST bias via the Bjerknes feedback (Li et al., 2015). The unrealistic present-day IOD–ISM correlation simulated by the majority of CMIP5 models may also be related to an overly strong control by ENSO (Li et al., 2017), likely leading to an underestimation of the projected future ISMR increase. Still, CMIP5 models project an increase in ISMR in a warmer climate with a reasonably strong consensus among models

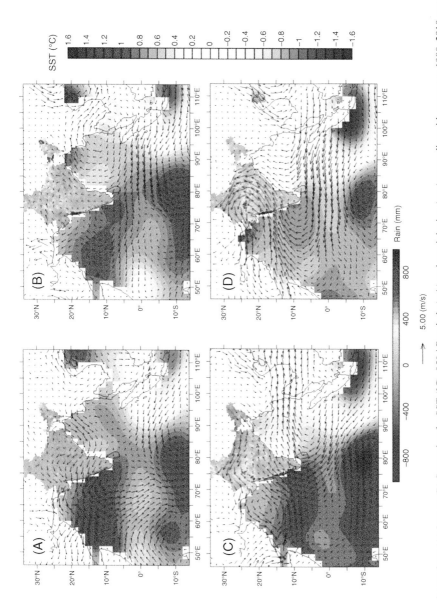

FIG. 8.5 2019 case from (A) June to (D) September for SST (°C). 850-hPa winds (m/s) and precipitation (mm) anomalies with respect to 1980–2010 mean climatology. SST data are taken from ERSSTv5 (Huang et al. 2017), 850-hPa wind vectors are obtained from ECMWF ERA5 reanalysis (Hersbach and Dee, 2016) and precipitation from daily gridded rainfall dataset over India (Pai et al. 2014).

(Jayashankar et al., 2015). CMIP5 models' projections tend to exhibit a positive IOD-like pattern in the tropical Indian Ocean with weaker (stronger) warming in the east (west) and an easterly wind trend (Zheng et al., 2010; 2013). The response is driven by the projected weakening of the Walker circulation in a warmer climate in the majority of models (Vecchi et al., 2006; Kociuba and Power, 2015).

In future projections, surface moisture increase dominates the changes in rainfall associated with the IOD, while IOD SST changes dominate the corresponding changes in the circulation, decreasing at a rate of 13.7%/°C (Huang et al., 2019). The ensemble spread in the IOD amplitude change is large (Ng et al., 2018), and it is related to that of the ENSO amplitude change (Hui and Zheng, 2018). The large spread in the IOD response to increasing GHGs with significant variations in the amplitude and skewness of the dipole and in climatological SST gradient is due to small differences in the mean thermocline depth induced by internal climate variability via the positive Bjerknes feedback (Ng et al., 2018).

The frequency of extreme IOD events is projected to increase under global warming conditions (Cai et al., 2014; see Chapter 21), with a persistence of the ENSO–IOD linkage in a warmer future world (Stuecker et al., 2017). The characteristics of the ENSO–IOD are likely to continue in the future, and given that ENSO and its predictability are modulated on decadal timescales (i.e., Wittenberg, 2009; Wittenberg et al., 2014; Karamperidou et al., 2014), the same should be expected for the IOD (Stuecker et al., 2017).

CMIP6 models largely improved in the simulation of the spatial and temporal pattern of the ISM (Gusain et al., 2020), especially over the Western Ghats and the foothills over the Himalayas, whereas a majority of the CMIP5 models underestimated rainfall over central and northern India (Jain et al., 2019). A subset of CMIP6 models (Table 8.1) confirms CMIP5 results, with a tendency toward larger IOD amplitude at the end of the 20th century and in the future, at least under the most extreme CMIP6 scenario SSP5-8.5 (Fig. 8.6A). SSP stands for shared socioeconomic pathways with five representing an economic vision of the future with relatively optimistic trends for human development but assuming an energy-intensive, fossil-fuel economy, and 8.5 correspondings to the forcing (in W/m^2) by 2100 (O'Neill et al., 2016). In summer (JJAS mean), SST regressed onto the IOD index project larger IOD-related anomalies over the Pacific and Indian Oceans (Fig. 8.6B and C). In the projection, precipitation and winds regressed onto the IOD index show modest positive anomalies over India and weaker easterlies along the Equator because of a weaker negative pole (Fig. 8.6D and E). This figure is just a flavor of the complex relationship as projected in the new generation of coupled climate models. For example, the methodology applied does not fully disentangle how the mean state changes and its role. A more systematic analysis and comparison of CMIP5 and CMIP6 experiments would be needed to fully understand differences and potential improvements. This is outside the scope of this chapter but it is under investigation in separated ongoing researches.

TABLE 8.1 List and some characteristics of CMIP6 models used.

Model name	Institute/ country	Atmosphere resolution (km)	Ocean resolution (km)	Earth system model
ACCESS-CM2	CSIRO-ARCCSS/ Australia	250	100	No
ACCESS-ESM1-5	CSIRO-ARCCSS/ Australia	250	100	Yes
BCC-CSM2-MR	BCC/China	100	50	No
CESM2	NCAR/United States	100	100	Yes
CESM2-WACCM	NCAR/United States	100	100	Yes
CNRM-CM6-1	CNRM-CERFACS/France	250	100	No
CNRM-CM6-1-HR	CNRM-CERFACS/France	100	25	No
CNRM-ESM2-1	CNRM-CERFACS/France	250	100	Yes
CanESM5	CCCma/ Canada	500	100	Yes
EC-Earth3	EC-Earth-Consortium/ Europe	100	100	No
EC-Earth3-Veg	EC-Earth-Consortium/ Europe	100	100	No
FGOALS-f3-L	CAS/China	100	100	No
FGOALS-g3	CAS/China	250	100	No
GFDL-ESM4	NOAA-GFDL/United States	100	50	Yes
HadGEM3-GC31-LL	MOHC-NERC/United Kingdom	250	100	No
INM-CM4-8	INM/Russia	100	100	No
INM-CM5-0	INM/Russia	100	50	No

(continued)

TABLE 8.1 (Cont'd)

Model name	Institute/ country	Atmosphere resolution (km)	Ocean reso- lution (km)	Earth sys- tem model
IPSL-CM6A-LR	IPSL/France	250	100	Yes
KACE-1-0-G	NIMS-KMA/ South Korea	250	100	No
MCM-UA-1-0	UA/United States	250	250	No
MIROC-ES2L	MIROC/Japan	500	100	Yes
MIROC6	MIROC/Japan	250	100	No
MPI-ESM1-2-HR	MPI-M DWD DKRZ/Germany	100	50	Yes
MPI-ESM1-2-LR	MPI-M AWI/ Germany	250	250	Yes
MRI-ESM2-0	MRI/Japan	100	100	Yes
NESM3	NUIST/China	250	100	No
NorESM2-LM	NCC/Norway	250	100	Yes
NorESM2-MM	NCC/Norway	100	100	Yes
UKESM1-0-LL	MOHC NERC NIMS-KMA NIWA/United Kingdom	250	100	Yes

More details about the models can be found at https://pcmdi.llnl.gov/CMIP6/.

8.5   Challenges and future perspectives

The main challenges in having a complete picture of the influence of IOD on the ENSO-monsoon teleconnection remains related to a full understanding of the IOD itself and to a full and agreed understanding of how the IOD is related to ENSO on one side, and its pure (i.e., independent from ENSO) relationship with ISM on the other. While some progress have been made in recent decades in simulating ENSO variability (Bellenger et al., 2014), many important issues remain open due to the available short observational record and the important biases affecting ISM and the Indian Ocean in state-of-the-art climate models (Li et al., 2015; Annamalai et al., 2017).

An exhaustive analysis of the IOD recorded past events would allow a categorization of the main characteristics of the processes at play, but the observed

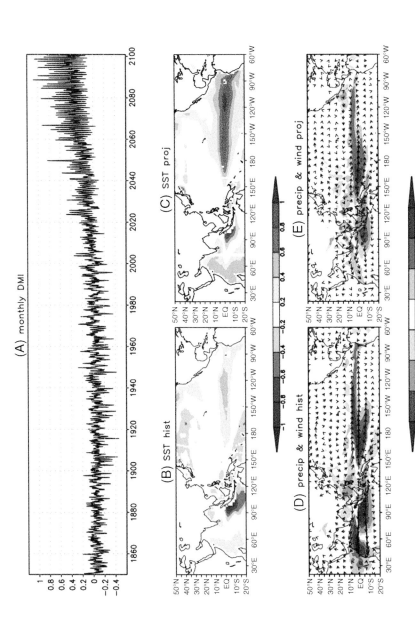

FIG. 8.6 (A) Monthly IOD index (anomalies, °C) for 20th and 21st centuries from a subset of CMIP6 models (Table 8.1). The index has been computed as in Fig. 8.2 (same areas difference and anomalies with respect to 1980–2010 mean climatology). The index has been computed for each model and then averaged to obtain the ensemble mean. (B, C) SST (°C, shaded) and (D, E) precipitation (mm/day, shaded) and 850-hPa wind (m/s, vectors) regressed on the IOD index (values for 1°C change in the index) for JJAS mean during the historical period and the future projection, respectively. One member for each model has been considered. For the 21st century, the SSP5-8.5 scenario has been used. In (C) and (E), the time series have been detrended before computing the regression, to keep out the trend from the related variability.

record remains short to have a statistically robust assessment. On the other hand, in state-of-the-art coupled climate models the simulation of the IOD, ISM, and related characteristics (including mean state and variability) still has large biases thus precluding a complete understanding of the processes at work. For example, it is not clear whether the weak IOD-monsoon relationship simulated in the models (Fig. 8.6) is realistic or not, due to the exaggerated IOD variability or to the overly strong control of ENSO simulated by current global climate models. Moreover, it has to be clarified whether poor simulations of other factors, like IOD-induced cross-equatorial flows, may be important. Similarly, a complete understanding of the coupling and feedback processes between the developing phase of the IOD and the summer monsoon rainfall over India, including the possible feedback on the development of the IOD in the subsequent season, is still missing. As a consequence, much more in-depth observational and modeling studies (including model's improvements) are clearly needed to understand the IOD effect on ISM and the relative roles of remote versus local forcing on the ISM–ENSO relationship as well as the role of internal atmospheric processes in modulating that relationship. In future climate projections, it would be useful to understand how changes in the simulated IOD properties contribute to the relative importance of thermodynamic and dynamic monsoon processes at play in global warming frameworks.

For all the points above, crucial keys are the collection of as much as possible observations during known IOD events, and how IOD and related properties are simulated in state-of-the-art climate models. The need for more observations in the Indian Ocean is particularly important because of its changes within the last warm decades (Hermes et al., 2019). For the simulation of IOD and related models' performance, more efforts should be dedicated to reducing systematic biases in coupled climate models and/or in performing ad-hoc sensitivity experiments in a large set of different coupled models to clarify the dynamics involved in IOD formation and related teleconnections. One possibility is to design coordinated international efforts with specific common experiments, likely following current CMIP frameworks (Zhou et al., 2016) or the CORE-II experience (Rahaman et al., 2020).

8.6 Conclusions

The IOD is one of the dominant modes of variability of the tropical Indian Ocean. It is recognized as having important teleconnections worldwide but here it has been considered in terms of its influence on the ISM and its relationship with ENSO. In particular, the literature focused on its active or passive role has been reviewed, evidencing how the influence of the IOD on the ISMR can be interpreted as having a direct impact through moisture transport over the Western Indian Ocean or modifications of the local Hadley cell, or alternatively in the framework of the TBO. Recently, more literature is available on the role of the IOD independent from ENSO, or even as trigger of ENSO itself. Still, combining modeling and observational studies, the precise relationship between IOD and ISM remains elusive, with or without ENSO.

Considering major ENSO and/or IOD events, it has been recorded how in 1997 the failure of the negative relationship between ENSO and the ISMR was associated with a positive IOD event developing that summer, or how the strong IOD of 1994 has been responsible for a stronger than normal monsoon that summer. Recently, the big IOD event recorded in 2019 and its evolution within the summer evidenced how the interactions between ISM, IOD, and ENSO are complex, and highly influenced by internal dynamical processes.

CMIP5 and CMIP6 models agree in projecting stronger IOD events in the future, but how this project on atmospheric anomalies over the Indo-Pacific region may not be fully consistent. A systematic analysis of the two model inter-comparisons sets is needed to fully understand the possible differences.

Conflict of interest statement

On behalf of all authors, the corresponding author states that there is no conflict of interest.

Acknowledgements

CINECA is acknowledged for providing resources for the storage of CMIP6 data via the "CMIP6_EC" DRES space. All CMIP6 simulations used are also freely available from the Earth System Grid Federation (https://esgf-node.llnl.gov/search/cmip6/). S.B.R. is supported by the Belmont Forum and JPI-Climate project INTEGRATE (An Integrated data-model study of interactions between tropical monsoons and extratropical climate variability and extremes) with funding by UK NERC grant NE/P006809/1. S.S. gratefully acknowledges the financial support given by the Ministry of Earth Sciences (MoES), Government of India to support the ST Radar facility at ACARR, CUSAT, Kochi. The Centre for Climate Change Research (CCCR) at Indian Institute of Tropical Meteorology (IITM) is fully funded by the Ministry of Earth Sciences, Government of India. P.T. is funded by Institut de Recherche pour le Développement (IRD, France).

References

Abram, N.J., Wright, N.M., Ellis, B., Dixon, B.C., Wurtzel, J.B., England, M.H., et al., 2020. Coupling of Indo-Pacific climate variability over the last millennium. Nature 579, 385–392. doi:10.1038/s41586-020-2084-4.

Abram, N.J., Gagan, M.K., Cole, J.E., Hantoro, W.S., Mudelsee, M., 2008. Recent intensification of tropical climate variability in the Indian Ocean. Nat. Geosci 1, 849–853. doi:10.1038/ngeo357.

Agrawal, N., Pandey, V.K., Shahi, N., 2017. ENSO-IOD changing relationship and its impact on Indian Summer Monsoon. VSRD Intern. VSRD Int. J. Tech. Non-Tech. Res. 8 (5), 165–174.

Allan, R., Chambers, D., Drosdowsky, W., Hendon, H., Latif, M., Nicholls, N., et al., 2001. Is there an Indian Ocean dipole and is it independent of the El Niño-Southern Oscillation? CLIVAR Exch. 6, 18–22.

Anil, N., Ramesh Kumar, M.R., Sajeev, R., Saji, P.K., 2016. Role of distinct flavors of IOD events on Indian summer monsoon. Nat. Hazards 82, 1317–1326. doi:10.1007/s11069-016-2245-9.

Annamalai, H., Taguchi, B., McCreary, J.P., Nagura, M., Miyama, T., 2017. Systematic errors in South Asian monsoon simulation: importance of equatorial Indian Ocean processes. J. Clim. 30, 8159–8178. doi:10.1175/JCLI-D-16-0573.1.

Annamalai, H., Liu, P., Xie, S.P., 2005. Southwest Indian Ocean SST variability: its local effect and remote influence on Asian monsoons. J. Clim. 18 (20), 4150–4167. doi:10.1175/JCLI3533.1.

Annamalai, H., Murtugudde, R., Potemra, J., Xie, S.P., Liu, P., Wang, B., 2003. Coupled dynamics over the Indian Ocean: spring initiation of the zonal mode. Deep-Sea Res. II (50), 2305–2330. doi:10.1016/S0967-0645(03)00058-4.

Ashok, K., Guan, Z., Yamagata, T., 2003. A look at the relationship between the ENSO and the Indian Ocean dipole. J. Meteor. Soc. Japan 81, 41–56. doi:10.2151/jmsj.81.41.

Ashok, K., Guan, Z., Saji, N.H., Yamagata, T., 2004. Individual and combined influences of ENSO and the Indian Ocean dipole on the Indian summer monsoon. J. Clim. 17, 3141–3155. doi:10.1175/1520-0442(2004)017<3141:IACIOE>2.0.CO;2.

Ashok, K., Guan, Z., Yamagata, T., 2001. Impact of Indian Ocean dipole on the relationship between the Indian monsoon rainfall and ENSO. Geophys. Res. Lett. 28, 4499–4502. doi:10.1029/2001GL013294.

Ashok, K., Saji, N.H., 2007. On impacts of ENSO and Indian Ocean dipole events on the subregional Indian summer monsoon rainfall. Nat. Haz. 42 (2), 273–285. doi:10.1007/s11069-006-9091-0.

Ashrit, R.G., Kumar, K.R., Kumar, K.K., 2001. ENSO-Monsoon relationships in a greenhouse warming scenario. Geophys. Res. Lett. 28, 1727–1730. doi:10.1029/2000GL012489.

Baquero-Bernal, A., Latif, M., Legutke, S., 2002. On dipole-like variability of sea surface temperature in the Indian Ocean. J. Clim. 15, 1358–1368. doi:10.1175/1520-0442(2002)015<1358:ODVOSS>2.0.CO;2.

Bazo, J., de las Nieves Lorenzo, M., Porfirio da Rocha, R., 2013. Relationship between monthly rainfall in NW Peru and tropical sea surface temperature. Adv. Meteor. doi:10.1155/2013/152875.

Behera, S.K., Krishnan, R., Yamagata, T., 1999. Unusual ocean–atmosphere conditions in the tropical Indian Ocean during 1994. Geophys. Res. Lett. 26, 3001–3004. doi:10.1029/1999GL010434.

Behera, S.K., Rao, S.A., Saji, N.H., Yamagata, T., 2003. Comments on "A cautionary note on the interpretation of EOFs". J. Clim. 16, 1094–1097. doi:10.1175/1520-0442(2003)016<1087:COACNO>2.0.CO;2.

Behera, S.K., Luo, J.J., Masson, S., Delecluse, P., Gualdi, S., Navarra, A., et al., 2005. Paramount impact of the Indian Ocean dipole on the East African short rains: a CGCM study. J. Clim. 18, 4514–4530. doi:10.1175/JCLI3541.1.

Behera, S.K, Luo, J.J., Masson, S., Rao, S.A., Sakuma, H., Yamagata, T., 2006. A CGCM study on the interaction between IOD and ENSO. J. Clim 19, 1608–1705. doi:10.1175/JCLI3797.1.

Behera, S.K., Ratnam, J.V., Masumoto, Y., Yamagata, T., 2012. Origin of extreme summers in Europe —the Indo-Pacific connection. Clim. Dyn. doi:10.1007/s00382-012-1524-8.

Behera, S.K., Yamagata, T., 2003. Influence of the Indian Ocean dipole on the Southern Oscillation. J. Meteor. Soc. Jpn 81 (1), 169–177. doi:10.2151/jmsj.81.169.

Behera, S.K., Ratnam, J.V., 2018. Quasi-asymmetric response of the Indian summer monsoon rainfall to opposite phases of the IOD. Sci. Rep. doi:10.1038/s41598-017-18396-6.

Bellenger, H., Guilyardi, E., Leloup, J., Lengaigne, M., Vialard, J., 2014. ENSO representation in climate models: from CMIP3 to CMIP5. Clim. Dyn. 42, 1999–2018. doi:10.1007/s00382-013-1783-z.

Black, E., Slingo, J., Sperber, K.R., 2003. An observational study of the relationship between excessively strong short rains in coastal East Africa and Indian Ocean SST. Mon. Wea. Rev. 31, 74–94. doi:10.1175/1520-0493(2003)131<0074:AOSOTR>2.0.CO;2.

Boschat, G., Terray, P., Masson, S., 2011. Interannual relationships between Indian summer monsoon and Indo-Pacific coupled modes of variability during recent decades. Clim. Dyn. 37, 1019–1043. doi:10.1007/s00382-010-0887-y.

Boschat, G., Terray, P., Masson, S., 2012. Robustness of SST teleconnections and precursory patterns associated with the Indian summer monsoon. Clim. Dyn. 38, 2143–2165. doi:10.1007/s00382-011-1100-7.

Cai, W., Wu, L., Lengaigne, M., Li, T., McGregor, S., Kug, J.-S., et al., 2019. Pantropical climate interactions. Science 363, 6430. doi:10.1126/science.aav4236.

Cai, W., Santoso, A., Wang, G., Weller, E., Wu, L., Ashok, K., et al., 2014. Increased frequency of extreme Indian Ocean Dipole events due to greenhouse warming. Nature 510 (7504), 254–258. doi:10.1038/nature13327.

Cai, W., Sullivan, A., Cowan, T., 2011. Interactions of ENSO, the IOD, and the SAM in CMIP3 models. J. Clim. 24 (6), 1688–1704. doi:10.1175/2010JCLI3744.1.

Cai, W., Sullivan, A., Cowan, T., 2009. Climate change contributes to more frequent consecutive positive Indian Ocean dipole events. Geophys. Res. Lett. 36, L23704. doi:10.1029/2009GL040163.

Chan, S., Behera, S., Yamagata, T., 2008. The Indian Ocean Dipole teleconnection to South America. Geophys. Res. Lett. 35, L14S12. doi:10.1029/2008GL034204.

Chen, W., Wang, L., Feng, J., Wen, Z., Ma, T., Yang, X., et al., 2019. Recent progress in studies of the variabilities and mechanisms of the East Asian monsoon in a changing climate. Adv. Atm. Sci. 36, 887–901. doi:10.1007/s00376-019-8230-y.

Cherchi, A., Navarra, A., 2013. Influence of ENSO and of the Indian Ocean Dipole on the Indian summer monsoon variability. Clim. Dyn. 41, 81–103. doi:10.1007/s00382-012-1602-y.

Cherchi, A., Gualdi, S., Behera, S., Luo, J.J., Masson, S., Yamagata, T., et al., 2007. The influence of tropical Indian Ocean SST on the Indian summer monsoon. J. Clim. 20, 3083–3105. doi:10.1175/JCLI4161.1.

Chowdary, J.S., Xie, S.P., Luo, J.J., Hafner, J., Behera, S., Masumoto, Y., et al., 2011. Predictability of Northwest Pacific climate during summer and the role of the tropical Indian Ocean. Clim. Dyn. 36, 607–621. doi:10.1007/s00382-009-0686-5.

Chowdary, J.S., Parekh, A., Kakatkar, R., Gnanaseelan, C., Sriniva, G., Singh, P., et al., 2016. Tropical Indian Ocean response to the decay phase of El Niño in a coupled model and associated changes in south and east-Asian summer monsoon circulation and rainfall. Clim. Dyn. 47, 831–844. doi:10.1007/s00382-015-2874-9.

Cretat, J., Terray, P., Masson, S., Sooraj, K.P., Roxy, M.K., 2017. Indian Ocean and Indian Summer Monsoon: relationships without ENSO in ocean-atmosphere coupled simulations. Clim. Dyn. 49, 1429–1448. doi:10.1007/s00382-016-3387-x.

Cretat, J., Terray, P., Masson, S., Sooraj, K.P., 2018. Intrinsic precursors and timescale of the tropical Indian Ocean Dipole: Insights from partially decoupled experiment. Clim. Dyn. 51, 1311–1352. doi:10.1007/s00382-017-3956-7.

Delman, A.S., Sprintall, J., McClean, J.L., Talley, L.D., 2016. Anomalous Java cooling at the initiation of positive Indian Ocean Dipole events. J. Geophys. Res. Oceans. doi:10.1002/2016JC011635.

Dey, R., Lewis, S.C., Abram, N.J., 2019. Investigating observed northwest Australian rainfall trends in coupled model intercomparison project phase 5 detection and attribution experiments. Int. J. Climatol 39, 112–127. doi:10.1002/joc.5788.

Doi, T., Behera, S.W., Yamagata, T., 2020. Predictability of the super IOD event in 2019 and its link with El Niño Modoki. Geophys. Res. Lett. 47, e2019GL086713. doi:10.1029/2019 GL086713.

Dommenget, D., Latif, M., 2002. A cautionary note on the interpretation of EOFs. J. Clim. 15, 216–225. doi:10.1175/1520-0442(2002)015<0216:ACNOTI>2.0.CO;2.

Dommenget, D., 2011. An objective analysis of the observed spatial structure of the tropical Indian Ocean SST variability. Clim. Dyn. 36, 2129–2145. doi:10.1007/s00382-010-0787-1.

Drbohlav, H.K.L., Gualdi, S., Navarra, A., 2007. A diagnostic study of the Indian Ocean dipole mode in El Niño and non-El Niño years. J. Clim. 20, 2961–2977. doi:10.1175/JCLI4153.1.

Du, Y., Cai, W., Wu, Y., 2013. A new type of the Indian Ocean Dipole since the mid-1970s. J. Clim. 26 (3), 959–972. doi:10.1175/JCLI-D-12-00047.1.

Du, Y., Xie, S.P., Huang, G., Hu, K., 2009. Role of air-sea interaction in the long persistence of El Niño-induced North Indian Ocean warming. J. Clim. 22 (8), 2023–2038. doi:10.1175/2008 JCLI2590.1.

Endris, H.S., Lennard, C., Hewitson, B., Dosio, A., Nikulin, G., Artan, G.A., 2019. Future changes in rainfall associated with ENSO, IOD and changes in the mean state over eastern Africa. Clim. Dyn. 52, 2029–2053. doi:10.1007/s00382-018-4239-7.

Fischer, A.S., Terray, P., Delecluse, P., Gualdi, S., Guilyardi, E., 2005. Two independent triggers for the Indian Ocean Dipole/zonal mode in a coupled GCM. J. Clim. 18, 3428–3449. doi:10.1175/JCLI3478.1.

Gadgil, S., Francis, P.A., Vinayachandran, P.N., 2019. Summer monsoon of 2019: understanding the performance so far and speculating about the rest of the season. Curr. Sci. 117 (5), 783–793.

Gadgil, S., Vinayachandran, P.N., Francis, P.A., Gadgil, S., 2004. Extremes of the Indian summer monsoon rainfall, ENSO and equatorial Indian Ocean oscillation. Geophys. Res. Lett. 31, L12213. doi:10.1029/2004GL019733.

Gadgil, S., Rajeevan, M., Nanjundiah, R., 2005. Monsoon prediction—why yet another failure? Curr. Sci. 84, 1713–1719.

Gadgil, S., Rajeevan, M., Francis, P.A., 2007. Monsoon variability: links to major oscillations over the equatorial Pacific and Indian oceans. Curr. Sci. 93, 182–194.

Gershunov, A., Schneider, N., Barnett, T., 2001. Low-frequency modulation of the ENSO-Indian monsoon rainfall relationship: signal or noise? J. Clim. 14, 2486–2492. doi:10.1175/1520-0442(2001)014<2486:LFMOTE>2.0.CO;2.

Goswami, B.N., 1998. Interannual variations of Indian summer monsoon in a GCM: external conditions versus internal feedbacks. J. Clim. 11, 501–521. doi:10.1175/1520-0442(1998)011<0501:IVOISM>2.0.CO;2.

Gualdi, S., Guilyardi, E., Navarra, A., Masina, S., Delecluse, P., 2003. The interannual variability in the tropical Indian Ocean as simulated by a CGCM. Clim. Dyn. 20, 567–582. doi:10.1007/s00382-002-0295-z.

Guan, Z., Ashok, K., Yamagata, T., 2003. Summer-time response of the tropical atmosphere to the Indian Ocean dipole sea surface temperature anomalies. J. Meteorol. Soc. Jpn 81, 531–561. doi:10.2151/jmsj.81.533.

Guan, Z., Yamagata, T., 2003. The unusual summer of 1994 in East Asia: IOD teleconnections. Geophys. Res. Lett. 30. doi:10.1029/2002GL016831.

Gusain, A., Ghosh, S., Karmakar, S., 2020. Added value of CMIP6 over CMIP5 models in simulating Indian summer monsoon rainfall. Atm. Res 232, 104680. doi:10.1016/j.atmosres.2019.104680.

Ham, Y.-G., Choi, J.-Y., Kug, J.-S., 2017. The weakening of the ENSO-Indian Ocean Dipole (IOD) coupling strength in recent decades. Clim. Dyn. 49 (1), 249–261. doi:10.1007/s00382-016-3339-5.

Han, W., Shinoda, T., Fu, L.L., McCreary, J.P., 2006. Impact of atmospheric intraseasonal oscillations on the Indian Ocean dipole during the 1990s. J. Phys. Oceanogr 111, 679–690. doi:10.1175/JPO2892.1.

Hastenrath, S., Nicklis, A., Greischar, L., 1993. Atmospheric-hydrospheric mechanisms of climate anomalies in the western equatorial Indian Ocean. J. Geophys. Res 98 (C11), 20219–20235. doi:10.1029/93JC02330.

Hastenrath, S., 2002. Dipoles, temperature gradients and tropical climate anomalies. Bull. Am. Meteor. Soc 83, 735–740. doi:10.1175/1520-0477(2002)083<0735:WLACNM>2.3.CO;2.

Hermes, J.C., Masumoto, Y., Beal, L.M., Roxy, M.K., Vialard, J., Andres, M., et al., 2019. A sustained ocean observing system in the Indian Ocean for climate related scientific knowledge and societal needs. Front. Mar. Sci. doi:10.3389/fmars.2019.00355.

Hersbach, H., Dee, D.J.E.N., 2016. ERA5 reanalysis is in production. ECMWF Newsletter 147 (7), 5–6.

Hong, C.C., Li, T., Ho, L., Chen, Y.C., 2010. Asymmetry of the Indian Ocean basin-wide SST anomalies. Roles of ENSO and IOD. J. Clim 23, 3563–3576. doi:10.1175/2010JCLI3320.1.

Hong, C.C., Lu, M.M., Kanamitsu, M., 2008. Temporal and spatial characteristics of positive and negative Indian Ocean dipole with and without ENSO. J. Geophys. Res. Atmos 113, D08107. doi:10.1029/2007JD009151.

Hossain, I., Rasel, H.M., Imteaz, M.A., Mekanik, F., 2020. Long-term seasonal rainfall forecasting using linear and non-linear modelling approaches: a case study for western Australia. Meteor. Atm. Phys 132, 131–141. doi:10.1007/s00703-019-00679-4.

Huang, B., Kinter, J.L., 2002. Interannual variability in the tropical Indian Ocean. J. Clim. 107 (C11). doi:10.1029/2001JC001278.

Huang, B., Thorne, P.W., Banzon, V.F., Boyer, T., Chepurin, G., Lawrimore, J.H., et al., 2017. Extended reconstructed sea surface temperature, version 5 (ERSSTv5): upgrades, validations, and intercomparisons. J. Clim. 30 (20), 8179–8205.

Huang, P., Zheng, X.T., Ying, J., 2019. Disentangling the changes in the Indian Ocean Dipole-related SST and rainfall variability under global warming in CMIP5 models. J. Clim. 32, 3803–3818. doi:10.1175/JCLI-D-18-0847.1.

Hui, C., Zheng, X.T., 2018. Uncertainty in Indian Ocean dipole response to global warming: the role of internal variability. Clim. Dyn. doi:10.1007/s00382-018-4098-2.

Ihara, C., Kushnir, Y., Cane, M.A., De La Peña, V.H., 2007. Indian summer monsoon rainfall and its link with ENSO and Indian Ocean climate indices. Int. J. Clim. 27, 179–187. doi:10.1002/joc.1394.

Izumo, T., de Boyer Montegut, C., Luo, J.-J., Behera, S.K., Masson, S., Yamagata, T., 2008. The role of the western Arabian Sea upwelling in Indian monsoon rainfall variability. J. Clim. 21, 5603–5623. doi:10.1175/2008JCLI2158.1.

Izumo, T., Vialard, J., Lengaigne, M., de Boyer Montégut, C., Behera, S.K, Luo, J.J., et al., 2010. Influence of the state of the Indian Ocean Dipole on the following year's El Niño. Nat. Geosci 3, 168–172. doi:10.1038/ngeo760.

Jain, S., Salunke, P., Mishra, S.K., Sahany, S., Choudhary, N., 2019. Advantage of NEX-GDDP over CMIP5 and CORDEX Data: Indian summer monsoon. Atm. Res 228, 152–160. doi:10.1016/j.atmosres.2019.05.026.

Jayasankar, C.B., Surendran, S., Rajendran, K., 2015. Robust signals of future projections of Indian summer monsoon rainfall by IPCC AR5 climate models: role of seasonal cycle and interannual variability. Geophys. Res. Lett. 42 (9), 3513–3520. doi:10.1002/2015GL063659.

Jourdain, N.C., Lengaigne, M., Vialard, J., Izumo, T., Sen Gupta, A., 2016. Further insights on the influence of the Indian Ocean dipole on the following year's ENSO from observations and CMIP5 models. J. Clim. 29, 637–658. doi:10.1175/JCLI-D-15-0481.1.

Kajtar, J.B., Santoso, A., England, M.H., Cai, W., 2016. Tropical climate variability: interactions across the Pacific, Indian, and Atlantic Oceans. Clim. Dyn. doi:10.1007/s00382-016-3199-z.

Karamperidou, C., Cane, M.A., Lall, U., Wittemberg, A.T., 2014. Intrinsic modulation of ENSO predictability viewed through a local Lyapunov lens. Clim. Dyn. 42, 253–270. doi:10.1007/s00382-013-1759-z.

Kayanne, H., Iijima, H., Nakamura, N., McClanahan, T.R., Behera, S., Yamagata, T., 2006. The Indian Ocean Dipole Index recorded in Kenyan coral annual density bands. Geophys. Res. Lett. 33, L19709. doi:10.1029/2006GL027168.

Kinter III, J.L., Miyakoda, K., Yang, S., 2002. Recent changes in the connection from the Asian monsoon to ENSO. J. Clim. 15, 1203–1215. doi:10.1175/1520-0442(2002)015<1203:RCITCF>2.0.CO;2.

Kirtman, B., Shukla, J., 2000. Influence of the Indian summer monsoon on ENSO. Quart. J. Roy. Meteor. Soc 126, 213–239. doi:10.1002/qj.49712656211.

Kociuba, G., Power, S.B., 2015. Inability of CMIP5 models to simulate recent strengthening of the Walker circulation: implications for projections. J. Clim. 28, 20–35. doi:10.1175/JCLI-D-13-00752.1.

Krishnan, R., Mujumdar, M., Vaidya, V., Ramesh, K.V., Satyan, V., 2003. The abnormal Indian summer monsoon of 2000. J. Clim. 16, 1177–1194.

Krishan, R., Ayantika, D.C., Kumar, V., Pokhrel, S., 2011. The long-lived monsoon depressions of 2006 and their linkage with the Indian Ocean Dipole. Int. J. Climatol 31, 1334–1352. doi:10.1002/joc.2156.

Krishnaswamy, J., Vaidyanathan, S., Rajagopalan, B., Bonell, M., Sankaran, M., Bhalla, R.S., Badiger, S., 2015. Non-stationary and non-linear influence of ENSO and Indian Ocean Dipole on the variability of Indian monsoon rainfall and extreme rain events. Clim. Dyn. 45, 175–184. doi:10.1007/s00382-014-2288-0.

Kug, J.-S., Ham, Y.-G., 2012. Indian Ocean feedback to the ENSO transition in a multi-model ensemble. J. Clim. 25, 6942–6957. doi:10.1175/JCLI-D-12-00078.1.

Kug, J.-S., Kang, I.-S., 2006. Interactive feedback between the Indian Ocean and ENSO. J. Clim. 19, 1784–1801. doi:10.1175/JCLI3660.1.

Kumar, K.K., Rajagopalan, B., Cane, M.A., 1999. On the weakening relationship between the Indian monsoon and ENSO. Science 284, 2156–2159. doi:10.1126/science.284.5423.2156.

Lau, N.C., Nath, M.J., 2003. Atmosphere–ocean variations in the Indo-Pacific sector during ENSO episode. J. Clim. 16, 3–20. doi:10.1175/1520-0442(2003)016<0003:AOVITI>2.0.CO;2.

Lau, N.-C., Nath, M.J., 2004. Coupled GCM simulation of atmosphere-ocean variability associated with zonally asymmetric SST changes in the tropical Indian Ocean. J. Clim. 17, 245–265. doi:10.1175/1520-0442(2004)017<0245:CGSOAV>2.0.CO;2.

Lau, N.C., Nath, M.J., 2012. A model study of the air–sea interaction associated with the climatological aspects and interannual variability of the South Asian summer monsoon development. J. Clim. 25, 839–857. doi:10.1175/JCLI-D-11-00035.1.

Lau, N.-C., Wang, B., 2006. Interactions between the Asian monsoon and the El Niño Southern Oscillation. In: Wang, B. (Ed.), The Asian Monsoon. Springer-Praxis, Chichester, pp. 479–511.

Li, G., Xie, S.P., Du, Y., 2015. Monsoon-induced biases of climate models over the tropical Indian Ocean. J. Clim. 28, 3058–3072. doi:10.1175/JCLI-D-14-00740.1.

Li, Z., Lin, X., Cai, W., 2017. Realism of modelled Indian summer monsoon correlation with the tropical Indo-Pacific affects projected monsoon changes. Sci. Rep 7 (1), 1–7. doi:10.1038/s41598-017-05225-z.

Li, T., Wang, B., Chang, C.P., Zhang, Y.S., 2003. A theory for the Indian Ocean dipole-zonal mode. J. Atmos. Sci 60, 2119–2135. doi:10.1175/1520-0469(2003)060<2119:ATFTIO>2 .0.CO;2.

Li, T., Liu, P., Fu, X., Wang, B., Meehl, G.A., 2006. Spatiotemporal structures and mechanisms of the Tropospheric Biennial Oscillation in the Indo-Pacific warm ocean regions. J. Clim. 19, 3070–3087. doi:10.1175/JCLI3736.1.

Loschnigg, J., Meehl, G.A., Webster, P.J., Arblaster, J.M., Compo, G.P., 2003. The Asian monsoon, the tropospheric biennial oscillation and the Indian Ocean dipole in the NCAR CSM. J. Clim. 16, 2138–2158. doi:10.1175/1520-0442(2003)016<1617:TAMTTB>2.0.CO;2.

Lu, B., Ren, H.-L., 2020. What caused the extreme Indian Ocean Dipole event in 2019? Geophys. Res. Lett. 47. doi:10.1029/2020GL087768 e2020GL087768.

Luo, J.J., Zhang, R., Behera, S.K., Masumoto, Y., Jin, F.F., Lukas, R., et al., 2010. Interactions between El Niño and extreme Indian Ocean dipole. J. Clim. 23, 726–742. doi:10.1175/200 9JCLI3104.1.

Manatsa, D., Behera, S.K., 2013. On the epochal strengthening in the relationship between rainfall of East Africa and IOD. J. Clim. 26, 5655–5673. doi:10.1175/JCLI-D-12-00568.1.

McPhaden, M.J., Nagura, N., 2014. Indian Ocean Dipole interpreted in terms of recharge oscillator theory. Clim. Dyn. 42, 1569–1586. doi:10.1007/s00382-013-1765-1.

Meehl, G.A., Arblaster, J.M., 2002. Indian monsoon GCM sensitivity experiments testing tropospheric biennial oscillation transition conditions. J. Clim. 15, 923–944. doi:10.1175/1520-0442(2002)015<0923:IMGSET>2.0.CO;2.

Meehl, G.A., Arblaster, J.M., Loschnigg, J., 2003. Coupled ocean-atmosphere dynamical processes in the tropical Indian and Pacific oceans and the TBO. J. Clim. 16, 2138–2158. doi:10.1175/2767.1.

Murtugudde, R., Beauchamp, J., Bussalacchi, A., Nerem, S., 1995. Secular sea-level change in the Indian Ocean: comparison of model result with TOPEX/Poseidon data. Trans. Am. Geophys. Union (EOS) 76, G32A–63211.

Murtugudde, R., Goswami, B.N., Busalacchi, A.J., 1998. Air sea interactions in the southern Indian Ocean and its relation to interannual variability of the monsoons over India, Proceedings of the International Conference on Monsoon and Hydrological Cycle, 184–188.

Murtugudde, R., Busalacchi, A.J., 1999. Internannual variability of the dynamics and thermodynamics of the tropical Indian Ocean. J. Clim. 12, 2300–2326. doi:10.1175/1520-0442(1999)012<2300:IVOTDA>2.0.CO;2.

Murtugudde, R., Busalacchi, A.J., McCreary, J.P., 2003. Comment on "Dipoles, temperature gradients and tropical climate anomalies". Bull. Am. Meteor. Soc 84, 1422–1423. doi:10.1175/BAMS-84-10-1422.

Nakamura, N., Kayanne, H., Iijima, H., McClanahan, T.R., Behera, S.K., Yamagata, T., 2009. Mode shift in the Indian Ocean climate under global warming stress. Geophys. Res. Lett. 36 (23), L23708. doi:10.1029/2009GL040590.

Ng, B., Cai, W., Cowan, T., Bi, D., 2018. Influence of internal climate variability on Indian Ocean dipole properties. Sci. Rep 8, 13500. doi:10.1038/s41598-018-31842-3.

O'Neill, B.C., Tebaldi, C., van Vuuren, D.P., Eyring, V., Friedlingstein, P., Hurtt, G., et al., 2016. The Scenario Model Intercomparison Project (ScenarioMIP) for CMIP6. Geosci. Model Dev 9, 3461–3482. doi:10.5194/gmd-9-3461-2016.

Pai, D.S., Sridhar, L., Rajeevan, M., Sreejith, O.P., Satbhai, N.S., Mukhopadhyay, B., 2014. Development of a new high spatial resolution (0.25 × 0.25) long period (1901–2010) daily gridded rainfall data set over India and its comparison with existing data sets over the region. Mausam 65 (1), 1–18.

Pepler, A., Timbal, B., Rakich, C., Coutts-Smith, A., 2014. Indian Ocean dipole overrides ENSO's influence on cool season rainfall across the eastern seaboard of Australia. J. Clim. 27, 3816–3826. doi:10.1175/JCLI-D-13-00554.1.

Rahaman, H., Srinivasu, U., Panickal, S., Durgadoo, J.V., Griffies, S.M., Ravichandran, M., et al., 2020. An assessment of the Indian Ocean mean state and seasonal cycle in a suite of interannual CORE-II simulations. Ocean Modell. doi:10.1016/j.ocemod.2019.101503.

Rao, K.G., Goswami, B.N., 1988. Interannual variations of sea surface temperature over the Arabian Sea and the Indian monsoon: a new perspective. Mon. Weath. Rev 116, 558–568.

Rao, S.A., Yamagata, T., 2004. Abrupt termination of Indian Ocean dipole events in response to intraseasonal disturbances. Geophys. Res. Lett. 31, L19306. doi:10.1029/2004GL020842.

Rao, S.A., Behera, S.K., Masumoto, Y., Yamagata, T., 2002. Interannual subsurface variability in the Tropical Indian Ocean with a special emphasis on the Indian Ocean Dipole. Deep-Sea Res 49, 1549–1572.

Rao, S.A., Luo, J.J., Behera, S.K., Yamagata, T., 2009. Generation and termination of Indian Ocean dipole events in 2003, 2006 and 2007. Clim. Dyn. 33, 751–767. doi:10.1007/s00382-008-0498-z.

Rasmusson, E.M., Carpenter, T.H., 1983. The relationship between eastern equatorial Pacific sea surface temperature and rainfall over India and Sri Lanka. Mon. Wea. Rev 111, 517–528. doi:10.1175/1520-0493(1983)111<0517:TRBEEP>2.0.CO;2.

Rayner, N.A., Parker, D.E., Horton, E.B., Folland, C.K., Alexander, L.V., Rowell, D.P., et al., 2003. Global analyses of sea surface temperature, sea ice, and night marine air temperature since the late nineteenth century. J. Geophys. Res 108 (D14), 4407. doi:10.1029/2002JD002670.

Reverdin, G., Cadet, D., Gutzler, D., 1986. Interannual displacements of convection and surface circulation over the equatorial Indian Ocean. Quart. J. Roy. Meteorol. Soc 112, 43–46. doi:10.1002/qj.49711247104.

Sahu, N., Behera, S.K., Yamashiki, Y., Takara, K., Yamagata, T., 2012. IOD and ENSO impacts on the extreme stream-flows of Citarum river in Indonesia. Clim. Dyn. doi:10.1007/s00382-011-1158-2.

Saji, N.H., 2018. The Indian Ocean Dipole. Oxf. Res. Encycl. Clim. Sci. doi:10.1093/acrefore/9780190228620.013.619.

Saji, N.H., Goswami, B.N., Vinaychandran, P.N., Yamagata, T., 1999. A dipole mode in the tropical Indian Ocean. Nature 401, 360–363. doi:10.1038/43854.

Santoso, A., England, M.H., Cai, W., 2012. Impact of Indo-Pacific feedback interactions on ENSO dynamics diagnosed using ensemble climate simulations. J. Clim. 25, 7743–7763. doi:10.1175/JCLI-D-11-00287.1.

Schott, F.A., Xie, S.-P., McCreary, J.P., 2009. Indian Ocean circulation and climate variability. Rev. Geophys 47, RG1002. doi:10.1029/2007rg000245.

Shukla, R.P., Huang, B., 2016. Interannual variability of the Indian summer monsoon associated with the air–sea feedback in the northern Indian Ocean. Clim. Dyn. 46, 1977–1990. doi:10.1007/s00382-015-2687-x.

Sikka, D.R., 1980. Some aspects of the large-scale fluctuations of summer monsoon rainfall over India in relations to fluctuations in the planetary and regional scale circulation parameters. J. Earth Syst. Sci 89, 179–195. doi:10.1007/BF02913749.

Slingo, J.M., Annamalai, H., 2000. 1997: The El Niño of the century and the response of the Indian summer monsoon. Mon. Wea. Rev 128, 1778–1797. doi:10.1175/1520-0493(2000)128<1778:TENOOT>2.0.CO;2.

Sreejith, O.P., Panickal, S., Rajeevan, M., 2015. An Indian precursor for Indian summer monsoon rainfall variability. Geophys. Res. Lett. 42 (21), 9345–9354. doi:10.1002/2015GL065950.

Srinivasan, J., Nanjundiah, R.S., 2002. The evolution of Indian summer monsoon in 1997 and 1983. Meteor. Atm. Phys 79 (3-4), 243–257. doi:10.1007/s007030300006.

Srivastava, A., Pradhan, M., Goswami, B.N., Rao, S.A., 2019. Regime shift of Indian summer monsoon rainfall to a persistent arid state: external forcing versus internal variability. Met. Atm. Phys 131, 211–224. doi:10.1007/s00703-017-0565-2.

Stuecker, M.F., Timmermanss, A., Jin, F.F., Chikamoto, Y., Zhang, W., Wittenberg, A.T., et al., 2017. Revisiting ENSO/Indian Ocean dipole phase relationships. Geophys. Res. Lett. 44, 2481–2492. doi:10.1002/2016GL072308.

Taschetto, A.S., Ambrizzi, T., 2012. Can Indian Ocean SST anomalies influence South American rainfall? Clim. Dyn. 38, 1615–1628. doi:10.1007/s00382-011-1165-3.

Terray, P., Dominiak, S., Delecluse, P., 2005. Role of the southern Indian Ocean in the transitions of the monsoon-ENSO system during recent decades. Clim. Dyn. 24, 169–195. doi:10.1007/s00382-004-0480-3.

Terray, P., Chauvin, F., Douville, H., 2007. Impact of southeast Indian Ocean sea surface temperature anomalies on monsoon-ENSO dipole variability in a coupled ocean-atmosphere model. Clim. Dyn. 28, 553–580. doi:10.1007/s00382-006-0192-y.

Terray, P., Kamala, K., Masson, S., Madec, G., Sahai, A.K., Luo, J.J., et al., 2012. The role of the intra-daily SST variability in the Indian monsoon variability and monsoon-ENSO–IOD relationships in a global coupled model. Clim. Dyn. 39, 729–754. doi:10.1007/s00382-011-1240-9.

Terray, P., Masson, S., Prodhomme, C., Roxy, M.K., Sooraj, K.P., 2016. Impacts of Indian and Atlantic oceans on ENSO in a comprehensive modeling framework. Clim. Dyn. 46, 2507–2533. doi:10.1007/s00382-015-2715-x.

Ummenhofer, C.C., Schwarzkopf, F.U., Meyers, G.A., Behrens, E., Biastoch, A., Boning, C.W., 2013. Pacific Ocean contribution to the asymmetry in eastern Indian Ocean variability. J. Clim. 26, 1152–1171. doi:10.1175/JCLI-D-11-00673.1.

Ummenhofer, C.C., Sen Gupta, A., Briggs, P.R., England, M.H., McIntosh, P.C., Meyers, G.A., et al., 2011. Indian and Pacific ocean influences on southeast Australian drought and soil moisture. J. Clim. 24, 1313–1336. doi:10.1175/2010JCLI3475.1.

Vecchi, G.A., Soden, B.J., Wittenberg, A.T., Held, I.M., Leetma, A., Harrison, M.J., 2006. Weakening of tropical Pacific atmospheric circulation due to anthropogenic forcing. Nature 441, 73–76. doi:10.1038/nature04744.

Walker, G.T., 1924. Correlation in seasonal variations of weather—a further study of world weather. Mon. Weather Rev. doi:10.1175/1520-0493(1925)53<252:CISVOW>2.0.CO;2.

Wang, H., Murtugudde, R., Kumar, A., 2016. Evolution of Indian Ocean dipole and its forcing mechanisms in the absence of ENSO. Clim. Dyn. 47, 2481–2500. doi:10.1007/s00382-016-2977-y.

Wang, H., Kumar, A., Murtugudde, R., Narapusetty, B., Seip, K., 2019. Covariations between the Indian Ocean dipole and ENSO: A modeling study. Clim. Dyn. 53, 5743–5761. doi:10.1007/s00382-019-04895-x.

Webster, P.J., Hoyos, C.D., 2010. Beyond the spring barrier? Nat. Geosci 3, 152–153. doi:10.1038/ngeo800.

Webster, P.J., Magana, V., Palmer, T.N., Shukla, J., Tomas, R.A., Yanai, M., et al., 1998. Monsoons: processes, predictability and the prospects for prediction. J. Geophys. Res 103, 14451–14510. doi:10.1029/97JC02719.

Webster, P.J., Moore, A.M., Loschnigg, J.P., Leben, R.R., 1999. Coupled ocean-atmosphere dynamics in the Indian Ocean during 1997-1998. Nature 401, 356–360. doi:10.1038/43848.

Weller, E., Cai, W., 2013. Asymmetry in the IOD and ENSO teleconnection in a CMIP5 model ensemble and its relevance to regional rainfall. J. Clim. 26 (14), 5139–5149. doi:10.1175/JCLI-D-12-00789.1.

Wieners, C.E., Dijkstra, H.A., de Ruijter, W.P.M, 2017a. The influence of the Indian Ocean on ENSO stability and flavor. J. Clim. 30, 2601–2620. doi:10.1175/JCLI-D-16-0516.1.

Wieners, C.E., Dijkstra, H.A., de Ruijter, W.P.M, 2017b. The influence of atmospheric convection on the interaction between the Indian Ocean and ENSO. J. Clim. 30, 10155–10178. doi:10.1175/JCLI-D-17-0081.1.

Wittenberg, A.T., 2009. Are historical records sufficient to constrain ENSO simulations? Geophys. Res. Lett. 36, L12702. doi:10.1029/2009GL038710.

Wittenberg, A.T., Rosati, A., Delworth, T.L., Vecchi, G.A., Zeng, F., 2014. ENSO modulations: Is it decadally predictable? J. Clim. 27, 2667–2681. doi:10.1175/JCLI-D-13-00577.1.

Wu, R., Kirtman, B., 2004. Impact of the Indian Ocean on the Indian summer monsoon-ENSO relationship. J. Clim. 17, 3037–3054. doi:10.1175/1520-0442(2004)017<3037:IOTIOO>2.0.CO;2.

Xie, S.P., Annamalai, H., Schott, F.A., McCreary, J.P., 2002. Structure and mechanisms of south Indian Ocean climate variability. J. Clim. 15 (8), 864–878. doi:10.1175/1520-0442(2002)015<0864:SAMOSI>2.0.CO;2.

Yamagata, T., Behera, S.K., Rao, S.A., Saji, N.H., 2003. Comments on "Dipoles, temperature gradients and tropical climate anomalies". Bull. Amer. Meteor. Soc 84, 1418–1422.

Yamagata, T., Behera, S.K., Luo, J.J., Masson, S., Jury, M.R., Rao, S.A., 2004Coupled Ocean-Atmospheric Variability in the Tropical Indian Ocean147. Earth Climate: The Ocean-Atmosphere Interaction, Geophys. Monogr., pp. 189–212, https://agupubs.onlinelibrary.wiley.com/doi/book/10.1029/GM147 Print ISBN: 9780875904122 Online ISBN: 9781118665947 doi: 10.1029/GM147.

Yang, J., Liu, Q., Xie, X.P., Liu, Z., Wu, L., 2007. Impact of the Indian Ocean SST basin mode on the Asian summer monsoon. Geophys. Res. Lett. 34, L02708. doi:10.1029/2006GL028571.

Yuan, Y., Li, C.Y., 2008. Decadal variability of the IOD-ENSO relationship. Chin. Sci. Bull 53, 1745–1752. doi:10.1007/s11434-008-0196-6.

Zhao, Y., Nigam, S., 2015. The Indian Ocean dipole: a monopole in SST. J. Clim. 28, 3–19. doi:10.1175/JCLI-D-14-00047.1.

Zhao, S., Jin, F.-F., Stuecker, M.F., 2019. Improved predictability of the Indian Ocean Dipole using seasonally modulated ENSO forcing forecasts. Geophys. Res. Lett. 46, 9980–9990. doi:10.1029/2091GL084196.

Zheng, X.T., Xie, S.P., Vecchi, G.A., Liu, Q., Hafner, J., 2010. Indian Ocean Dipole response to global warming: analysis of ocean-atmosphere feedbacks in a coupled model. J. Clim. 23, 1240–1253. doi:10.1175/2009JCLI3326.1.

Zheng, X.T., Xie, S.P., Du, Y., Liu, L., Huang, G., Liu, Q., 2013. Indian Ocean Dipole response to global warming in the CMIP5 multi-model ensemble. J. Clim. 26, 6067–6080. doi:10.1175/JCLI-D-12-00638.1.

Zhou, Q., Duan, W., Mu, M., Feng, R., 2015. Influence of positive and negative Indian Ocean Dipoles on ENSO via the Indonesian throughflow: results from sensitivity experiments. Adv. Atm. Sci 32, 783–793. doi:10.1007/s00376-014-4141-0.

Zhou, T., Turner, A.G., Kinter, J.L., Wang, B., Qian, Y., Chen, X., et al., 2016. GMMIP (v1.0) contribution to CMIP6: global monsoons model inter-comparison project. Geosci. Model Dev 9, 3589–3604. doi:10.5194/gmd-9-3589-2016.

Zubair, L., Rao, S.A., Yamagata, T., 2003. Modulation of Sri Lankan Maha rainfall by the Indian Ocean Dipole. Geophys. Res. Lett. 30 (2). doi:10.1029/2002GL015639.

Chapter 9

Influence of South Tropical Indian Ocean dynamics on the Indian summer monsoon

Yan Du[a,b,c], Zesheng Chen[a,b], Ying Zhang[a,b], Kaiming Hu[d], Xiaotong Zheng[e], Weidong Yu[f]

[a]State Key Laboratory of Tropical Oceanography, South China Sea Institute of Oceanology, Chinese Academy of Sciences, Guangzhou, China, [b]Southern Marine Science and Engineering Guangdong Laboratory, Guangzhou, China, [c]University of Chinese Academy of Sciences, Beijing, China, [d]Center for Monsoon System Research, Institute of Atmospheric Physics, Chinese Academy of Sciences, Beijing, China, [e]Physical Oceanography Laboratory and Key Laboratory of Ocean-Atmosphere Interaction and Climate in Universities of Shandong, Ocean University of China, Qingdao, China, [f]School of Atmosphere Sciences, Sun Yat-Sen University, Guangzhou, China

9.1   Introduction

El Niño-Southern Oscillation (ENSO) is the dominant mode of air–sea interaction in the equatorial Pacific with profound impacts on the tropical Indian Ocean (TIO). It has been well documented that El Niño, the positive phase of ENSO, is followed by TIO sea surface temperature (SST) warming with about one season lag (Klein et al., 1999; Lau and Nath, 2000, 2003; Alexander et al., 2002; Chowdary and Gnanaseelan, 2007).

Among three tropical oceans, the Indian Ocean is the unique one where the annual mean winds along the equator flow eastward (Fig. 9.1A). The mean equatorial thermocline is flat and deep due to the weak westerly winds (Fig. 9.1B), thus limiting the effect of thermocline displacement on SST variability (Schott et al., 2009). In contrary to the popular belief that the TIO cannot develop the interannual variability on its own, the SST was perceived as a passive response to ENSO having limited feedback on the atmosphere (e.g., Latif and Barnett, 1995). However, with the discovery of Indian Ocean dipole (IOD) mode (Saji et al., 1999; Webster et al., 1999; also see Chapter 8) and Indo-Western Pacific Ocean capacitor mode (IPOC, Xie et al., 2009, 2010, 2016; Yang et al., 2007; Kosaka et al., 2013), both meteorologists and oceanographers have realized that the TIO can be energetic and affect the Indo-Pacific climate (Ashok et al., 2001; Li and Mu, 2001; Xie et al., 2002; Clark et al., 2003; Saji and Yamagata, 2003; Manatsa et al., 2008; Du et al., 2011; Chen et al., 2019).

Indian Summer Monsoon Variability: El Niño-teleconnections and beyond.
DOI: https://doi.org/10.1016/B978-0-12-822402-1.00013-2

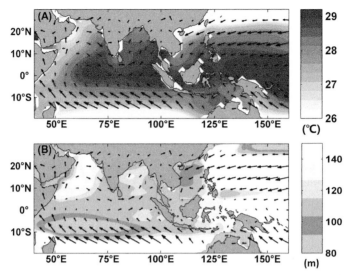

FIG. 9.1   The annual mean SST (°C, A) and thermocline (D20, m, B) over Indo-Pacific region, superimposed with annual mean surface winds (m/s). "*SST*, sea surface temperature."

Typically, ENSO is phase-locked to boreal winter, the eastern Pacific SST anomalies develop in boreal summer, peak in boreal winter, and decay rapidly in boreal spring in the following year (Fig. 9.2). Whereas, the TIO SST warming persists from boreal spring to summer (Fig. 9.2) and has a climatic impact on the Indo-Pacific region (Yang et al., 2007; Chen et al., 2016; Xie et al., 2016). The Indian Ocean warming was deemed to be basin-wide and caused by El Niño-induced surface heat flux anomalies. However, recent studies have revealed that the Indian Ocean warming is mechanically distinct among sub-basins, and a series of advances have been made in explaining the SST warming among the TIO sub-basins in post-ENSO years (Izumo et al., 2008; Du et al., 2009; Xie et al., 2010, 2016; Chen et al., 2019; Kakatkar et al. 2020).

The wind shear between the equatorial westerlies and southeasterly trade winds induces an open-ocean upwelling in the southwest TIO, uplifting the local thermocline (Fig. 9.1B). Remotely, the southwest TIO thermocline ridge is influenced by the Rossby waves that emanate from the eastern TIO (Masumoto and Meyers, 1998; Rao and Behera, 2005; Yokoi et al., 2008; Tozuka et al., 2010; Gnanaseelan and Vaid, 2010; Chakravorty et al., 2014; Sayantani and Gnanaseelan, 2015) and from Pacific (Vaid et al., 2007). Over the southwest TIO, thermocline ridge influences the subsurface Rossby waves to interact with the atmosphere through strong feedback of SST (Xie et al., 2002; Kakatkar et al. 2020).

In this chapter, we discuss how the south TIO ocean dynamics shape the SST warming over the TIO and describe its climatic impacts on the north Indian Ocean, especially for Indian summer monsoon.

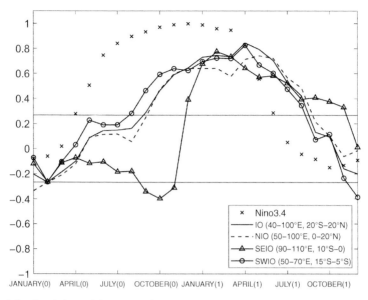

FIG. 9.2 Correlations with NDJ(0) Niño3.4 SST index with Indian Ocean SST anomalies for different regions, the values in two straight lines denote the 95% confidence level according to the students' *t* test. "*SST*, sea surface temperature."

9.2   Data and methods

This section introduces observational and reanalysis datasets. The Extended Reconstructed Sea Surface Temperature (ERSST) version 5 dataset is derived from the International Comprehensive Ocean-Atmosphere Dataset (ICOADS) based on the latest datasets and improved method. It is available at $2° \times 2°$ resolution in the global and from 1854 to the present (Huang et al., 2017). The sea level anomaly is obtained from Archiving, Validation, and Interpretation of Satellites Oceanographic (AVISO) data from the CNES data distribution center. AVISO is a merged multisatellite (ERA-1/2, T/P, GFO, Envisat, and Jason-1/2) altimetry product, which spans from 1993 to the present with a resolution of $0.25° \times 0.25°$. The surface wind and wind stress data are based on the NCEP/ NCAR Reanalysis 1 for the period 1993–2018 (Kalnay et al., 1996). The GPCP version 2.2 provides a monthly combined precipitation dataset with a resolution of $2.5° \times 2.5°$ from 1979 to the present (Adler et al., 2003; Huffman et al., 2009). The objectively analyzed subsurface temperature and salinity fields from the ENACT/ENSEMBLES version 4 (EN4) are used to calculate the mixed layer, barrier layer, and thermocline depths and detect the response of the ocean vertical stratification to ENSO impact (Good et al., 2013; Levitus et al., 2012). The mixed layer depth is defined as the depth where the potential density changes by $0.8°C$ relative to SST (Kara et al., 2000). The depth of $20°C$ (D20) is used to represent the thermocline depth (e.g., Saji et al., 2006). The layer between the bottom of the mixed layer and isothermal layer is referred to as the barrier layer (e.g., Godfrey and Lindstrom, 1989; Sprintall and Tomczak, 1992). The monthly temperature and salinity fields from the EN4 database have a regular $1°$ horizontal grid and 42 levels in the vertical (Good et al., 2013). All datasets are available from 1979 to 2018.

Our presentation relies on regression and correlation analyses with respect to the Niño3.4 SST index averaged during November–December(0)–January(1) [NDJ(0)]. The Niño3.4 index is calculated as SST anomalies averaged in $5°S$–$5°N$, $170°$–$120°W$ (Trenberth, 1997). Here, numerals "0" and "1" in parentheses denote the developing and decay years of El Niño, respectively.

9.3 Characteristics of TIO warming and its climatic influence

This section represents the characteristics of SST warming over TIO and discusses the mechanism for the TIO warming and associated climatic impacts.

9.3.1 The Southwest Indian Ocean

The southwest TIO SST warming occurs at El Niño mature phase and persists until El Niño decaying phase (Fig. 9.2). Surface heat flux change cannot fully explain the SST warming in the southwest TIO (Figs. 9.3 and 9.6; Klein et al., 1999), while later studies highlighted the importance of oceanic Rossby waves (Xie et al., 2002; Huang and Kinter, 2002; Murtugudde et al., 2000; Behera and Yamagata, 2001). Southwest TIO SST variability is mainly caused by thermocline fluctuation, which is forced by ENSO or IOD or both (Xie et al., 2002; Yu et al., 2005; Tozuka et al., 2010; Ma et al., 2014; Chakravorty et al. 2014).

During an IOD and/or ENSO event, anomalous anticyclonic wind anomaly pattern develops in the southeast Indian Ocean (Fig. 9.6A and E), such wind stress curl anomalies and the shear-induced by the equatorial easterlies lead to local Ekman pumping and force downwelling oceanic Rossby waves (Sayantani and Gnanaseelan, 2015) that propagate westward (Figs. 9.4 and 9.5). The thermocline feedback in the southwest TIO is strong (Huang and Kinter, 2002; Xie et al., 2002). The late spring SST anomalies over the thermocline ridge are mainly caused by off-equatorial oceanic Rossby waves that are generated from the south TIO (Figs. 9.4 and 9.5). These Rossby waves are mainly forced by wind stress curl anomalies and resultant Ekman pumping anomalies during the mature phase of El Niño (Fig. 9.5), rather than by the oceanic Kelvin waves exciting from the eastern boundary of TIO (Masumoto and Meyers 1998). The downwelling Rossby waves deepen the thermocline, resulting in significant warming in the subsurface ocean. Meanwhile, the mixed layer is deepened by the Ekman downwelling, and a thick barrier layer forms (Fig. 9.5A). The subsurface warming and barrier layer move westward followed by the westward-propagating downwelling Rossby waves and influence the mixed layer temperature in the thermocline dome in the southwest TIO (Figs. 9.3A and 9.4B; Chowdary et al., 2009). During MAM(1), the subsurface warming apparently reaches the

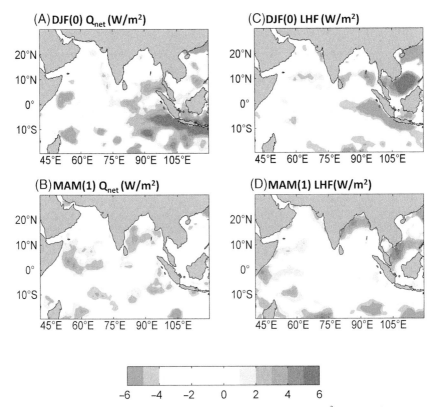

FIG. 9.3   Net heat flux (Qnet) and latent heat flux (LHF) anomalies (W/m^2), expressed as regression upon the NDJ(0) Niño −3.4 index during December–February(0, A and C) and March–May(1, B and D). "*SST*, sea surface temperature."

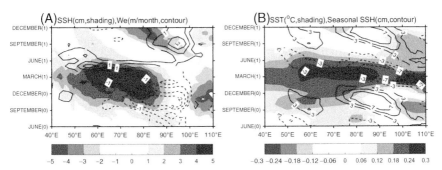

FIG. 9.4   Longitude-time sections of (A) sea surface height (SSH) and (B) SST anomalies over the southern TIO (averaged meridionally in 5°–15°S), expressed as regression upon the NDJ(0) Niño3.4 index. Superimposed are (A) Ekman pumping (contour intervals: 0.5 m/month), and (B) satellite SSH climatology (contour intervals: 2 cm), with the zero contour omitted and negative values dashed. "*SST*, sea surface temperature; *TIO*, tropical Indian Ocean."

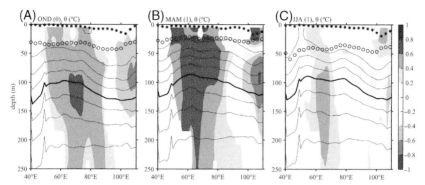

FIG. 9.5 (A) Regression of EN4 potential temperature anomalies (°C, shading averaged meridionally in 5°–15° S) upon the NDJ(0) Niño3.4 index. Only the values at the 95% confidence level or higher are displayed. Also, superimposed are climatology of temperature (contour), mixed-layer depth (open circle), and barrier layer (asterisk). The interval of the climatic temperature is 2°C and the 20°C isotherm is thickened, (B) and (C) are the same as (A) but for the spring (MAM1) and summer (JJA1) of next year, respectively.

sea surface (Fig. 9.5B), indicating that thermocline feedback has been initiated. Further, the warming in the mixed layer, thermocline, and sea surface weakens in the following season (Fig. 9.5C). Moreover, the thermal advection of the Indonesian Throughflow (ITF) also contributes to the variability of SST in the southwest TIO (Yokoi et al., 2008; Zhou et al., 2008; Du et al., 2008; Tozuka et al., 2010). The thermocline dome can be expected to become shallower if the ITF shut down (Hirst and Godfrey, 1993).

9.3.2 Somalia–Oman upwelling

In boreal spring and summer, the SST in the Western Arabian Sea coastal zone cools down, mainly due to the Somalia-Oman upwelling. This upwelling starts in late spring, peaks in summer, mostly caused by the strong low-level southwesterly jet (Findlater, 1969) and consequent offshore oceanic Ekman transport (Montégut et al., 2007; Murtugudde et al., 2007; Izumo et al., 2008). It brings cold and nutrient-rich water into the ocean mixed layer (Brock and McClain, 1992; Liao et al., 2016).

Since the Somalia–Oman upwelling strongly cools the SST in the western Arabian Sea during boreal summer, its variations can affect the SST in the North Indian Ocean, and hence the Indian Summer Monsoon variability (Shukla, 1975). When the southwesterly winds along the Somalia–Oman coasts are weaker than normal during boreal spring, they cause a decrease in offshore Ekman transport, coastal upwelling, and latent heat flux. These in turn warm the mixed layer in the Western Arabian Sea. Izumo et al. (2008) used both observations and a coupled atmosphere–ocean general circulation model to verify that the variations of the Somalia–Oman upwelling have significant impacts on the moisture transport toward the Indian Subcontinent. Specifically, a decrease in

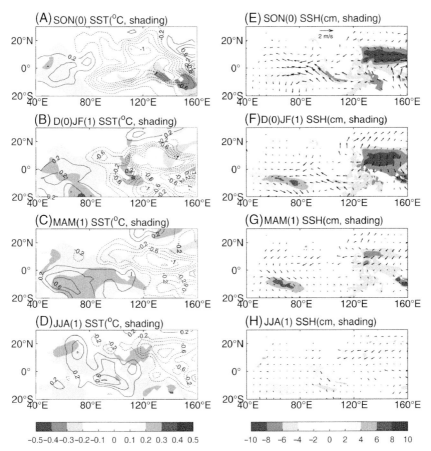

FIG. 9.6   Seasonally SST (°C, shading) and precipitation (mm/day) anomalies (left), as well as SSH (cm, shading) and surface wind velocity (m/s, vector) anomalies (right), expressed as regression upon the NDJ(0) Niño3.4 index. "*SST*, sea surface temperature."

Somalia–Oman upwelling increases the SST along the coasts and thus enhances local evaporation and water vapor transport toward the Indian Western Ghats. As a result, monsoon rainfall strengthens along the west coast of India (Izumo et al., 2008).

9.3.3   The North Indian Ocean

Du et al. (2009) found that the north Indian Ocean features a peculiar double peak SST warming following El Niño (Fig. 9.2). The first peak of SST warming appears in boreal winter (Fig. 9.2) and is caused by surface heat flux adjustments, including solar radiation and wind-induced latent heat flux (Fig. 9.3), while the second peak appears in post-El Niño summer (Fig. 9.2) and cannot be directly explained by El Niño. Indeed, it is caused by air–sea interaction over the north Indian Ocean, which is anchored by the south TIO warming (Xie et al., 2002; Du et al., 2009; Chen et al., 2019).

The southwest TIO warming enhances local convection and induces asymmetric wind anomalies pattern in spring (Chakravorty et al. 2013) following El Niño (Fig. 9.6E). The asymmetric wind anomalies pattern features anomalous northeasterly winds north and northwesterly winds south of the equator (Fig. 9.6E). The asymmetrical pattern is also found in precipitation (Chakravorty et al. 2013). The precipitation increases over the anomalously warm south Indian Ocean and decreases over the relatively cool north Indian Ocean (Fig. 9.6C; Wu et al., 2008; Wu and Yeh, 2010; Du et al., 2009, 2013; Chen et al., 2019).

During winter and following spring, the southwest TIO SST warming induces anomalous northerly winds across the equator (Fig. 9.6E and F). Due to the Coriolis effect, the wind anomalies off the equator turn northeasterly over the north Indian Ocean. Before the summer monsoon onset, the anomalous northeasterly winds over the north Indian Ocean intensify the climatological northeasterly winter monsoon and cool the North Indian Ocean SST. As the climatological wind turns into the southwesterly monsoon in May, the anomalous northeasterly winds contribute to the second peak of SST warming in the North Indian Ocean by reducing latent heat flux (Du et al., 2009). Therefore, the second SST warming of the North Indian Ocean in the early summer essentially results from the cross-equator ocean–atmosphere interaction triggered by the slow-propagating downwelling Rossby waves south of the equator (Izumo et al., 2008; Du et al., 2009, 2013; Xie et al., 2016; Chen et al., 2019).

9.3.4   Impacts on Indian summer monsoon

From December to May, the Indian Ocean Intertropical Convergence Zone (ITCZ) is located over the south TIO. Because of the relatively warm SST in the south TIO at that time, small SST variations can modulate air–sea heat flux exchange. Therefore, the anomalous SST warming over the southwest TIO enhances local convection, which is verified both in observations (Fig. 9.6B and C) and Atmospheric general circulation model experiments (Annamalai et al., 2005; Chen et al., 2019). The southwest TIO SST warming, which is driven by ocean dynamics, induces a "C-shaped" wind anomalies pattern (Chakravorty et al., 2013) with northeasterly (northwesterly) winds north (south) of the equator (Fig. 9.6C). These northeasterly wind anomalies over the North Indian Ocean weaken the prevailing southwesterly winds during the Indian summer monsoon season (Chowdary et al., 2019). As a result, the northwestward migration of the ITCZ is delayed. In early summer, the southwest TIO SST warming effect causes a significant delay in Indian summer monsoon onset by a week (Annamalai et al., 2005).

Even though the mean southwesterly winds over the Arabian Sea are weakened in post-El Niño years, the moisture transport into the Indian Subcontinent seems to be stronger than normal due to excess evaporation in a warmer north Indian Ocean (Izumo et al., 2008). Moisture convergence is enhanced over the windward side of the Indian Western Ghats and south of Himalaya due to the orographic effect, maintaining positive rainfall anomalies over the western coast of India (Fig. 9.6D).

9.4 Discussion and summary

In this chapter, we reviewed the mechanisms that induce and sustain the SST warming over the south TIO after El Niño. Indeed, the TIO SST warming features considerable regional variations. During El Niño mature phase, both anomalous solar radiation and latent heat flux warm most of the TIO, except for the southwest TIO where the mean thermocline is shallow. The oceanic Rossby waves forced by El Niño/IOD, through interacting with the shallow thermocline, dominate the SST warming over the southwest TIO (Fig. 9.7A). Because of strong thermocline feedback in the southwest thermocline ridge, the local SST warming triggers air–sea interaction across the equator (Fig. 9.7A). Thus, the north Indian Ocean features a double-peak SST warming with an

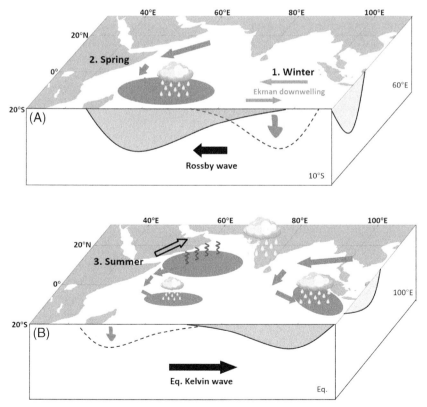

FIG. 9.7 Schematic diagrams of South TIO impacts on the Indian summer monsoon. (A) MAM; (B) JJA. "*TIO*, tropical Indian Ocean."

even larger second peak. During winter and early spring, the ITCZ over TIO locates south of the equator. The southwest TIO SST warming enhances local convection and prevents the northward migration of ITCZ, delaying the summer monsoon onset by about one week. Besides, it also induces a "C-shaped" wind anomalies pattern with anomalous northeasterly (northwesterly) winds in the north (south) of the equator during boreal spring, when El Niño decays quickly as well as its direct impact on TIO (Fig. 9.7A). The northeasterly wind anomalies along the Somalia–Oman coast weaken the coastal upwelling and give rise to the SST warming over the Arabian Sea. Around May, the northeasterly wind anomalies, triggering by the southwest TIO SST warming maintained by downwelling Rossby waves, decrease the southwesterly monsoonal winds over the north Indian Ocean (Fig. 9.7A). Although the mean southwesterly winds in post-El Niño year are weaker than normal, a warmer north Indian Ocean with more evaporation strengthens the monsoonal moisture transport into the Indian Subcontinent (see Chapter 5). Due to the orographic effect, more moisture convergence and resultant rainfall appear over the windward side of the Western Ghats and south of Himalaya (Fig. 9.7).

During the developing and mature phases of El Niño, the El Niño-forced oceanic Rossby waves play a dominant role in inducing the southwest TIO SST warming. The approach of westward propagating oceanic Rossby waves near the thermocline ridge results in deepening the thermocline and also contributes to the local SST warming (Fig. 9.7A). The oceanic Rossby waves that finally arrive at the western boundary of the TIO are reflected as oceanic Kelvin waves (Chen et al., 2019). The Kelvin waves propagate eastward along the equator and eventually reach the eastern boundary of TIO (Fig. 9.7B). As a result, the thermocline off Sumatra and Java deepens, and the summer monsoon-induced coastal upwelling is suppressed, setting a favorable condition for the southeast TIO SST warming (Figs. 9.6H and 9.7B). Such ocean wave dynamics are obvious in strong El Niño cases, such as 1997/1998 and 2015/2016 events (Chen et al., 2019). The southeast TIO warming also enhances local convection and forces asymmetric wind anomaly pattern over the eastern TIO (Fig. 9.7B), as the southwest TIO SST warming does in the western TIO (Fig. 9.5D and H). Associated easterly wind anomalies response sustains the SST warming in the Bay of Bengal and the South China Sea by altering the latent heat flux after the onset of the Asian summer monsoon (Jing et al., 2011; Yang et al., 2015; Chen et al., 2018, 2019). This chapter highlights that the anomalous south TIO warming driven by ocean dynamics has a considerable influence on the Indian summer monsoon by modulating the TIO air–sea interactions.

Funding

This work is jointly supported by the National Natural Science Foundation of China (41830538, 42090042, 41805057), the National Key R&D Projects of China (2019YFA0606703), the Chinese Academy of Sciences (XDA15020901,

133244KYSB20190031, ZDRW-XH-2019-2), the Guangdong Natural Science Foundation (2018A030310023, 2020A1515010361), the Southern Marine Science and Engineering Guangdong Laboratory (Guangzhou) (GML2019ZD0303, GML2019ZD0306, 2019BT02H594).

Acknowledgements

We acknowledge the NOAA/OAR/ESRL PSD, Boulder, Colorado, USA, for providing the monthly ERSSTv5, the monthly NCEP/NCAR reanalysis, and Global Precipitation Climatology Project data (http://www.esrl.noaa.gov/psd/), the CNES for providing the AVISO data (http://www.aviso.altimetry.fr/en/data/data-access.html), and the Met Office Hadley Centre for providing the EN4 data (https://www.metoffice.gov.uk/hadobs/en4/).

References

Adler, R.F., Huffman, G.J., Chang, A., Ferraro, R., Xie, P.P., Janowiak, J., et al., 2003. The version-2 global precipitation climatology project (GPCP) monthly precipitation analysis (1979.present). J. Hydrometeorol. 4 (6), 1147–1167. doi:10.1175/1525-7541(2003)004<1147:Tvgpcp>2.0.Co;2.

Alexander, M.A., Blade, I., Newman, M., Lanzante, J.R., Lau, N.C., Scott, J.D., 2002. The atmospheric bridge: the influence of ENSO teleconnections on air-sea interaction over the global oceans. J. Clim. 15 (16), 2205–2231. doi:10.1175/1520-0442(2002)015<2205:Tabtio>2.0.Co;2.

Annamalai, H., Liu, P., Xie, S.P., 2005. Southwest Indian Ocean SST variability: its local effect and remote influence on Asian monsoons. J. Clim. 18 (20), 4150–4167. doi:10.1175/jcli3533.1.

Ashok, K., Guan, Z.Y., Yamagata, T., 2001. Impact of the Indian Ocean Dipole on the relationship between the Indian monsoon rainfall and ENSO. Geophys. Res. Lett. 28 (23), 4499–4502. doi:10.1029/2001gl013294.

Behera, S.K., Yamagata, T., 2001. Subtropical SST dipole events in the southern Indian ocean. Geophys. Res. Lett. 28 (2), 327–330. doi:10.1029/2000gl011451.

Brock, J.C., McClain, C.R., 1992. Interannual variability in phytoplankton blooms observed in the northwestern Arabian Sea during the southwest monsoon. J. Geophys. Res. 97 (C1), 733–750. doi:10.1029/91jc02225.

Chakravorty, S., Chowdary, J.S., Gnanaseelan, C., 2013. Spring asymmetric mode in the tropical Indian Ocean: role of El Niño and IOD. Clim Dyn 40, 1467–1481. doi:10.1007/s00382-012-1340-1.

Chakravorty, S., Gnanaseelan, C., Chowdary, J.S., Luo, J.J., 2014. Relative role of El Niño and IOD forcing on the southern tropical Indian Ocean Rossby waves. J. Geophys. Res. 119, 5105–5122. doi:10.1002/2013JC009713.

Chen, Z., Wen, Z., Wu, R., Lin, X., Wang, J., 2016. Relative importance of tropical SST anomalies in maintaining the Western North Pacific anomalous anticyclone during El Nio to La Nia transition years. Clim. Dyn. 46 (3-4), 1027–1041. doi:10.1007/s00382-015-2630-1.

Chen, Z., Du, Y., Wen, Z., Wu, R., Wang, C., 2018. Indo-Pacific climate during the decaying phase of the 2015/16 El Nio: role of southeast tropical Indian Ocean warming. Clim. Dyn. 50 (11-12), 4707–4719. doi:10.1007/s00382-017-3899.z.

Chen, Z., Du, Y., Wen, Z., Wu, R., Xie, S.-P., 2019. Evolution of South Tropical Indian Ocean warming and the climatic impacts following strong El Nino events. J. Clim. 32 (21), 7329. doi:10.1175/jcli-d-18-0704.1.

Chowdary, J.S., Gnanaseelan, C., 2007. Basin-wide warming of the Indian Ocean during El Niño and Indian Ocean dipole years. Int. J. Clim 27 (11), 1421–1438.

Chowdary, J.S., Gnanaseelan, C., Xie, S.P., 2009. Westward propagation of barrier layer formation in the 2006-07 Rossby wave event over the tropical southwest Indian Ocean. Geophys. Res. Lett. 36 (4). doi:10.1029/2008gl036642.

Chowdary, J.S., Patekar, D., Srinivas, G., Gnanaseelan, C., Parekh, A., 2019. Impact of the Indo-Western Pacific Ocean Capacitor mode on South Asian summer monsoon rainfall. Clim. Dyn. 53 (3-4), 2327–2338. doi:10.1007/s00382-019.04850-w.

Clark, C.O., Webster, P.J., Cole, J.E., 2003. Interdecadal variability of the relationship between the Indian Ocean zonal mode and East African coastal rainfall anomalies. J. Clim. 16 (3), 548–554. doi:10.1175/1520-0442(2003)016<0548:Ivotrb>2.0.Co;2.

Du, Y., Qu, T.D., Meyers, G., 2008. Interannual variability of the sea surface temperature off Java and Sumatra in a global GCM. J. Clim. 21, 2451–2465. doi:10.1175/2007JCLI1753.1.

Du, Y., Xie, S.-P., Huang, G., Hu, K., 2009. Role of Air-Sea Interaction in the Long Persistence of El Nino-Induced North Indian Ocean Warming. J. Clim. 22 (8), 2023–2038. doi:10.1175/2008jcli2590.1.

Du, Y., Yang, L., Xie, S.P., 2011. Tropical Indian Ocean Influence on Northwest Pacific Tropical Cyclones in Summer Following Strong El Nino. J. Clim. 24 (1), 315–322. doi:10.1175/2010JCLI3890.1.

Du, Y., Xie, S.-P., Yang, Y.L., Zheng, X.T., Liu, L., Huang, G., 2013. Indian Ocean variability in the CMIP5 multimodel ensemble: the basin mode. J. Clim. 26, 7240–7266. doi:10.1175/JCLI-D-12-00678.1.

Findlater, J., 1969. A major low-level air current near the Indian Ocean during the northern summer. Quart. J. Roy. Meteor. Soc. 95 (404), 362–380. doi:10.1002/qj.49709540409.

Gnanaseelan, C., Vaid, B.H., 2010. Interannual variability in the biannual Rossby waves in the tropical Indian Ocean and its relation to Indian Ocean dipole and El Niño forcing. Ocean Dyn. 60, 27–40. doi:10.1007/s10236-009.0236-z.

Good, S.A., Martin, M.J., Rayner, N.A., 2013. EN4: quality controlled ocean temperature and salinity profiles and monthly objective analyses with uncertainty estimates. J. Geophys. Res. 118 (12), 6704–6716. doi:10.1002/2013jc009067.

Godfrey, J.S., Lindstrom, E.J., 1989. The heat budget of the equatorial western Pacific surface mixed layer. J. Geophys. Res.: Oceans 94 (C6), 8007–8017. doi:10.1029/JC094iC06p08007.

Hirst, A.C., Godfrey, J.S., 1993. The role of Indonesian throughflow in a global ocean GCM. J. Phys. Oceanogr. 23 (6), 1057–1086. doi:10.1175/1520-0485(1993)023<1087:WDTITA>2.0.CO;2.

Huang, B., Thorne, P.W., Banzon, V.F., Boyer, T., Chepurin, G., Lawrimore, J.H., et al., 2017. Extended reconstructed sea surface temperature, Version 5 (ERSSTv5): upgrades, validations, and intercomparisons. J. Clim. 30 (20), 8179.8205. doi:10.1175/jcli-d-16-0836.1.

Huang, B.H., Kinter, J.L., 2002. Interannual variability in the tropical Indian Ocean. J. Geophys. Res. 107 (C11). doi:10.1029/2001jc001278 20-21-20-26

Huffman, G.J., Adler, R.F., Bolvin, D.T., Gu, G., 2009. Improving the global precipitation record: GPCP Version 2.1. Geophys. Res. Lett. 36 (17). doi:10.1029/2009gl040000.

Izumo, T., de Boyer Montégut, C., Luo, J.-J., Behera, S., Yamagata, T., 2008. The role of the western Arabian Sea upwelling in Indian monsoon rainfall variability. J. Climate 21, 5603–5623. doi:10.1175/2008JCLI2158.1.

Jing, Z.Y., Qi, Y.Q., Du, Y., 2011. Upwelling in the continental shelf of northern South China Sea associated with 1997–1998 El Niño. J. Geophys. Res. 116, C02033. doi:10.1029/2010JC006598.

Kakatkar, R., Gnanaseelan, C., Chowdary, J.S., 2020. Asymmetry in the tropical Indian Ocean subsurface temperature variability. Dyn. Atm. Oceans 90, 101142. doi:10.1016/j.dynat-moce.2020.101142.

Kara, A.B., Rochford, P.A., Hurlburt, H.E., 2000. An optimal definition for ocean mixed layer depth. J.Geophys. Res. 105 (C7), 16803–16821. doi:10.1029/2000JC900072.

Kalnay, E., Kanamitsu, M., Kistler, R., Collins, W., Deaven, D., Gandin, L., et al., 1996. The NCEP/NCAR 40-year reanalysis project. Bull. Am. Meteorol. Soc. 77 (3), 437–471. doi:10.1175/1520-0477(1996)077<0437:Tnyrp>2.0.Co;2.

Klein, S.A., Soden, B.J., Lau, N.C., 1999. Remote sea surface temperature variations during ENSO: evidence for a tropical atmospheric bridge. J. Clim. 12 (4), 917–932. doi:10.1175/1520-0442(1999)012<0917:Rsstvd>2.0.Co;2.

Kosaka, Y., Xie, S.-P., Lau, N.-C., Vecchi, G.A., 2013. Origin of seasonal predictability for summer climate over the Northwestern Pacific. Proceedings of the National Academy of Sciences of the United States of America 110(19), 7574-7579. doi:10.1073/pnas.1215582110.

Latif, M., Barnett, T.P., 1995. Interactions of the tropical oceans. J. Clim. 8 (4), 952–964. doi:10.1175/1520-0442(1995)008<0952:Iotto>2.0.Co;2.

Lau, N.C., Nath, M.J., 2003. Atmosphere-ocean variations in the Indo-Pacific sector during ENSO episodes. J. Clim. 16 (1), 3–20. doi:10.1175/1520-0442(2003)016<0003:Aoviti>2.0.Co;2.

Levitus, S., Antonov, J.I., Boyer, T.P., Baranova, O.K., Garcia, H.E., Locarnini, R.A., et al., 2012. World ocean heat content and thermosteric sea level change (0-2000 m), 1955-2010. Geophys. Res. Lett. 39 (10). doi:10.1029/2012gl051106.

Li, C.Y., Mu, M.Q., 2001. The influence of the Indian Ocean dipole on atmospheric circulation and climate. Adv. Atmos. Sci. 18 (5), 831–843. doi:10.1007/BF03403506.

Liao, X., Zhan, H., Du, Y., 2016. Potential new production in two upwelling regions of the western Arabian Sea: estimation and comparison. J. Geophys. Res. 121 (7), 4487–4502. doi:10.100 2/2016jc011707.

Ma, J., Du, Y., Zhan, H., Liu, H., Wang, J., 2014. Influence of oceanic Rossby waves on phy-toplankton production in the southern tropical Indian Ocean. J. Marine Syst. 134, 12–19. doi:10.1016/j.jmarsys.2014.02.003.

Manatsa, D., Chingombe, W., Matarira, C.H., 2008. The impact of the positive Indian Ocean dipole on Zimbabwe droughts. Int. J. Climatol. 28 (15), 2011–2029. doi:10.1002/joc.1695.

Masumoto, Y., Meyers, G., 1998. Forced Rossby waves in the southern tropical Indian Ocean. J. Geophys. Res. 103 (C12), 27589–27602. doi:10.1029/98jc02546.

Montégut, C.B., Vialard, J., Shenoi, S.S., Shankar, D., Durand, F., Ethé, C., Madec, G., 2007. Sim-ulated seasonal and interannual variability of the mixed layer heat budget in the northern Indian Ocean. J. Climate 20 (13), 3249–3268.

Murtugudde, R., McCreary, J.P., Busalacchi, A.J., 2000. Oceanic processes associated with anoma-lous events in the Indian Ocean with relevance to 1997-1998. J. Geophys. Res. 105 (C2), 3295–3306. doi:10.1029/1999jc900294.

Murtugudde, R., Seager, R., Thoppil, P., 2007. Arabian Sea response to monsoon variations. Pale-oceanography 22 (4). doi:10.1029/2007pa001467.

Rao, S.A., Behera, S.K., 2005. Subsurface influence on SST in the tropical Indian Ocean: struc-ture and interannual variability. Dyn. Atmos. Oceans 39(1-2), 103-135. doi:10.1016/j.dynat-moce.2004.10.014.

Saji, N.H., Goswami, B.N., Vinayachandran, P.N., Yamagata, T., 1999. A dipole mode in the tropical Indian Ocean. Nature 401 (6751), 360–363. doi:10.1038/43855.

Saji, N.H., Yamagata, T., 2003. Possible impacts of Indian Ocean Dipole mode events on global climate. Clim. Res. 25 (2), 151–169. doi:10.3354/cr025151.

Saji, N.H., Xie, S.P., Yamagata, T., 2006. Tropical Indian Ocean variability in the IPCC twentieth-century climate simulations. J. Clim. 19 (17), 4397–4417. doi:10.1175/JCLI3847.1.

Sayantani, O., Gnanaseelan, C., 2015. Tropical Indian Ocean subsurface temperature variability and the forcing mechanisms. Clim. Dyn. 44, 2447–2462. doi:10.1007/s00382-014-2379.y.

Schott, F.A., Xie, S.-P., McCreary Jr, J.P., 2009. Indian Ocean circulation and climate variability. Rev. Geophys. 47 (1). doi:10.1029/2007rg000245.

Shukla, J., 1975. Effect of Arabian sea-surface temperature anomaly on Indian summer monsoon: a numerical experiment with the GFDL model. J. Atmos. Sci. 32 (3), 503–511. doi:10.1175/1520-0469(1975)032<0503:Eoasst>2.0.Co;2.

Sprintall, J., Tomczak, M., 1992. Evidence of the barrier layer in the surface layer of the tropics. J. Geophys. Res.: Oceans 97 (C5), 7305–7316. doi:10.1029/92JC00407.

Tozuka, T., Yokoi, T., Yamagata, T., 2010. A modeling study of interannual variations of the Seychelles Dome. J. Geophys. Res. 115 (C4). doi:10.1029/2009jc005547.

Trenberth, K., 1997. The definition of El Niño. Bull. Am. Meteor. Soc. 78, 2771–2777. doi:10.1175/1520-0477 (1997)078<2771:TDOENO>2.0.CO;2.

Vaid, B.H., Gnanaseelan, C., Polito, P.S., Salvekar, P.S., 2007. Influence of pacific on Southern Indian Ocean Rossby waves. Pure Appl. Geophys. 164, 1765–1785. doi:10.1007/s00024-007-0230-7.

Webster, P.J., Moore, A.M., Loschnigg, J.P., Leben, R.R., 1999. Coupled ocean-atmosphere dynamics in the Indian Ocean during 1997-98. Nature 401 (6751), 356–360. doi:10.1038/43848.

Wu, R., Kirtman, B.P., Krishnamurthy, V., 2008. An asymmetric mode of tropical Indian Ocean rainfall variability in boreal spring. J. Geophys. Res. 113 (D5). doi:10.1029/2007jd009316.

Wu, R., Yeh, S.-W., 2010. A further study of the tropical Indian Ocean asymmetric mode in boreal spring. J. Geophys. Res. 115 (D8). doi:10.1029/2009jd012999.

Xie, S.P., Annamalai, H., Schott, F.A., McCreary, J.P., 2002. Structure and mechanisms of South Indian Ocean climate variability. J. Clim. 15 (8), 864–878. doi:10.1175/1520-0442(2002)015<0864:Samosi>2.0.Co;2.

Xie, S.-P., Hu, K., Hafner, J., Tokinaga, H., Du, Y., Huang, G., et al., 2009. Indian Ocean capacitor effect on Indo-western Pacific climate during the summer following El Nino. J. Clim. 22, 730–747. doi:10.1175/2008JCLI2544.1.

Xie, S.-P., Du, Y., Huang, G., Zheng, X.T., Tokinaga, H., Hu, K., et al., 2010. Decadal shift in El Nino influences on Indo-western Pacific and East Asian climate in the 1970s. J. Clim. 23 (12), 3352–3368. doi:10.1175/2010JCLI3429.1.

Xie, S.-P., Kosaka, Y., Du, Y., Hu, K., Chowdary, J.S., Huang, G., 2016. Indo-western Pacific ocean capacitor and coherent climate anomalies in post-ENSO summer: a review. Adv. Atmos. Sci. 33 (4), 411–432. doi:10.1007/s00376-015-5192-6.

Yang, J., Liu, Q., Xie, S.-P., Liu, Z., Wu, L., 2007. Impact of the Indian Ocean SST basin mode on the Asian summer monsoon. Geophys. Res. Lett. 34 (2). doi:10.1029/2006gl028571.

Yang, Y.L., Xie, S.-P., Du, Y., Tokinaga, H., 2015. Interdecadal difference of interannual variability characteristics of South China Sea SSTs associated with ENSO. J. Clim. 28, 7145–7160. doi:10.1175/JCLI-D-15-0057.1.

Yokoi, T., Tozuka, T., Yamagata, T., 2008. Seasonal variation of the Seychelles Dome. J. Clim. 21 (15), 3740–3754. doi:10.1175/2008jcli1957.1.

Yu, W.D., Xiang, B.Q., Liu, L., Liu, N., 2005. Understanding the origins of interannual thermocline variations in the tropical Indian Ocean. Geophys. Res. Lett. 32 (24). doi:10.1029/2005gl024327.

Zhou, L., Murtugudde, R., Jochum, M., 2008. Seasonal influence of Indonesian throughflow in the southwestern Indian Ocean. J. Phys. Oceanogr. 38 (7), 1529–1541. doi:10.1175/2007JPO3730.1.

Chapter 10

Atlantic Niño—Indian summer monsoon teleconnections

Ramesh Kumar Yadav

Indian Institute of Tropical Meteorology, Pune, India

10.1 Introduction

The rainfall in India peaks during boreal summer and contributes about 80% of the annual rainfall in the months of June through September (JJAS), called the Indian summer monsoon (ISM). It is part of the South Asian monsoon system and is characterized by strong, large-scale monsoonal circulations. It is one of the largest global phenomena of the general circulation that not only affects the life of millions of people of India but has its impacts on the other parts of the global weather and climate. The temporal and spatial variation of ISM rainfall plays a decisive role in causing large-scale droughts and floods in India, seriously affecting the agriculture production and the agrarian economy of the country (Saha et al., 1979). It also plays an important role in the water management and the economic planning of the country. The spatial distribution of summer monsoon rainfall over the Indian region is not homogeneous. North-East India and the Western Ghats receive maximum rainfall, followed by Central India (Fig. 10.1). Due to different land-ocean configuration and orography (especially the Tibetan plateau), the South Asian monsoon brings a maximum rainfall over Northeast India (Fig. 10.1) with the lower-level convergent (cyclonic) flow (850-hPa positive vorticity) and upper-troposphere divergent (anticyclonic) flow (200-hPa negative vorticity; Fig. 10.2). The enhanced divergence (negative vorticity) over Northeast India is well known as the Tibetan High. This enlarged upper-troposphere divergence of the Tibetan High reinforces the outbreak of convective activities over there. Quite often, the maximum convection, located in Northeast India, follows the movement of the Tibetan High.

The ISM is a fully coupled land–atmosphere–ocean system. It is linked with the local air–sea interactions and the remote climatic phenomena, for example, El-Niño-Southern Oscillation (ENSO), the Indian Ocean Dipole, western equatorial Pacific sea surface temperature (SST) (Li et al., 2017), and the Atlantic Niño. Similar to the Indian rainfall, the Atlantic Niño typically peaks in boreal

Indian Summer Monsoon Variability: El Niño-teleconnections and beyond.
DOI: https://doi.org/10.1016/B978-0-12-822402-1.00005-3
197

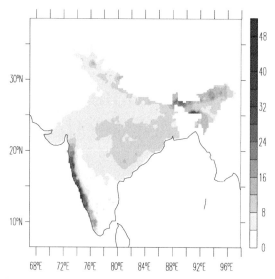

FIG. 10.1 Spatial pattern of seasonal mean climatology (period: 1982:2019) of rainfall (mm/day) of India for the season June–September (JJAS).

summer. The warmest temperature in the tropical Atlantic is on average found in the western basin and to the north of the equator, coincide with the position of the intertropical convergence zone (ITCZ). The wind field is dominated by the convergence of the north-easterly and south-easterly trade wind that meets at zonally oriented ITCZ north of the equator (Lubbecke and McPhaden, 2013). The Atlantic Niño is the interannual SST variability in the cold tongue region of the east tropical Atlantic and along the southwest African coast (Zebiak, 1993). The canonical Atlantic Niño mode is governed by feedbacks involving ocean dynamics and ocean–atmosphere interactions (Lubbecke and McPhaden, 2017; Carton et al., 1996), referred to as the Bjerknes feedback (Bjerknes, 1969), very similar to ENSO: a warm anomaly in the eastern equatorial basin, called the positive phase of Atlantic Niño, results in a relaxation of the trade winds to its west, which leads to reduced upwelling and a deepening of the thermocline in the east that causes further warming, and vice-versa for cold phase (Zebiak, 1993; Carton and Huang, 1994). The cold and warm event of Atlantic Niño is symmetric with respect to amplitude, spatial patterns, and temporal evolution, whereas that of ENSO in the Pacific is asymmetric (Lubbecke and McPhaden, 2017)

The warming of the Atlantic cold tongue causes a southward shift of ITCZ, enhancing atmospheric convection over the tropical East Atlantic and West Africa (Nobre and Shukla, 1996; Janicot et al., 1998; Yadav, 2017a, Yadav et al., 2018, 2020). The active ITCZ with intense convection and tropospheric heating modulates the weather and climate remotely over the Indian subcontinent and Pacific (Kucharski et al., 2007, 2008, 2009; Barimalala et al., 2012;

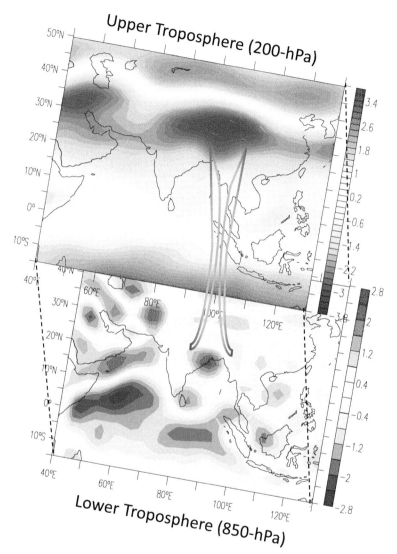

FIG. 10.2   Spatial pattern of seasonal mean climatology (period: 1982:2019) of vorticity at 200- and 850-hPa levels representing upper (upper panel) and lower (lower panel) troposphere, respectively, for the season June–September (JJAS). The *cray arrows* illustrate the intense vertical motion.

Yadav, 2017a, Yadav et al., 2018). Several studies (Kucharski et al., 2007, 2008, and 2009; Barimalala et al., 2012) showed that the warmer east-equatorial Atlantic SST force low-level convergent flow, brings moist air, in the immediate surrounding coastal regions of Africa, and causes increased rainfall there. A Gill–Matsuno type quadrupole response (Gill, 1980, Matsuno, 1966) to the Atlantic Niño with a low-level anticyclone located over India weakens the ISM circulation. Atlantic Niño can also induce changes in the Indian Ocean SST, especially along the coast of Africa and in the West Indian Ocean basin. A recent study (Yadav, 2017a; Yadav et al., 2018, 2020) suggested that the ISM rainfall could be influenced by the east equatorial Atlantic SST via Eurasian waves as the atmospheric bridge. The intensified ITCZ over the Atlantic and West Africa due to positive SST anomaly provokes meridional stationary wave poleward. This creates anomalous positive and negative geopotential heights (GPHs) over subtropics (Mediterranean) and extratropics (Northwest Europe), respectively, with anomalous negative GPH over Northwest Europe and vice-versa for negative SST anomaly. The anomalous GPH on Northwest Europe acts as a center of action for the propagation of a Rossby wave train to the North–West of India, intensifying the Tibetan High westward, which incite the monsoon activity over India (Yadav, 2017a; Yadav et al., 2018, 2020).

The chapter revisits the link between the Atlantic Niño and the ISM. This study discussed the mechanism by which the ITCZ modulated by the Atlantic Niño affects the upper-troposphere GPHs through the meridional stationary wave, which further modifies the Central Asia GPH by the Eurasian wave, consequently shifting the Tibetan High and hence ISM rainfall.

10.2 Methods and data

The SST data is obtained from OISST v2.0 (Optimum Interpolation SST, Reynolds, 1993) for the period 1982–2019. NCEP/NCAR reanalysis data for the same period have been used (Kalnay et al., 1996). For rainfall over the Indian landmass, the monthly gridded rainfall at $0.25° \times 0.25°$ resolutions for the same period is used from the India Meteorological Department (IMD) (Pai et al., 2015).

Throughout the chapter, the summer season is defined as the average of June through September (JJAS). A total of 38 years, JJAS boreal summer season from 1982 to 2019 has been considered for this study. Empirical orthogonal function (EOF) analysis is performed on the linearly detrended JJAS seasonal gridded SST comprising regions of tropical Atlantic (50°W–15°E, 20°S–10°N) to study the interannual variability of the dominant mode of Atlantic Niño. The corresponding PCs are used as the basis for understanding their interannual variability. Contiguous correlation, and regression of SST, rainfall, 200-hPa GPH, and wind fields (u and v) have been examined. All the data series have been linearly detrended before carrying out the correlation and regression analysis.

The Rossby wave activity fluxes have been calculated to show the existence and propagation of the Rossby wave in the mean flow over the Eurasian region. The horizontal flux of Rossby wave formulas are as follows:

$$W_x = \frac{p\cos\varphi}{2|U|}\left(\begin{array}{c} \dfrac{U}{a^2\cos^2\varphi}\left[\left(\dfrac{\partial\psi'}{\partial\lambda}\right)^2 - \psi'\dfrac{\partial^2\psi'}{\partial\lambda^2}\right] + \\ \dfrac{V}{a^2\cos\varphi}\left[\dfrac{\partial\psi'}{\partial\lambda}\dfrac{\partial\psi'}{\partial\varphi} - \psi'\dfrac{\partial^2\psi'}{\partial\lambda\partial\varphi}\right] \end{array} \right)$$

$$(10.1)$$

$$W_y = \frac{p\cos\varphi}{2|U|} \left(\begin{array}{l} \dfrac{U}{a^2\cos\varphi}\left[\dfrac{\partial\psi'}{\partial\lambda}\dfrac{\partial\psi'}{\partial\varphi} - \psi'\dfrac{\partial^2\psi'}{\partial\lambda\partial\varphi}\right] + \\[2ex] \dfrac{V}{a^2}\left[\left(\dfrac{\partial\psi'}{\partial\varphi}\right)^2 - \psi'\dfrac{\partial^2\psi'}{\partial\varphi^2}\right] \end{array} \right) \tag{10.2}$$

where $W_{x,y}$ is the wave flux in the horizontal x and y direction, a is earth's radius, (ϕ, λ) are latitude and longitude, respectively, Geostrophic stream function is defined as $\psi = \phi/f$, ϕ is geopotential and f (= 2 Ω sin ϕ) the Coriolis parameter with the earth's rotation rate Ω, ϕ' is perturbation stream function, U is the zonal wind, V is meridional wind, $|U|$ is wind magnitude, and p is pressure normalized by 1000 hPa (=pressure/1000 hPa). In theory, the flux **W** is independent of the wave phase and parallel to the local group velocity of planetary waves. The flux tends to diverge out of forcing regions. The details of the individual terms of **W** are explained by Takaya and Nakamura, (2001).

10.3 Atlantic Niño and ISM rainfall

The coherent seasonality of summertime Atlantic Niño and ISM is suggestive of a possible link between them (Kucharski et al., 2007, 2008, 2009; Barimalala et al., 2012; Yadav, 2017a; Yadav et al., 2018, 2020), but it has remained unclear whether and how the Atlantic Niño affects ISM. Therefore, it becomes important to explore the influence of the Atlantic Niño on ISM. For this, EOF analysis has been performed on the linearly detrended JJAS seasonal SST in the tropical Atlantic region over the domain (50°W–15°E, 20°S–10°N), for the period 1982–2019. The first leading EOF (EOF1), which accounts for 52.57% of the variance, represents the Atlantic Niño pattern (Fig. 10.3A) and the respective principal component, PC1, as an Atlantic Niño index, referred hereafter as ANI (Fig. 10.3B) is obtained to study the influence of Atlantic Niño on ISM. The EOF1 SST pattern (Fig. 10.3A) shows the highest loading (largest positive SST anomalies) over the south of east equatorial Atlantic. The correlation coefficient (CC) between ANI and high-resolution IMD rainfall (Fig. 10.3C) shows significant positive CC over North-Eastern India (East UP, Bihar, Jharkhand, West Bengal, and parts of Assam and Meghalaya) and Peninsular India. The anomalous positive rainfall observed over the peninsular India (Fig. 10.3C) is mainly associated with La Niña (negative phase of ENSO), since Atlantic Niño is associated with weak La Niña event in the Pacific (Yadav et al., 2018; Yadav, 2017a). Strong easterly wind anomalies over southern peninsular India from Central Pacific via warm western Pacific transport moisture and lead to excess rainfall in the former region (Fig. 1C of Yadav et al., 2018). Further, in response to anomalous warming over the equatorial west Pacific, a weak cyclonic circulation over the southern tip of India as a part of the Matsuno–Gill (Gill, 1980; Matsuno, 1966) pattern supports enhanced rainfall locally (Srinivas et al. 2018).

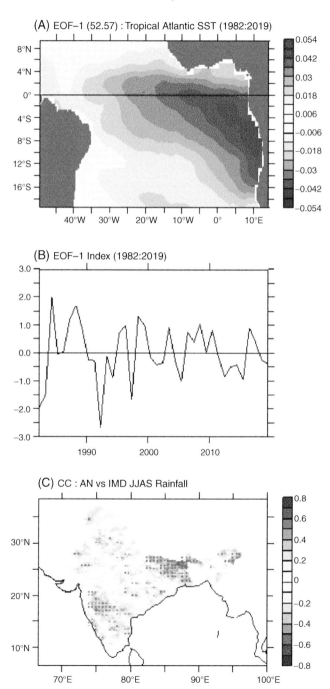

FIG. 10.3   (A) The first mode of empirical orthogonal function (EOF1) of seasonal JJAS SST anomalies (°C; shaded) for the period 1982–2019 over the equatorial Atlantic region, (B) respective principal component, PC1, and (C) contiguous correlation between PC1 and IMD rainfall over India. "*IMD*, India Meteorological Department."

The CC of ANI with 925- and 200-hPa divergence and regression of divergent zonal and meridional winds onto ANI are shown in Fig. 10.4. The lower-level (925-hPa) shows strong convergence (negative divergence anomaly) north of equatorial east Atlantic at the position of ITCZ, inferring active ITCZ. In contrary, the upper-troposphere (200-hPa) shows the opposite mirror image to the lower-troposphere with strong divergence (positive divergence anomaly). The pattern resembles strong convection (as depicted by light cray arrows, Fig. 10.4) supported by strong lower-level convergence with the compensatory enlarged upper-troposphere divergence (Fig. 10.4). As we know, the Atlantic Niño has warm anomalies in the east equatorial Atlantic associated with the lower-level westerly wind anomalies over the west equatorial Atlantic, relaxing the trade wind along the equator. This results in weak upwelling over west coast of Africa with deepening of the thermocline that leads to warming of SST in the east equatorial Atlantic. The warming of SST in the east equatorial Atlantic intensify the ITCZ over east Atlantic and west Africa.

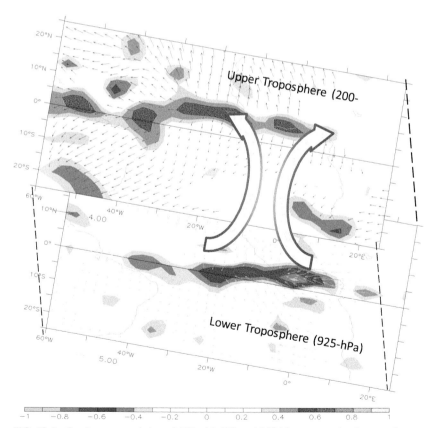

FIG. 10.4   Contiguous correlation of ANI with 925- and 200-hPa, representing lower and upper troposphere, respectively, divergence (color shaded) and regression of divergent winds (arrows). The 925- and 200-hPa divergent winds are shown by blue and red, respectively.

Fig. 10.5 shows climatology of 200-hPa GPH (black contours) and Rossby Wave Activity Flux (Cray Arrows) and CC between ANI and GPH (color shade, the significant area is shown by the grey dotted). The CC between ANI and 200 hPa GPH (Fig. 10.5) shows significant positive CC over North Africa, Mediterranean Sea, Central Asia, and North India, and negative CC over Northwest Europe. The negative anomaly over tropical East Atlantic and tropical Africa, and positive anomaly over North Africa and Mediterranean illustrate the poleward shift of the upper-troposphere climatological maximum GPH of North Africa. Since, the amplification of the tropical heat source owing to the active ITCZ over East Atlantic and West Africa, enhance the upper-troposphere outflow/divergence, which in turn had pushed the upper-troposphere GPH poleward over North Africa, creating negative GPH anomalies over tropical east Atlantic and Africa and positive GPH anomalies over North Africa and Mediterranean. Moreover, the successive meridional negative, positive, and negative GPH anomalies from the tropics to extratropics located at the equatorial Africa, Mediterranean, and Northwest Europe, respectively, clearly display the stationary wave owing to the enhancement of the upper-troposphere outflow/divergence of the active Atlantic and Africa ITCZ (Yadav, 2017a; Yadav et al., 2018, 2020). The Rossby wave activity flux (Takaya and Nakamura,

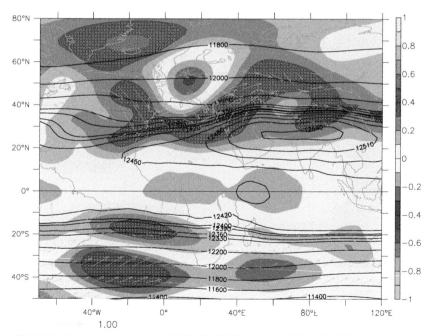

FIG. 10.5 Contiguous correlation of ANI with 200-hPa geopotential height (GPH; color shaded) and superimposed by climatology of GPH (black contours) and horizontal component of wave activity flux (*cray arrows*). The *gray dots* indicate the GPH signals are significant at 95% confidence level.

2001) (Fig. 10.5, cray arrows) shows the preferred region of the propagation of the circum-global Rossby wave train zonally oriented from the North Atlantic toward the Central Asia via Northwest Europe. The negative GPH anomaly formed over Northwest Europe owing to the Atlantic Niño warming, imposes positive GPH anomaly at Central Asia and north India by this Eurasian Rossby wave-train dispersion, as illustrated in Fig. 10.5. Overall, the warm phase of Atlantic Niño is associated with active ITCZ over East Atlantic and West Africa, upper-troposphere negative GPH anomaly over Northwest Europe, positive GPH anomalies over Mediterranean and Central Asia, north-westward shifting of Tibetan High, and excess rainfall over North-Eastern India.

It is found that the Atlantic Niño is simultaneously associated with the rainfall over the north-eastern part of India during its peak rainy season (JJAS). Therefore, the box 82.5°E–90.5°E, 22°N–28.25°N (Fig. 10.6A), representing North-Eastern India (east UP, Bihar, Jharkhand, West Bengal, and parts of Assam and Meghalaya), has been considered for constructing an index, named hereafter as NEISR (Fig. 10.6B). The simultaneous CC between this NEISR index and gridded rainfall over Indian landmass (Fig. 10.6A) does not show any big patch of significant rainfall except North-Eastern India. This suggests that the north-eastern part is a homogeneous region and independent from other parts of the ISM rainfall (Yadav and Roxy, 2019; Yadav et al., 2020). The CC between NEISR and ANI is 0.67, significant at 99.9% confidence level. The CC between NEISR and SST over Atlantic Niño region (Fig. 10.6C) displays the canonical positive phase of Atlantic Niño with a dipole-like structure with significant positive correlation in the east equatorial and south–east equatorial Atlantic, and negative correlation in the extratropical South-West Atlantic, known as south Atlantic Ocean dipole (Nnamchi et al., 2016). This clearly shows the strong relationship between NEISR and Atlantic Niño.

Fig. 10.7 is similar to Fig. 10.5, but for NEISR. The pattern shows similar to Fig. 10.5. The CC between NEISR and 200 hPa GPH (Fig. 10.7) shows significant positive CC over the east Mediterranean Sea and Central Asia and negative CC over Northwest Europe. The positive and negative GPH anomaly over Central and Eastern Asia infers the north-westward shift of Tibetan High. This means that the shift of Tibetan High north-westward is favorable for intense convection over North-Eastern India. The shifting of Tibetan High is associated with upper-troposphere divergence inciting intense convection to occur over there (see Chapter 16). The consecutive negative, positive, and negative anomalies over tropical east Atlantic, east Mediterranean, and Northwest Europe, respectively, indicate the meridional stationary wave owing to Atlantic Niño. This clearly establishes the relationship between Atlantic Niño and NEISR. The Rossby wave activity flux (Takaya and Nakamura, 2001) (Fig. 10.7, cray arrows) shows the preferred region of the propagation of Rossby wave train zonally oriented from the North Atlantic toward the Central Asia via Northwest Europe (Ding and Wang, 2005, 2007; Yadav, 2016, 2017a, b; Yadav et al., 2018, 2020). The negative GPH anomaly formed over Northwest

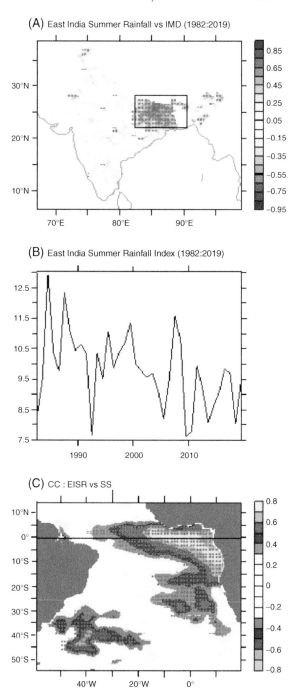

FIG. 10.6   (A) Contiguous correlation between NEISR (highlighted by black box) and IMD rainfall over India. (B) Time-series of NEISR for the period 1982–2019. (C) Contiguous correlation between NEISR and SST over the tropical Atlantic. The *gray dots* in (A) & (C) indicates 95% significance. "*IMD*, Indian Meteorological Department."

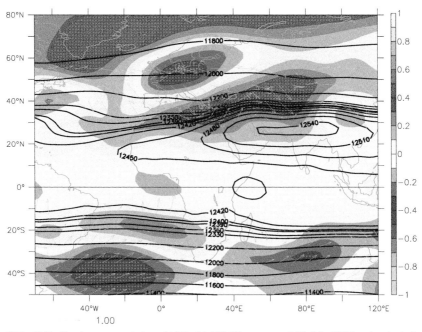

FIG. 10.7   Contiguous correlation of ANI with 200-hPa geopotential height (GPH; color shaded) and superimposed by climatology of GPH (black contours) and horizontal component of wave activity flux (cray arrows). The *gray dots* indicate the GPH signals are significant at 95% confidence level. This is shown for NEISR.

Europe by the Atlantic Niño warming has imposed consecutive positive and negative GPH anomalies at Central Asia and the East Asia, respectively, by this Eurasian Rossby wave-train dispersion. The positive and negative GPH anomalies formed over Central and east Asia respectively shift the Tibetan High north-westward. The upper-troposphere divergence is associated with Tibetan High incite intense convection over North-Eastern India and consequently shifts the maximum convection, climatologically situated over Northeast India during the summer season, toward North-Eastern India. In summary, the positive phase of Atlantic Niño, the zonal dipole structure over Eurasia with negative and positive GPH anomaly over Northwest Europe and Central Asia, respectively, and the shifting of Tibetan High north-westward are favorable for rainfall in North-Eastern India.

10.4   Discussion and summary

In this chapter, the influence of Atlantic Niño on ISM rainfall during boreal summer (JJAS) is revisited and discussed. Both the ISM and Atlantic Niño, primarily peaks during boreal summer season. The observational analysis revealed that positive Atlantic Niño supports intensification of the ITCZ, enhances the upper-troposphere divergence, and pushes the climatological maximum GPH of north Africa poleward. This generates the meridional stationary wave with successive negative, positive, and negative GPH anomalies over tropical Africa, Mediterranean, and Northwest Europe, respectively. The anomalous GPH over Northwest Europe falls on the path of the zonally oriented mid-latitude/extratropical Rossby wave train toward the Eurasian region, which reinforces the positive GPH anomaly over Central Asia. This positive GPH anomaly shifts the Tibetan High north-westward from its climatological position, thereby shifting the upper-troposphere large-scale divergence. The upper-troposphere large-scale divergence incites the monsoon activity over the north of East India, by shifting the maximum convection of South Asia to North-Eastern India.

A study by Yadav et al. (2018) has shown the AGCM simulations forced with observed tropical SST (Hurrell et al., 2003) for the period of 1955–2001 is able to capture the meridional stationary wave structure from North Africa to Northwest Europe. Further, this hypothesis has been tested by carrying out an idealized coupled model experiment in which warm SST anomalies are prescribed over the Atlantic Niño region. The model experiment clearly showed a meridional wave pattern in GPH anomalies between tropics and extra-tropics. Model experiment further confirmed that the Atlantic Niño (as the SST had been kept same as in control run over rest of the globe) warming generates the stationary wave meridionally, similar to that of observations. Overall, despite having some limitations, the model experiment was able to represent the atmospheric adjustments associated with the Atlantic Niño, supporting the hypothesis that the Atlantic Niño could generate meridional stationary wave over Africa and Europe, which further influences the midlatitude wave pattern and the Tibetan High. A more detailed analysis of the influence of individual Atlantic Niño events on the Indian region in the summer season is desirable. Most importantly, the role of Atlantic Niño in modulating the ENSO-monsoon teleconnections is an open question and essentially needs further attention.

Acknowledgement

All the data have been downloaded from free internet and their data sources are duly acknowledged. Computational and graphical analyses required for this study have been completed with the free software NCL and Ferret.

References

Barimalala, R., Bracco, A., Kucharski, F., 2012. The representation of the South Tropical Atlantic teleconnection to the Indian Ocean in the AR4 coupled models. Clim. Dyn. 38, 1147–1166.

Bjerknes, J., 1969. Atmospheric teleconnections, from the equatorial Pacific. Mon. Weather Rev. 97, 163–172.

Carton, J.A., Cao X., Benjamin S. G., Da Silva A. M., 1996. Decadal and interannual SST variability in the tropical Atlantic. J. Phys. Oceanogr. 26, 1165–1175.

Carton, J.A., Huang, B., 1994. Warm events in the tropical Atlantic. J. Phys. Oceanogr. 24, 888–903.

Ding, Q., Wang, B., 2007. Intraseasonal teleconnection between the summer Eurasian wave train and the Indian Monsoon. J. Clim. 20, 3751–3767.

Ding, Q., Wang, B., 2005. Circumglobal teleconnection in the northern hemisphere summer. J. Clim. 18, 3483–3505.

Gill, A.E., 1980. Some simple solutions for heat-induced tropical circulations. Q. J. R. Meteorol. Soc. 106, 447–462.

Hurrell, J.W., Kushnir, Y., Ottersen, G., Visbeck, M., 2003. An overview of the North Atlantic Oscillation: climatic significance and environmental impact. Geophys. Monogr., 1–35.

Janicot, S., Harzallah, A., Fontaine, B., Moron, V., 1998. West African monsoon dynamics and eastern equatorial Atlantic and Pacific SST Anomalies. J. Clim. 11, 1874–1882.

Kalnay E., Kanamitsu M., Kistler R., Collins W., Deaven D., Gandin L., Iredel M., Saha S., White G., Woollen J., Zhu Y., Chelliah M., Ebisuzaki W., Higgins W., Janowiak J., Mo K.C., Ropelewski C., Wang J., Leetmaa A., Reynolds R., Jenne R., Joseph D., 1996. The NCEP/NCAR 40-year reanalysis project. Bull. Am. Meteor. Soc. 77: 437–471.

Kucharski, F., Bracco A., Yoo J., Molteni F., 2007. Low-frequency variability of the Indian Monsoon-ENSO relationship and the Tropical Atlantic: the weakening of the 1980s and 1990s. J. Clim. 20, 4255–4266.

Kucharski, F., Bracco A., Yoo J., Tompkins A., Feudale L., Ruti P., Dell'Aquila A., 2009. A simple Gill-Matsuno-type mechanism explains the tropical atlantic influence on African and Indian Monsoon rainfall. Q. J. Roy. Meteor. Soc. 135, 569–579.

Kucharski, F., Bracco A., Yoo J., Molteni F., 2008. Atlantic forced component of the Indian monsoon interannual variability. Geophys. Res. Lett. 35, L04706. doi:10.1029/2007GL033037.

Li, G., Xie, S.-P., He, C., Chen, Z., 2017. Western Pacific emergent constraint lowers projected increase in Indian summer monsoon rainfall. Nat. Clim. Change 7, 708–712.

Lübbecke, J.F., McPhaden, M.J., 2017. Symmetry of the Atlantic Niño mode. Geophys. Res. Lett. 44, 965–973.

Lübbecke, J.F., McPhaden, M.J., 2013. A comparative stability analysis of Atlantic and Pacific Niño modes. J. Clim. 26, 5965–5980.

Matsuno, T., 1966. Quasi-geostrophic motions in the equatorial area. J. Meteorol. Soc. Jpn. 44, 25–43.

Nnamchi, H.C., Li, J.P., Kucharski, F., Kang, I.-S., Keenlyside, N.S., Chang, P., et al., 2016. An equatorial–extratropical dipole structure of the Atlantic Niño. J. Clim. 29, 7295–7311.

Nobre, P., Shukla, J., 1996. Variations of sea surface temperature, wind stress, and rainfall over the tropical Atlantic and South America. J. Clim. 9, 2464–2479.

Pai, D.S., Sridhar L., Badwaik M.R., Rajeevan M., 2015. Analysis of the daily rainfall events over India using a new long period (1901–2010) high resolution (0.25° × 0.25°) gridded rainfall data set. Clim. Dyn. 45(3–4), 755–776.

Reynolds, R.W., 1993. Impact of Mount Pinatubo aerosols on satellite-derived sea surface temperatures. J. Clim. 6, 768–774.

Saha, K.R., Mooley, D.A., Saha, S., 1979. The Indian monsoon and its economic impact. Geo. J. 3, 171–178.

Srinivas, G., Chowdary, J.S., Kosaka, Y., Gnanaseelan, C., Parekh, A., Prasad, K.V., 2018. Influence of the Pacific–Japan pattern on Indian summer monsoon rainfall. J. Clim 15 (10), 3943–3958 31.

Takaya, K., Nakamura, H., 2001. A formulation of a phase-independent wave-activity flux for stationary and migratory quasigeostrophic eddies on a zonally varying basic flow. J. Atmos. Sci. 58 (6), 608–627.

Yadav, R.K., 2016. On the relationship between Iran surface temperature and north-west India summer monsoon rainfall. Int. J. Climatol. 36 (13), 4425–4438. doi:10.1002/joc.4648.

Yadav, R.K., 2017a. On the relationship between east equatorial Atlantic SST and ISM through Eurasian wave. Clim. Dyn. 48, 281–295. doi:10.1007/s00382-016-3074-y.

Yadav, R.K., 2017b. Mid-latitude Rossby wave modulation of the Indian summer monsoon. Q. J. R. Meteorol. Soc. 143, 2260–2271. doi:10.1002/qj.3083.

Yadav, R.K., Srinivas, G., Chowdary, J.S., 2018. Atlantic Niño modulation of the Indian summer monsoon through Asian jet. npj Clim. Atmos. Sci. 1 (23), 1–11. doi:10.1038/s41612-018-0029-5.

Yadav, R.K., Roxy, M.K., 2019. On the relationship between north India summer monsoon rainfall and east equatorial Indian Ocean warming. Global Planet. Change 179, 23–32. doi:10.1016/j.gloplacha.2019.05.001.

Yadav, R.K., Simon Wang, S.-Y., Wu, C.-H., Gillies, R.R., 2020. Swapping of the Pacific and Atlantic Niño influences on north central India summer monsoon. Clim. Dyn. 54 (9), 4005–4020. doi:10.1007/s00382-020-05215-4.

Zebiak, S.E., 1993. Air–sea interaction in the equatorial Atlantic region. J. Clim. 6, 1567–1586.

Chapter 11

Teleconnections between tropical SST modes and Indian summer monsoon in observation and CMIP5 models

Indrani Roy[a], Ramesh Kripalani[b]

[a]*University College London (UCL), IRDR, London, United Kingdom,* [b]*Indian Institute of Tropical Meteorology (MoES), Pune, India (retired)*

11.1 Introduction

The Indian summer monsoon (ISM) plays a critical role in the well-being of a billion people. ISM serves a very important part of India's socioeconomic infrastructure as it receives about 80% of the total rainfall during this season. Society is so finely tuned to its rains that its variation has profound impacts on country's industry, agriculture, and economy. As India has the second largest population in the world, it also has consequences for global wealth generation. Variability of ISM is strongly influenced by sea surface temperature (SST) of tropics, among which El Niño Southern Oscillation (ENSO) and the Indian Ocean Dipole (IOD) are two dominant modes of climate variability.

ENSO, the most dominant tropospheric variability around tropical Pacific is found to be strongly coupled with the ISM (Azad and Rajeevan, 2016; Kumar, 2006; Ashok, 2001; Roy, 2018a; Li et al., 2017; Hrudya et al., 2020b; Sikka, 1980; Rasmussen and Carpenter, 1983). Excess rainfall years in India are usually aligned with cold events of ENSO or La Niña years, while drought years match with El Niño. SST anomaly in the Pacific in opposite phases of ENSO are shown in Fig. 11.1A. Studies also detected some complementary effects of the IOD on ISM (Ashok et al., 2001). However, such linkages were shown could be sensitive to the reference period chosen (Roy and Collins, 2015; Roy, 2018b).

In the tropical Indian Ocean (IO), one of the leading modes of interannual variability is the IOD. IOD index is measured by the difference in SST anomaly between the western equatorial IO and the south eastern equatorial IO. It has a profound impact on the climate and weather of India, Australia, and East Africa.

Indian Summer Monsoon Variability: El Niño-teleconnections and beyond.
DOI: https://doi.org/10.1016/B978-0-12-822402-1.00012-0

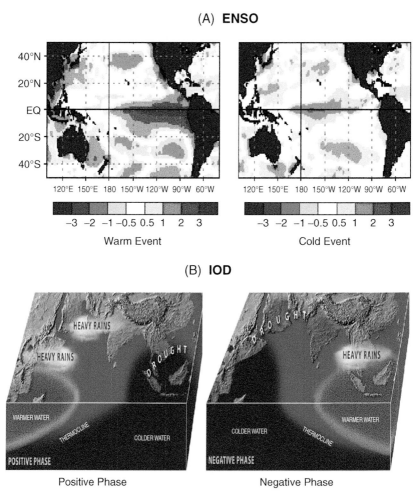

FIG. 11.1 Opposite phases of the (A) ENSO showing reverse sea surface temperature anomaly in the tropical Pacific [http://www.cpc.noaa.gov/products/analysis_ monitoring/ ensocycle/] and (B) Indian Ocean Dipole and effects [http://www.whoi.edu/] (Accessed 31 December 2020). "*ENSO*, El Niño southern oscillation."

Different characteristics of IOD and its impact are discussed elaborately in various studies (Saji et al. 1999, 2003; Webster et al., 1999; Hrudya et al., 2020b). During the negative phase of IOD, SST is anomalously cold in the western equatorial region of IO, while warm in the eastern side and the reverse is true for the positive phase (Fig. 11.1B). During the positive (negative) phase of the IOD, Darwin in Australia experiences drought (heavy rain) which is also associated with cold (warm) surrounding ocean water. On the other hand, in India or East Africa, there is heavy rainfall in the positive IOD phase and less rainfall in the negative phase. The impact of IOD is mainly seen around countries adjacent to the IO; whereas, ENSO though originated in the tropical Pacific Ocean, have profound impacts in most parts of the globe (Brown et al., 2009; Cai et al., 2009; Roy, 2018a; Tedeschi and Collins, 2016; Preethi et al., 2015). It has become common to discuss two types of ENSO. One is dominated by the variability of SST centered in the Eastern Tropical Pacific, commonly known as the East Pacific (EP) type or Canonical ENSO, and the other dominated by strong SST variability around the central tropical Pacific, commonly known as Central Pacific (CP) type or ENSO Modoki (Ashok et al., 2007; Kao et al., 2009; Kug et al., 2009).

The teleconnection patterns associated with these two ENSO types indicate different impacts on various parts of the world (Australia: Cai and Cowan, 2009; Brown et al., 2009; South America: Tedeschi and Collins, 2016; Tedeschi et al., 2013; India: Ashok et al., 2007; Roy et al., 2017; Roy et al., 2019). The formation mechanisms of these two ENSO types are also different; for the Canonical case, the thermocline plays a dominant role, while for Modoki, zonal advection and mid-latitude interactions are proposed as more important (Kao et al. 2009; Yu et al. 2011). Different solar decadal influence on the mechanisms of these two types of ENSO is also recently explored (Roy and Kripalani, 2019a).

Another new type of ENSO named the "mega-ENSO," has been identified (Wang et al., 2013; Zhang et al., 2017 and Wu and Zhang, 2015). It has a pattern similar to the interdecadal Pacific oscillation but involves both interannual and multidecadal SST variations. Many researchers also discussed other types of ENSO events and their impact on ISM. For example, Zhang et al. (2016) analyzed different impacts of atypical and typical developing ENSOs on the ISM. The current study, however, focuses mainly on Canonical and Modoki types of ENSO.

Various modeling groups around the world have coordinated climate model experiments, known as Coupled Model Intercomparison Project (CMIP5) (Taylor et al. 2012). Experiments with various radiative forcing scenarios were also examined in hypothetical future climate change situations (Riahi et al. 2011). Representative Concentration Pathways 8.5 (RCP 8.5) scenarios are the highest forcing situations as examined. These model simulations may be used to better understand the ENSO–ISM teleconnection during various types of ENSO events and changes under climate change (Li et al., 2017; Jourdain et al., 2013; Roy and Tedeschi, 2016; Roy et al., 2017; Roy et al., 2019; Azad and Rajeevan, 2016).

The ISM represents a large-scale heat source along the equator around the intertropical convergence zone that covers Central North East (CNE) India. Following the linear theory, it is linked to both the regional Hadley cell as well as the Walker Circulation (Gill, 1980), and hence the main focus of many recent analyses was on the CNE region (Roy et al. 2017; Bollasina et al. 2011; Goswami et al. 2006). Earlier works studied extratropical ENSO teleconnections in CMIP5 models (Hurwitz et al., 2014; Charlton-Perez, 2013) separating models as the high top (H) and low top (L). H models have upper lids up to the Stratopause (1 hPa) and it is believed that these models with polar vortex features may capture polar annular mode patterns better (Osprey et al., 2013; Seviour et al., 2016; Charlton-Perez, 2013). ISM—southern annular mode teleconnection in observations was also identified, where ENSO Modoki plays the dominant part (Prabhu et al., 2016). Studies further separated CMIP5 models based on H or L to detect whether it is possible to segregate a subset of models performing better over the others (Roy et al., 2017, 2019).

Changes of ISM in the future had been analyzed (Azad and Rajeevan, 2016; Zou and Zhou, 2015; Roy, 2017) that suggested a significant change from the historical period. Azad and Rajeevan (2016) used 20 different CMIP5 models and studied ISM and ENSO teleconnection in a future scenario and suggested a shift to a shorter period from 2.5 to 3 years in the future from 3- to 5-year period band of variability of the historical period. How ISM–ENSO teleconnection will evolve in the future, especially when the focus is on the various subcategories of ENSO were also explored in recent studies (Roy et al. 2019).

Various thermodynamic scaling arguments had been put forward to explain future changes in ISM (Held and Soden, 2006). Those included "wet get wetter" (Chou et al., 2009) and "warmer-get-wetter" (Xie et al., 2010; John et al., 2009) concept. Recent studies suggest that changes in circulation are crucial (Seager et al., 2010) and hence it is appropriate to use techniques to decompose ISM rainfall changes associated with both circulation and moisture. Studies (Roy et al., 2019) employed such a technique to explain the changes in ISM rainfall under climate change considering the effect of both "thermodynamical" and "dynamical" part (Chung et al., 2014; Seager et al., 2012). It discussed changes in the behavior of ENSO–ISM teleconnection during differently flavored ENSO situations (Huang and Xie, 2015; Huang, 2014).

As ENSO plays a crucial role in the variation of ISM, improved understanding of ENSO could be a step forward on advanced ISM prediction. The work of Cappondonti et al. (2015) nicely discussed the current state of understanding on ENSO diversity and identified major areas of knowledge gaps. Studies pointed out that more attention is needed on the tropical/extratropical teleconnections associated with ENSO (Roy et al. 2018). It discussed that bias correction in models could be an important area of model improvements and highlighted the strengths and weaknesses of CMIP5 models in different oceanic sectors that not only included Pacific but also Atlantic. The link between changes in ENSO and the mean state of the Pacific climate would be an important area that needs to be explored to understand the varied nature of ENSO in the future (Cai et al. 2015). Interestingly, Yeh et al. (2018) discussed that future changes of ENSO teleconnection do not currently indicate strong agreement among models.

Disruption of ISM ENSO teleconnection in later decades was discussed (Hrudya et al., 2020a; Seetha et al., 2020; Kripalani and Kulkarni, 1997; Krishna Kumar et al., 1999) with particular emphasis on the latter two decades of the 20th century (Roy, 2018b; Roy and Kripalani, 2019a, 2019b). Most models are unable to reproduce this observed change in the ENSO–monsoon relationship. Instead the models have a strong inverse relationship between monsoon and

ENSO, similar to the observations in the earlier period (Preethi et al., 2010; Rajeevan et al., 2012). Those studies have also shown that the large bias in SST over the IO and the imperfect relationship of monsoon with ENSO and IOD are some of the possible reasons for the lower skill of the models in simulating monsoons during the recent period, compared to the previous period. Other studies further indicated that Pacific Ocean, as well as North Atlantic, also played roles in the change in the pattern of ISM teleconnection (Roy, 2020, 2018b). The role of natural factors was addressed which suggested that the mean state of the Pacific climate and regions of the North Atlantic served crucial parts. CMIP5 model results, in these connections, were shown poorly representing observed influences of natural factors (Roy, 2018b; Roy and Kripalani, 2019b). A hypothesized mechanism is proposed for such a disruption in ISM– ENSO teleconnection which indicated that regional Hadley circulation played a major role (Roy, 2018b).

Overall, this chapter presents a brief overview of our general understanding of teleconnections between some important tropical SST modes and ISM in observation and CMIP5 models. It further indicates areas of biases in CMIP5 models and provides directions for reducing uncertainties in simulating ISM.

11.2   Tropical ocean SST teleconnections to ISM

11.2.1   ISM-IOD teleconnection: Observations

IOD develops in boreal summer (JJA) while peaks in fall (SON) (Saji et al., 1999; Webster et al., 1999). In recent decades two extreme positive IOD occurred, one during 1997 and the other in 2019. During 1997, extreme IOD resulted in drought and bushfire in Indonesia and Australia, whereas flooding in the Eastern part of Africa (Saji et al., 1999; Webster et al., 1999). Another extreme IOD happened in 2019 and it is believed that it has a relevance to widespread Australian bush fire during that year. ISM in 2019, June through September was an excess monsoon (110% of long-period average), whereas the previous excess monsoon was during 1994. There is a positive skewness of the IOD, where positive events tend to be stronger in amplitude than negative events (Cai et al., 2012; Ogata et al., 2012). Australia's millennium drought may be explained in light of tendencies of IOD events (Ummenhofer et al., 2009). Not only IOD, but IOD-ENSO combined behavior is also important and can influence regional rainfall patterns (see Chapter 8). Teleconnections of ENSO, ENSO Modoki, and IOD with rainfall over different parts of the world, especially the East African rainfall was discussed by Preethi et al. (2015). It also showed opposite impacts during some seasons. ENSO and IOD exert an offsetting impact on ISM, with El Niño tend to lower ISM rainfall, whereas a positive IOD tends to increase that (Hrudya et al., 2020b; Li et al., 2017; Ashok et al., 2004).

Using reanalysis and model, correspondence between ENSO, IOD, and ISM was studied in detail (Cherchi and Navarra, 2013; Behera et al., 2006; Krishnaswamy et al., 2014). Fig. 11.2 represents a schematic depicting the

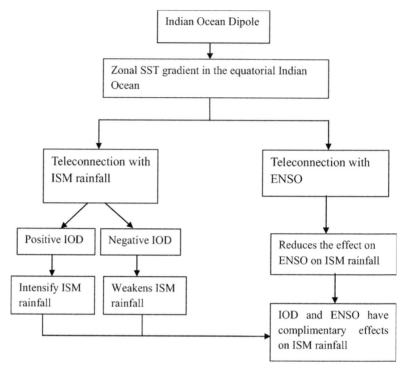

FIG. 11.2 A schematic showing the IOD–ENSO and IOD–ISM relationships (Hrudya et al., 2020b). "*ENSO*, El Niño southern oscillation; *IOD*, Indian Ocean Dipole; *ISM*, Indian summer monsoon."

observed IOD–ENSO and IOD–ISM relationships. Variability in the tropical IO has also found to be strongly connected with the phase of both ENSO and IOD (Chakravorty et al. 2013). Studies also suggested that decadal variability in tropical IO SST influences the monsoon dynamics, that involve Hadley and Walker circulation and modulates ISM (Vibhute et al., 2020; Roy and Collins, 2015). From the observed record, ENSO–IOD correlation is found positive and significant since later decades of the 20th century which may correspond with either weak/strong monsoon–ENSO relationship or strong/weak monsoon–IOD relationship (Cherchi and Navarra, 2013). It is consistent with the fact that over the same time, the teleconnection between ISM–ENSO weakened (Kripalani and Kulkarni, 1997; Krishna Kumar et al., 1999), whereas the relationship strengthened for ISM–IOD (Ashok et al. 2001).

In Table 11.1, IOD years are computed from values averaged in SON exceeding 1 standard deviation (std) from the mean (see details in Cherchi and Navarra, 2013). Those mostly agree with previous classifications (Yuan and Yin, 2008; Saji and Yamagata, 2003, among others). ENSO years are based on November-December-January values exceeding 0.5 std. All La Niña/El

TABLE 11.1 List of La Niña, El Niño, negative, and positive Indian Ocean Dipole years.

	List of years
El Niño	1951 **1957**, 1963, **1965**, 1968, 1969, **1972**, 1976, 1977, **1982**, 1986, 1987, **1991**, **1994**, **1997**, 2002
La Niña	**1949**, 1950, 1954, **1955**, 1956, 1964, 1967, **1970**, 1971, **1973**, **1975**, **1984**, **1988**, 1995, 1998, **1999**, 2000
positive IOD	1961, 1963, 1967, 1972, 1982, 1987, 1991, 1994, 1997, 2002
negative IOD	1956, 1958, 1960, 1964, 1968, 1974, 1975, 1992, 1996, 1998

Cherchi and Navarra (2013)

Niño years in the table agree with the US National Weather Service classification (http://www.cdc.noaa.gov/products/analysis_monitoring/ensostuff/ensoyears.html). Years exceeding 1 std are in bold. At least half the time potentially predictable monsoon years coincide with a positive IOD event or an El Niño (Cherchi and Navarra, 2013). In fact, for all the cases the two events co-occurred; whereas, the correspondence is weaker for negative IOD and La Niña events.

11.2.2   ISM–IOD teleconnection: model results

Community Earth System Model Large Ensemble (CESM-LE) and CMIP5 models simulated well the seasonal phase-locking of IOD (Liu et al., 2014; Hui et al., 2018). A modeling study conducted IODs response under increased greenhouse gas scenario and found models have limitations to account for the role of internal climate variability (Ng et al., 2018). A small difference in the depth of mean thermocline, which is caused by internal climate variability, generates significant variations in the skewness, amplitude of IOD, and the climatological zonal SST gradient. They used the simulations from the CESM-LE model for present-day cases as well as for the future. The performance of 21 CMIP5 models in the simulation of IOD is evaluated and compared with the CMIP3 version (Liu et al. 2014). It is noticed that the ENSO and IOD amplitudes are closely related in models, e.g., if a model generates a weak/strong ENSO, it is likely to simulate a weak/strong IOD too.

Analyzing 34 CMIP5 models, IOD, ENSO, and ISM were analyzed in both historical and future scenarios (Li et al. 2017). For observation, they considered the period 1950–1999. Due to an overly strong control by ENSO, the majority of models simulate an unrealistic IOD and ISM rainfall (ISMR) correlation (Fig. 11.3). In the simulated present-day scenario, a positive IOD is associated with a reduction in ISM rainfall (Fig. 11.3C), which is different from what we discussed in Figs 11.1B and 11.2. Models with a larger IOD amplitude produce a greater ISM–IOD negative correlation. Such an unrealistic

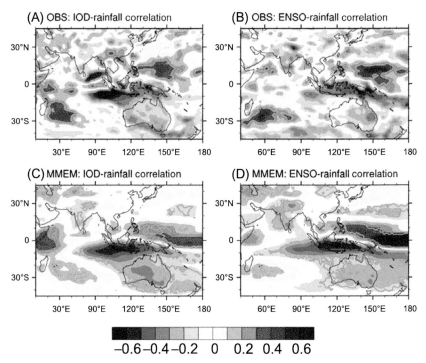

FIG. 11.3 Present-day rainfall correlations with IOD (left) and ENSO (right) for observation (top, A and B) and using the multimodel ensemble mean (MMEM) (bottom, C and D). In (A) and (B) areas within the green contours indicate regions where the correlation is statistically significant above the 95% level. In (C) and (D) the green contours show areas where the sign of the correlation is consistent over 80% of CMIP5 models (Li et al., 2017). "*ENSO*, El Niño southern oscillation; *IOD*, Indian Ocean Dipole."

correlation of the historical period is relevant to the future projection of ISM and will induce an underestimation/overestimation in the projected rainfall. The impact on ISMR from the ENSO and IOD tends to offset and was seen in model simulations as well as in observations. For example, the El Niño of 1997 was strongest in the last century, which alone would reduce ISMR. However, the influence of ENSO on ISMR was overwhelmed by the concurrent 1997 extreme positive IOD event, which was liable for a moderate increase in ISMR. There is no significant rainfall response to IOD events seen in Fig. 11.3A, as the dominant and concurrent effect from ENSO suppresses the impact from IOD. Positive correlation in observation for IOD is only noticed around north eastern region of India. Rainfall responses to ENSO, however, indicates a broadly consistent and significant region, i.e., a reduction in ISM rainfall during El Niño and vice versa for La Niña. (Fig. 11.3B). Uncertainties in the projection of ISM can partly be induced by present-day simulation of ENSO, IOD, their connection and correlations with ISM

(Li et al. 2017). However, there is some consistency around CNE region for either rainfall ENSO teleconnection or rainfall IOD teleconnection through the sign for IOD is different (Fig. 11.3).

11.2.3   ISM–ENSO teleconnection

Similar to IOD, ENSO also develops in boreal summer (JJA) though peaks in boreal winter (DJF) (Wang and Fiedler, 2006). Jourdain et al. (2013) discussed spatial pattern of ISM rainfall using CMIP5 (Fig. 11.4) and CMIP3 models and addressed ISM ENSO teleconnection. Climatological spatial pattern suggests that there are more rainfall in parts of Eastern India and also along regions of Western Ghat from South India. It is followed by the CNE region, which also receives high rainfall. Some CMIP5 models capture such spatial patterns reasonably well and those model results with their names (total 9 in number) and are shown in Fig. 11.4 (right). Recent studies, however, indicated that there is disagreement among models in various ISM-related features (Roy, 2017). For example, in the CNE region of India, a decreasing trend is noticed for ISM in observation during the latter decades of the last century (Bollasina et al., 2011; Goswami et al., 2006) and in one chosen model NOAA GFDL CM3 (Bollasina et al. 2011, shown in Fig. 11.5). That is one feature of ISM among others that suggests inconsistency among CMIP5 models (Roy, 2017). Regions from CNE as used by Bollasina et al. (2011) and Goswami et al. (2006) are marked by CI (76°–87°E, 20°–28°N) and CII (74.5°E–86.5°E and 16.5°N–26.5°N), respectively, in Fig. 11.6. Precipitation time series around the Indian subcontinent vary widely among models and analyses with various future scenarios indicate that the Indian subcontinent shows much larger uncertainty, compared to that from the whole world (Roy, 2017).

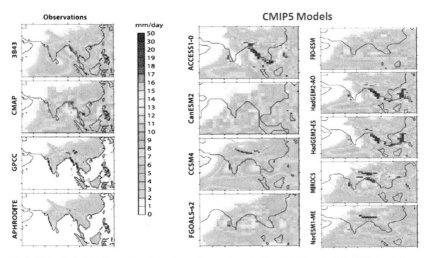

FIG. 11.4   Rainfall climatology in various observations (left) and in historical CMIP5 simulations (right) (after, Jourdain et al. 2013).

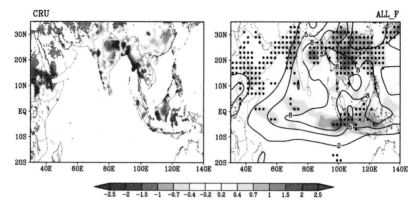

FIG. 11.5 A decreasing trend of ISM rainfall in CNE region of India in observation (CRU data) shown left and the model NOAA GFDL CM3 using all forcing (natural and anthropogenic) shown right (Bollasina et al., 2011). "*CNE*, Central North East; *ISM*, Indian summer monsoon."

11.2.4   ISM in various ENSO phases: historical period

To address various ENSO phases, different locations of the Pacific Ocean need to be identified. Those regions can be marked (Fig. 11.6A) as follows: Canonical region (5°N–5°S, 90°W–140°W), region A (10°S–10°N, 165°E–140°W), region B (15°S–5°N, 110°W–70°W) and region C (10°S–20°N, 125°E–145°E).

Following SST anomalies (SSTAs) in those regions, various definitions for ENSO are used (Ashok et al., 2007; Kao et al., 2009; Kug et al., 2009; Tedeschi et al., 2013):

- ENSO Modoki Index (EMI).
  EMI = (region A SSTA) – 0.5(region B SSTA) – 0.5(region C SSTA).
- ENSO Modoki (ENM/LNM): EMI is greater/less than $0.7\sigma_M$ and region A SSTA is greater than $0.7\sigma_A$. σ_M is the standard deviation (std) of the EMI and σ_A is the std of the region A SSTAs.
- ENSO Canonical (ENC/LNC): SSTA in the Canonical region is greater/less than $0.7\sigma_C$, σ_C is std of SSTAs in that region.
- Mixed ENSO Canonical and Modoki (ENCM/LNCM): SSTAs satisfy both the Canonical as well as Modoki conditions.

Recent studies discussed ISM–ENSO teleconnection in various ENSO phases (Roy et al., 2017, 2019; Roy and Tedeschi, 2016). Correlation and compositing techniques (Ashok et al., 2007; Tedeschi et al., 2013) were mainly used, while the trend was initially removed from the data. Regional teleconnection in specified locations of India (as shown in Fig. 11.6B and C) was also addressed. For correlation analyses, correlation between ISM rainfalls with the SST of canonical region, region A and the EMI index were computed and respective spatial plots were analyzed. The level of significance for the correlation analysis was tested using Student's t test. For compositing, the average value of various meteorological parameters (precipitation, meridional, and zonal wind at 850 mb and 200 mb) was calculated for chosen Indian regions, in various ENSO phases.

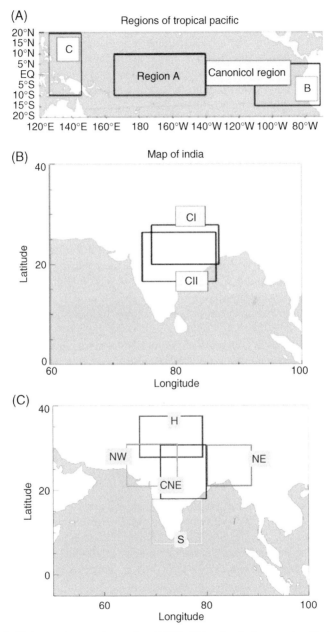

FIG. 11.6 Various regions used to explore ENSO–ISM regional teleconnection. Regions of the tropical Pacific used for defining different types of ENSO (A); map of India showing CI and CII region from CNE (B); map of India subdivided in five different arbitrary chosen regions—CNE, Hilly region (H), North West (NW), North East (NE), and Southern region (S) (C). "*CNE*, Central North East; *ENSO*, El Niño southern oscillation; *ISM*, Indian summer monsoon."

Years of ENSO phases were calculated following the definition of respective phases. Those were then subtracted from the average of the whole record to indicate the anomaly value of that parameter in the respective ENSO phases. For significant testing, the hypergeometric test (Meyer, 1997; Ropelewski and Halpert, 1987) is applied which is also used in other studies (Tedeschi et al., 2013; Grimm, 2004). Compositing is useful to better capture asymmetries in response to El Niño from La Niña phases.

Results showed that not all regions in India are affected by ENSO (Roy, 2017; Roy et al., 2017). So, two regional teleconnections were focused: (1) a positive rainfall signal around "hilly" and CNE region of India during La Niña (and vice versa for El Niño) and (2) a similar signal for regions of southern peninsular India (regions shown in Fig. 11.6C). Correlation study indicated that more than 50% of the CMIP5 models captured these two regional teleconnections, while the first signal is captured by more than 80% of models (Roy et al. 2017). Using compositing technique, it is found that most models again agree on the sign of regional teleconnection around the CNE and hilly region of India. That suggests the robustness of ENSO teleconnection in that region. Separating models based on High Top or Low Top did not make much difference. Even many CMIP6 models are unable to represent the regional summer rainfall patterns over India associated with ENSO (Mahendra et al. 2020).

Using CMIP5 model outputs in different ENSO phases, studies further investigated regional teleconnection in five arbitrarily chosen regions of the Indian subcontinent (Roy and Tedeschi, 2016), namely, CNE region, Hilly region (H), North West (NW), North East (NE), and Southern India (S) (Fig. 11.6C). To overcome issues of resolution in CMIP5 models, each region was kept at least 10 degrees wide, both longitude and latitude-wise. Local wind fields and remote influences from the tropical Pacific were considered and results were compared with observations/reanalysis to pinpoint areas of agreement and shortcomings. Meridional and zonal components of wind at 200 mb and 850 mb levels were explored.

Results among models and observations are more consistent for zonal velocity at 200 hPa (u-200) in CNE region (Fig. 11.7). It is also true for CI and CII regions (Roy et al. 2017). In CNE region, precipitation is generally negatively correlated with the local zonal eastward velocity at 200 hPa (u-200) in CMIP5 models and observation (Fig. 11.7). For La Niña, more precipitation, in general, is seen for almost all models, though less for El Niño. Moreover, during El Niño, u-200 is positive in almost all cases, while for La Niña it is negative which indicates a change in the direction of Walker circulation. However, analyses with meridional component of wind (v) at 200 mb (v-200) indicate circulation of the upper branch of Hadley cells in regions CNE and Southern India, though suggest the best agreement among models in comparison with other fields, but there are some deviations from observations, indicating missing mechanisms in models.

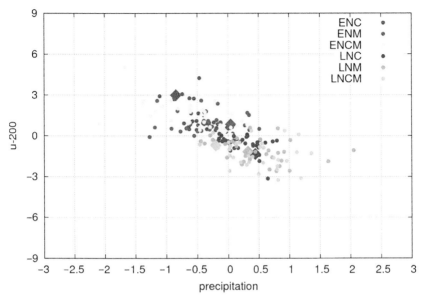

FIG. 11.7 Composite anomaly (JJA) plot around CNE region (18°N–31°N, 86°E–75°E) for ISM precipitation (mm/day) versus local zonal eastward velocity at 200 hPa (u-200) (m/s) in different phases of El Niño and La Niña in CMIP5 models. Composites for ENC, ENM, and ENCM are shown by red, pink ,and yellow, whereas that of LNC, LNM, and LNCM are shown by blue, green, and cyan, respectively. Data of observation/reanalyses are shown by the same colored large diamond. If models are in agreement with observations, then those diamonds are masked in the schematics. Results are similar if either CI or CII regions are used instead of CNE region (Roy and Tedeschi, 2016). "*ENC*, Canonical El Niño; *ENCM*, Canonical and Modoki El Niño; *ENM*, El Niño Modoki; *ISM*, Indian summer monsoon; *LNC*, *Canonical La Niña*; *LNCM*, Canonical and Modoki La Niña; *LNM*, La Niña Modoki."

Fig. 11.8 shows composite anomaly plot (JJA) for model ensemble mean in nine selected models (models shown in Fig. 11.4) for ISM (top) and SST (bottom) in different El Niño and La Niña cases. Results are also similar using all CMIP5 models. Significant regions at 95% level are outlined. Around CNE, there is less rain for all EN phases, while more rain for LN. Moreover, for SST there is warming around tropical Pacific for all EN cases while cooling for LN phases. Thus, there is a clear connection (anticorrelation) between tropical Pacific SST and ISM around CNE region of India in various ENSO phases (Fig. 11.8) (Roy et al. 2017). It is also consistent with the direction of the wind in Walker circulation (Fig. 11.7) and matches with observation too.

11.2.5   ISM in various ENSO phases: historical vs RCP scenario

Teleconnections between ISM and ENSO in CMIP5 simulations were explored by comparing historical period with future scenario (Roy et al. 2019). The

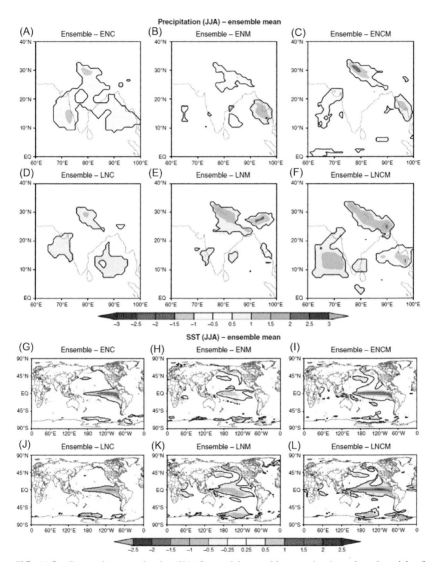

FIG. 11.8 Composite anomaly plot (JJA) for model ensemble mean in nine selected models of ISM precipitation (mm/day) (top, A–F) and SST (°C) (bottom, G–L). Composites are shown in different El Niño (A–C, G–I) and La Niña cases (D–F, J–L) (Roy et al., 2017). "*ISM*, Indian summer monsoon; *SST*, sea surface temperature."

correlation coefficients in the RCP and historical scenario between tropical Pacific SST and ISM are shown in two regions of CNE, namely, CI and CII (Fig. 11.9).

Almost all models show a negative correlation for the RCP as well as historical scenarios. Clustering around the one-to-one line is clearly noticed.

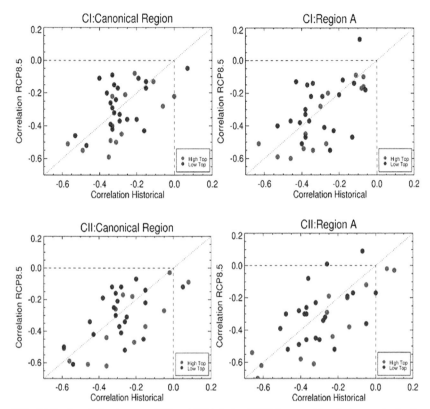

FIG. 11.9 Correlation between SST and ISM during RCP8.5 versus historical scenario. In the right, SST is chosen in region (A), whereas in the left canonical region is chosen. ISM rainfall is considered in two different regions of CNE, CI (top) and CII (bottom). In both scenarios, if model results show a negative correlation those are confined in a box of dashed lines (Roy et al., 2019). "*CNE*, Central North East; *ISM*, Indian summer monsoon; *SST*, sea surface temperature."

A tighter relationship, with lesser spread among models, is apparent when instead of SST of region A, the canonical region is used and also for CI instead of CII, respectively. These correlations are relatively insensitive to the choice of SST indices, (either Niño4 and Niño3 instead of region A and canonical region, respectively). High Top or Low Top models do not make many differences.

A subset of models was selected, based on their ability to simulate the spatial pattern of ISM in the historical period (Jourdain et al. 2013, also shown in Fig.11.4), and those are used in the CNE region, to examine future predictions (Fig. 11.10). Results were also similar using all models. For mixed canonical Modoki events and pure canonical ENSO events, the teleconnection is spatially extended in the future over most of India. Interestingly for pure ENSO Modoki events, that teleconnection disappears, and practically no influence is

El Niño

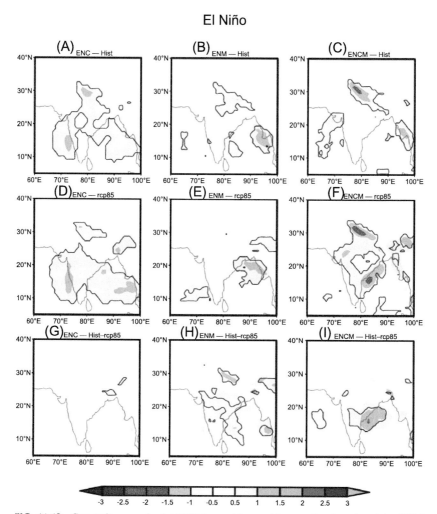

FIG. 11.10 Composite anomaly plot (JJA) for model ensemble mean in selected models of ISM (mm/day) for El Niño and La Niña cases. ISM composites in historical (A–C), RCP8.5 (D–F) and historical-RCP8.5 scenarios (G–I) are shown. Results are similar combining all models. Ensemble mean for various ENSO phases are shown [ENC/ LNC (A, D, G), ENM/ LNM (B, E, H) and ENCM/LNCM (C, F, I)]. Significant regions are marked by a black outline (Roy et al., 2019). "*ENC*, Canonical El Niño; *ENCM*, Canonical and Modoki El Niño; *ENM*, El Niño Modoki; *ENSO*, El Niño southern oscillation; *ISM*, Indian summer monsoon; *LNC*, Canonical La Niña; *LNCM*, Canonical and Modoki La Niña; *LNM*, La Niña Modoki."

detected in the future in any parts of India (Fig. 11.10). A rainfall decomposition technique revealed a battle between changes in moisture change which act to strengthen the rainfall teleconnection and changes in circulation which act to weaken it (Roy et al. 2019).

La Niña

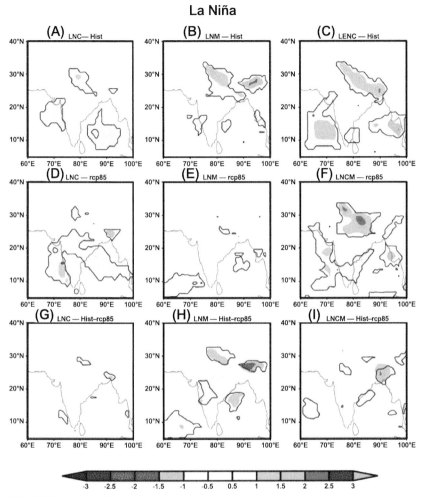

FIG. 11.10   (continued)

11.3   Mechanisms: role of natural factors

Studies addressed on mechanisms and discussed the importance on an improved understanding of the tropic and mid-latitude connection (Prabhu et al., 2016; Roy and Kripalani, 2019a; Roy, 2018b). Moreover, Roy and Kripalani (2019a) analyzed the role of natural factors, mainly the sun on ISM in a holistic way by exploring various major tropospheric and stratospheric modes. Initiated by solar decadal variability it depicted coupling in the ocean–atmosphere system, by formulating a flow chart, that involved ENSO (Fig. 11.11). Possible mechanisms for ENSO Modoki (or CP ENSO),

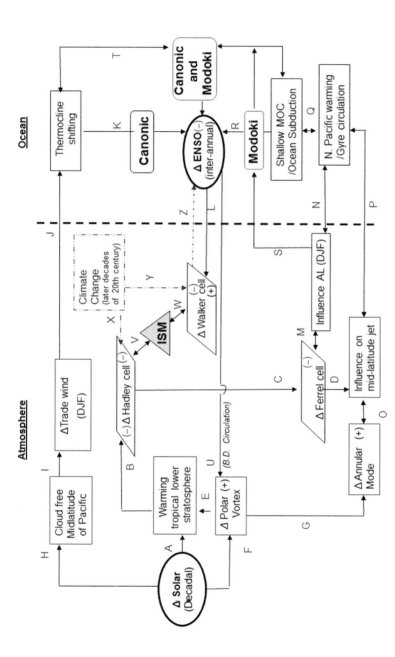

FIG. 11.11 A flow chart is presented depicting the role of the sun in atmosphere and ocean coupling mainly during DJF. Possible mechanisms for ENSO Modoki, Canonic ENSO, and Canonic–Modoki ENSO are suggested. Three major variability, namely, solar, QBO, and ENSO are shown by oval outlines, whereas major circulations, responsible for modulating the effect are shown by nonrectangular parallelograms. Pathways of signals are marked by "A" – "Z," e.g., Brewer Dobson (B.D.) circulation via ENSO can influence Polar vortex is shown by "U." *Dash-dotted lines* indicate how ENSO-related teleconnection (e.g., ISM) can be altered in the latter decades of the last century. Directions of change in behavior are shown by "+" (for increase) or "–" (for decrease) (Roy and Kripalani, 2019a). *DJF*, December-January-February; *ENSO*, El Niño southern oscillation; *ISM*, Indian summer monsoon.

Canonic (or EP) ENSO, and Canonic–Modoki (EP and CP) ENSO were proposed considering their relevance to the decadal-scale variation of mid-latitude jets, Hadley and Walker circulation. The upper stratospheric polar vortex features were included too. Some results of solar signature were presented which could possibly trigger different types of ENSO, agreeing with proposed pathways of the flow chart. Since the 1970s, there was a changing behavior of the ENSO (Ashok and Yamagata, 2009; Yeh et al., 2009), which was also addressed.

Mechanisms of ENSO–ISM teleconnection and a change in the later period were addressed even in many earlier studies (Kripalani and Kulkarni, 1997; Krishna Kumar et al., 1999, etc). ENSO–ISM relationship had significant changes in the past three multidecadal epochs (early: 1931–1960, middle: 1961–1990, and recent: 1991–2015) (Seetha et al., 2020). The rainfall during the early epoch was above normal, whereas in the last two, it was relatively dry. The ISMR–ENSO connection and its deviation are more distinct in CNE India. During the early epoch, the effect of La Niña was more dominant which can be attributed to the excess rainfall, whereas the ISM during the consecutive two multidecadal epochs is below normal and may be attributed to the strong El Niño events. The changes in equatorial Walker and regional Hadley circulations during the three epochs are linked to the changes in rainfall and other circulation-related parameters (Seetha et al. 2020).

Some studies further focused later two decades of the last century where CP ENSO became more persistent and frequent (Ashok and Yamagata, 2009; Yeh et al. 2009) and disruption in ISM–ENSO connection was more pronounced (Roy, 2018b; Roy and Kripalani, 2019b; Kumar et al. 1999). There was also a common shifting point for regional monsoon–ENSO relationship around the 1970s with a recovery in the late 1990s (Yim et al., 2013). Roy (2018b) separated out a period 1976–1996 that covered two full solar cycles, where two explosive volcanos erupted during active phases of strong solar cycles. The similar period also matched the period of abrupt global warming. A hypothesized mechanism was proposed that could be generated by explosive volcanos and initiated via a preferential alignment of the NAO phase (Fig. 11.12). Inciting extratropical Rossby wave to modulate the Aleutian Low (AL) had an influence on CP ENSO. The usual effect of East-West (E-W) Walker circulation on ISM was partly overtaken by North-South (N-S) Hadley circulation. Winter NAO and Eurasian snow played an important role to modulate N-S Hadley circulation (see Chapter 12). Disruption of ENSO and ISM teleconnection during that period of abrupt warming and a subsequent recovery thereafter can be explained from that angle. Interestingly, the ensemble of CMIP5 model, and also individual models, fails to conform with such observation. These issues were further elaborated by Roy and Kripalani (2019b). Observation suggested that the regional Hadley circulation, via NAO in the northern hemisphere and IOD in the southern hemisphere, had roles in the change in ISM behavior (Roy and Collins, 2015). More recently, importance of western north Pacific low-level circulation influence on modulating

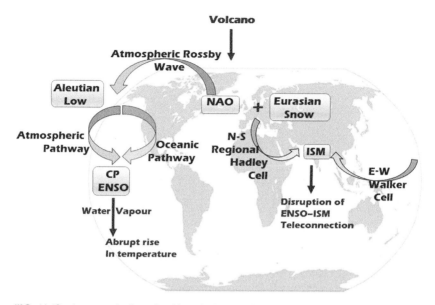

FIG. 11.12 A proposed schematic of hypothesized mechanisms for a change in ISM–ENSO teleconnection during 1976–1996 is presented. It is initiated by explosive volcanos, those erupted during active phases of strong solar cycles. That influences various places [Aleutian Low (AL) and Eurasian snow cover] and modulates various modes (ISM, NAO, and CP ENSO), shown by the blue box. From North Atlantic, AL is impacted via atmospheric Rossby waves. That signal is transported to trigger CP ENSO through Oceanic (*green arrow*) and atmospheric pathways (*pink arrow*). Increased CP ENSO raises atmospheric water vapor and can cause an abrupt rise in global temperature (the effect is shown by red). Combination of Eurasian Snow and NAO strongly modulates N-S regional Hadley circulation (*yellow arrow*) to impact ISM. A disruption in usual ENSO–ISM teleconnection is thus seen (shown by red) which normally occurs via E-W Walker circulation (*yellow arrow*) (Roy, 2018b). "*CP ENSO*, Central Pacific ENSO; *ISM*, Indian summer monsoon; *NAO*, North Atlantic oscillation."

ENSO–ISM teleconnections is highlighted (Mahendra et al., 2020; Ramu et al., 2020). Such features though well captured in the observation but shown missing in models (Roy and Kripalani, 2019b). Analyses on global hydrological cycle suggest that the differences among models mainly originate in regional levels and could be related to inconsistency in representing regional teleconnection features. Interestingly, in terms of global thermodynamic scaling arguments, all models perform reasonably well (Roy and Kripalani, 2019b).

11.4 Discussion and summary

The variability of ISM has major impacts on the socioeconomic structure of the country and this study focuses on the main factors that cause such variability. Teleconnection between ISM with two dominant tropical SST modes, namely, the ENSO and IOD are the key focus here. Results are presented and discussed using observed/ reanalyses data and CMIP5 simulations in historical as well as future scenarios. Climatology and some features of ISM though captured well in some CMIP5 models, but there are large disagreements among models in various ISM-related features. Time series of ISM in future scenarios indicate that there is much larger uncertainty, compared to the rainfall of the whole world.

ENSO and IOD exert an offsetting impact on ISM, with El Niño tend to lower ISM rainfall, whereas a positive IOD tends to increase that. There is a positive skewness of the IOD where positive events tend to be stronger in amplitude than negative events. Moreover, the ENSO–IOD connection is positive and significant since the latter decades of the last century and it may correspond with either weak (strong) monsoon–ENSO relationship or strong (weak) monsoon–IOD relationship. The period when ISM–ENSO teleconnection weakened the teleconnection between ISM–IOD strengthened. Model results suggested that seasonal phase-locking of IOD is simulated well though models have limitations to account for the role of internal climate variability. If a model simulates a weak/strong ENSO, it is shown to generate a weak/strong IOD too. Due to an overly strong control by ENSO, the majority of CMIP5 models simulate an unrealistic IOD–ISMR correlation. In the simulated present-day scenario, a positive IOD is associated with a reduction in ISM rainfall. Such unrealistic correlation of the historical period is relevant to the future projection of ISM and will cause an underestimation/overestimation in the projected rainfall. Uncertainties in the projection of ISM can partly be induced by present-day simulation of ENSO, IOD, their connection, and their correlations with ISM.

For ENSO and ISM teleconnection, especially three different categories, i.e., Canonical, Modoki, and mixed Canonical, and Modoki cases are focused. Correlation and compositing technique both indicated that CMIP5 models capture regional teleconnections in CNE well in models as in observations. Even 80% of models agree on the sign of regional teleconnection around the CNE, suggesting the robustness of ENSO teleconnection in that region. In the CNE region, precipitation is generally negatively correlated with the local zonal eastward velocity at 200 hPa (u-200) in CMIP5 models and matches with observation. However, for the meridional velocity at 200 hPa (v-200), models though suggest highly consistent behavior but deviates from observation. In the historical period for all ENSO subcategories, there is a clear connection between the Walker circulation, tropical Pacific SST, and ISM rainfall around the CNE India in CMIP5 models and observation/reanalyses.

Using a subset of selective models, ISM and the ENSO teleconnection were further analyzed for a future scenario and compared with historical cases in various ENSO events. Interestingly, for future cases, the situation changes. For mixed canonical Modoki events and pure canonical ENSO events, that teleconnection from regions of CNE India is spatially extended over most of India in the future, though disappears for pure ENSO Modoki events. Practically no influence is detected in future for pure Modoki cases in any part of India. However, in terms of ISM–ENSO teleconnection, when the focus is only on tropical Pacific SST, almost all models show a negative correlation in regions of CNE India (CI and CII) not only for historical scenarios but also for future cases. These correlations are relatively insensitive to the choice of various SST indices (either Niño4 and Niño3 instead of SST of region A and canonical region, respectively). Lesser spread among models is apparent when SST of the canonical region is used instead of region A and also if CI from CNE India is considered instead of CII. A rainfall decomposition technique revealed a battle between changes in moisture which acts to strengthen the rainfall teleconnection and changes in circulation which acts to weaken it.

Focusing on mechanisms, tropic, mid-latitude connection, and the role of natural factors were addressed. Formulating a flow chart, coupling in the atmosphere and ocean is presented, which is initiated by solar decadal-scale variability and involved different types of ENSO. Considering relevance to the decadal-scale variation of Hadley, Walker circulation and mid-latitude jets, possible mechanisms for ENSO Modoki (or CP ENSO), Canonical (or EP) ENSO, and Canonical–Modoki (or CP and EP) ENSO were proposed. Finally, disruption of ISM–ENSO teleconnection in the latter two decades of the 20th century is addressed. One hypothesized mechanism is proposed that could be originated in the North Atlantic and modulates regional Hadley circulation, which partly overtakes influence from the Walker circulation on ISM.

Overall, this study analyzed results of observation and models to improve our understanding of the interaction between some important tropical SST modes and ISM. It would be useful to advance future prediction skills of ISM rainfall and reduce uncertainty in models.

Conflict of Interest

There is no conflict of interest.

Funding

This work did not receive any funding.

Acronyms

AL	Aleutian Low.
CESM-LE	Community Earth System Model Large Ensemble.
CMIP5	Coupled Model Inter-comparison Project Phase 5.
CNE	Central North East.
CP ENSO	Central Pacific ENSO.
DJF	December-January-February.
E-W	East West.
EP ENSO	East Pacific ENSO.
ENC	Canonical El Niño.
ENCM	Canonical and Modoki El Niño.
ENM	El Niño Modoki.
ENSO	El Niño Southern Oscillation.

H	High Top.
IO	Indian Ocean.
IOD	Indian Ocean Dipole.
ISM	Indian Summer Monsoon.
ISMR	Indian Summer Monsoon Rainfall.
JJA	June-July-August.
L	Low Top.
LNC	Canonical La Niña.
LNCM	Canonical and Modoki La Niña.
LNM	La Niña Modoki.
MMEM	Multi-Model Ensemble Mean.
MOC	Meridional Overturning Circulation.
NAO	North Atlantic Oscillation.
N-S	North-South.
RCP	Representative Concentration Pathway.
SON	September-October-November.
SST	Sea Surface Temperature.

References

Ashok, K., Guan, Z., Yamagata, T., 2001. Impact of Indian Ocean dipole on the relationship between the Indian monsoon rainfall and ENSO. Geophys. Res. Lett. 28 (23), 4499–4502.

Ashok, K., Guan, Z., Saji, N.H., Yamagata, T., 2004. Individual and combined influences of ENSO and the Indian Ocean Dipole on the Indian summer monsoon. J. Clim. 17, 3141–3155.

Ashok, K., Behera, S.K., Rao, S.A., Weng, H., Yamagata, T., 2007. El Niño Modoki and its possible teleconnection. J. Geophys. Res. 112, C11007. doi:10.1029/2006JC003798.

Ashok, K., Yamagata, T, 2009. Climate change the El Niño with a difference. Nature 461, 481–484. doi:10.1038/461481a.

Azad, S., Rajeevan, M., 2016. Possible shift in the ENSO-Indian monsoon rainfall relationship under future global warming. Sci. Rep. 6, 20145.

Behera, S.K., Luo, J.J., Masson, S., Rao, S.A., Sakuma, H., 2006. A GCM study on the interaction between IOD and ENSO. J. Clim. 19, 1688–1705.

Bollasina, M.A., et al., 2011. Anthropogenic aerosols and the weakening of the South Asian summer monsoon. Science 502. doi:10.1126/science.1204994 334.

Brown, J.N., McIntosh, P.C., Pook, M.J., Risbey, J.S., 2009. An investigation of the links between ENSO flavors and rainfall processes in Southeastern Australia. Mon. Weather Rev. 137, 3786–3795.

Cai, W., Cowan, T., 2009. La Niña Modoki impacts Australia autumn rainfall variability. Geophys. Res. Lett. 36 L12805doi:2009GL037885.

Cai, W., Qiu, Y., 2012. An observation-based assessment of nonlinear feedback processes associated with the Indian Ocean Dipole. J. Clim. 26, 2880–2890.

Cai, et al., 2015. ENSO and greenhouse warming. Nat. Clim. Change 5, 849–859.

Cappondonti, A., et al., 2015. Understanding ENSO diversity. Bull. Am. Meteor. Soc. 96, 921–938. doi:10.1175/BAMS-D-13-00117.1.

Chakravorty, S., Chowdary, J.S., Gnanaseelan, C., 2013. Spring asymmetric mode in the tropical Indian Ocean: role of El Nino and IOD. Clim. Dyn. 40, 1467–1481. doi:10.1007/s00382-012-1340-1.

Charlton-Perez, A.J., 2013. On the lack of stratospheric dynamical variability in low-top versions of the CMIP5 models. J. Geophys. Res. Atmos. 118 (6). doi:10.1002/jgrd.50125.

Cherchi, A., Navarra, A., 2013. Influence of ENSO and of the Indian Ocean Dipole on the Indian summer monsoon variability. Clim. Dyn. 41, 81–103 https://doi.org/10.1007/s00382-012-1602-y.

Chou, C., Neelin, J., Chen, C., Tu, J., 2009. Evaluating the "rich-get-richer" mechanism in tropical precipitation change under global warming. J. Clim. 22, 1982–2005.

Chung, C.T.Y., Power, S.B., Arblaster, J.M., Rashid, H.A., Roff, G.L., 2014. Nonlinear precipitation response to El Niño and global warming in the Indo-Pacific. Clim. Dyn. 42, 1837–1856. doi:10.1007/s00382-013-1892-8.

Gill, A.E., 1980. Some simple solutions of heat induced tropical circulations. Q. J. R. Meteorol. Soc 106, 447–462.

Goswami, B.N., Venugopal, V., Sengupta, D., Madhusoodanan, M.S., Xavier, P.K., 2006. Increasing trend of extreme rain events over india in a warming environment. Science 314, 1442.

Grimm, A.M., 2004. How do La Niña events disturb the summer monsoon system in Brazil? Clim. Dyn. 22, 123–138.

Held, I.M., Soden, B.J., 2006. Robust responses of the hydrological cycle to global warming. J. Clim. 19, 5686.

Hrudya, P.H., Varikoden, H., Vishnu, R., et al., 2020a. Changes in ENSO-monsoon relations from early to recent decades during onset, peak and withdrawal phases of Indian summer monsoon. Clim. Dyn. 55, 1457–1471. doi:10.1007/s00382-020-05335-x.

Hrudya, P.H., Varikoden, H., Vishnu, R., 2020b. A review on the Indian summer monsoon rainfall, variability and its association with ENSO and IOD. Meteorol. Atmos. Phys. doi:10.1007/s00703-020-00734-5.

Huang, P., 2014. Regional response of annual-mean tropical rainfall to global warming. Atmos. Sci. Lett. 15, 103–109. doi:10.1002/asl2.475.

Huang, P., Xie, S.-P., 2015. Mechanisms of change in ENSO-induced tropical Pacific rainfall variability in a warming climate. Nat. Geosci. 8, 922–926. doi:10.1038/ngeo2571.

Hui, C., Zheng, X., 2018. Uncertainty in Indian Ocean Dipole response to global warming: the role of internal variability. Clim. Dyn. 51, 3597–3611. doi:10.1007/s00382-018-4098-2.

Hurwitz, M.M., et al., 2014. Extra-tropical atmospheric response to ENSO in the CMIP5 Models. Clim. Dyn. doi:10.1007/s00382-014-2110-z.

John, O., Allan, R.P., Soden, J., 2009. How robust are observed and simulated precipitation responses to tropical ocean warming? Geophys. Res. Lett. 36, L14702.

Jourdain, N.C., Sen Gupta, A., Taschetto, A.S., Ummenhofer, C.C., Moise, A.F., Ashok, K., 2013. The Indo-Australian monsoon and its relationship to ENSO and IOD in reanalysis data and the CMIP3/CMIP5 simulations. Clim. Dyn. 41, 3073–3102. doi:10.1007/s00382-013-1676-1.

Kao, H.-Y., Yu, J.-Y., 2009. Contrasting eastern-Pacific and central-Pacific types of El Nino. J. Clim. 22, 615–632.

Kripalani, R.H., Kulkarni, A., 1997. Climatic impact of El Niño/La Niña on the Indian monsoon: a new perspective. Weather 52, 39–46.

Krishna Kumar, K., Rajagopalan, B., Cane, M.A., 1999. On the weakening relationship between the Indian monsoon and ENSO. Science 284, 2156–2159.

Krishnaswami, J., Vaidyanathan, S., Rajagopalan, B., Bonnel, M., Sankaran, M., Bhalla, R.S., et al., 2014. Non-stationary and non-linear influence of ENSO and Indian Ocean Dipole on Indian summer monsoon rainfall and extreme rain events. Clim. Dyn. doi:10.1007/s00382-014-2288-0.

Kug, J.-S., Jin, F.-F., An, S.-I., 2009. Two types of El Niño events: cold tongue El Niño and warm pool El Niño. J. Clim. 22, 1499–1515.

Krishna Kumar, K., Rajagopalan, B., Cane, M.A., 1999. On the weakening relationship between the Indian monsoon and ENSO 284 (5423), 2156–2159. doi:10.1126/science.284.5423.2156.

Kumar, K.K., Rajagopalan, B., Hoerling, M., Bates, G., Cane, M., 2006. Unraveling the mystery of Indian monsoon failure during El Niño. Science 314, 115–119.

Li, Z., Lin, X., Cai, W., 2017. Realism of modelled Indian summer monsoon correlation with the tropical Indo-Pacific affects projected monsoon changes. Sci. Rep. 7 (1), 4929. doi:10.1038/s41598-017-05225-z https://doi.org/.

Liu, L., et al., 2014. Indian Ocean variability in the CMIP5 multi-model ensemble: the zonal dipole mode. Clim. Dyn 43, 1715–1730.

Mahendra, N., Chowdary, J.S., Darshana, P., Sunitha, P., Parekh, A., Gnanaseelan, C., 2020. Inter-decadal modulation of interannual ENSO-Indian Summer monsoon rainfall teleconnections in observations and CMIP6 models: regional patterns. Int. J. Climatol. doi:10.1002/joc.6973.

Meyer, P.L., 1970. Introductory Probability and Statistical Applications. Addison-Wesley, Boston, MA.

Ng, B., Cai, W., Cowan, T., et al., 2018. Influence of internal climate variability on Indian Ocean Dipole properties. Sci. Rep. 8, 13500. doi:10.1038/s41598-018-31842-3.

Ogata, T., Xie, S.-P., Lan, J., Zheng, X.-T., 2012. Importance of ocean dynamics for the skewness of the Indian Ocean Dipole Mode. J. Clim. 26, 2145–2159.

Osprey, S.M., Gray, L.J., Hardiman, S.C., Butchart, N., Hinton, T.J., 2013. Stratospheric variability in twentieth-century CMIP5 simulations of the met office climate model: high top versus low top. J. Clim. 26, 1595–1606.

Prabhu, A., Kripalani, R.H., Preethi, B., Pandithurai, G., 2016. Potential role of the February–March southern annular mode on the Indian summer monsoon rainfall: a new perspective. Clim. Dyn 47 (3), 1161–1179.

Preethi, B., Kripalani, R.H., Krishna Kumar, K., 2010. Indian summer monsoon rainfall variability in global coupled ocean-atmospheric models. Clim. Dyn. 35, 1521–1539.

Preethi, B., Sabin, T., Adedoyin, J., et al., 2015. Impacts of the ENSO Modoki and other tropical Indo-Pacific climate-drivers on African Rainfall. Sci. Rep. 5, 16653. doi:10.1038/srep16653.

Rajeevan, M., Unnikrishnan, C.K., Preethi, B., 2012. Evaluation of the ENSEMBLES multi-model seasonal forecasts of Indian summer monsoon variability. Clim. Dyn. 38, 2257–2274.

Ramu, D.A., Chowdary, J.S., Pillai, P.A., Sradhara, S.N.S., Koteswararao, K., Ramakrishna, S.S.V.S., 2020. Impact of El Niño Modoki on Indian summer monsoon rainfall: role of western north Pacific circulation in observations and CMIP5 models. Int. J. Climatol. 40, 2117–2133.

Rasmussen, E.M., Carpenter, T.H., 1983. The relationship between eastern equatorial Pacific SST and rainfall over India and Sri Lanka. Mon. Weather Rev. 111, 517–552.

Riahi, K., Rao, S., Krey, V., Cho, C., Chirkov, V., Fischer, G., et al., 2011. RCP-8.5—a scenario of comparatively high greenhouse gas emission. Clim. Change 109, 33–57. doi:10.1007/s10584-011-0149-y.

Ropelewski, C.F., Halpert, M.S., 1987. Global and regional scale precipitation patterns associated with the El Niño-Southern Oscillation. Mon. Weather Rev. 115, 1606–1626.

Roy, I., Collins, M., 2015. On Identifying the role of Solar variability and the El Niño Southern Oscillation on Indian summer monsoon rainfall. Atmos. Sci. Lett. 16 (2), 162–169. doi:10.1002/asl2.547.

Roy, I., Tedeschi, R.G., 2016. 'Influence of ENSO on regional ISM precipitation—local atmospheric Influences or remote influence from Pacific. Atmosphere 7, 25. doi:10.3390/atmos7020025.

Roy, I., Tedeschi, R.G., Collins, M., 2017. 'ENSO teleconnections to the Indian summer monsoon in observations and models. Int. J. Climatol. 37 (4), 1794–1813. doi:10.1002/joc.4811.

Roy, I., 2017. Indian summer monsoon and El Niño southern oscillation in CMIP5 models: a few areas of agreement and disagreement. Atmosphere 8 (8), 154. doi:10.3390/atmos8080154.

Roy, I., 2018. 'Climate Variability and Sunspot Activity—Analysis of the Solar Influence on Climate'. Springer Nature 18 chapters, 216 pages, ISBN 978-3-319-77107-6, doi:10.1007/978-3-319-77107-6.

Roy, I, 2018b. Addressing on abrupt global warming, warming trend slowdown and related features in recent decades. Front. Earth Sci. 6, 136. doi:10.3389/feart.2018.00136.

Roy, I., Gagnon, A.S., Siingh, D., 2018. 'Evaluating ENSO teleconnections using observations and CMIP5 models. Theor. Appl. Climatol. doi:10.1007/s00704-018-2536-z 2018.

Roy, I., Kripalani, R., 2019a. 'The role of natural factors (Part 1): addressing on mechanism of different types of ENSO, related teleconnections and solar influence'. Theor. Appl. Climatol., 1–12. doi:10.1007/s00704-018-2597-z.

Roy, I., Kripalani, R., 2019b. The role of natural factors (Part 2): Indian summer monsoon in climate change period—Observation and CMIP5 models. Theor. Appl. Climatol. doi:10.1007/s00704-019-02864-2.

Roy, I., Tedeschi, R.G., Collins, M., 2019. ENSO teleconnections to the Indian summer monsoon under changing climate. Int. J. Climatol. doi:10.1002/joc.5999.

Roy, I., 2020. Major climate variability and natural factors in boreal winter. Pure Appl. Geophys. doi:10.1007/s00024-020-02522-z.

Saji, N.H., Goswami, B.N., Vinayachandran, P.N., Yamagata, T., 1999. A dipole mode in the tropical Indian Ocean. Nature 401, 360–363.

Saji, N.H., Yamagata, T., 2003. Possible impacts of Indian Ocean dipole mode events on global climate. Clim. Res. 25, 151–169.

Seager, R., Naik, N., Vecchi, G.A., 2010. Thermodynamic and dynamic mechanisms for large-scale changes in the hydrological cycle in response to global warming. J. Clim. 23, 4651–4668. doi:10.1175/2010JCLI3655.1.

Seager, R., Naik, N., Vogel, L., 2012. Does global warming cause intensified interannual hydroclimate variability? J. Clim 25, 3355–3372. doi:10.1175/JCLI-D-11-00363.1.

Seetha, C.J., et al., 2020. Significant changes in the ENSO-monsoon relationship and associated circulation features on multidecadal timescale. Clim. Dyn. 54, 1491–1506.

Seviour, W.J.M., Gray, L.J., Mitchell, D.M, 2016. Stratospheric polar vortex splits and displacements in the high-top CMIP5 climate models. J. Geophys. Res.: Atmos. 121 (4), 1400–1413.

Sikka, D.R., 1980. Some aspects of large-scale fluctuations of summer monsoon rainfall over India in relation to fluctuations in the planetary and regional scale circulation parameters. Proc. Indian Acad. Sci. (Earth Planet Sci.) 89, 179–195.

Taylor, K.E., Stouffer, R.J., Meehl, G.A., 2012. An overview of CMIP5 and the experimental design. Bull. Am. Meteor. Soc. 93, 485–498. doi:10.1175/BAMS-D-11-00094.1.

Tedeschi, R.G., Cavalcanti, I.F.A., Grimm, A.M., 2013. Influences of two types of ENSO on South American precipitation. Int. J. Climatol. 33, 1382–1400. doi:10.1002/joc.3519.

Tedeschi, R.G., Collins, M., 2016. The influence of ENSO on South American precipitation during austral summer and autumn in observations and models. Int. J. Climatol. 36 (2), 618–635.

Ummenhofer, C.C., et al., 2009. What causes southeast Australia's worst droughts? Geophys. Res. Lett. 36, L04706.

Vibhute, A., Halder, S., Singh, P., Parekh, A., Chowdary, J.S., Gnanaseelan, C., 2020. Decadal variability of tropical Indian Ocean sea surface temperature and its impact on the Indian summer monsoon. Theor. Appl. Climatol. 141, 551–566. doi:10.1007/s00704-020-03216-1.

Wang, C., Fiedler, P.C., 2006. ENSO variability and the eastern tropical Pacific: a review. Progr. Oceanogr. 69, 239–266 2006.

Wang, B., Liu, J., Kim, H.J., Webster, P.J., Yim, S.Y., Xiang, B.Q., 2013. Northern Hemisphere summer monsoon intensified by mega-El Niño/southern oscillation and Atlantic multidecadal oscillation. Proc. Natl. Acad. Sci. USA 14, 5347–5352.

Webster, P.J., Moore, A.M., Loschnigg, J.P., Lebel, R.R., 1999. Coupled oceanic-atmospheric dynamics in the Indian Ocean during 1997–1998. Nature 401, 356–360.

Wu, Z., Zhang, P., 2015. Interdecadal variability of the mega-ENSO–NAO synchronization in winter. Clim. Dyn 45, 1117–1128.

Yeh, S., Kug, J., Dewitte, B., Kwon, M., Kirtman, B., Jin, F, 2009. El Niño in a changing climate. Nature 461, 511–514. doi:10.1038/nature08316.

Yim, S.Y., Wang, B., Liu, J., Wu, Z., 2013. A comparison of regional monsoon variability using monsoon indices. Clim. Dyn. 43, 1423–1437.

Yu, J.-Y., Kim, S.T., 2011. Relationships between extratropical sea level pressure variations and the Central-Pacific and Eastern-Pacific types of ENSO. J. Clim. 24, 708–720.

Yuan, Y., Yin, L.C., 2008. Decadal variability of the IOD-ENSO relationship. Chin. Sci. Bull. 53, 1745–1752.

Zhang, L., Wu, Z., Zhou, Y., 2016. Different impacts of typical and atypical ENSO on the Indian summer rainfall: ENSO-developing phase. Atmos.-Ocean 54, 440–456.

Zhang, P., Z., Wu, H., Chen, (2017). Interdecadal variability of the ENSO–North Pacific atmospheric circulation in winter. Atmos.-Ocean, 55, 110–120. doi: 10.1080/07055900.2017.1291411.

Zou, L., Zhou, T., 2015. Asian summer monsoon onset in simulations and CMIP5 projections using four Chinese climate models. Adv. Atmos. Sci. 32, 794. doi:10.1007/s00376-014-4053-z.

Part III

Subtropical and Extratropical teleconnections to Indian Summer Monsoon

Chapter 12

Eurasian snow and the Asian summer monsoon

Song Yang[a,b], Mengmeng Lu[a], Renguang Wu[b,c]

[a]*School of Atmospheric Sciences and Guangdong Province Key Laboratory for Climate Change and Natural Disaster Studies, Sun Yat-sen University, Guangzhou, China,* [b]*Southern Marine Science and Engineering Guangdong Laboratory (Zhuhai), Guangdong, China,* [c]*Key Laboratory of Geoscience Big Data and Deep Resource of Zhejiang Province, School of Earth Sciences, Zhejiang University, Hangzhou, China*

12.1 Introduction

Snow is an important element of the land surface process that together with the oceanic process drives the variability of the Asian monsoon. It is among the earliest predictors applied in seasonal monsoon prediction (Blanford, 1884) and perhaps even climate prediction in general. Since the time of Blanford, cold-season Eurasian snow cover (ESC) has been considered as an important influencing factor of the variations of both the Indian summer monsoon (ISM) and the East Asian summer monsoon (EASM) (Normand, 1953; Hahn and Shukla, 1976; Dey and Bhanu Kumar, 1983; Yeh et al., 1983; Barnett et al., 1989; Yang and Xu, 1994; Yang and Lau, 1998; Fasullo, 2004; Halder and Dirmeyer, 2017; Zhang et al., 2019; Lu et al., 2020a). Extensive studies have demonstrated various relationships between snow and monsoon and led to a better understanding of the popular inverse ESC–ISM relationship. However, snow–monsoon relationships are complex due to the sophisticated spatial-temporal variations of both snow and monsoon (Fasullo, 2004). For example, snow variations are not always in phase over the western, central, and eastern Eurasian continent (e.g., Yim et al., 2010). Even within a particular spatial domain, the starting and ending time of snow events of the year and the total seasonal snow amount vary substantially on different time scales. Furthermore, snow–monsoon relationships are not stable, undergoing apparent interdecadal variations, and modulations due to changing background warming (climate) (Zhang et al., 2019).

In this chapter, the observed and model-simulated features of the multiple-timescale relationship between Eurasian cold-season snow and Asian summer monsoon are discussed. Regional features of snow–monsoon relations including the features for the ISM and the EASM and the role of snow over the Tibetan Plateau (TP) are particularly addressed.

Indian Summer Monsoon Variability: El Niño-teleconnections and beyond.
DOI: https://doi.org/10.1016/B978-0-12-822402-1.00017-X
241

12.2   Eurasian snow and the Indian monsoon

Snow modulates surface radiation, moisture, and energy budget and influences the land surface and lower-tropospheric thermal states (e.g., Namias, 1985; Walsh et al., 1985; Cohen and Rind, 1991; Karl et al., 1993; Groisman et al., 1993). Anomalous snow alters these thermal states through the snow-albedo effect and snow-hydrological effect (e.g., Barnett et al., 1989; Yasunari et al., 1991; Xu and Dirmeyer, 2013). In the snow-albedo effect, excessive snow cover reflects more solar radiation reaching the surface and thus reduces the amount of solar radiation absorbed by the surface. In the snow-hydrological effect, more solar radiation is consumed in melting snow and the wetter land surface leads to more surface evaporation. These changes in surface heat fluxes cause variability in surface air temperature, which, in turn, modulates atmospheric circulation both locally and remotely.

Since Blanford (1884) found an inverse relationship between the Himalayan winter snow and ISM rainfall, substantial investigations have been conducted to understand how Eurasian snow is linked to and affects the ISM including the associated physical processes (Walker, 1910; Hahn and Shukla, 1976; Dey and Bhanu Kumar, 1983; Dickson, 1984; Khandekar, 1991; Parthasarathy and Yang, 1995; Yang, 1996; Sankar-Rao et al., 1996; Matsuyama and Masuda, 1998; Kripalani et al., 1996, 2003a; Bamzai and Shukla, 1999; Kripalani and Kulkarni, 1999; Robock et al., 2003; Fasullo, 2004; Dash et al., 2005; Singh and Oh, 2005; Mamgain et al., 2010; Peings and Douville, 2010; Senan et al., 2016; Amita et al., 2017; Halder and Dirmeyer, 2017). While the inverse relationship between winter (December–February) ESC and subsequent ISM rainfall was largely confirmed (e.g., Hahn and Shukla, 1976; Dickson, 1984), it was also recognized that the low-level higher pressure perturbation over central Asia and high-level lower pressure anomaly over India following more winter ESC weakened the following ISM (Sankar-Rao et al., 1996). In response to the Western Eurasian snow depth or accumulation anomalies, the extratropical atmospheric circulation during winter through spring (March–May) undergoes significant changes, which serve as the precursors of ISM variations (Dash et al., 2005; Mamgain et al., 2010). ESC also influences the advance of ISM apparently (Dey and Bhanu Kumar, 1983), although the start date of Eurasian snowfall is not obviously related to the onset date of ISM (Mamgain et al., 2010). Positive snow depth anomaly over the TP in early April can delay the onset of the monsoon (Senan et al., 2016). The Eurasian snow strongly affects the ISM in the early months of the monsoon season such as the first two months after monsoon onset due to the snow-hydrological effect, which is relatively stronger over eastern Eurasia, while the snow-albedo effect is more prevalent over Western Eurasia (Halder and Dirmeyer, 2017). Atmospheric model experiments showed that the response of monsoon to cold-season land surface process including snow condition is mainly in May and June, and the monsoon response in July–September is much smaller (Yang and Lau, 1998; Fig. 12.1).

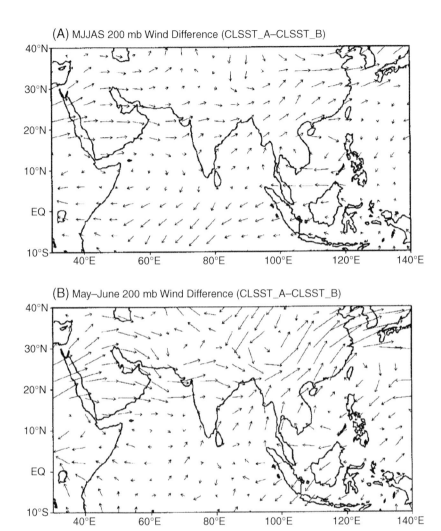

FIG. 12.1 (A) Composite patterns of the difference in 200-mb winds between more snow and soil moisture experiment and climatology ("control") experiment. (A) is for May–September mean, and (B) is for May and June average. The change in summer monsoon circulation associated with land surface processes (A) is accounted for mainly by early summer mean (B), with little contribution from the change in late summer. (from Yang and Lau, 1998)

The ESC–ISM relationship is characterized by many regional features (Parthasarathy and Yang, 1995; Bamzai and Shukla, 1999; Robock et al., 2003; Fasullo, 2004). For example, the monsoon rainfall over Western and Central India is more negatively related to ESC (Parthasarathy and Yang, 1995), and ISM rainfall is more negatively affected by snow cover over Western Eurasia

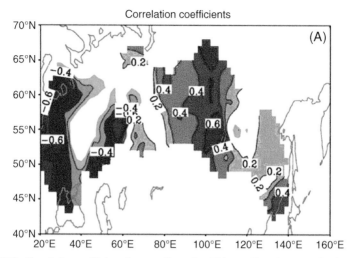

FIG. 12.2 Correlation coefficients between December–February Eurasian snow depth anomalies and the following ISM rainfall during 1951–1994. The ISM rainfall is related negatively to the snow over western Eurasia, but positively to the snow over central-eastern Eurasia (from Dash et al., 2005). "*ISM*, Indian summer monsoon."

(Bamzai and Shukla, 1999). Positive correlation even occurs between the winter snow depth over Eastern Eurasia and ISM rainfall (Kripalani and Kulkarni, 1999). As seen from Fig. 12.2 (from Dash et al., 2005), ISM rainfall variation is also associated with a west-east dipole pattern of the Eurasian snow anomalies (also see Kripalani and Kulkarni, 1999; Ye and Bao, 2001).

Numerical experiments are believed to provide a better understanding of the physical processes through which Eurasian winter–spring snow affects the Asian summer monsoon especially in terms of the snow-albedo and snow-hydrological effects (Barnett et al., 1989; Yasunari et al., 1991; Zwiers, 1993; Vernekar et al., 1995; Douville and Royer, 1996; Ose, 1996; Yang and Lau, 1998; Dong and Valdes, 1998; Watanabe and Nitta, 1998; Ferranti and Molteni, 1999; Bamzai and Marx, 2000; Corti et al., 2000; Becker et al., 2001; Gong et al., 2004; Dash et al., 2006; Turner and Slingo, 2011; Saha et al., 2013). According to Barnett et al. (1989), the snow-related atmospheric signals from changing Eurasian snow distribution with fixed snow depth cannot persist, but disappear shortly after snowmelt, indicating a small snow-albedo effect on the summer monsoon. However, the snow-hydrological effect resulted from snowfall and snow accumulation change leads to prominent changes in soil temperature, soil moisture, surface evaporation, and atmospheric circulation, and their delayed impact on the summer monsoon. This result about the difference between the snow-albedo and snow-hydrological effects has been confirmed by numerous studies (Yasunari et al., 1991; Vernekar et al., 1995; Lau and Bua,1998; Yang and Lau, 1998). Based on the snow-hydrological effect, the feedback of snow on the atmosphere can be negative or positive (Lau and Bua, 1998), depending on the conditions of the atmosphere and the land surface.

The inverse relationship between the ESC/depth and the ISM rainfall has been confirmed in numerical model studies. Dash et al. (2006) conducted two sensitivity experiments with high and low Eurasian snow depth. The different fields of mean monsoon circulation in the two experiments confirm the influence of Eurasian snow depth on the ISM rainfall. Based on the out-of-phase snow variations over the TP and northwest part of the Eurasian continent before a weak ISM, Turner and Slingo (2011) conducted two sensitivity experiments with snow anomalies added in the above two regions, respectively. Their model results indicate a main contribution of the TP snow anomalies in weakening the ISM. Saha et al. (2013) performed a sensitivity experiment with reduced snowfall over the Eurasian region. The reduced snowfall induces a increase in winter and spring snow depth, which, in turn, leads to increase in the ISM rainfall.

The ESC–ISM relationship can be modulated by El Niño-Southern Oscillation (ENSO) and tends to diminish during ENSO years (Sankar-Rao et al., 1996; Yang, 1996; Matsuyama and Masuda, 1998; Fasullo, 2004; Peings and Douville, 2010). More winter ESC can be followed by stronger, instead of weaker, ISM during El Niño years (Yang, 1996). Stratified analysis of the ESC–ISM relationship also confirms that the inverse snow–ISM relationship is obvious only during non-ENSO years (Fasullo, 2004). Model experiments further demonstrate that the snow–ISM relationship may exist independent of ENSO, indicating the independent impact of Eurasian snow on ISM (Ferranti and Molteni, 1999; Turner and Slingo, 2011).

Furthermore, the relationship between Eurasian snow and ISM has experienced secular changes (Kripalani et al., 2003a; Peings and Douville, 2010; Zhang et al., 2019). According to Kripalani et al. (2003b), a switch of the negative relationship between winter/spring Western Himalayan snow cover and the ISM rainfall to a positive relationship occurred in the 1990s, which was plausibly related to the changes in snow cover extent and snow depth due to global warming. Peinngs and Douville (2010) examined the sliding correlation of the ISM rainfall with entire ESC in both winter and spring, and with the western ESC in winter and the central ESC in spring, and detected apparent interdecadal changes in the snow–monsoon relationship. The correlation between Western Eurasian winter snow and ISM rainfall experienced a change from negative to positive in the late 1990s. The relationship between Central Eurasian spring snow and ISM rainfall changed its sign in the early 1990s and became strongly positive after the late 1990s. Zhang et al. (2019) showed that the inverse relationship of central Eurasian spring snow cover with ISM rainfall was obvious during 1967–1990, but disappeared during 1991–2015 (Fig. 12.3). The above change in the snow–ISM rainfall relationship was believed to be related to the reduced lagged snow-hydrological effect due to land surface warming. This dis-

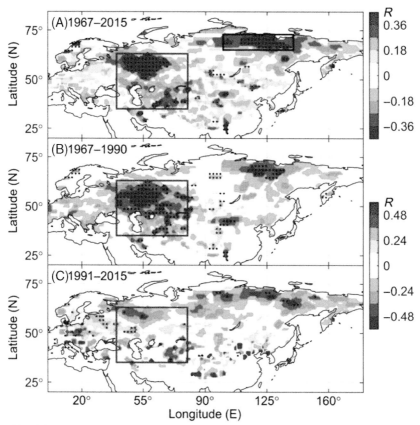

FIG. 12.3 Correlation of Eurasian spring snow cover fraction with all-Indian summer monsoon rainfall for 1967–2015 (A), 1967–1990 (B), and 1991–2015 (C). *Black dots* indicate the statistically significant values above the 95% confidence level. *Red boxes* represent central Eurasia (40°–80°E, 35°–65°N) and the *black box* in (A) denotes northeast Eurasia (100°–140°E, 65°–73°N). Note the significantly weakened snow–monsoon relationship within the *red box* in (C) (from Zhang et al., 2019).

appearance of snow–monsoon relationship was also attributed to the weakened regulation of central Eurasian spring snow cover on the summer mid-tropospheric temperature over the Iranian Plateau, and hence the land-ocean thermal contrast, after 1990. The interdecadal change in snow–monsoon relationship can also be affected by other factors. During the negative phase of the Atlantic Multi-decadal Oscillation, the North Atlantic SST anomaly may lead to more snow over the TP through atmospheric teleconnection and reduce the local atmospheric heating during spring and summer, which, in turn, weakens the ISM (Goswami et al., 2006; Feng and Hu, 2008; Shen et al., 2011). This interdecadal change in the snow–monsoon relationship may partly explain the inconsistency in the relationship between ISM and the preceding ESC in previous studies (Liu and Yanai, 2001; Robock et al., 2003; Fasullo, 2004).

Given that both snow and sea surface temperature (SST) anomalies can affect the Asian monsoon, their relative roles in monsoon variations are also studied, mostly based on experiments using sophisticated climate models (e.g., Yang and Lau, 1998; Dash et al., 2019). According to Yang and Lau (1998), SST anomalies make a more important combination to monsoon variability; however, the effect of land surface process including the forcing of ground wetness measured by snow amount and soil moisture content cannot be neglected, especially in the early stage of the monsoon. On the other hand, the sensitivity experiments with varying observed SST but climatological snow depth show that the simulated ISM rainfall differs from the observed in the absence of initial snow depth even though observed SST is prescribed as the boundary condition (Dash et al., 2019). The response of ISM rainfall to the varying snow depth anomaly in the presence of climatological SST is stronger than that to the varying SST anomaly in the presence of climatological snow.

12.3 Eurasian snow and the East Asian monsoon

Compared to those of the ISM, the spatial distributions and temporal variations of the EASM rainfall and circulation are more complicated (Tao and Chen, 1987; Ha et al., 2012). The EASM covers a region with a wider latitude range, which leads to more complex rainfall and circulation features, and thus an index of all EASM rainfall, constructed by averaging the rainfall over the entire monsoon region as for the ISM index, could hardly measure the variability of EASM.

Early studies of ground-based data revealed an in-phase relationship between Eurasian cold-season snow and Southern China presummer rainfall (Chen and Yan, 1979, 1981), and more extensive satellite-derived ESC was negatively related to the Northern China summer monsoon (Zhao, 1984). Analysis of longer records of ESC and rainfall observation indicates that the ESC is correlated negatively with the rainfalls over Western, Northeastern, and Central China, but positively with the rainfalls over Southern and Northern China (Yang and Xu, 1994; see Fig. 12.4). Particularly, the ESC is most significantly linked to the rainfalls over Southern and Northern China, compared to the other domains examined. More studies have been conducted to understand the relationship in the past two decades (Zhang, 1986; Kripalani et al., 2002; Liu and Yanai, 2002; Zhang et al., 2017; Zhang et al., 2018; Luo and Wang, 2019; Lu et al., 2020a). The variability of ESC can also be measured by several dominant modes and these modes are linked to EASM differently (Yim et al., 2010). The most dominant mode of spring ESC, measuring the relatively uniform variability of the snow, exhibits a weak link to EASM rainfall, while the second most dominant mode, depicting a zonal dipole of snow variability between Central-Western Eurasia and Eastern Eurasia (measuring an out-of-phase interannual variation of snow between the two domains), is related to the East Asian rainfall more strongly. A dipole pattern with less spring snow cover over Eastern Eurasia and more snow over Central-Western Eurasia is accompanied by more summer rainfall over

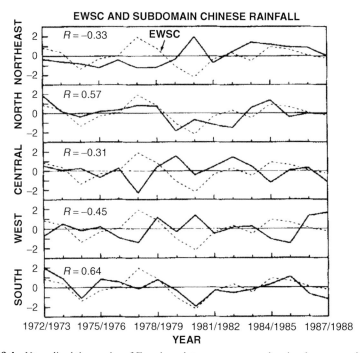

FIG. 12.4 Normalized time series of Eurasian winter snow cover and regional summer rainfalls over Northeastern, Northern, Central, Western, and Southern China. Correlation coefficients are also shown for the snow–monsoon relationships for the various regions, respectively (from Yang and Xu, 1994).

Northern East Asia including Korea, Northeastern China, and Southern Japan. Overall, compared to the ESC–ISM relationship, snow–monsoon association is more complex and exhibits larger temporal and spatial variations in the East Asian monsoon region, leading to a seemingly weaker ESC–EASM link (see review in Lu et al., 2020b).

As for the ESC–ISM link, the snow-albedo effect and snow-hydrological effect are also the two important mechanisms for explaining the relationship between Eurasian snow and EASM. Emphasizing the importance of the snow-hydrological effect via snowmelt and latent heat variations, Zhang et al. (2017) analyzed the variability of Eurasian spring snow decrement, measured by the difference in snow water equivalent between February and May, and its relationship with EASM rainfall. A dipole pattern of spring snow decrement with positive values over the Lake Baikal region and negative values over Siberia to the west excites an anomalous Rossby wave pattern over the extratropics (Fig. 12.5). The anomalous wave pattern modulates the atmospheric circulation over East Asia and causes more (less) summer rainfall over northeastern (southern) China. Importantly, from spring to summer, persistent atmospheric features mainly occur in the snow-concerned regions (see the two boxes in Fig. 12.5)

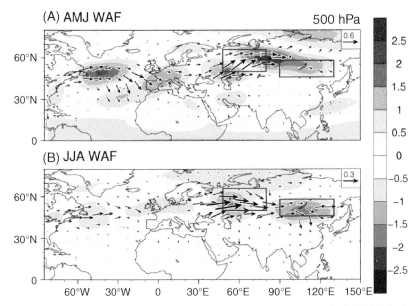

FIG. 12.5 Waveguides in response to Eurasian spring snow decrement, one in the high latitudes, and the other in subtropical regions, as depicted by 500-hPa stream function (shadings, in m^2/s) and the associated wave activity flux (vectors, in m^2/s^2) anomalies regressed onto a snow decrement index defined by the two boxes (see text for detail). (A) is for late spring (April–June) and (B) for summer (June–August) (from Zhang et al., 2017).

but not over the North Atlantic and West Europe. Moreover, the Eurasian snow anomalies and Arctic sea ice anomalies synergistically regulate atmospheric teleconnection patterns and cause EASM rainfall variations (Amita et al., 2018; Zhang et al., 2018).

Accumulated evidence has indicated that the snow over TP also plays an important role in regulating the variability of EASM (Guo and Wang, 1986; Wu and Qian, 2000; Liu et al., 2014). Because of its high altitude, snow exists over certain TP regions all year round and its climate impact exhibits an apparent seasonality (Xiao et al., 2016; Bao and You, 2019; Xiao et al., 2019). The snow-albedo effect may be dominant over the snow-hydrological effect in cold seasons (winter and spring months), but the opposite appears in warmer seasons (Yasunari et al., 1991; Souma and Wang, 2010; Turner and Slingo, 2011; Wu et al., 2012a). When more snow accumulates over the TP, mostly central-eastern TP, monsoon rainfall increases over the middle and lower branches of the Yangtze River and decreases over South China and North China (Zhang and Tao, 2001; Xu et al., 2012). According to Wang et al. (2018), more snow cover over the western TP is accompanied by an obvious upper-level wave pattern with an anomalous cyclone at both upper and lower levels over Northeast China (Figs. 12.6A and B). The anomalous southwesterlies to the south transport more

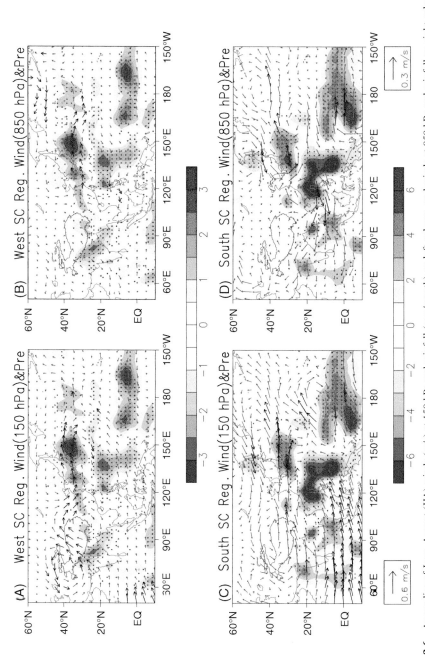

FIG. 12.6 Anomalies of June–August (JJA) winds (m/s) at 150 hPa and rainfall (mm/month) on left column, and winds at 850 hPa and rainfall on right column, obtained by linear regression against JJA interannual snow cover index in western TP region (70°E–79°E, 32.5°N–40.5°N; (A) and (B)) and Southern TP region (80°E–99°E, 26.5°N–31.5°N; (C) and (D)). *Dotted areas* denote significant rainfall anomalies (at the 95% confidence level), and *black vectors* denote significant wind anomalies (at the 90% confidence level). The scale for wind vectors is shown at bottom left for (A) and (C), and bottom right for (B) and (D) (from Wang et al., 2018). *"TP, Tibetan Plateau."*

moist air from the lower latitudes, leading to excessive rainfall from the Yangtze River to the east of Japan. On the other hand, more snow cover over the southern TP is accompanied by anomalous upper-level westerlies over the tropical Indian Ocean with an elongated anomalous cyclone extending from the TP to the North Pacific (Fig. 12.6C). An anomalous lower-level cyclone covers East Asia (Fig. 12.6D), inducing heavy rainfall from the Yangtze River to the east of Japan. The anomalies of East Asian wind and rainfall corresponding to more snow cover over the western TP and the southern TP seem similar; however, the wind anomalies are more significant over mid-latitude Asia for western TP snow but over the tropical Indian Ocean for more southern TP snow. The effect of western TP snow is through the mid-latitude Asian wave pattern, and the influence of southern TP snow follows several paths including a meridional overturning circulation between the TP and the North Indian Ocean, a zonal overturning circulation between the North Indian Ocean and the Philippine Sea, and a meridional wave pattern along the East Asian coast (Liu et al., 2014).

The other distinguishing feature of the East Asian monsoon is its winter component (Lau and Li, 1984; Ding, 1994; Yang et al., 2002; Jhun and Lee, 2004; Chen et al., 2005; Chang et al., 2006; Gao and Yang, 2009; Li and Yang, 2010; Hu and Yang, 2015; Li and Yang, 2017). Eurasian snow, the Arctic Oscillation, the Siberian high, and East Asian upper-tropospheric jet stream, mid-tropospheric trough, and surface cold surge are among or are related to, the coupled systems of the winter monsoon (Gong et al., 2001; Park et al., 2011; Li et al., 2017). Recently, Lu et al. (2020a) used the NCAR Community Earth System Model (CESM 1.2.2) to conduct a sensitivity experiment with a winter snowfall rate of 1.1 mm/day added in Northern China and Southern Mongolia ($35°N–47.5°N/70°E–125°E$). Compared with the control experiment, the East Asian winter monsoon intensifies in response to the snowfall perturbation (Fig. 12.7). The surface continental high and oceanic low strengthen, with anomalous northerly wind and thus decreased surface temperature along the East Asian coastal region. An anomalous mid-tropospheric wave train forms in the extratropics, with a deepened East Asian trough. At the upper troposphere, the East Asia and western Pacific westerly jet stream becomes stronger.

A stronger winter monsoon is accompanied by a weak EASM, such as decreased summer rainfall over Central-Southern China (Chen et al., 2000). The remote effect of Pacific and Indian Ocean SST anomalies was claimed as an important factor in establishing the relationship between the winter and summer monsoons (Wang and Wu, 2012; Chen et al., 2013). It was also recognized that the Eurasian snow plays a bridging role in the link between the East Asian winter monsoon and EASM. According to Lu et al. (2020a), the strengthening of upper tropospheric westerlies as a response to enhanced snow (Fig. 12.7) is accompanied by anomalous surface convergence in northeastern Asia and anomalous divergence related to the compensating circulation over subtropical East Asia and the western Pacific. Therefore, decreased rainfall occurs over subtropical East Asia and western Pacific region (Fig. 12.8). This robust feature,

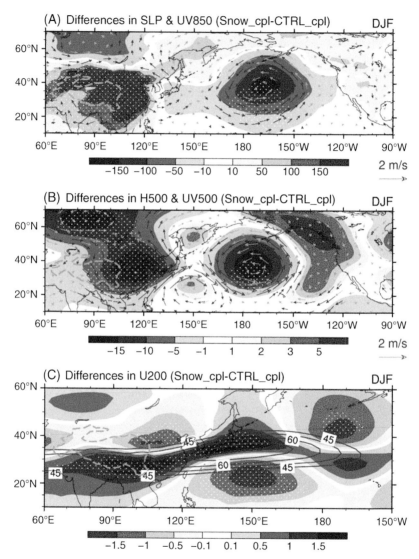

FIG. 12.7 Differences in winter (A) sea level pressure (shadings, in Pa) and 850-hPa horizontal winds (vectors, in m/s), (B) 500-hPa geopotential height (shadings, in meter) and horizontal winds (vectors), and (C) 200-hPa zonal wind (shadings) between sensitivity experiment Snow_cpl and control run CTRL_cpl. In (C), the purple contours outline the climatological mean of 200-hPa zonal wind in CTRL_cpl. *White dots* indicate the values above the 90% confidence level. The topography of the TP above 1500 m is marked with *dash green line* (from Lu et al., 2020). "*TP*, Tibetan Plateau."

appearing in CESM experiments of both 50-year integrations and 100-year integrations, demonstrates the relationship between more Eurasian winter snow, stronger East Asian winter monsoon, and weaker EASM.

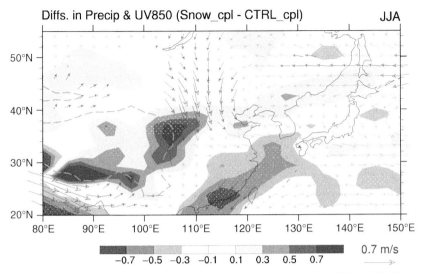

FIG. 12.8 Same as Fig. 12.7, but for summer 850-hPa winds (vectors) and precipitation (shadings, in mm/day) (modified from Lu et al., 2020).

As in its effect on ESC–ISM relationship, ENSO modulates the relationship between Eurasia snow and the East Asian monsoon (Wu and Kirtman, 2007), although snow also affects ENSO and ENSO–EASM relationship (e.g., Wu et al., 2012b). Indeed, ENSO can first cause changes in Eurasian winter land surface conditions including snow variability, which then exerts an impact on the following summer monsoon. This process has been referred to as ENSO's "indirect effect" on the large-scale Asian summer monsoon (Yang and Lau, 1998).

The variations of ENSO, snow, winter monsoon, and summer monsoon as well as their inter-relationships, experience changes on interdecadal time scales (Chen and Wu, 2000; Zhang et al., 2004; Kwon et al., 2007; Ye et al., 2015; Jia and Ge, 2017; Jia et al., 2018), including those modulated by the Pacific Decadal Oscillation (Chen et al., 2013). Recently, particular effort has also been devoted to understanding the interdecadal change in TP snow and its effect on the EASM (You et al., 2011; Duan et al., 2018). The TP snow has decreased since the 1990s, enhancing surface warming and thus land–sea thermal contrast (Si and Ding, 2013). Correspondingly, the summer monsoon rain band over East Asia shifts northward.

12.4 Summary and discussion

Inverse relationship between winter-spring snow and summer monsoon occurs in many Asian monsoon regions, especially the commonly known inverse ESC–ISM relationship. The inverse relationship is attributed to the preceding change in extratropical Eurasian snow, which causes changes in land surface processes including conditions of surface albedo, soil moisture, temperature, temperature gradients, and consequently atmospheric circulation and others. The persistency of the land surface processes, which can also be enhanced by atmosphere–land interaction (Yang et al., 2004; Henderson et al., 2018; Lu et al., 2020a), is mostly needed to explain the lag relationship between cold-season snow and the following summer monsoon.

However, the features of the snow–monsoon relationships vary from one region to another, and these complex regional features have attracted substantial interest of investigations in the past decades. While the ISM is negatively and more strongly related to the western Eurasian snow, eastern Eurasian snow affects the EASM more significantly. Positive correlation can even be seen between eastern Eurasian snow and the ISM. Furthermore, the ESC is negatively related to the following summer monsoon rainfalls over most of India and parts of East Asia, such as West, Central, and Northeast China. However, it is positively related to the summer monsoon rainfall over South China and North China. In addition, accumulated evidence has shown that the TP snow strongly affects the Asian summer monsoon, especially the EASM.

The ESC, the Asian monsoon, and snow–monsoon relationships are also temporally unstable, undergoing interdecadal variations, and changing under the global warming backgrounds including the long-term changes in aerosols and carbon dioxide in the atmosphere (e.g., Hu et al., 2003; Bollasina et al., 2011; Annamalai et al., 2013; Krishnan et al., 2016; Kitoh, 2017; Zhang et al., 2019; Ha et al., 2020; Wang et al., 2020 and references therein). This feature, which is not discussed in detail in this chapter, can be found in numerous studies with long records of observations and outputs from various versions of the Coupled Model Intercomparison Project models. Based on the perturbed physics ensemble simulations of the Hadley Centre climate model, the inverse relationship between Eurasian snow and the ISM still exists in the warming world (Panda et al., 2016). Analysis of the snow and rainfall data also indicates that the generally inverse relationship between the central Eurasian spring snow cover and the ISM rainfall has disappeared since 1990 (Zhang et al., 2019), owing to the loss of the snow's regulation on the temperature over the Iranian Plateau and thus the land–sea thermal contrast that affects monsoon variation. Nevertheless, cautions should always be taken when distinguishing features between the long-term change under global warming background and the interdecadal variation of the snow–monsoon relationship, which may partially explain the seemingly inconsistent snow–monsoon relationships among various investigations. Since the variability of snow over the TP affects significantly the extreme weather and climate events in China and the TP snow changes apparently under the global warming background, You et al. (2020) called for a need of investigations of monsoon response to snow variation in the warming climate.

The development of climate forecast systems provides a useful tool for the dynamical prediction of snow–monsoon relationship and coupled global models have demonstrated certain skills. For example, the climate forecast system of the U.S. National Centers for Environmental Prediction shows a reasonable skill in predicting ESC and snow water equivalent (Zuo et al., 2015; He et al., 2016). The predictions initiated in April can capture the relationship between Eurasian spring snow and the summer rainfall over China.

Funding

This study was supported by the National Key R&D Program of China (Grant 2019YFC1510400), the National Natural Science Foundation of China (Grants 91637208 and 41721004), the Second Tibetan Plateau Scientific Expedition and Research Program (STEP) (2019QZKK0208), and the Jiangsu Collaborative Innovation Center for Climate Change.

Acknowledgement

Comments from two anonymous reviewers are helpful for improving the overall quality of the manuscript.

References

Amita, P., Oh, J., Kim, I.-W., Kripalani, R.H., Mitra, A.K., Pandithurai, G., 2017. Summer monsoon rainfall variability over North East regions of India and its association with Eurasian snow, Atlantic sea surface temperature and Arctic Oscillation. Clim. Dyn 49, 2545–2556. doi:10.1007/s00382-016-3445-4.

Amita, P., Oh, J., Kim, I.-W., Kripalani, R.H., Pandithurai, G., 2018. SMMR-SSM/I derived Greenland sea ice variability: links with Indian and Korean monsoons. climate dyn 50, 1023–1043. doi:10.1007/s00382-017-3659-0.

Annamalai, H., Hafner, J., Sooraj, K.P., Pillai, P., 2013. Global warming shifts the monsoon circulation, drying South Asia. J. Clim. 26, 2701–2718.

Bao, Y., You, Q., 2019. How do westerly jet streams regulate the winter snow depth over the Tibetan Plateau? Clim. Dyn 53, 353–370.

Bamzai, A.S., Shukla, J., 1999. Relation between Eurasian snow cover, snow depth, and the Indian summer monsoon: an observational study. J. Clim. 12, 3117–3132.

Bamzai, A.S., Marx, L., 2000. COLA AGCM simulation of the effect of anomalous spring snow over Eurasia on the Indian summer monsoon. Quart. J. Roy. Meteor. Soc. 126, 2575–2584.

Barnett, T.P., Dümenil, L., Schlese, U., Roeckner, E., Latif, M., 1989. The effect of Eurasian snow cover on regional and global climate variations. J. Atmos. Sci. 46, 661–686.

Becker, B.D., Slingo, J.M., Ferranti, L., Molteni, F., 2001. Seasonal predictability of the Indian summer monsoon: what role do land surface conditions play? Mausam 52, 47–62.

Blanford, H.F., 1884. On the connexion of the Himalaya snowfall with dry winds and seasons of drought in India. Proc. Roy. Soc. London 37, 3–22.

Bollasina, M.A., Ming, Y., Ramaswamy, V., 2011. Anthropogenic aerosols and the weakening the South Asian summer monsoon. Science 334, 502–505.

Chang, C.-P., Wang, Z., Hendon, H., 2006. The Asian winter monsoon. In: Wang, B. (Ed.), The Asian Monsoon. Springer, Berlin, pp. 89–127.

Chen, L.-T., Yan, Z.-X., 1979. Impact of Himalayan winter-spring snow cover on atmospheric circulation and southern China rainfall in the rainy season (in Chinese). In: Yangtze River Regulating Office (Ed.), Collected Papers on Medium- and Long-Term Hydrological and Meteorological Forecasts. Water Conservancy and Power Press, Beijing, pp. 185–194 Vol. 1.

Chen, L.-T., Yan, Z.-X., 1981. A statistical study of the impact of Himalayan winter-spring snow cover anomalies on the early summer monsoon (in Chinese). In: Yangtze River Regulating Office (Ed.), Collected Papers on Medium- and Long-Term Hydrological and Meteorological Forecasts. Water Conservancy and Power Press, Beijing, pp. 133–141 Vol. 2.

Chen, L.-T., Wu, R., 2000. Interannual and decadal variations of snow cover over Qinghai-Xizang Plateau and their relationships to summer monsoon rainfall in China. Adv. Atmos. Sci. 17, 18–30.

Chen, W., Graf, H., Huang, R., 2000. The interannual variability of East Asian winter monsoon and its relation to the summer monsoon. Adv. Atmos. Sci. 17, 48–60.

Chen, W., Yang, S., Huang, R., 2005. Relationship between stationary planetary wave activity and the East Asian winter monsoon. J. Geophys. Res. 110, D14110.

Chen, W., Feng, J., Wu, R., 2013. Roles of ENSO and PDO in the link of the East Asian winter monsoon to the following summer monsoon. J. Clim. 26, 622–635.

Cohen, J., Rind, D., 1991. The effect of snow cover on the climate. J. Clim. 4, 689–706.

Corti, S., Molteni, F., Brankovic, C., 2000. Predictability of snow depth anomalies over Eurasia and associated circulation patterns. Quart. J. Roy. Meteor. Soc. 126, 241–262.

Dash, S.K., Singh, G.P., Shekhar, M.S., Vernekar, A.D., 2005. Response of the Indian summer monsoon circulation and rainfall to seasonal snow depth anomaly over Eurasia. Clim. Dyn 24, 1–10.

Dash, S.K., Parth Sarthi, P., Panda, S.K., 2006. A study on the effects of Eurasian snow on the summer monsoon circulation and rainfall using a spectral GCM. Int. J. Climatol. 26, 1017–1025.

Dash, S.K., Paliwal, M., Panda, S.K., Karri, S., 2019. Comparison of Indian summer monsoon rainfall anomalies in response to changes in snow depths and SSTs in a GCM. Mausam 70, 71–86.

Dey, B., Bhanu Kumar, O.S.R.U., 1983. Himalayan winter snow cover area and summer monsoon rainfall over India. J. Geophys. Res. 88, 5471–5474.

Dickson, R.R., 1984. Eurasian snow cover versus Indian monsoon rainfall—an extension of the Hahn-Shukla results. J. Clim. Appl. Meteor. 23, 171–173.

Ding, Y., 1994. Monsoon over China. Kluwer Academic, Norwell, MA, p. 420.

Dong, B.W., Valdes, P.J., 1998. Modeling the Asian summer monsoon rainfall and Eurasian winter/spring snow mass. Quart. J. Roy. Meteor. Soc. 124, 2567–2596.

Douville, H., Royer, J.F., 1996. Sensitivity of the Asian summer monsoon to an anomalous Eurasian snow cover within the Meteo-France GCM. Clim. Dyn 12, 449–466.

Duan, A., Xiao, Z., Wang, Z., 2018. Impacts of the Tibetan Plateau winter/spring snow depth and surface heat source on Asian summer monsoon: A review. Chin. J. Atmos. Sci. 44, 755–766 In Chinese.

Fasullo, J., 2004. A stratified diagnosis of the Indian monsoon-Eurasian snow cover relationship. J. Climate 17, 1110–1122.

Feng, S., Hu, Q., 2008. How the North Atlantic multidecadal oscillation may have influenced the Indian summer monsoon during the past two millennia. Geophys. Res. Lett. 35, L01707.

Ferranti, L., Molteni, F., 1999. Ensemble simulations of Eurasian snow-depth anomalies and their influence on the summer Asian monsoon. Quart. J. Roy. Meteor. Soc. 125, 2597–2610.

Gao, H., Yang, S., 2009. A severe drought event in northern China in winter 2008-09 and the possible influences of La Nina and Tibetan Plateau. J. Geophys. Res. 114, D24104 doi:10.102 9/2009JD012430.

Gong, D., Wang, S., Zhu, J., 2001. East Asian winter monsoon and Arctic oscillation. Geophys. Res. Lett. 28, 2073–2076.

Gong, G., Entekhabi, D., Cohen, J., Robinson, D., 2004. Sensitivity of atmospheric response to modeled snow anomaly characteristics. J. Geophys. Res. 109, D06107. doi:10.1029/2003 JD004160.

Goswami, B.N., Madhusoodanan, M.S., Neema, C.P., Sengupta, D., 2006. A physical mechanism for North Atlantic SST influence on the Indian summer monsoon. Geophys. Res. Lett. 33, L02706.

Groisman, P.Y., Karl, T.R., Knight, R.W., 1993. Observed impact of snow cover on the heat balance over the continental spring temperature. Science 263, 198–200.

Guo, Q.-Y., Wang, J.-J., 1986. The snow cover on Tibetan Plateau and its effect on the monsoon over East Asia (in Chinese). Plateau Meteor 5, 116–124.

Ha, K.-J., Kim, B.-H., Chung, E.-S., Chan, J.L.C., Chang, C.-P., 2020. Major factors of monsoon rainfall changes: natural versus anthropogenic forcing. Environ. Res. Lett. doi:10.1088733/1748-9326/ab7767.

Ha, K.-J., Heo, K.-Y., Lee, S.-S., Yun, K.-S., Jhun, J.-G., 2012. Variability in the East Asian monsoon: a review. Meteorol. Appl. 19, 200–215.

Hahn, D.G., Shukla, J., 1976. An apparent relationship between Eurasian snow cover and Indian monsoon rainfall. J. Atmos. Sci. 33, 2461–2462.

Halder, S., Dirmeyer, P.A., 2017. Relation of Eurasian snow cover and Indian summer monsoon rainfall: importance of the delayed hydrological effect. J. Clim. 30, 1273–1289.

He, Q., Zuo, Z., Zhang, R., Yang, S., Wang, W., Zhang, R., et al., 2016. Prediction skill and predictability of Eurasian snow cover fraction in the NCEP Climate Forecast Version 2 reforecasts. Int. J. Climatol. 36, 4071–4084.

Henderson, G.R., Peings, Y., Furtado, J.C., Kushner, P.J., 2018. Snow–atmosphere coupling in the Northern Hemisphere. Nat. Clim. Change 8, 954–963.

Hu, C., Yang, S., 2015. An optimal index for measuring the effect of East Asian winter monsoon on China winter temperature. Clim. Dyn 45, 2571–2589.

Hu, Z.-Z., Yang, S., Wu, R., 2003. Long-term climate variations in China and global warming signals. J. Geophys. Res. 108 (D19), 4614. doi:10.1029/2003JD003651.

Jhun, J.-G., Lee, E.-J., 2004. A new East Asian winter monsoon index and associated characteristics of the winter monsoon. J. Climate 17, 711–726.

Jia, X.-J., Ge, J-W., 2017. Interdecadal changes in the relationship between ENSO, EAWM, and the wintertime precipitation over China at the end of the twentieth century. J. Clim. 30, 1923–1937.

Jia, X.-J., Cao, D.-R., Ge, J.-W., Wang, M., 2018. Interdecadal change of the impact of Eurasian snow on spring precipitation over Southern China. J. Geophys. Res. Atmos. 123, 10073–10089.

Karl, T.R., Groisman, P.Y., Knight, R.W., Heim Jr., R.R., 1993. Recent variations of snow cover and snowfall in North America and their relation to precipitation and temperature variations. J. Climate 6, 1327–1344.

Khandekar, M.L., 1991. Eurasian snow cover, Indian monsoon and El Nino/Southern Oscillation—a synthesis. Atmos. Ocean 29, 636–647.

Kitoh, A., 2017. The Asian monsoon and its future change in climate models: A review. J. Meteor. Soc. Japan 95, 7–33. doi:10.2151/jmsj.2017-002.

Kripalani, R.H., Kulkarni, A., 1999. Climatology and variability of historical Soviet snow depth data: some new perspectives in snow-Indian monsoon teleconnections. Clim. Dyn 15, 475–489.

Kripalani, R.H., Singh, S.V., Vernekar, A.D., Thapliyal, V., 1996. Empirical study on Nimbus-7 snow mass and Indian summer monsoon rainfall. Int. J. Climatol. 16, 23–24.

Kripalani, R.H., Kim, B.-J., Oh, J.-H., Moon, S.-E., 2002. Relationship between Soviet snow and Korean rainfall. Int. J. Climatol. 22, 1313–1325.

Kripalani, R.H., Kulkarni, A., Sabade, S.S., 2003a. Western Himalayan snow cover and Indian monsoon rainfall. A re-examination with INSAT and NCEP/NCAR data. Theor. Appl. Climatol. 74, 1–18.

Kripalani, R.H., Kulkarni, A., Sabade, S.S., 2003b. Indian monsoon variability in a global warming scenario. Nat. Hazards 29, 189–206.

Krishnan, R., Sabin, T.P., Vellore, R., Mujumdar, M., Sanjay, J., et al., 2016. Deciphering the desiccation trend of the South Asian monsoon hydroclimate in a warming world. Clim. Dyn 47, 1007–1027.

Kwon, M.-H., Jhun, J.-G., Ha, K.-J., 2007. Decadal change in East Asian summer monsoon circulation in the mid-1990s. Geophys. Res. Lett. 34, L21706.

Lau, K.-M, Bua, W.R., 1998. Mechanisms of monsoon–southern oscillation coupling: insights from GCM experiments. Clim. Dyn 14, 759–779.

Lau, K.-M., Li, M., 1984. The monsoon of East Asia and its global associations—a survey. Bull. Am. Meteor. Soc. 65, 114–125.

Li, Q., Yang, S., Wu, T., Liu, X., 2017. Sub-seasonal dynamical prediction of East Asian cold surges. Weather Forecast. 32, 1675–1694.

Li, Y., Yang, S., 2010. A dynamical index for the East Asian winter monsoon. J. Clim. 23, 4255–4262.

Li, Y., Yang, S., 2017. Feedback attributions to the dominant modes of East Asian winter monsoon variations. J. Clim. 30, 905–920.

Liu, G., Wu, R., Zhang, Y., Nan, S., 2014. The summer snow cover anomaly over the Tibetan Plateau and its association with simultaneous precipitation over the Meiyu-Baiu region. Adv. Atmos. Sci. 31, 755–764.

Liu, X., Yanai, M., 2001. Relationship between the Indian monsoon rainfall and the tropospheric temperature over the Eurasian continent. Quart. J. Roy. Meteor. Soc. 127, 909–937.

Liu, X., Yanai, M., 2002. Influence of Eurasian spring snow cover on Asian summer rainfall. Int. J. Climatol. 22, 1075–1089.

Lu, M., Kuang, Z., Yang, S., Li, Z., Fan, H., 2020a. A bridging role of winter snow over northern China and southern Mongolia in linking the East Asian winter and summer monsoons. J. Clim. 33, 9849–9862.

Lu, M., Wu, R., Yang, S., Wang, Z., 2020b. Relationships between Eurasian cold-season Snows and Asian summer monsoons: regional characteristics and seasonality. Trans. Atmos. Sci. 43, 93–103 In Chinese.

Luo, X., Wang, B., 2019. How autumn Eurasian snow anomalies affect east Asian winter monsoon: a numerical study. Clim. Dyn 52, 69–82.

Mamgain, A., Dash, S.K., Sarthi, P.P., 2010. Characteristics of Eurasian snow depth with respect to Indian summer monsoon rainfall. Meteorol. Atmos. Phys. 110, 71–83.

Matsuyama, H., Masuda, K., 1998. Seasonal/interannual variations of soil moisture in the former USSR and its relationship to Indian summer monsoon rainfall. J. Clim. 11, 652–658.

Namias, J., 1985. Some empirical evidence for the influence of snow cover on temperature and precipitation. Mon. Wea. Rev. 113, 1542–1553.

Normand, C., 1953. Monsoon seasonal forecasting. Quart. J. Roy. Meteor. Soc. 79, 463–473.

Ose, T., 1996. The comparison of the simulated response to the regional snow mass anomalies over Tibet, Eastern Europe, and Siberia. J. Meteor. Soc. Japan 74, 845–866.

Panda, S.K., Dash, S.K., Bhaskaran, B., Pattnayak, K.C., 2016. Investigation of the snow monsoon relationship in a warming atmosphere using Hadley Centre climate model. Global Planet. Change 147, 125–136.

Park, T.-W., Ho, C.-H., Yang, S., 2011. Relationship between the Arctic Oscillation and cold surges over East Asia. J. Clim. 24, 68–83.

Parthasarathy, B., Yang, S., 1995. Relationships between regional Indian summer monsoon rainfall and Eurasian snow cover. Adv. Atmos. Sci. 12, 143–150.

Peings, Y., Douville, H., 2010. Influence of the Eurasian snow cover on the Indian summer monsoon variability in observed climatologies and CMIP3 simulations. Clim. Dyn 34, 643–660. doi:10.1007/s00382-009-0565-0.

Robock, A., Mu, M., Vinnikov, K., Robinson, D., 2003. Land surface conditions over Eurasia and Indian summer monsoon rainfall. J. Geophys. Res. 108, 4131. doi:10.1029/2002 JD002286.

Saha, S.K., Pokhrel, S., Chaudhari, H.S., 2013. Influence of Eurasian snow on Indian summer monsoon in NCEP CFSv2 free run. Clim. Dyn 41, 1801–1815.

Sankar-Rao, M., Lau, K.-M., Yang, S., 1996. On the relationship between Eurasian snow cover and the Asian summer monsoon. Int. J. Climatol. 16, 605–616.

Senan, R., Orsolini, Y.J., Weisheimer, A., Vitart, F., Balsamo, G., Stockdale, T.N., et al., 2016. Impact of springtime Himalayan–Tibetan Plateau snowpack on the onset of the Indian summer monsoon in coupled seasonal forecasts. Clim. Dyn 47, 2709–2725. doi:10.1007/s00382-016-2993-y.

Shen, C.-M., Wang, W.-C., Zeng, G., 2011. Decadal variability in snow cover over the Tibetan Plateau during the last two centuries. Geophys. Res. Lett. 38, L10703.

Si, D., Ding, Y., 2013. Decadal change in the correlation pattern between the Tibetan Plateau winter snow and the East Asian summer precipitation during 1979-2011. J. Climate 26, 622–634.

Singh, G.P., Oh, J.-H., 2005. Study on snow depth anomaly over Eurasia, Indian rainfall and circulations. J. Meteor. Soc. Japan 83, 237–250.

Souma, K, Wang, Y., 2010. A comparison between the effects of snow albedo and infiltration of melting water of Eurasian snow on East Asian summer monsoon rainfall. J. Geophys. Res. Atmos. 115, D02115.

Tao, S., Chen, L., 1987. A review of recent research on the East Asian summer monsoon in China. In: Chang, C.-P., Krishnamurti, T.N. (Eds.), Monsoon Meteorology. Oxford University Press, Oxford, pp. 60–92.

Turner, A.G., Slingo, J.M., 2011. Using idealized snow forcing to test teleconnections with the Indian summer monsoon in the Hadley Centre GCM. Clim. Dyn 36, 1717–1735.

Vernekar, A.D., Zhou, J., Shukla, J., 1995. The effect of Eurasian snow cover on the Indian monsoon. J. Clim. 8, 248–266.

Walker, G.T., 1910. On the meteorological evidence for supposed changes of climate in India. Mem. Indian Meteor. 21, 1–21.

Walsh, J.E., Jasperson, W.H., Ross, B., 1985. Influences of snow cover and soil moisture on monthly air temperature. Mon. Weather Rev. 113, 756–768.

Wang, B., Biasutti, M., Byrne, M., Castro, C., Chang, C.-P., Cook, K., et al., 2021. Monsoon climate change assessment. Bull. Amer. Meteor. Soc. 102, E1–E19.

Wang, L., Wu, R., 2012. In-phase transition from the winter monsoon to the summer monsoon over East Asia: Role of the Indian Ocean. J. Geophys. Res. 117, D11112.

Wang, Z.-B., Wu, R., Chen, S.-F., Huang, G., Liu, G., Zhu, L.-H., 2018. Influence of western Tibetan Plateau summer snow cover on East Asian summer rainfall. J. Geophys. Res. 123, 2371–2385. doi:10.1002/2017JD028016.

Watanabe, M., Nitta, T., 1998. Relative impacts of snow and sea surface temperature anomalies on an extreme phase in the winter atmospheric circulation. J. Climate 11, 2837–2857.

Wu, T.-W., Qian, Z.-A., 2000. Further analyses of the linkage between winter and spring snow depth anomaly over Qinghai-Xizang Plateau and summer rainfall of eastern China (in Chinese). Acta Meteor. Sin. 58, 570–581.

Wu, R., Kirtman, B.P., 2007. Observed relationship of spring and summer East Asian rainfall with winter and spring Eurasian snow. J. Climate 20, 1285–1304.

Wu, Z., Jiang, Z., Li, J., Zhong, S., Wang, L., 2012a. Possible association of the western Tibetan Plateau snow cover with the decadal to interdecadal variations of northern China heatwave frequency. Climate Dyn 39, 2393–2402.

Wu, Z., Li, J., Jiang, Z., Ma, T., 2012b. Modulation of the Tibetan Plateau snow cover on the ENSO teleconnections: from the East Asian summer monsoon perspective. J. Climate 25, 2481–2489.

Xiao, Z., Duan, A., 2016. Impacts of Tibetan Plateau snow cover on the interannual variability of the East Asian summer monsoon. J. Climate 29, 8495–8514.

Xiao, Z., Duan, A., Wang, Z., 2019. Atmospheric heat sinks over the western Tibetan Plateau associated with snow depth in late spring. Int. J. Climatol. 39, 5170–5180.

Xu, L., Dirmeyer, P., 2013. Snow–atmosphere coupling strength. Part II: Albedo effect versus hydrological effect. J. Hydrometeorol. 14, 404–418.

Xu, X., Guo, J., Koike, T., Liu, Y., Shi, X., Zhu, F., et al., 2012. Downstream Effect" of winter snow cover over the eastern Tibetan Plateau on climate anomalies in East Asia. J. Meteor. Soc. Japan 90, 113–130.

Yang, S., 1996. ENSO-snow-monsoon associations and seasonal-interannual predictions. Int. J. Climatol. 16, 125–134.

Yang, S., Lau, K.-M., 1998. Influences of sea surface temperature and ground wetness on Asian summer monsoon. J. Climate 11, 3230–3246.

Yang, S., Lau, K.-M., Jim, K.-M., 2002. Variations of the East Asian jet stream and Asian-Pacific-American winter climate anomalies. J. Climate 15, 306–325.

Yang, S., Lau, K.-M., Yoo, S.-H., Kinter, J.L., Miyakoda, K., Ho, C.-H., 2004. Upstream subtropical signals preceding the Asian summer monsoon circulation. J. Climate 17, 4213–4229.

Yang, S., Xu, L., 1994. Linkage between Eurasian winter snow cover and regional Chinese summer rainfall. Int. J. Climatol. 14, 739–750.

Yasunari, T., Kitoh, A., Tokioka, T., 1991. Local and remote responses to excessive snow mass over Eurasia appearing in the northern spring and summer climate—a study with the MRI-GCM. J. Meteor. Soc. Japan 69, 473–487.

Ye, H.-C, Bao, C.-H., 2001. Lagged teleconnections between snow depth in northern Eurasia, rainfall in Southeast Asia and sea-surface temperatures over the tropical Pacific. Int. J. Climatol. 21, 1607–1621.

Ye, K., Wu, R., Liu, Y., 2015. Interdecadal change of Eurasian snow, surface temperature, and atmospheric circulation in the late 1980s. J. Geophys. Res. Atmos. 120, 2738–2753. doi:10.1002/2015JD023148.

Yeh, T.C., Wetherald, R.T., Manabe, S., 1983. A model study of the short-term climatic and hydrologic effects of sudden snow-cover removal. Mon. Wea. Rev. 111, 1013–1024.

Yim, S.-Y., Jhun, J.-G., Lu, R., Wang, B., 2010. Two distinct patterns of spring Eurasian snow cover anomaly and their impacts on the East Asian summer monsoon. J. Geophys. Res. 115, D22113.

You, Q., Kang, S., Ren, G., Fraedrich, K., Pepin, N., Yan, Y., Ma, L., 2011. Observed changes in snow depth and number of snow days in the eastern and central Tibetan Plateau. Climate Res 46, 171–183.

You, Q., Wu, T., Shen, L., Pepin, N., Zhang, L., Jiang, Z., et al., 2020. Review of snow cover variation over the Tibetan Plateau and its influence on the broad climate system. Earth-Sci. Rev. 201, 103043.

Zhang, R.H., Zhang, R.N., Zuo, Z., 2017. Impact of Eurasian spring snow decrement on East Asian summer precipitation. J. Clim. 30, 3421–3437.

Zhang, R.N., Sun, C., Li, W., 2018. Relationship between the interannual variations of Arctic sea ice and summer Eurasian teleconnection and associated influence on summer precipitation over China. Chin. J. Geophys. 61, 91–105 In Chinese.

Zhang, T., Wang, T., Krinner, G., Wang, X., Gasser, T., Peng, S., et al., 2019. The weakening relationship between Eurasian spring snow cover and Indian summer monsoon rainfall. Sci. Adv. 5, eaau8932.

Zhang, X.-C., 1986. Statistical correlation between snow cover of Eurasia in winter-spring and rainfall and temperature of eastern China in summer. Sci. Bull. 31, 1412–1417.

Zhang, Y., Li, T., Wang, B., 2004. Decadal change of the spring snow depth over the Tibetan Plateau: the associated circulation and influence on the East Asian summer monsoon. J. Clim. 17, 2780–2793.

Zhao, Q., 1984. Eurasian snow cover and the East Asian monsoon. Meteorology 7, 27–29 in Chinese.

Zuo, Z., Yang, S., Zhang, R., Xiao, D., Guo, D., Ma, L., 2015. Response of summer rainfall over China to spring snow anomalies over Siberia in the NCEP CFSv2 reforecast. Quart. J. Roy. Meteor. Soc. 141, 939–944.

Zwiers, F.W., 1993. Simulation of the Asian summer monsoon with the CCC GCM-1. J. Clim. 6, 469–486.

Chapter 13

Coupling of the Indian, western North Pacific, and East Asian summer monsoons

Yu Kosaka

Research Center for Advanced Science and Technology, The University of Tokyo, Tokyo, Japan

13.1 Introduction

The Indian monsoon is a subsystem of the continental-scale Asian monsoon. The Asian summer monsoon also includes the western North Pacific (WNP) and East Asian summer monsoons (Murakami and Matsumoto, 1994). The WNP summer monsoon forms as the Intertropical Convergence Zone migrates northward to the Philippine Sea where the monsoon trough develops (Fig. 13.1A). The East Asian summer monsoon is characterized by a quasistationary rainband, called the Meiyu-Baiu rainband, which extends zonally from Southeastern China to the east of Japan (Fig. 13.1A). Both are subject to strong interannual variability of precipitation (Fig. 13.1B). Climatologically, the Meiyu-Baiu rainband gradually advances poleward from mid-May to mid-July, and suddenly weakens in mid to late July. This seasonal transition occurs in parallel with the so-called "convection jump" in the WNP summer monsoon, where the convective region abruptly shifts northward (Ueda et al., 1995; Zhou et al., 2016). The convection jump induces surface anticyclone in the midlatitude WNP through a teleconnection and leads to the disappearance of the Meiyu-Baiu front. Timing of the convection jump thus affects the East Asian summer monsoon, giving rise to interannual covariation of the WNP and East Asian summer monsoons.

How the WNP and East Asian summer monsoons are linked to the Indian summer monsoon is a matter of debate. While external forcing such as changes in greenhouse gas concentrations and orbital parameters can force multiple monsoons simultaneously and cause their covariation (Biasutti et al., 2018; Seth et al., 2019). The focus of this chapter is to demonstrate the interlinkages brought by internal coupling processes between monsoons, such as those described above for the WNP, East Asian summer monsoons, and Indian summer monsoon. Yet, external influence, such as teleconnections associated with

Indian Summer Monsoon Variability: El Niño-teleconnections and beyond.
DOI: https://doi.org/10.1016/B978-0-12-822402-1.00002-8

263

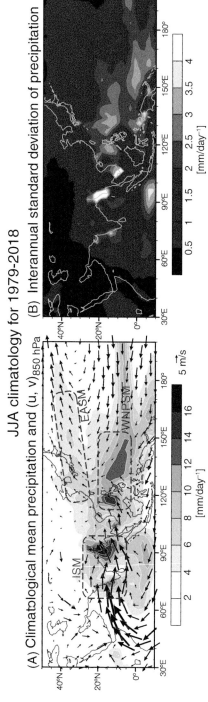

FIG. 13.1 JJA climatology for 1979–2018. (A) Climatological mean precipitation (shading) and 850 hPa wind velocity (arrows). Boxes with purple dashed lines roughly indicate the monsoons considered in this chapter. WNPSM, the western North Pacific summer monsoon. EASM, the East Asian summer monsoon. (B) Interannual standard deviation of precipitation. Based on CMAP and JRA-55, Japanese 55-year Reanalysis. "*CMAP*, Climate Prediction Center Merged Analysis of Precipitation; *JJA*, June-July-August."

El Niño-Southern Oscillation (ENSO), potentially drives those coupling processes and is also within the scope of this chapter.

Ha et al. (2018) review and reassess linkages between the Indian and East Asian summer monsoons. They found that their correlation is overall moderate or insignificant due to compensating influences that lead to their positive (e.g., decaying ENSO) and negative (e.g., developing ENSO) correlations. Fig. 13.2 examines interannual associations of the Indian summer monsoon precipitation with several climate variables in a broader domain, based on observational and reanalysis datasets. Here, instead of averaging over the entire Indian summer monsoon domain, rainfall anomalies over two regions of South Asian areas are used to examine climate variability associated with the Indian summer monsoon, based on the known relationship with teleconnection patterns, as described below (Ding and Wang, 2007; Srinivas et al., 2018). Anomalous precipitation in Northern India is associated with prominent wave-like circulation anomalies in the upper troposphere, which is reminiscent of the Silk Road teleconnection pattern (Enomoto, 2004), superposed on westerly deceleration and acceleration along 30°N and 50°N, respectively (Fig. 13.2B). An associated anticyclonic anomaly over the Korean Peninsula features an equivalent barotropic structure with a slight poleward phase tilt with height, collocated with surface warm anomalies (Figs. 13.2A and B). Dry anomalies are also found over the East China Sea, South Korea, and Japan, while precipitation anomalies are weak and mostly insignificant in China. These midlatitude signatures will be reviewed in Section 13.4. Meanwhile, anomalous rainfall increase in Southern India is linked to dry and surface anticyclonic anomalies in the WNP summer monsoon region (Fig. 13.2C), as reviewed in Section 13.3. Together with anomalously high sea surface temperature (SST) in the North Indian Ocean, these anomalies are key features of the Indo-western Pacific Ocean Capacitor (IPOC) mode (Xie et al., 2016), while a La Niña signal is also evident in the equatorial Pacific (Fig. 13.2D).

As demonstrated in Fig. 13.2, teleconnection patterns are essential for coupling among the Asian monsoons. The present chapter reviews key teleconnection patterns that mediate those inter-monsoon linkages. The striking difference in the teleconnection patterns between Figs. 13.2A, B and C, D highlights sensitivity of the associated teleconnections to subregional features of monsoon variability. In order to synthesize preceding studies on potential inter-monsoon linkages, the discussion is based mainly on analyses made for this review in a consistent manner, rather than fully depending on figures of preceding studies, which can yield inconsistencies due to different methods, datasets, seasons, and analysis periods. Besides, a large ensemble simulations of a high resolution atmospheric general circulation model (AGCM), which enables to separate atmospheric internal and SST-forced variability is used and to obtain statistically significant results.

The model and observational datasets used for the original figures in this chapter are described in Section 13.2. Section 13.3 reviews a tropical teleconnection

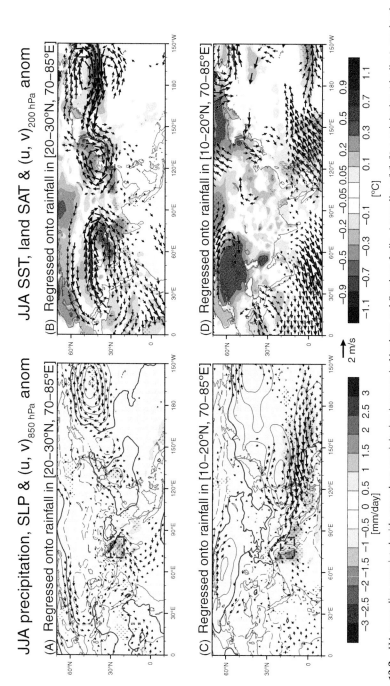

FIG. 13.2 JJA anomalies associated with anomalous precipitation increase in Northern and Southern India. Anomalies of (left) precipitation (shading), sea level pressure (contours for every 0.2 hPa; zero contour thickened), 850 hPa wind velocity (arrows), (right) SST, land surface air temperature (shading), and 200 hPa wind velocity (arrows) regressed onto precipitation anomalies averaged over (A, B) 20°–30°N, 70°–85°E and (C, D) 10°–20°N, 70°–85°E as indicated by dashed magenta boxes in the left panels. Based on precipitation from CMAP, SST from HadISST, and sea level pressure, wind velocity, and land surface air temperature from JRA-55. Stippling indicates statistical significance for anomalies shown in shading with the confidence level of 95% based on t test. Arrows are shown only where either zonal or meridional components are significant at 95% confidence level, based on t test. "*CMAP*, Climate Prediction Center Merged Analysis of Precipitation; *JJA*, June-July-August; *SST*, sea surface temperature."

pathway that can link the WNP summer monsoon variability with the Indian summer monsoon. Westward influence is transmitted through an atmospheric Rossby wave, which corresponds to one lobe of the Pacific-Japan teleconnection pattern and a key component of the IPOC mode. Section 13.4 examines eastward influence from the Indian summer monsoon to East Asian monsoon via the eastward Silk Road teleconnection along with the midlatitude Asian jet. External drivers of these teleconnections are assessed in Section 13.5, followed by concluding remarks in Section 13.6.

13.2 Data

Besides adopting figures from preceding studies, this chapter discusses original analyses based on observational and reanalysis datasets and model simulations. The Japanese 55-year Reanalysis (JRA-55; Kobayashi et al., 2015) with a horizontal resolution of 1.25°, US Climate Prediction Center Merged Analysis of Precipitation (CMAP; Xie and Arkin, 1997) with 2.5° resolution, and Hadley Centre Sea Ice and SST (HadISST; Rayner et al., 2003) version 1.1 with 1° resolution are utilized in this chapter.

Large ensemble simulations with a high-resolution AGCM, called Database for Policy Decision Making for Future Climate Change (d4PDF; Mizuta et al., 2017) are also used. A 100-member ensemble simulations with the Meteorological Research Institute AGCM version 3.2 with 60 km horizontal resolution are driven by historical SST and sea ice from Centennial Observation-Based Estimates of SST version 2 (COBE-SST2; Hirahara et al., 2014) are used. In addition to initial conditions, the ensemble is perturbed by small perturbations artificially introduced to SST within the range of observational errors. Thus one can find the significance of SST anomalies in regression analyses applied to inter-member spread. Nevertheless, these artificial SST perturbations are tiny, and thus we can regard the ensemble mean and deviations from it (hereafter "inter-member variability") as externally-driven and atmospheric internal variability components, respectively. The large ensemble enables us to clearly separate these components and obtain robust signals. Variability of concatenated all ensemble members (hereafter "all-member variability") includes both the externally driven and atmospheric internal components and can be directly compared with observed variability.

Additionally, a partial coupling experiment called "NoENSO" with the Geophysical Fluid Dynamics Laboratory Climate Model version 2.1 (CM2.1; Delworth et al., 2006) was also used. The experimental design is identical to the experiment analyzed in Kosaka et al. (2013), i.e., the SST is restored to model climatology in the tropical eastern Pacific in 15°S–15°N, from the dateline eastward to the American coast, with 5° buffer zones north, south, and west. Outside of this restoring region, the model ocean and atmosphere are fully coupled. Radiative forcing is fixed at preindustrial levels. The model is integrated for 310 years, and the first 10 years are discarded as the spin-up.

All analyses of these observational and reanalysis datasets and d4PDF are limited to 1979-2018 with the focus on interannual variability of June-July-August (JJA) fields after linear detrending throughout the analysis period.

13.3 The tropical pathway

13.3.1 The WNP summer monsoon variability

The WNP summer monsoon features, strong precipitation associated with the Intertropical Convergence Zone extending from the South China Sea to the east of the Philippines (Fig. 13.1A). Toward this region, the monsoon westerlies from the Northern Indian Ocean and the trade winds along the equatorial periphery of the North Pacific subtropical high confluence in the lower troposphere. These winds turn northward to midlatitude East Asia and feed moisture to the East Asian summer monsoon.

Variability of WNP summer monsoon precipitation is associated with a Matsuno (1966), Gill (1980) type Rossby wave-like circulation anomalies (Figs. 13.3A and B). Enhanced precipitation accompanies anomalous cyclonic circulation in the lower troposphere with a slight northward displacement. Embedded in the climatologically confluent flow, the zonally elongated circulation anomalies gain kinetic energy through barotropic energy conversion (Kosaka and Nakamura, 2006). Besides, the surface anomalous cyclone intensifies the surface wind and thus evaporation, which positively feeds back to the anomalous precipitation while lowering SST (Kosaka and Nakamura 2006). Although the latter leads to a weakly negative local interannual correlation between SST and precipitation in the tropical WNP in summer (Wang et al., 2005), the energy conversion and moist feedback energize strong interannual variability of the WNP summer monsoon.

Anomalous diabatic heating associated with a WNP summer monsoon anomaly further drives a poleward teleconnection pattern called the Pacific–Japan (PJ) pattern (Nitta, 1987; Kosaka and Nakamura, 2006, 2010), also known as the East Asia–Pacific pattern (Huang and Sun, 1992). The PJ pattern features a meridional dipole of precipitation and lower-tropospheric circulation and forms a see-saw relationship between the WNP and East Asian summer monsoons (Figs. 13.3A and B).

13.3.2 Westward influence through atmospheric Rossby waves

In response to the enhanced WNP summer monsoon convection, the anomalous lower tropospheric cyclone extends westward as phase propagation of Rossby waves, with westerly anomalies on its southern flank reaching the Arabian Sea (Fig. 13.3A). This circulation feature is well simulated in d4PDF (Fig. 13.3C). The westward-extension of the anomalous circulation can affect the Indian summer monsoon (Fig. 13.3; Chowdary et al., 2013). Consistent with Chowdary et al. (2013) and Srinivas et al. (2018), there is an increase

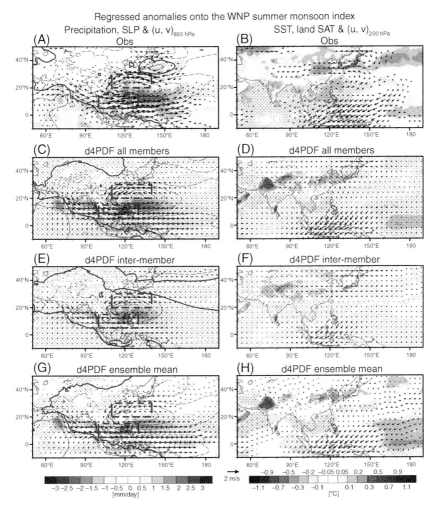

FIG. 13.3   JJA anomalies associated with the WNP summer monsoon intensification. Anomalies of (left) precipitation (shading), sea level pressure (contours for every 0.2 hPa; zero contour thickened), 850 hPa wind velocity (arrows), (right) SST, land surface air temperature (shading), and 200 hPa wind velocity (arrows) regressed onto the WNP summer monsoon index defined as 850 hPa zonal wind velocity difference between (5°–15°N, 100°–130°E) and (20°–30°N, 110°–140°E) as indicated by dashed magenta boxes in the left panels. Based on (A, B) precipitation from CMAP, SST from HadISST, and sea level pressure, wind velocity and land surface air temperature from JRA-55, and (C, D) all-member, (E, F) intermember and (G, H) ensemble-mean variability of d4PDF. Stippling indicates statistical significance for anomalies shown in shading with the confidence level of 95% based on *t* test. Arrows are shown only where either zonal or meridional components are significant at 95% confidence level, based on *t* test. "*CMAP*, Climate Prediction Center Merged Analysis of Precipitation; *JJA*, June-July-August; *SST*, sea surface temperature; *WNP*, western North Pacific."

in precipitation over Central-Eastern India extending to the Northern Bay of Bengal and a decrease spanning the Arabian Sea, Southern India, and the Southern Bay of Bengal (Figs. 13.3A and C). Precipitation also decreases around the Ganges Delta (Figs. 13.3A and C). While these precipitation anomalies are only marginally significant in observations (Fig. 13.3A), d4PDF well reproduces them with high statistical significance (Fig. 13.3C), suggesting robustness of those signals.

Srinivas et al. (2018) proposed two pathways for this WNP-Indian summer monsoon linkage. First, the westward-extending anomalous cyclone changes moisture transport and thereby convection in South Asia. Precipitation enhancement in Central-Eastern India and decrease to its south are likely induced by surface Ekman convergence to the north and divergence to the south of anomalous surface westerlies. Precipitation decrease around the Ganges Delta is presumably due to anomalous northeasterlies which counteracts background southerlies and associated moisture influx. While in observations the significant anomalies of precipitation are confined to southern India (Fig. 13.3A), the large ensemble simulations robustly identify the inter-monsoon linkage (Fig. 13.3C). In particular, the inter-member spread of d4PDF ensemble underpins these atmospheric direct influences (Fig. 13.3E). Furthermore, similarity of the inter-member and ensemble mean anomalies (Figs. 13.3E and G) suggests that a similar atmospheric influence plays an important role in the SST-forced variability.

Second, the atmospheric circulation anomalies change Indian Ocean SST. The stronger monsoon westerlies cool SST in the northern Indian Ocean, which in turn changes evaporation and atmospheric circulation. While in observations, the SST anomalies associated with the WNP summer monsoon index include response to ENSO (see Section 13.5), the cold SST anomalies are evident in the NoENSO experiment (Fig. 13.4), supporting the role of surface wind anomalies in forcing SST (Kosaka et al., 2013; Xie et al., 2016). Reduction of evaporation in response to cooler SST in the northern Indian Ocean weakens moisture supply to the Indian summer monsoon region, resulting in a precipitation decrease (Chowdary et al., 2013; Srinivas et al., 2018). The cooler Indian Ocean can further excite the cold tropospheric Kelvin wave, which is suggested to feedback to the enhanced WNP summer monsoon and form the IPOC mode (Kosaka et al., 2013; Xie et al., 2016).

13.4   The midlatitude pathway

In midlatitudes, jet streams act as a waveguide for Rossby waves and facilitate teleconnections extending in a zonal direction (Branstator, 2002). In boreal summer, the Asian jet waveguide forms along the northern rim of the South Asian (or Tibetan) high in the upper troposphere. The teleconnection pattern along the summertime Asian jet waveguide is called the Silk Road pattern (Enomoto, 2004; Sato and Takahashi, 2006; Kosaka et al., 2009). This wave train pattern

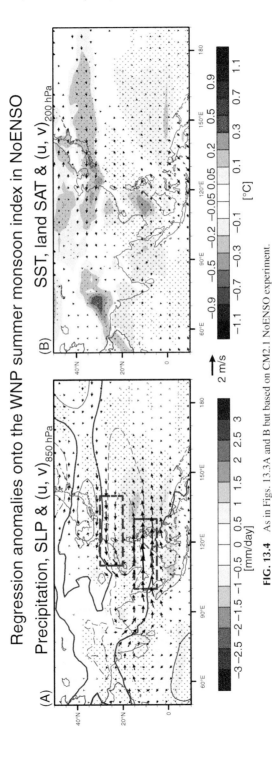

FIG. 13.4 As in Figs. 13.3A and B but based on CM2.1 NoENSO experiment.

sometimes extends further downstream to form the circumglobal teleconnection pattern (Ding and Wang, 2005). Since the focus here is on the Eurasian sector, we use the name of the Silk Road pattern.

13.4.1 The silk road pattern

Fig. 13.5 shows the Silk Road pattern extracted through an empirical orthogonal function (EOF) analysis of interannual variability of JJA 200 hPa meridional wind velocity in 25º–50ºN, 30º–130ºE in d4PDF. The method is similar to Kosaka et al. (2009) but applied to the seasonal mean-field with the domain slightly adjusted to better capture the jet stream in d4PDF. Yet, the resulting patterns are insensitive to the meridional range of the EOF domain. To examine arbitrary zonal phases of the wave train, we use both the leading and second EOF modes (EOF1 and EOF2, respectively). They respectively explain 31.8% and 23.6% of the variance integrated over the domain and are well separated under the rule of North et al. (1982), indicating the dominance of a particular zonal phase corresponding to EOF1. In JRA-55, this zonal phase anchoring is even clearer (with variance fractions of 36.4% and 20.9 %), but in different phases compared to d4PDF (not shown).

Despite seasonal averaging, the potential predictability estimated following Rowell et al. (1995) are as low as 10.9% and 2.7% for EOF1 and EOF2, respectively, in d4PDF. This result suggests that the Silk Road pattern is essentially an atmospheric internal variability and its potential predictability is low (but still non-zero), consistent with Kosaka et al. (2012). However, this result may be contaminated by model biases. Potential role of SST variability on the Silk Road pattern will be revisited in Section 13.5.3.

As documented in previous studies, the Silk Road pattern features a stationary Rossby wave train trapped by the Asian jet in the upper troposphere, with a zonal wavenumber of about 6 (Figs. 13.5A and B; Ding and Wang, 2005; Chen and Huang, 2012; Stephan et al., 2019). Vertically, it features an equivalent barotropic structure, but with a slight westward phase tilt with height (Figs. 13.5C–F), thus leading to poleward heat transport (not shown). Kosaka et al. (2009) and Chen et al. (2013) argue that baroclinic energy conversion associated with the tilting structure efficiently maintains the Silk Road pattern against dissipation. An additional contribution from barotropic energy conversion has been pointed out (Sato and Takahashi, 2006; Kosaka et al., 2009). These results imply that the Silk Road pattern is a dynamical mode inherent to the Asian jet.

13.4.2 Interaction with the Asian summer monsoons

Consistent with the mode-like characteristic, the Silk Road pattern can develop from various sources, such as anomalous diabatic heating in various locations (Yasui and Watanabe, 2010; Chen and Huang, 2012; Lin and Lu, 2016) and

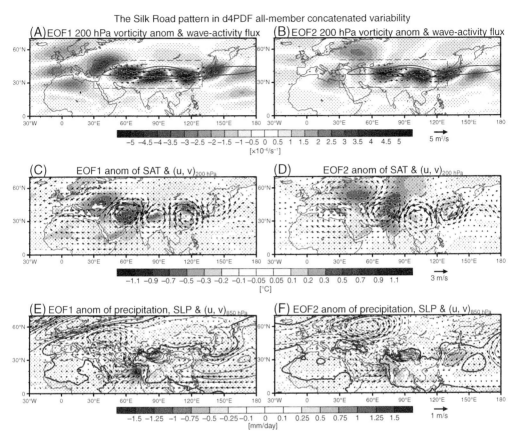

FIG. 13.5 JJA anomalies associated with the Silk Road pattern in d4PDF. Anomalies of (A, B) 200 hPa vorticity (shading), (C, D) SST and land surface air temperature (shading) and 200 hPa wind velocity (arrows), (E, F) precipitation (shading), sea level pressure (contours for every 0.2 hPa; zero contour thickened) and 850 hPa wind velocity (arrows) regressed onto the (left) leading and (right) second principal components of 200 hPa meridional wind velocity within (25°–50°N, 30°–130°E) as indicated by dashed magenta boxes in (A, B). Arrows and contours in (A, B) represent the associated wave activity flux of Takaya and Nakamura (2001) and climatological zonal wind velocity at 25 and 30 m/s, respectively. Based on all-member variability of d4PDF. Stippling indicates statistical significance for anomalies shown in shading with the confidence level of 95% based on *t* test. In (C–F), arrows are shown only where either zonal or meridional components are significant at 95% confidence level, based on *t* test. "*JJA*, June-July-August; *SST*, sea surface temperature."

blocking in Europe (Kosaka et al., 2012; Lau and Kim, 2012). Indian summer monsoon variability is also known to force (Ding and Wang, 2005; Lin, 2009; Saeed et al., 2011) or interact with (Ding and Wang, 2007; Stephan et al., 2019) the Silk Road pattern. Indeed, in d4PDF, both EOF1 and EOF2 feature significantly enhanced precipitation in Northern India, Pakistan, and the Eastern Arabian Sea, with a stronger signal in association with EOF1 (Figs. 13.5E and F). The associated positive diabatic heating anomalies excite a warm Rossby wave response extending northwestward, which interacts with the background Asian jet and induces a stationary Rossby wave train (Enomoto et al., 2003; Lin, 2009), as in the monsoon–desert mechanism (Rodwell and Hoskins, 1996). In Fig. 13.5C, the precipitation increase is apparently aided by warmer SST in the Arabian Sea, consistent with Chen and Huang (2012). This may partly explain higher potential predictability in EOF1 compared to EOF2.

The Silk Road pattern is one of key teleconnection patterns that affect summer climate in East Asia (Hsu and Lin, 2007; Kosaka et al., 2011; Chen and Huang, 2012; Shimpo et al., 2019; Thompson et al., 2019). Climatologically, the Meiyu-Baiu rainband, which characterizes the East Asian summer monsoon, forms along dynamically induced upwelling due to mid-tropospheric temperature advection (Sampe and Xie, 2010) together with upper-tropospheric vorticity advection (Horinouchi, 2014). This ascent uplifts moisture to the mid-troposphere and sets favorable conditions for convection (Yokoyama et al., 2017). Circulation anomalies induced by the Silk Road pattern thus shift the rainband and alter rainfall intensity. The Silk Road pattern also changes the extent and strength of the WNP subtropical high (or the Bonin high) and causes anomalously or cool summers to East Asia (Wakabayashi and Kawamura, 2004; Yasunaka and Hanawa, 2006).

It is thus plausible that the Silk Road teleconnection pattern can link variations of the Indian and East Asian summer monsoons (see Chapter 14). Indeed, Krishnan and Sugi (2001) found such covariation from century-long station observations, with the associated upper-tropospheric circulation anomalies similar to the Silk Road pattern. In Fig. 13.5, EOF1 features a precipitation decrease in eastern China, Korean peninsula, the East China Sea and western Japan, and precipitation increase east of Japan. EOF2 is associated with similar zonal dipole anomalies of precipitation but with a zonal phase shift. These results highlight that subregional anomalies dominate in East Asian summer monsoon changes induced by the Silk Road pattern.

A recent study by Takemura and Mukougawa (2020) further suggests that the Silk Road pattern affects the WNP summer monsoon variability and triggers the PJ pattern (Fig. 13.6). Although their focus is on the subseasonal events that involve nonlinear wave breaking, modulations of occurrence in the subseasonal variability are likely manifested in interannual covariation of the Silk Road pattern and WNP summer monsoon. This poses a possible extension of the midlatitude pathway through which the Indian summer monsoon variability affects the WNP summer monsoon. Obviously, robustness of this inter-monsoon influence requires further examination.

13.5 External drivers of the intermonsoon linkages

13.5.1 Concurrent ENSO and the Indian Ocean Dipole mode

Typically, El Niño's equatorial Pacific SST anomalies begin to develop in late boreal spring, mature in December and dissipate by subsequent boreal summer.

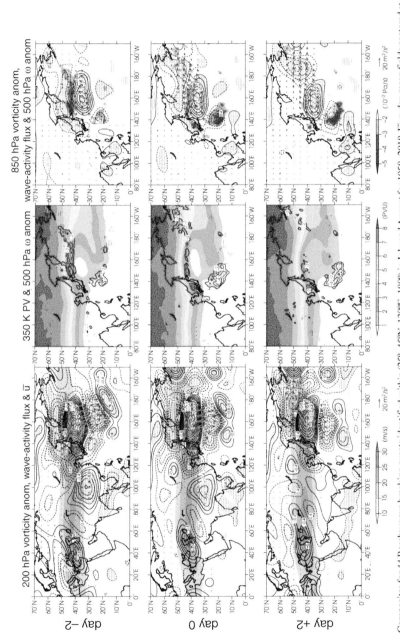

FIG. 13.6 Composites for 44 Rossby wave breaking events identified within (25°–45°N, 130°E–180°) in July and August for 1958–2018. Five-day mean fields centered at (top row) 2 days before the peak, (middle row) the peak day, and (right raw) 2 days after the peak of Rossby wave breaking. (left and right columns) Vorticity anomalies at (left) 200 and (right) 850 hPa (contoured for every 2×10^{-6} s^{-1}; zero contours omitted). (middle column) potential vorticity at 350 K isentropic surface (shading), and (middle and right columns) 500 hPa pressure velocity anomalies (contoured for $-2, -3, -4, \ldots \times 10^{-2}$ Pa s^{-1} in the left panels and shading in the right panels). Arrows in left and right panels show a wave activity flux of Takaya and Nakamura (2001) at respective levels. Shading in left panels represents climatological zonal wind velocity at 200 hPa. Stippling indicates statistical significance for anomalies shown by contours at the confidence level of 95%. Adopted from Takemura and Mukougawa (2020) (their Fig. 3d-l) under CC BY 4.0.

ENSO influences the Asian summer monsoon both in its developing and decaying years. The present and subsequent subsections review the ENSO influence in its developing and decaying year, respectively. Here, our focus is mainly on the influence of El Niño, which is overall stronger and more robust than La Niña, though further research will be beneficial on the asymmetric influence of El Niño and La Niña on monsoons.

In response to concurrent El Niño, the WNP summer monsoon circulation strengthens, with precipitation increase limited to south of 10°N (Fig. 13.7; Wang et al., 2013). Indeed, the WNP summer monsoon index is positively correlated with equatorial Pacific SST (Fig. 13.3). The circulation response is considered primarily as a response to anomalous diabatic heating on the central equatorial Pacific and cooling over the Maritime Continent (Naoi et al., 2020). As seen in the regression analysis onto the WNP summer monsoon index (Fig. 13.3), the associated anomalous cyclone and westerly anomalies on its southern flank extends westward to the Indian subcontinent in the lower troposphere, with precipitation increase around the head Bay of Bengal and decrease to its south.

Over the Indian subcontinent, precipitation anomalies are rather weak (Fig. 13.7). This is due to a compensation of opposing influences from ENSO and the co-occurring Indian Ocean Diploe (IOD), with El Niño acting to decrease the Indian summer monsoon rainfall while the positive IOD acting to increase it (Ashok et al., 2001). These influences are reviewed in Chapter 8 of this book. The IOD mode can develop independently of ENSO, and not all ENSO events accompany the IOD events. Saji and Yamagata (2003) report that 11 out of 19 IOD events for 1958-1997 did not accompany ENSO in the same polarity. Elucidating the individual influence of ENSO and the IOD mode on the inter-monsoon association is left for future research.

13.5.2   Decaying ENSO and the IPOC mode

In the IPOC mode, tropical Indian Ocean basin-wide SST anomalies interact with the WNP summer monsoon through an equatorial Kelvin wave that penetrates into the western Pacific, and through a westward-propagating Rossby wave induced by the anomalous diabatic heating in the tropical WNP (Xie et al., 2016). In the NoENSO simulation (Fig. 13.4), the IPOC mode associated with WNP summer monsoon intensification features tripolar precipitation anomalies in South Asia with precipitation increase over northeastern India and decrease over southern India and the Ganges Plain (e.g., Chowdary et al., 2019). Despite some displacements due to model biases, these anomalies are common to observed IPOC anomalies (Fig. 13.3A and B; Section 13.3.2). Since decaying ENSO is considered to be the main driver of the IPOC mode (Kosaka et al., 2013; Xie et al., 2016), the observed IPOC mode reflects the strong influence of ENSO. Indeed, the precipitation anomaly pattern is consistent with that observed in the summer of ENSO-decay years (with sign flipped; Fig. 13.6;

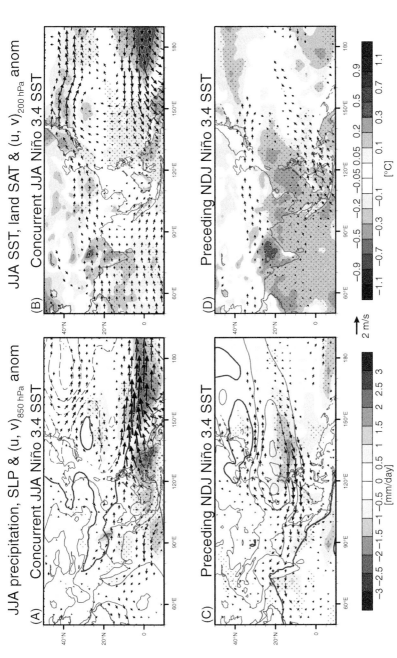

FIG. 13.7 JJA anomalies associated with ENSO. Anomalies of (left) precipitation (shading), sea level pressure (contours for every 0.2 hPa; zero contour thickened), 850 hPa wind velocity (arrows), (right) SST, land surface air temperature (shading), and 200 hPa wind velocity (arrows) regressed onto Niño 3.4 SST in (A, B) concurrent JJA and (C. D) preceding November-December-January. Based on precipitation from CMAP, SST, from HadISST, and sea level pressure, wind velocity and land surface air temperature from JRA-55. Stippling indicates statistical significance for anomalies shown in shading with the confidence level of 95% based on *t* test. Arrows are shown only where either zonal or meridional components are significant at 95% confidence level, based on *t* test. "*CMAP*, Climate Prediction Center Merged Analysis of Precipitation; *ENSO*, El Niño-Southern Oscillation; *JJA*, June-July-August."

also Chowdary et al., 2013; Kosaka et al., 2020; Chapter 5 of this book). The similarity of the Indian summer monsoon anomalies between NoENSO and in observed ENSO-decay years suggests that the IPOC mode is the key driver of the interaction between the WNP and Indian summer monsoons. Chowdary et al. (2013) found that the influence of the Indian Ocean SST anomalies on summer rainfall in Southern and Western India is stronger in ENSO-forced than unforced IPOC events since the Indian Ocean basin-wide SST anomalies are broader and stronger in the former case.

In the boreal summer of 2020, the IPOC mode contributed to enhanced and weaker precipitation in the East Asian and WNP summer monsoons, respectively (Takaya et al., 2020). It is suggested that the extreme positive IOD event in the preceding boreal autumn to winter contributed to the IPOC development through ocean Rossby waves in the tropical South Indian Ocean (Takaya et al., 2020). This potential IOD–IPOC linkage requires a further examination in the context of historical variability.

13.5.3 External drivers of the silk road pattern

Yasui and Watanabe (2010) examined the Silk Road pattern in a dry nonlinear atmospheric model forced by historical diabatic heating taken from a reanalysis dataset. They found that anomalous diabatic heating over the Mediterranean, North America, and equatorial Africa can effectively excite the Silk Road pattern as a linear wave response. In addition, nonlinear atmospheric dry dynamics such as blocking is also a candidate of the driver. In 2010, the extremely strong Silk Road pattern developed from a blocking formed in eastern Europe (Kosaka et al., 2012). This blocking caused extreme heat wave and a lot of wildfires in eastern Europe (Grumm, 2011). In its downstream, unusually heavy monsoon rainfall hit Pakistan and Northwestern India under the Silk Road pattern, causing record-breaking floods (Hong et al., 2011). Further downstream, the Silk Road pattern was a major cause of a record heat wave event in Korea and Japan. This sequence of extreme weather events illustrates that upstream atmospheric perturbations drive covariation of the Indian and East Asian summer monsoons. However, the statistical relationship between the Silk Road pattern and European blocking activity remains to be examined.

Recent studies found decadal-to-interdecadal components of the Silk Road pattern (Lin et al., 2016; Stephan et al., 2019) with a shift in the dominant zonal phase (Wu et al., 2016) and a broader meridional extent (Wang et al., 2017) compared to its interannual counterpart. Based on model simulations, Lin et al. (2016) and Monerie et al. (2018) argue that the Atlantic multidecadal variability (AMV) (also called the Atlantic Multidecadal Oscillation; Kerr, 2000; Knight et al., 2005) contributes to the interdecadal Silk Road pattern (Fig. 13.8). Stephan et al. (2019) further dug into the involved processes and concluded that AMV can modulate Indian summer monsoon rainfall and thereby drive the interdecadal Silk Road pattern. Yet, as apparent in Fig. 13.8, the response

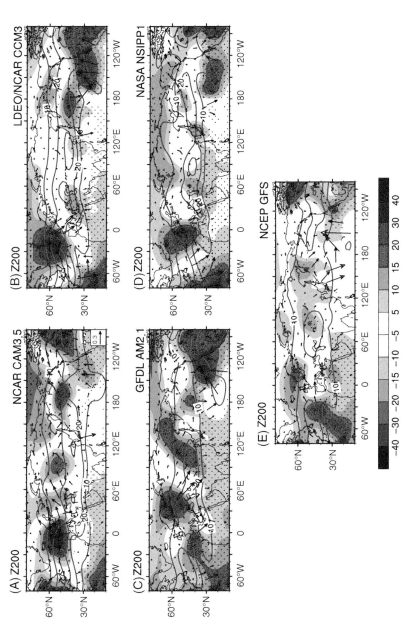

FIG. 13.8 AGCM responses to AMV SST anomalies in the North Atlantic in JJA. The thirty-year difference of 200 hPa geopotential height (shading, unit: m) between sensitivity and control experiments in (A) NCAR CAM3.5, (B) LDEO/NCAR CCM3, (C) GFDL AM2.1, (D) NASA NSIPP1, and (E) NCEP GFS. Arrows represent corresponding wave activity flux with a scale in (A) (unit: m^2/s^2). The control experiment is driven by climatological SST, while in the sensitivity experiment observed AMV North Atlantic SST anomalies are superposed on it. Contours show climatological 200 hPa zonal wind velocity (for 10, 20, and 30 m/s) in the control experiments. Stippling indicates statistical significance for geopotential height anomalies with the confidence level of 90% based on t test. Adopted from Lin et al. (2016) under CC BY 4.0. "AGCM, atmospheric general circulation model: AMV, Atlantic Multidecadal Variability: JJA, June–July–August; SST, sea surface temperature."

patterns against AMV SST anomalies prescribed to AGCMs are quite different especially in their zonal phases, illustrating a lack of robustness. It, therefore, remains inconclusive how robustly the AMV can drive the Silk Road pattern and thus induce the Indian and East Asian summer monsoon linkage.

13.6 Concluding remarks

This chapter has reviewed possible internal processes and external drivers of inter-monsoon linkages among the Indian, WNP, and East Asian summer monsoons. Fig. 13.9 shows a visual abstract of this chapter. The key teleconnections include the westward propagation of Rossby waves from the WNP summer monsoon anomalies (as part of the IPOC mode), the PJ pattern, and the Silk Road pattern. Reviewed several external phenomena that can drive the inter-monsoon linkages by exciting the teleconnections. There are other phenomena or forcing that can drive their covariation on interannual and longer time scales, such as forcing from volcanic and anthropogenic aerosols and greenhouse gas concentration changes (Biasutti et al., 2018; Seth et al., 2019).

An important implication obtained from this review is that the inter-monsoon linkages are dominated by subregional features in each of the monsoons. Specifically, an increase of WNP summer monsoon rainfall induces tripolar rainfall anomalies in the Indian summer monsoon region (e.g., Chowdary et al. 2019). Influence of the Silk Road pattern features a zonal dipole in East Asian summer monsoon precipitation anomalies, whose pattern depends on the zonal phase of the Silk Road wave train. WNP summer monsoon precipitation anomalies associated with concurrent ENSO are meridionally orthogonal to those regressed onto the WNP summer monsoon index. Furthermore, studies note seasonality of these influences. For instance, Indian summer monsoon precipitation anomalies associated with the IPOC mode varies considerably from June to August (Chowdary et al., 2017; Kosaka et al., 2020). Wavelength of the Silk

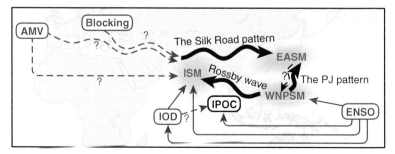

FIG. 13.9 A visual summary of this chapter. Monsoons are shown in blue, and internal processes linking the monsoons are indicated in black. External drivers are represented in red. Since the IPOC mode involves WNP summer monsoon variability in its internal component, it is denoted in black. A question mark is added to the linkage that requires further research for robustness. "*EASM*, East Asian summer monsoon; *IPOC*, Indo-western Pacific Ocean Capacitor; *ISM*, the Indian summer monsoon; *WNPSM*, WNP summer monsoon."

Road pattern is different between July and August due to seasonal changes of the background westerlies (Kosaka et al., 2009). Seasonal changes of climatological precipitation and circulation can further modulate the influence of the teleconnection patterns. Thus, variability of an entire monsoon activity represented by, e.g., seasonal-mean precipitation averaged over the entire monsoon region can be a consequence of strong compensation of subregional and subseasonal features. Examining relationships among predefined indices thus may fail in capturing key anomalies associated with inter-monsoon linkages. Superposition of influences from multiple drivers would further complicate the interpretation of such superficial analyses. Rather, this review illustrates the importance of a process-based study of the inter-monsoon linkages.

Funding

The author is supported by Japan Society for the Promotion of Science (Grant number JP18H01278, JP18H01281, and JP19H05703), by Japanese Ministry of Education, Culture, Sports, Science and Technology through Integrated Research Program for Advancing Climate Models (JPMXD0717935457), by the Environmental Research Development Fund (JPMEERF20192004) of the Environmental Restoration and Conservation Agency of Japan, and by Japan Science and Technology Agency through Belmont Forum CRA "InterDec."

References

Ashok, K., Guan, Z., Yamagata, T, 2001. Impact of the Indian Ocean dipole on the relationship between the Indian monsoon rainfall and ENSO. Geophys. Res. Lett. 28, 4499–4502. doi:1 0.1029/2001GL013294.

Biasutti, M., Voigt, A., Boos, W.R., Braconnot, P., Hargreaves, J.C., Harrison, S.P., et al., 2018. Global energetics and local physics as drivers of past, present and future monsoons. Nat. Geosci. 11, 392–400. doi:10.1038/s41561-018-0137-1.

Branstator, G., 2002. Circumglobal teleconnections, the jet stream waveguide, and the North Atlantic oscillation. J. Clim. 15, 1893–1910. doi:10.1175/1520-0442(2002)015<1893:CTTJS W>2.0.CO;2.

Chen, G., Huang, R., 2012. Excitation mechanisms of the teleconnection patterns affecting the July precipitation in Northwest China. J. Clim. 25, 7834–7851. doi:10.1175/JCLI-D-11-00684.1.

Chen, G., Huang, R., Zhou, L., 2013. Baroclinic instability of the silk road pattern induced by thermal damping. J. Atmos. Sci. 70, 2875–2893. doi:10.1175/JAS-D-12-0326.1.

Chowdary, J.S., Gnanaseelan, C., Chakravorty, S., 2013. Impact of Northwest Pacific anticyclone on the Indian summer monsoon region. Theor. Appl. Climatol. 113, 329–336. doi:10.1007/ s00704-012-0785-9.

Chowdary, J.S., Harsha, H.S., Gnanaseelan, C., Srinivas, G., Parekh, A., Pillai, P., et al., 2017. Indian summer monsoon rainfall variability in response to differences in the decay phase of El Niño. Clim. Dyn. 48, 2707–2727. doi:10.1007/s00382-016-3233-1.

Chowdary, J.S., Patekar, D., Srinivas, G., Gnanaseelan, C., Parekh, A., 2019. Impact of the Indo-Western Pacific Ocean capacitor mode on South Asian summer monsoon rainfall. Clim. Dyn. **53**, 2327–2338. doi:10.1007/s00382-019-04850-w.

Delworth, T.L., Broccoli, A.J., Rosati, A., Stouffer, R.J., Balaji, V., Beesley, J.A., et al., 2006. GFDL's CM2 global coupled climate models. Part I: formulation and simulation characteristics. J. Clim. 19, 643–674. doi:10.1175/JCLI3629.1.

Ding, Q., Wang, B, 2005. Circumglobal teleconnection in the northern hemisphere summer. J. Clim. 18, 3483–3505. doi:10.1175/JCLI3473.1.

Ding, Q., Wang, B, 2007. Intraseasonal teleconnection between the Summer Eurasian wave train and the Indian monsoon. J. Clim. 20, 3751–3767. doi:10.1175/JCLI4221.1.

Enomoto, T., 2004. Interannual variability of the bonin high associated with the propagation of Rossby waves along the Asian Jet. J. Meteorol. Soc. Japan 82, 1019–1034. doi:10.2151/jmsj.2004.1019.

Enomoto, T., Hoskins, B.J., Matsuda, Y., 2003. The formation mechanism of the Bonin high in August. Q. J. R. Meteorol. Soc. 129, 157–178. doi:10.1256/qj.01.211.

Gill, A.E., 1980. Some simple solutions for heat-induced tropical circulation. Q. J. R. Meteorol. Soc. 106, 447–462. doi:10.1002/qj.49710644905.

Grumm, R.H., 2011. The Central European and Russian heat event of July–August 2010. Bull. Am. Meteorol. Soc. 92, 1285–1296. doi:10.1175/2011BAMS3174.1.

Ha, K.-J., Seo, Y.-W., Lee, J.-Y., Kripalani, R.H., Yun, K.-S, 2018. Linkages between the South and East Asian summer monsoons: a review and revisit. Clim. Dyn. 51, 4207–4227. doi:10.1007/s00382-017-3773-z.

Hirahara, S., Ishii, M., Fukuda, Y., 2014. Centennial-scale sea surface temperature analysis and its uncertainty. J. Clim. 27, 57–75. doi:10.1175/JCLI-D-12-00837.1.

Hong, C.-C., Hsu, H.-H., Lin, N.-H., Chiu, H., 2011. Roles of European blocking and tropical-extratropical interaction in the 2010 Pakistan flooding. Geophys. Res. Lett. 38, L13806. doi:10.1029/2011GL047583.

Horinouchi, T., 2014. Influence of upper tropospheric disturbances on the synoptic variability of precipitation and moisture transport over summertime East Asia and the Northwestern Pacific. J. Meteorol. Soc. Japan 92, 519–541. doi:10.2151/jmsj.2014-602.

Hsu, H.-H., Lin, S.-M., 2007. Asymmetry of the tripole rainfall pattern during the East Asian summer. J. Clim. 20, 4443–4458. doi:10.1175/JCLI4246.1.

Huang, R., Sun, F., 1992. Impacts of the Tropical Western Pacific on the East Asian summer monsoon. J. Meteorol. Soc. Japan 70, 243–256. doi:10.2151/jmsj1965.70.1B_243.

Kerr, R.A., 2000. A North Atlantic climate pacemaker for the centuries. Science 288, 1984–1985. doi:10.1126/science.288.5473.1984.

Knight, J.R., Allan, R.J., Folland, C.K., Vellinga, M., Mann, M.E., 2005. A signature of persistent natural thermohaline circulation cycles in observed climate. Geophys. Res. Lett. 32, L20708. doi:10.1029/2005GL024233.

Kobayashi, S., Ota, Y., Harada, Y., Ebita, A., Moriya, M., Onoda, H., et al., 2015. The JRA-55 Reanalysis: general specifications and basic characteristics. J. Meteorol. Soc. Japan 93, 5–48. doi:10.2151/jmsj.2015-001.

Kosaka, Y., Chowdary, J.S., Xie, S.-P., Min, Y.-M., Lee, J.-Y., 2012. Limitations of seasonal predictability for summer climate over East Asia and the Northwestern Pacific. J. Clim. 25, 7574–7589. doi:10.1175/JCLI-D-12-00009.1.

Kosaka, Y., Nakamura, H., 2006. Structure and dynamics of the summertime Pacific-Japan teleconnection pattern. Q. J. R. Meteorol. Soc. 132, 2009–2030. doi:10.1256/qj.05.204.

Kosaka, Y., Nakamura, H., 2010. Mechanisms of meridional teleconnection observed between a summer monsoon system and a subtropical anticyclone. Part I: the Pacific-Japan pattern. J. Clim. 23, 5085–5108. doi:10.1175/2010JCLI3413.1.

Kosaka, Y., Nakamura, H., Watanabe, M., Kimoto, M., 2009. Analysis on the dynamics of a wave-like teleconnection pattern along the summertime Asian jet based on a reanalysis dataset and climate model simulations. J. Meteorol. Soc. Japan 87, 561–580. doi:10.2151/jmsj.87.561.

Kosaka, Y., Takaya, Y., Kamae, Y., 2020. The Indo-western Pacific Ocean capacitor effectTropical and Extratropical Air-Sea Interactions. Elsevier, Amsterdam ed. S. Behera.

Kosaka, Y., Xie, S.-P., Lau, N.-C., Vecchi, G.A., 2013. Origin of seasonal predictability for summer climate over the Northwestern Pacific. Proc. Natl. Acad. Sci. U.S.A. 110, 7574–7579. doi:10.1073/pnas.1215582110.

Kosaka, Y., Xie, S.-P., Nakamura, H., 2011. Dynamics of interannual variability in summer precipitation over East Asia. J. Clim. 24, 5435–5453. doi:10.1175/2011JCLI4099.1.

Krishnan, R., Sugi, M., 2001. Baiu rainfall variability and associated monsoon teleconnections. J. Meteorol. Soc. Japan 79, 851–860. doi:10.2151/jmsj.79.851.

Lau, W.K.M., Kim, K.-M, 2012. The 2010 Pakistan flood and Russian heat wave: teleconnection of hydrometeorological extremes. J. Hydrometeorol. 13, 392–403. doi:10.1175/JHM-D-11-016.1.

Lin, H., 2009. Global extratropical response to diabatic heating variability of the Asian summer monsoon. J. Atmos. Sci. 66, 2697–2713. doi:10.1175/2009JAS3008.1.

Lin, J.-S., Wu, B., Zhou, T.-J., 2016. Is the interdecadal circumglobal teleconnection pattern excited by the Atlantic multidecadal oscillation? Atmos. Ocean. Sci. Lett. 9, 451–457. doi:10.1080/16742834.2016.1233800.

Lin, Z., Lu, R., 2016. Impact of summer rainfall over southern-central Europe on circumglobal teleconnection. Atmos. Sci. Lett. 17, 258–262. doi:10.1002/asl.652.

Matsuno, T., 1966. Quasi-geostrophic motions in the equatorial area. J. Meteorol. Soc. Japan 44, 25–43. doi:10.2151/jmsj1965.44.1_25.

Mizuta, R., Murata, A., Ishii, M., Shiogama, H., Hibino, K., Mori, N., et al., 2017. Over 5,000 Years of Ensemble Future Climate Simulations by 60-km Global and 20-km Regional Atmospheric Models. Bull. Am. Meteorol. Soc. 98, 1383–1398. doi:10.1175/BAMS-D-16-0099.1.

Monerie, P.-A., Robson, J., Dong, B., Dunstone, N., 2018. A role of the Atlantic Ocean in predicting summer surface air temperature over North East Asia? Clim. Dyn. 51, 473–491. doi:10.1007/s00382-017-3935-z.

Murakami, T., Matsumoto, J., 1994. Summer monsoon over the Asian Continent and Western North Pacific. J. Meteorol. Soc. Japan 72, 719–745. doi:10.2151/jmsj1965.72.5_719.

Naoi, M., Kamae, Y., Ueda, H., Mei, W., 2020. Impacts of seasonal transitions of ENSO on atmospheric river activity over East Asia. J. Meteorol. Soc. Japan 98, 655–668. doi:10.2151/jmsj.2020-027.

Nitta, T., 1987. Convective activities in the Tropical Western Pacific and their impact on the Northern Hemisphere summer circulation. J. Meteorol. Soc. Japan 65, 373–390. doi:10.2151/jmsj1965.65.3_373.

North, G.R., Bell, T.L., Cahalan, R.F., Moeng, F.J., 1982. Sampling errors in the estimation of empirical orthogonal functions. Mon. Weather Rev. 110, 699–706. doi:10.1175/1520-0493(1982)110<0699:SEITEO>2.0.CO;2.

Rayner, N.A., Parker, D.E., Horton, E.B., Folland, C.K., Alexander, L.V., Rowell, D.P., et al., 2003. Global analyses of sea surface temperature, sea ice, and night marine air temperature since the late nineteenth century. J. Geophys. Res. 108, 4407. doi:10.1029/2002JD002670.

Rodwell, M.J., Hoskins, B.J., 1996. Monsoons and the dynamics of deserts. Q. J. R. Meteorol. Soc. 122, 1385–1404. doi:10.1002/qj.49712253408.

Rowell, D.P., Folland, C.K., Maskell, K., Ward, M.N., 1995. Variability of summer rainfall over tropical north Africa (1906–92): observations and modelling. Q. J. R. Meteorol. Soc. 121, 669–704. doi:10.1002/qj.49712152311.

Saeed, S., Müller, W.A., Hagemann, S., Jacob, D., Mujumdar, M., Krishnan, R., 2011. Precipitation variability over the South Asian monsoon heat low and associated teleconnections. Geophys. Res. Lett. 38. L08702. doi:10.1029/2011GL046984.

Saji, N.H., Yamagata, T, 2003. Structure of SST and surface wind variability during Indian Ocean Dipole mode events: COADS observations. J. Clim. 16, 2735–2751. doi:10.1175/1520-0442(2003)016<2735:SOSASW>2.0.CO;2.

Sampe, T., Xie, S.-P., 2010. Large-scale dynamics of the Meiyu-Baiu rainband: environmental forcing by the westerly Jet. J. Clim. 23, 113–134. doi:10.1175/2009JCLI3128.1.

Sato, N., Takahashi, M., 2006. Dynamical processes related to the appearance of quasi-stationary waves on the subtropical jet in the midsummer northern hemisphere. J. Clim. 19, 1531–1544. doi:10.1175/JCLI3697.1.

Seth, A., Giannini, A., Rojas, M., Rauscher, S.A., Bordoni, S., Singh, D., et al., 2019. Monsoon responses to climate changes—connecting past, present and future. Curr. Clim. Chang. Reports 5, 63–79. doi:10.1007/s40641-019-00125-y.

Shimpo, A., Takemura, K., Wakamatsu, S., Togawa, H., Mochizuki, Y., Takekawa, M., et al., 2019. Primary factors behind the heavy rain event of July 2018 and the subsequent heat wave in Japan. Sci. Online Lett. Atmos. 15A, 13–18. doi:10.2151/sola.15A-003.

Srinivas, G., Chowdary, J.S., Kosaka, Y., Gnanaseelan, C., Parekh, A., Prasad, K.V.S.R, 2018. Influence of the Pacific–Japan Pattern on Indian Summer Monsoon Rainfall. J. Clim. 31, 3943–3958. doi:10.1175/JCLI-D-17-0408.1.

Stephan, C.C., Klingaman, N.P., Turner, A.G., 2019. A mechanism for the recently increased interdecadal variability of the silk road pattern. J. Clim. 32, 717–736. doi:10.1175/JCLI-D-18-0405.1.

Takaya, K., Nakamura, H., 2001. A formulation of a phase-independent wave-activity flux for stationary and migratory Quasigeostrophic Eddies on a Zonally varying basic flow. J. Atmos. Sci. 58, 608–627. doi:10.1175/1520-0469(2001)058<0608:AFOAPI>2.0.CO;2.

Takaya, Y., Ishikawa, I., Kobayashi, C., Endo, H., Ose, T., 2020. Enhanced Meiyu-Baiu rainfall in early summer 2020: aftermath of the 2019 super IOD event. Geophys. Res. Lett. 47, e2020GL090671. doi:10.1029/2020GL090671.

Takemura, K., Mukougawa, H., 2020. Dynamical relationship between quasi-stationary Rossby wave propagation along the Asian Jet and Pacific-Japan Pattern in Boreal summer. J. Meteorol. Soc. Japan 98, 169–187. doi:10.2151/jmsj.2020-010.

Thompson, V., Dunstone, N.J., Scaife, A.A., Smith, D.M., Hardiman, S.C., Ren, H.-L., et al., 2019. Risk and dynamics of unprecedented hot months in South East China. Clim. Dyn. 52, 2585–2596. doi:10.1007/s00382-018-4281-5.

Ueda, H., Yasunari, T., Kawamura, R., 1995. Abrupt seasonal change of large-scale convective activity over the western pacific in the northern summer. J. Meteorol. Soc. Japan 73, 795–809. doi:10.2151/jmsj1965.73.4_795.

Wakabayashi, S., Kawamura, R., 2004. Extraction of major teleconnection patterns possibly associated with the anomalous summer climate in Japan. J. Meteorol. Soc. Japan 82, 1577–1588. doi:10.2151/jmsj.82.1577.

Wang, B., Ding, Q., Fu, X., Kang, I.-S., Jin, K., Shukla, J., et al., 2005. Fundamental challenge in simulation and prediction of summer monsoon rainfall. Geophys. Res. Lett. 32, L15711. doi:10.1029/2005GL022734.

Wang, B., Xiang, B., Lee, J.-Y., 2013. Subtropical High predictability establishes a promising way for monsoon and tropical storm predictions. Proc. Natl. Acad. Sci. 110, 2718–2722. doi:10.1073/pnas.1214626110.

Wang, L., Xu, P., Chen, W., Liu, Y., 2017. Interdecadal Variations of the Silk Road pattern. J. Clim. 30, 9915–9932. doi:10.1175/JCLI-D-17-0340.1.

Wu, B., Lin, J., Zhou, T., 2016. Interdecadal circumglobal teleconnection pattern during boreal summer. Atmos. Sci. Lett. 17, 446–452. doi:10.1002/asl.677.

Xie, P., Arkin, P.A., 1997. Global precipitation: a 17-year monthly analysis based on gauge observations, satellite estimates, and numerical model outputs. Bull. Am. Meteorol. Soc. 78, 2539–2558. doi:10.1175/1520-0477(1997)078<2539:GPAYMA>2.0.CO;2.

Xie, S.-P., Kosaka, Y., Du, Y., Hu, K., Chowdary, J.S., Huang, G., 2016. Indo-western Pacific ocean capacitor and coherent climate anomalies in post-ENSO summer: a review. Adv. Atmos. Sci. 33, 411–432. doi:10.1007/s00376-015-5192-6.

Yasui, S., Watanabe, M., 2010. Forcing processes of the summertime circumglobal teleconnection pattern in a dry AGCM. J. Clim. 23, 2093–2114. doi:10.1175/2009JCLI3323.1.

Yasunaka, S., Hanawa, K., 2006. Interannual summer temperature variations over Japan and their relation to large-scale atmospheric circulation field. J. Meteorol. Soc. Japan 84, 641–652. doi:10.2151/jmsj.84.641.

Yokoyama, C., Takayabu, Y.N., Horinouchi, T., 2017. Precipitation characteristics over east Asia in early summer: effects of the subtropical jet and lower-tropospheric convective instability. J. Clim. 30, 8127–8147. doi:10.1175/JCLI-D-16-0724.1.

Zhou, W., Xie, S.P., Zhou, Z.Q., 2016. Slow preconditioning for the abrupt convective jump over the northwest Pacific during summer. J. Clim. 29, 8103–8113. doi:10.1175/JCLI-D-16-0342.1.

Chapter 14

Teleconnection along the Asian jet stream and its association with the Asian summer monsoon

Lin Wang[a], Peiqiang Xu[a], Jasti S. Chowdary[b]
[a]*Center for Monsoon System Research, Institute of Atmospheric Physics, Chinese Academy of Sciences, Beijing, China,* [b]*Indian Institute of Tropical Meteorology, Ministry of Earth Sciences, Pune, India*

14.1   Introduction

Teleconnection is a fundamental component of the climate system that refers to the climate variability links between geographically separated regions. In addition to the dipolar teleconnections, such as the Southern Oscillation (Walker and Bliss, 1932) and the North Atlantic Oscillation (Hurrel et al., 2003), and the wave-like teleconnections propagating along the great circles, such as the Pacific-North America pattern (Wallace and Gutzler, 1981), there is a third type of teleconnection that is trapped in the jet stream with the wave-like pattern. Teleconnections of this type include the Silk Road pattern (SRP) in boreal summer (e.g., Lu et al., 2002; Enomoto et al., 2003), the circumglobal teleconnection (CGT) in boreal winter (Branstator, 2002) and summer (Ding and Wang, 2005), and also the British–Baikal Corridor (BBC) pattern in boreal summer (Xu et al., 2019, 2020), among others. They owe their existence to the waveguide effect of jet streams (Ambrizzi and Hoskins, 1993). That is, certain band-like regions in the atmosphere with a local maximum of the meridional gradient of absolute vorticity or potential vorticity can form the atmospheric waveguide and trap the propagation of Rossby waves (Branstator, 1983, 2002; Hoskin and Ambrizzi, 1993; Ambrizzi et al., 1995; Branstator and Teng, 2017). As a result, the stationary Rossby waves trapped in the waveguide form teleconnection patterns that have a large zonal scale and small meridional scale often referred to as the waveguide teleconnection (Hsu and Lin, 1992; Branstator, 2002; Lu et al., 2002; Enomoto et al., 2003; Hu et al., 2018; Chowdary et al., 2019a; Teng and Branstator, 2019).

The SRP and CGT are two waveguide teleconnections along the Asian subtropical jet in boreal summer (e.g., Lu et al., 2002; Enomoto et al., 2003; Ding and Wang, 2005; Wang et al., 2017). They have almost identical centers of

Indian Summer Monsoon Variability: El Niño-teleconnections and beyond.
DOI: https://doi.org/10.1016/B978-0-12-822402-1.00009-0

action over Eurasia, but the CGT may extend further downstream to the Pacific and North America (Zhou et al., 2019; 2020). The high similarity between them often leads to the mixed usage of SRP and CGT in the literature (e.g., Yasui and Watanabe, 2010; Hong and Lu, 2016). Recent studies attempt to distinguish them from each other (e.g., Zhou et al., 2019), but this chapter will use the terminology SRP/CGT because disentangling the two is not the purpose of this chapter. In contrast, it gives a brief review of the interactions between the SRP/CGT and the Asian summer monsoon. Such a topic is essential because it helps to understand the variability of both the Asian summer monsoon and the SRP/CGT. It will be shown that variations of the SRP/CGT can alter the precipitation of the Indian summer monsoon (ISM) and the East Asian summer monsoon (EASM) from intraseasonal to interdecadal timescales on the one hand (Enomoto et al., 2003; Ding and Wang, 2007; Li et al., 2017; Wang et al., 2017) and that the latent heat released from the ISM and the EASM precipitation plays a crucial role in the formation and maintenance of the SRP/CGT on the other hand (Ding and Wang, 2005; Ding et al., 2011; Zhou et al., 2019, 2020).

14.2 Influences of the SRP/CGT on the ISM and EASM

The ISM and EASM are two distinct components of the Asian summer monsoon that have different formation and variability mechanisms (Wang, 2006). Nevertheless, their variabilities are closely related to each other on several timescales. For example, when the summer mean precipitation is above average over the ISM region on the interannual timescale, the precipitation is usually above normal over North China and below normal over the Yangtze River Valley and South Japan (Kripalani and Singh, 1993; Kripalani and Kulkarni, 1997; Zhang, 2001; Krishnan and Sugi, 2001). This ISM–EASM linkage has been noticed for decades, and it is partly attributed to the SRP/CGT and its influence on the east-west shift of the South Asian high (Wu, 2002; 2017; Enomoto et al., 2003). In climatology, the South Asian high is a dominant circulation system in the upper troposphere surrounding the Tibetan Plateau (Fig. 14.1A and B; Hoskins and Rodwell, 1995; Duan and Wu, 2005; Boos and Kuang, 2010). ISM and EASM are located at its western and eastern flank, respectively (Fig. 14.1C). Therefore, the east-west shift of the South Asian high induced by the SRP/CGT activity can interact with EASM and ISM simultaneously. The SRP/CGT has four centers of actions over East Europe, Caspian Sea, Mongolia, and the Korean Peninsula, respectively (Fig. 14.2A and B), all of which are equivalent barotropic and tilt slightly westward with height in the troposphere, with the maximum amplitude being located near the tropopause (Fig. 14.2C and D). Its second and third centers of action are located in the western and eastern portion of the South Asian high. Hence, the SRP/CGT can shift the South Asian high westward in its positive phase (Fig. 14.2A and B). The resultant large-scale upper-tropospheric divergence would induce anomalous upward motion over the northern Indian subcontinent and North East Asia, facilitating enhanced

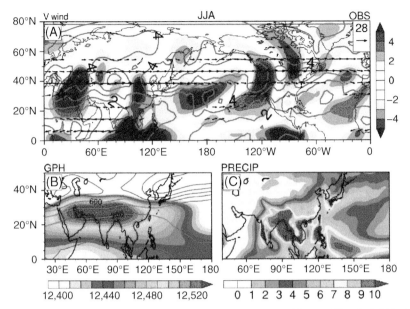

FIG. 14.1 The climatology of summer (June to August) mean (A) 200 hPa meridional wind (shading, unit: m/s), its standard deviation [contour, contour interval (CI) = 1 m/s starting from 0], and the total wind (vector, unit: m/s, only vectors whose zonal velocity are greater than 10 m/s are shown), (B) 200 hPa geopotential height (shading, unit: gpm) and its standard deviation (contour, CI = 200 gpm starting from 200 gpm), and (C) precipitation (shading, unit: mm/day). Climatology is defined as the average from 1979 to 2019. (A) and (B) are based on ERA5 dataset, and (C) is based on GPCP dataset.

upward motion and precipitation in these regions and leading to in-phase variations of precipitation between the ISM regions and North China (Fig. 14.2E and F). Meanwhile, the climatology of the atmospheric circulation is featured by easterly vertical shear of zonal wind over the ISM region in boreal summer. In the presence of this mean flow configuration, the SRP/CGT-induced enhanced ISM precipitation and latent heating may force strong circulation anomalies in the lower troposphere (Wang and Xie, 1996; Xie and Wang, 1996) and lead to above-normal water vapor convergence over the ISM region in the boundary layer (Ding and Wang, 2005). This process would further reinforce the ISM precipitation by supplying abundant water vapor to the ISM regions.

The above-mentioned influences of the SRP/CGT on the ISM and EASM can also be found on the interdecadal (Wang et al., 2017) and intraseasonal (Li et al. 2017) timescales. On the interdecadal timescale, the SRP/CGT has two regime shifts in 1972 and 1997, respectively (Wang et al., 2017). Its associated wave-like structure is similar to that on the interannual timescale except for its larger meridional scale (Fig. 14.2F vs Fig. 14.3). It accounts for approximately 40% of the precipitation reduction after 1997 over the ISM region and North

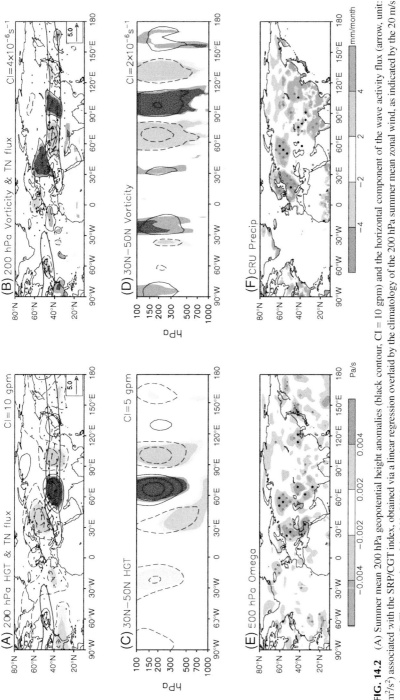

FIG. 14.2 (A) Summer mean 200 hPa geopotential height anomalies (black contour, CI = 10 gpm) and the horizontal component of the wave activity flux (arrow, unit: m²/s²) associated with the SRP/CGT index, obtained via a linear regression overlaid by the climatology of the 200 hPa summer mean zonal wind, as indicated by the 20 m/s purple contour. (C) The regression of the 30°–50°N averaged summer mean geopotential height (black contour, CI = 5 gpm) onto the SRP/CGT index. (B) and (D) are the same as (A) and (C), but geopotential height is replaced with relative vorticity [black contour, CI=4 (±2, ±6, ±10,…) × 10⁻⁶ s⁻¹ for(b) and CI = 2 (±1, ±3, ±5,…) × 10⁻⁶ s⁻¹ for (D)]. Regression of the summer mean (E) 500 hPa vertical velocity [shading interval (SI)=0.002 Pa/s], and (F) precipitation (SI = 2 mm month⁻¹) onto the SRP/CGT index. The light and dark shading in (A)–(D) and the gray and black dots in (E)–(F) indicate the 95% and 99% confidence levels based on a two-tailed Student's *t* test, respectively. The regression was applied to the period 1979–2015. This figure is adapted from Wang et al. (2017). "*SRP*, Silk Road pattern; *CGT*, Circumglobal teleconnection."

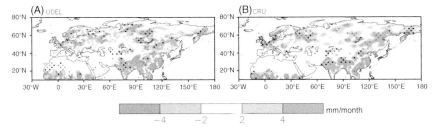

FIG. 14.3 Regression of the interdecadal component of the summer mean precipitation (unit: mm/month) from (A) UDEL v4.01 and (B) CRU TS v4.00 datasets onto the interdecadal SRP index from ERA-20C dataset during the period 1920–2010. Grey and black dots indicate the 95% and 99% confidence levels based on a two-tailed Student's t test, respectively. The figure is adapted from Wang et al. (2017).

China (Wang et al., 2017), highlighting the role of the SRP/CGT in the inter-decadal variations of the ISM and EASM. On the intraseasonal timescale, the SRP/CGT can alter the position of the western North Pacific subtropical high (WNPSH) and influence the EASM (Enomoto et al., 2003; also see Chapter 13). An example is the 2016 summer (Li et al. 2017) when the strong 2015/16 El Niño was in its decaying phase. The summer rainfall over the Yangtze River Valley is usually above normal after an El Niño winter because the WNPSH usually shifts westward and forms an anomalous lower-tropospheric anticyclone over the western North Pacific in the decaying phase of El Niño (Wang et al., 2000; Xie et al., 2009). However, the SRP/CGT was extremely strong and in its negative phase in August 2016. It resulted in an anticyclonic circulation anomaly over mid-latitude East Asia with the anomalous easterly flow over the middle-to-lower reaches of the Yangtze River Valley in the lower troposphere. This easterly flow weakened the water vapor transport induced by the WNPSH and caused weaker precipitation over the Yangtze River Valley. As a result, the precipitation anomalies over the Yangtze River Valley in the 2016 summer were positive and much weaker than those in a similar decaying phase of the El Niño in 1998 (Li et al. 2017). In the case of ISM rainfall, using historical data of more than 140 years, Chowdary et al. (2017) reported that in post-El Niño summers the Indian subcontinent receives low rainfall in June, and rainfall gradual increase by September. However, in 2016 ISM rainfall showed abnormal month-to-month variability with June rainfall being 89% (below normal) of its long period average, 107% (above normal) in July, 91% (below normal) in August, and 97% (normal) in September. Abnormal reduction in rainfall over the Indian subcontinent in August 2016 is associated with the anomalous moisture transport from ISM region to WNP region, in contrast to July, due to the presence of local cyclogenesis over WNP region (Chowdary et al., 2019b). The strong SRP/CGT altered the position of the anticyclone over WNP and caused changes in ISM rainfall in August 2016.

14.3   Influences of the ISM and EASM on the SRP/CGT

The SRP/CGT is an intrinsic atmospheric mode that can be generated solely by the internal dynamics of the atmosphere (Sato and Takahashi, 2006; Kosaka et al., 2009), but it can also be modulated by various atmospheric external forcing such as the latent heat anomalies associated with anomalous ISM (e.g., Ding and Wang, 2005; Ding et al., 2011) and EASM (e.g., Wei et al., 2015; Zhou et al., 2020) precipitation. When the ISM precipitation is above normal, the ISM-induced diabatic heating would excite a Matsuno–Gill type response (Ding and Wang, 2005). The associated anticyclonic Rossby wave response is located to the northwest of the ISM region (Fig. 14.4A). It overlaps the SRP/CGT's center of action near the Caspian Sea and contributes substantially to the excitation and maintenance of the SRP/CGT (Ding and Wang, 2005; Ding et al., 2011). This picture can also be understood from another point of view. That is, the upper-tropospheric divergence induced by the enhanced ISM precipitation forms a Rossby wave source (Sardeshmukh and Hoskins, 1988) in the subtropics near the Indian subcontinent. Although this Rossby wave source is located to the south of the subtropical jet, it can be advected northward to the latitude of the subtropical jet by the upper-tropospheric divergent wind and serves as an essential source of the SRP/CGT. Combining the review in Section 14.2, it can be seen that there is a positive feedback between the ISM precipitation and SRP/CGT. On the one hand, enhanced ISM could excite the positive phase of the SRP/CGT with an anomalous upper-tropospheric anticyclone near the Caspian Sea (Ding and Wang, 2005). The positive phase of the SRP/CGT facilitates enhanced ISM precipitation by inducing enhanced ascending motion and water vapor convergence over the ISM region. Hence, this positive feedback between the ISM and SRP/CGT is regarded as a crucial mechanism to excite, maintain, and anchor the SRP/CGT. Moreover, the variability of ISM is closely related to the El Niño-Southern Oscillation (ENSO). Therefore, the ISM is also a bridge to link the influence of ENSO to the SRP/CGT (Ding et al., 2011).

The EASM is located in the downstream portion of the SRP/CGT, and its influences on the SRP/CGT were not recognized until recent years. The Meiyu band is the primary precipitation band of the EASM, and it extends from the Yangtze River Valley northeastward to Japan (e.g., Ding and Chan, 2005; Huang et al., 2012). On the one hand, the EASM precipitation in the western portion of the Meiyu band is crucial for the formation of the SRP/CGT structure over the eastern portion of the Eurasian continent. Data diagnosis and numerical experiments with a simplified atmospheric general circulation model suggest that when the Meiyu precipitation is below normal, the resultant reduced latent heat release along the Yangtze River Valley can excite an anomalous anticyclone to its east and an anomalous cyclone to its west in the upper troposphere (Wei et al., 2015). This configuration indicates a northwestward shift of the South Asian high and a positive phase of the SRP/CGT. On the other hand, the precipitation in the eastern portion of the Meiyu band (i.e., surrounding Japan)

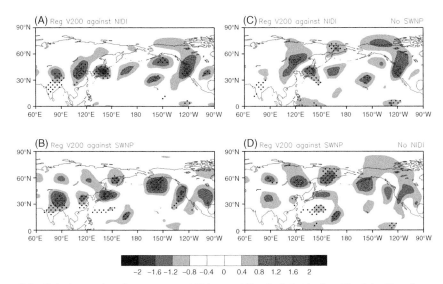

FIG. 14.4   Regression of summer mean 200 hPa meridional winds (shading, SI = 0.4 m/s) against averaged precipitation over the (A) Northern India (NIDI; 20°–35°N, 70°–85°E), and (B) subtropical western north Pacific Ocean (SWNP; 25°–45°N, 125°–150°E) for 1979–2013. Stippling indicates statistical significance exceeding the 99% confidence level based on the two-tailed Student's *t* test. (C) is the same as (A), but with the influence of SWNP precipitation removed using partial regression method. (D) is the same as (B), but with the influence of NIDI precipitation removed using partial regression method. The figure is adapted from Zhou et al. (2020).

is crucial for the SRP/CGT structure over the North Pacific. The variations of the precipitation surrounding Japan (referred to as P_{Japan} hereafter) are closely associated with an SRP/CGT-like Rossby wave train over the North Pacific (Fig. 14.4B). When the variability of P_{Japan} is removed from the ISM precipitation, the ISM-related SRP/CGT structure is weakened over the North Pacific (Fig. 14.4C). When the variability of the ISM precipitation is removed from P_{Japan}, in contrast, the P_{Japan} related atmospheric circulation resembles the SRP/CGT over the North Pacific (Fig. 14.4D). These results imply an independent effect of P_{Japan} on the SRP/CGT wave train over the North Pacific, which is further confirmed by numerical experiments based on a simplified atmospheric general circulation model (Zhou et al., 2020). Therefore, the EASM precipitation along the Meiyu band contributes substantially to the formation and maintenance of the SRP/CGT not only over the eastern portion of the Eurasian continent but also over its downstream regions such as the North Pacific.

14.4   Some remaining issues

Section 14.3 shows that the ISM and EASM precipitation is essential in the excitation and maintenance of the SRP/CGT via inducing anomalous diabatic heating. However, this is not the whole story, and several issues remain to be addressed. An important issue is to what extent the ISM and EASM precipitation contribute to the SRP/CGT. The magnitude of the ISM-related Rossby wave source in the upper troposphere is small compared to the climatology, so the efficiency of the ISM in exciting the SRP/CGT is yet to understand (Sato and Takahashi, 2006). For example, no distinct shift in the probability distribution functions of daily ISM or EASM precipitation anomalies is found when the SRP/CGT is in either a developing or mature phase in the 12,000-year simulation from the Community Atmospheric Model version 3 (Teng et al., 2013). Only a weak relationship of the SRP/CGT with the ISM (Beverley et al., 2019) and ENSO (Kosaka et al., 2012) is found in the hindcast data from the seasonal prediction systems, although the spatial structure of the SRP/CGT can be well captured. In contrast, the anomalous diabatic heating over South Europe (Lin et al., 2017) or East Europe (Cen et al., 2020) is suggested to be more efficient to excite the SRP/CGT than the ISM does (Fig. 14.5). Besides, diabatic heating anomalies over the North Indian Ocean (Chen et al., 2012), the West Africa

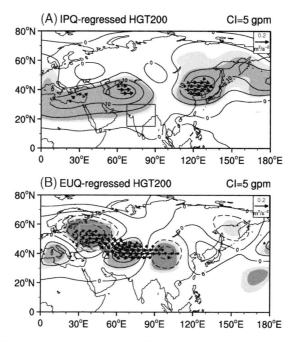

FIG. 14.5 Regression of the summer mean 200 hPa geopotential height (black contour, CI = 5 gpm) and wave activity flux anomalies on the regional mean diabatic heating <Q1> index over the (A) Indian subcontinent (15°–30°N, 67.5°–90°E) and (B) East Europe (42.5°–60°N, 30°–55°E). <Q1> is the heat source integrated vertically from 1000 hPa to 100 hPa. The dark (light) shading in (A, B) indicates geopotential height anomalies significant at the 99% (95%) confidence level. The boxes shown in (A, B) are the region used to calculate the Q1 index in the respective figure. Wave activity fluxes are omitted when both directions are less than 0.5 m²/s² in (A, B). The figure is taken from Cen et al. (2020).

monsoon region (Wang et al., 2012), the eastern Mediterranean (Yasui and Watanabe, 2010), and even the Great Plain (Teng et al., 2019) can also excite the SRP/CGT. All these results urge a further investigation of the atmospheric external forcing for the SRP/CGT.

A second issue is the stationarity of the SRP/CGT in linking the ISM and EASM. The associations between the ISM and EASM precipitation show strong interdecadal variations (e.g., Wu, 2017; Ha et al., 2018), so is the SRP/CGT (Wang et al., 2017). It is necessary to clarify whether the time-varying ISM–EASM linkage is induced by the interdecadal variations of the SRP/CGT or other factors. It may also be interesting to explore how the ISM–EASM relationship will change in a warming world. Addressing this issue may provide further insight into understanding the ISM–EASM linkage.

In addition to the SRP/CGT that is trapped in the subtropical Asian jet stream, a teleconnection named the BBC pattern along the polar front jet was identified recently (Xu et al., 2019). This BBC pattern is located more northward than the SRP/CGT, but it can also influence the temperature and rainfall over Northeast China and along with the Meiyu band (Xu et al., 2019). This coincidence implies that the BBC pattern may have some linkage with the SRP/CGT because the variations of the Meiyu precipitation can influence the SRP/CGT (Wei et al., 2015; Zhou et al., 2020; See Chapter 16). The authors' ongoing work found that some of the intraseasonal variations of the SRP/CGT are preceded by the BBC pattern, and some are not. Further study is needed in the future to clarify their relationship.

14.5   Summary

The waveguide teleconnection pattern along the Asian subtropical jet stream in boreal summer, i.e., SRP/CGT, and its associations with the Asian summer monsoon is briefly reviewed in this chapter. The interactions between the SRP/CGT and the Asian summer monsoon are emphasized. On the one hand, the SRP/CGT can shift the east-west movement of the South Asian high because two centers of action of the SRP/CGT are located in the western and eastern portion of the South Asian high, respectively. As a result, the SRP/CGT can induce anomalous vertical motion over the ISM and EASM regions and anomalous lower-tropospheric water vapor convergence on timescales extending from intraseasonal to interdecadal. The resultant circulation anomalies further lead to an in-phase precipitation anomaly over the northern Indian subcontinent and North China and precipitation of the opposite sign along the Meiyu band. On the other hand, the anomalous precipitation-related latent heat release over the ISM and EASM regions can feedback on the SRP/CGT and contribute to the formation and maintenance of the SRP/CGT. It has been well known that the anomalous latent heat release related to ISM can excite a Matsuno–Gill type Rossby wave response to the northwest of the ISM region. The latter overlaps a center of action of the SRP/CGT and thereby helps to excite and maintain the SRP/CGT. Moreover, it is recently recognized that the anomalous latent heat release of the EASM is also crucial for the

SRP/CGT. The anomalous latent heat release in the western portion of the Meiyu band contributes to the formation of the SRP/CGT structure over the eastern portion of the Eurasian continent via shifting the zonal position of the South Asian high. Meanwhile, the anomalous latent heat release in the eastern portion of the Meiyu band can excite a downstream Rossby wave train over the North Pacific and thereby contribute to the circumglobal structure of the SRP/CGT. Despite the above understanding, challenges remain to be addressed, including the extent to which the ISM and EASM precipitation contribute to the formation of the SRP/CGT, the stationarity of the SRP/CGT in linking the ISM and EASM, the relationship between the BBC pattern and the SRP/CGT, and many others.

Funding

This work is supported by the National Natural Science Foundation of China (41721004, 41925020, 42005057), the Fellowship of China Postdoctoral Science Foundation (2020M670418), and Special Research Assistant Project of Chinese Academy of Sciences. J.S.C. thank the Director, IITM for support and infrastructure.

References

Ambrizzi, T., Hoskins, B.J., Hsu, H.H., 1995. Rossby-wave propagation and teleconnection patterns in the Austral winter. J. Atmos. Sci. 52, 3661–3672.

Beverley, J.D., Woolnough, S.J., Baker, L.H., Johnson, S.J., Weisheimer, A., 2019. The northern hemisphere circumglobal teleconnection in a seasonal forecast model and its relationship to European summer forecast skill. Clim. Dyn. 52, 3759–3771.

Boos, W.R., Kuang, Z., 2010. Dominant control of the South Asian monsoon by orographic insulation versus plateau heating. Nature 463, 218–222.

Branstator, G., 1983. Horizontal energy propagation in a barotropic atmosphere with meridional and zonal structure. J. Atmos. Sci. 40, 1689–1708.

Branstator, G., 2002. Circumglobal teleconnections, the jet stream waveguide, and the North Atlantic oscillation. J. Clim. 15, 1893–1910.

Branstator, G., Teng, H., 2017. Tropospheric waveguide teleconnections and their seasonality. J. Atmos. Sci. 74, 1513–1532.

Cen, S., Chen, W., Chen, S., Liu, Y., Ma, T., 2020. Potential impact of atmospheric heating over East Europe on the zonal shift in the South Asian high: the role of the Silk Road teleconnection. Sci. Rep. 10, 6543.

Chen, G., Huang, R., 2012. Excitation mechanisms of the teleconnection patterns affecting the july precipitation in Northwest China. J. Clim. 25, 7834–7851.

Chowdary, J.S., Hu, K., Srinivas, G., Kosaka, Y., Wang, L., Rao, K.K., 2019a. The Eurasian jet streams as conduits for East Asian monsoon variability. Curr. Clim. Change Rep. 5, 233–244.

Chowdary, J., Srinivas, G., Du, Y., Gopinath, K., Gnanaseelan, C., Parekh, A., et al., 2019b. Month-to-month variability of Indian summer monsoon rainfall in 2016: role of the Indo-Pacific climatic conditions. Clim. Dyn. 52, 1157–1171 https://doi.org/10.1007/s00382-018-4185-4.

Chowdary, J., Harsha, H.S., Gnanaseelan, C., Srinivas, G., Parekh, A., Pillai, P., et al., 2017. Indian summer monsoon rainfall variability in response to differences in the decay phase of El Niño. Clim. Dyn. 48, 2707–2727. doi:10.1007/s00382-016-3233-1.

Ding, Y., Chan, J., 2005. The East Asian summer monsoon: an overview. Meteorol. Atmos. Phys. 89, 117–142.

Ding, Q.H., Wang, B., 2005. Circumglobal teleconnection in the northern hemisphere summer. J. Clim. 18, 3483–3505.

Ding, Q., Wang, B., 2007. Intraseasonal teleconnection between the summer Eurasian wave train and the Indian monsoon. J. Clim. 20, 3751–3767.

Ding, Q.H., Wang, B., Wallace, J.M., Branstator, G., 2011. Tropical-extratropical teleconnections in boreal summer: observed interannual variability. J. Clim. 24, 1878–1896.

Duan, A., Wu, G., 2005. Role of the Tibetan Plateau thermal forcing in the summer climate patterns over subtropical. Asia. Clim. Dyn. 24 (7–8), 793–807.

Enomoto, T., Hoskins, B.J., Matsuda, Y., 2003. The formation mechanism of the Bonin high in August. Quart. J. Roy. Meteorol. Soc. 129, 157–178.

Ha, K.-J., Seo, Y.-W., Lee, J.-Y., Kripalani, R.H., Yun, K.-S., 2018. Linkages between the South and East Asian summer monsoons: a review and revisit. Clim. Dyn. 51, 4207–4227.

Hoskins, B.J., Ambrizzi, T., 1993. Rossby-wave propagation on a realistic longitudinally varying flow. J. Atmos. Sci. 50, 1661–1671.

Hoskins, B.J., Rodwell, M.J., 1995. A model of the Asian summer monsoon. I. the global-scale. J. the Atmos. Sci. 52, 1329–1340.

Hong, X., Lu, R., 2016. The meridional displacement of the Summer Asian Jet, Silk Road pattern, and tropical SST anomalies. J. Clim. 29, 3753–3766.

Hsu, H.H., Lin, S.H., 1992. Global teleconnections in the 250-Mb Streamfunction Field During the Northern-Hemisphere winter. Mon. Weather Rev. 120, 1169–1190.

21Huang, R., Chen, J., Wang, L., Lin, Z., 2012. Characteristics, processes, and causes of the spatio-temporal variabilities of the East Asian monsoon system. Adv. Atmos. Sci. 29, 910–942.

Hurrell, J.W., Kushnir, Y., Ottersen, G., Visbeck, M., 2003. An overview of the North Atlantic Oscillation. North Atlantic Oscillation: Climatic Significance and Environmental Impact, pp. 1–35.

Hu, K., Huang, G., Wu, R., Wang, L., 2018. Structure and dynamics of a wave train along the wintertime Asian jet and its impact on East Asian climate. Clim. Dyn. 51, 4123–4137.

Kosaka, Y., Nakamura, H., Watanabe, M., Kimoto, M., 2009. Analysis on the dynamics of a wave-like teleconnection pattern along the summertime asian jet based on a reanalysis dataset and climate model simulations. J. Meteorol. Soc. Japan 87, 561–580.

Kosaka, Y., Chowdary, J.S., Xie, S.-P., Min, Y.-M., Lee, J.-Y., 2012. Limitations of seasonal predictability for summer climate over East Asia and the Northwestern Pacific. J. Clim. 25, 7574–7589.

Kripalani, R.H., Singh, S., 1993. Large scale aspects of India-China summer monsoon rainfall. Adv. Atmos. Sci. 10, 71–84.

Kripalani, R.H., Kulkarni, A., 1997. Rainfall variability over South–East Asia—connections with Indian monsoon and Enso extremes: new perspectives. Int. J. Climatol.: A J. Roy. Meteorol. Soc. 17, 1155–1168.

Krishnan, R., Sugi, M., 2001. Baiu rainfall variability and associated monsoon teleconnections. J. Meteorol. Soc. Japan 79, 851–860.

Li, C., Chen, W., Hong, X., Lu, R., 2017. Why was the strengthening of rainfall in summer over the Yangtze River Valley in 2016 less pronounced than that in 1998 under similar preceding El Niño Events?—role of midlatitude circulation in August. Adv. Atmos. Sci. 34, 1290–1300.

Lin, Z., Liu, F., Wang, B., Lu, R., Qu, X., 2017. Southern European rainfall reshapes the early-summer circumglobal teleconnection after the Late 1970s. Clim. Dyn. 48, 3855–3868.

Lu, R.Y., Oh, J.H., Kim, B.J., 2002. A teleconnection pattern in upper-level meridional wind over the North African and Eurasian Continent in Summer. Tellus Series a-Dyn. Meteorol. Oceanogr. 54, 44–55.

Sardeshmukh, P.D., Hoskins, B.J., 1988. The generation of global rotational flow by steady idealized tropical divergence. J. Atmos. Sci. 45, 1228–1251.

Sato, N., Takahashi, M., 2006. Dynamical processes related to the appearance of quasi-stationary waves on the subtropical jet in the midsummer northern hemisphere. J. Clim. 19, 1531–1544.

Teng, H., Branstator, G., 2019. Amplification of waveguide teleconnections in the boreal summer. Curr. Clim. Change Rep. 5, 421–432.

Teng, H., Branstator, G., Tawfik, A.B., Callaghan, P., 2019. Circumglobal response to prescribed Soil Moisture over North America. J. Clim. 32, 4525–4545.

Teng, H., Branstator, G., Wang, H., Meehl, G.A., Washington, W.M., 2013. Probability of Us heat waves affected by a subseasonal planetary wave pattern. Nat. Geosci. 6, 1056–1061.

Wallace, J.M., Gutzler, D.S., 1981. Teleconnections in the geopotential height field during the northern hemisphere winter. Mon. Weather Rev. 109, 784–812.

Walker, G.T., Bliss, E.W., 1932. World weather. V. Mem. Roy. Meteorol. Soc. 4, 53–84.

Wang, B., Wu, R.G., Fu, X.H., 2000. Pacific-East Asian teleconnection: how does ENSO affect East Asian climate? J. Clim. 13, 1517–1536.

Wang, B., 2006. The Asian Monsoon. Springer-Verlag Berlin, Heidelberg. doi:10.1007/3-540-37722-0.

Wang, B., Xie, X.S., 1996. Low-frequency equatorial waves in vertically Sheared Zonal Flow .1. Stable Waves. J. Atmos. Sci. 53, 449–467.

Wang, L., Xu, P., Chen, W., Liu, Y., 2017. Interdecadal Variations of the Silk Road Pattern. J. Clim. 30, 9915–9932.

Wang, H., Wang, B., Huang, F., Ding, Q., Lee, J.Y., 2012. Interdecadal Change of the boreal summer circumglobal teleconnection (1958–2010). Geophys. Res. Lett., 39.

Wei, W., Zhang, R.H., Wen, M., Kim, B.J., Nam, J.C., 2015. Interannual variation of the South Asian high and its relation with Indian and East Asian summer monsoon rainfall. J. Clim. 28, 2623–2634.

Wu, R.G., 2002. A mid-latitude Asian circulation anomaly pattern in boreal summer and its connection with the Indian and East Asian summer monsoons. Int. J. Climatol. 22, 1879–1895.

Wu, R.G., 2017. Relationship between Indian and East Asian summer rainfall variations. Adv. Atmos. Sci. 34, 4–15.

Xie, X., Wang, B., 1996. Low-frequency equatorial waves in vertically sheared zonal flow. Part II: unstable waves. J. Atmos. Sci. 53, 3589–3605.

Xie, S.-P., Hu, K., Hafner, J., Tokinaga, H., Du, Y., Huang, G., Sampe, T., 2009. Indian Ocean capacitor effect on Indo-Western Pacific climate during the Summer FOLLOWING El Nino. J. Clim. 22, 730–747.

Xu, P., Wang, L., Chen, W., 2019. The British–Baikal Corridor: A Teleconnection Pattern Along the Summertime Polar Front Jet over Eurasia. J. Clim. 32, 877–896.

Xu, P., Wang, L., Chen, W., Chen, G., Kang, I.-S., 2020. Intraseasonal variations of the British-Baikal Corridor pattern. J. Clim. 33, 2183–2200.

Yasui, S., Watanabe, M., 2010. Forcing processes of the summertime circumglobal teleconnection pattern in a dry AGCM. J. Clim. 23, 2093–2114.

Zhang, R.H., 2001. Relations of water vapor transport from Indian monsoon with that over East Asia and the summer rainfall in China. Adv. Atmos. Sci. 18, 1005–1017.

Zhou, F., Zhang, R., Han, J., 2019. Relationship between the circumglobal teleconnection and Silk Road pattern over Eurasian continent. Sci. Bull. 64, 374–376.

Zhou, F., Zhang, R., Han, J., 2020. Influences of the East Asian summer rainfall on circumglobal teleconnection. J. Clim. 33, 5213–5221.

Chapter 15

South Asian summer monsoon and subtropical deserts

K P Sooraj[a], Pascal Terray[b], Annalisa Cherchi[c]

[a]*Centre for Climate Change Research, Indian Institute of Tropical Meteorology, Ministry of Earth Sciences (IITM-MoES), Pune, India*, [b]*Sorbonne Universites (UPMC, Univ Paris 06)-CNRS-IRD-MNHN, LOCEAN Laboratory, Paris, France*, [c]*National Research Council of Italy, Institute of Atmospheric Sciences and Climate (CNR-ISAC), Bologna, Italy, Istituto Nazionale di Geofisica e Vulcanologia, INGV, Bologna, Italy*

15.1   Introduction

Monsoon and desert regions coexist at the subtropical latitudes of the African–Asian continent (e.g., Rodwell and Hoskins, 1996; Warner, 2004). The mutual association between these two contrasting climates has been studied in the past as well as in the recent times (e.g., Ramage, 1966; Charney, 1975; Charney et al., 1977; Shukla and Mintz, 1982; Sud and Fennessy, 1982; Smith, 1986a, 1986b; Mooley and Paolino, 1988; Sud et al., 1988; Parthasarathy et al., 1992; Yang et al., 1992; Webster, 1994; Rodwell and Hoskins, 1996, 2001; Sikka, 1997; Claussen, 1997; Bonfils et al., 2000; Douville et al., 2001; Xue et al., 2004; Yasunari et al., 2006; Wang, 2006; Wu et al., 2009; Biasutti et al., 2009; Lavaysse et al., 2009; Xue et al., 2010; Bollasina and Nigam, 2011a, 2011b; Bollasina and Ming, 2013; Tyrlis et al., 2013; Cherchi et al., 2014; Vinoj et al., 2014; Shekhar and Boos, 2017; Sooraj et al., 2019). As desertification is a fundamental process of the ongoing climate change (Cook and Vizy, 2015; Zhou, 2016; Wei et al., 2017) and climate projections of the South Asian monsoon remain uncertain (e.g., IPCC, 2013; Sabeerali et al., 2015; Annamalai et al., 2015; Sooraj et al., 2015; Krishnan et al., 2016; Kitoh, 2017; Singh and Achutarao, 2018; Wang et al., 2020), a renewed interest to understand the monsoon–desert relationships have grown. Furthermore, these two climate systems are affected by severe temperature, rainfall, and radiation biases in current climate models (e.g., Bollasina and Ming, 2013; Levine et al., 2013; Sperber et al., 2013; Roehrig et al., 2013; Prodhomme et al., 2014; Sandeep and Ajayamohan, 2015; Samson et al., 2016; Haywood et al., 2016; Terray et al., 2018) and improving their representation is an important aspect to have better climate forecasts and projections (e.g., Cherchi et al., 2014; Terray et al., 2018; Sooraj et al., 2019).

Indian Summer Monsoon Variability: El Niño-teleconnections and beyond.
DOI: https://doi.org/10.1016/B978-0-12-822402-1.00015-6

Against the backdrop of this, this chapter is an attempt to provide a comprehensive overview of the mutual relationships between these two contrasting climates thereby highlighting the underlying mechanisms. The chapter is organized as follows: Section 15.2 describes the salient climatological characteristics of the monsoon–desert system and highlights the historical background on the existence of subtropical deserts over African–Asian regions at the same latitude as the South Asian monsoon. Section 15.3 focuses on the south Asian monsoon (i.e., Indian Summer Monsoon, ISM) influence on the hot subtropical deserts, thus presenting the monsoon–desert paradigm using observation, reanalyses, and coupled General Circulation Models (GCMs). Section 15.4 reviews the literature on the potential role of deserts in modulating the ISM. Finally, Section 15.5 encapsulates the main highlights as drawn from this Chapter review.

15.2 The monsoon–desert system: background settings

The monsoon–desert system over African–Asian region is characterized by a sharp rainfall gradient during the boreal summer (e.g., from June to September) with heavy rainfall in the monsoon regions of the northern hemisphere (NH), but only sporadic and low rainfall in the adjacent arid regions (see Fig. 15.1A). The contrasts between the two climates are further corroborated when considering other important parameters such as surface albedo or net radiation budget at the Top of the Atmosphere (TOA; see Fig. 15.1B and C).

The subtropical desert regions, with bright sandy surface terrains, clear sky conditions, high temperature, reduced soil moisture, and lack of vegetation, are characterized by relatively high surface albedo (see Fig. 15.1B; Sikka, 1997; Warner, 2004; Terray et al., 2018; Sooraj et al., 2019). Fig. 15.1B shows two important examples for subtropical deserts: the Arabian-Iran-Thar desert located just to the west of the ISM system (Sikka 1997) and the Sahara just to the north of the West African Monsoon (WAM) (Lavaysse et al., 2009). In this chapter, these high albedo regions lying across North Africa and West Asia (i.e., the geographical region bounded by 20°W–75°E, 15°–40°N; see Fig. 15.1B) are collectively referred hereafter as "subtropical deserts" (or simply "arid regions").

As per the traditional theory dating back to the 1970s, the subtropical desert climates are associated with the descending branch of the Hadley circulation over the subtropics of the NH (Warner, 2004). However, NH Hadley circulation is weakest during boreal summer, which is completely out of synchronization with some of the observed summer climates features along the NH subtropics (e.g., the coexistence of both moist monsoon and desert climates at the same latitude). Consequently, this traditional view is now not well accepted in the literature (Yang et al., 1992; Rodwell and Hoskins, 1996; Wang, 2006; Wu et al., 2009). In a pioneering study, Charney (1975) made an attempt to explain the enhancement of subsidence over subtropical desert of the NH during boreal summer, by invoking a mechanism called biosphere-albedo feedback. As per their study, the pronounced reduction in vegetation cover (i.e., over-grazing) over subtropical desert

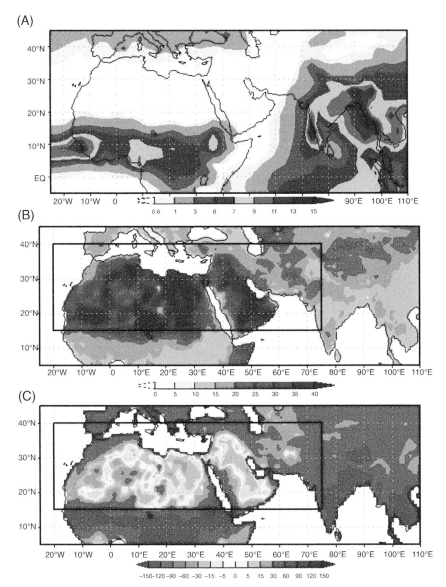

FIG. 15.1 Climatological map of (A) rainfall (mm/day), (B) land surface albedo (%), and (C) net radiation budget at TOA (W/m²), for boreal summer period (June–September). In (A) rainfall climatology is computed for the 1986–2014 period from Global Precipitation Climatology Project (GPCP version 2.1; Huffman et al., 2009). In (B) and (C), albedo and radiation climatology is computed for the 2000–2018 period from the Clouds and the Earth's Radiant Energy System Energy Balanced and Filled (CERES-EBAF edition 4.0; Kato et al., 2018). In (B) and (C), the region highlighted in the black rectangle refers to "subtropical deserts." "*TOA*, top of the atmosphere."

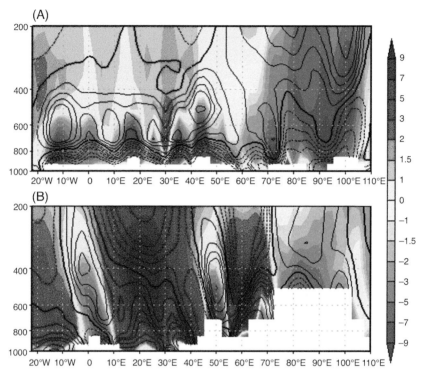

FIG. 15.2 Vertical cross-section of boreal summer atmospheric circulation in terms of vertical component of velocity (10^{-2} Pa/s, shading) and horizontal wind divergence (10^{-6} s^{-1}. contours) during July, along a pressure-longitude plane averaged over two latitude bands (A) 20°N and (B) 40°N. The circulation fields are taken from the ERA-Interim reanalysis (Dee et al., 2011) and the climatology is computed for the 1986–2014 period. The negative (dashed) and positive (continuous) contours correspond, respectively, to absolute magnitudes of 1, 2, 3, and 4 units. The zero contours are highlighted in thick black color. Negative (positive) shading implies ascending (descending), while negative (positive) contours imply convergence (divergence). The presentation using July climatology follows the observational conjecture that it corresponds to the peak of monsoon activity (e.g., Tyrlis et al., 2013).

regions increases surface albedo, thus causing enhanced radiative cooling (see Fig. 15.1C). This radiative loss is balanced by enhanced descent (see Fig. 15.2), which in turn results in reduced rainfall, thus leading to a further decrease in vegetation cover (i.e., desertification amplification). However, their mechanism ignored the possible influence of horizontal heat advection and hence was found to be inappropriate over the NH subtropical deserts (e.g., Hoskins, 1986). Later, Yang et al. (1992) and Webster (1994) proposed another concept involving a closed "Walker type" circulation linking convection over South Asia to the subsidence over arid regions to its west. Nevertheless, subsequent studies (i.e., the seminal work of Rodwell and Hoskins, 1996, 2001; Hoskins, 1986) dispelled this notion, as their studies showed no signature of a closed overturning circulation. Based on the scale analysis of the thermodynamic equation, Rodwell and

Hoskins (1996) also argued for the importance of horizontal heat advection for the existence of these subtropical desert regions. This currently recognized paradigm is often referred to in the literature as the "monsoon–desert mechanism," and is further described in Section 15.3.

In contrast to the subtropical desert, the vegetated land surface over the South Asian landmass shows a reduced surface albedo and the ISM region is radiatively surplus highlighting the role of clouds and moisture on the radiation balance (see Fig. 15.1B and C; Sooraj et al., 2019). Furthermore, the ISM land region and the Bay of Bengal are characterized by strong ascending motion extending throughout the troposphere (see Fig. 15.2A), thus exemplifying the monsoon-induced stratification due to the ISM rainfall and its associated diabatic heating (Fig. 15.1A). The corresponding low-level convergence is consistent with the thermo-dynamical response due to this diabatic heating (Fig. 15.2A; e.g., Neelin and Held, 1987).

The transition zone between this ISM convection center and the hot subtropical deserts to its west (i.e., extending from Mediterranean to west Asian landmass) is characterized by the existence of several regional heat lows with lower-level convergence and ascending motions (i.e., below 600-hPa) capped aloft by upper-level subsidence and divergence (i.e., mostly confined between 200 and 600-hPa) as displayed in Fig. 15.2A. Fig. 15.2B shows the intensification of these descending motions (extending throughout the troposphere) over west Asia and eastern Mediterranean regions at the northern boundary of the domain, thus highlighting the pronounced subsidence over these regions. Fig. 15.2A and B highlights that the southern and northern regions of the hot subtropical deserts show totally different vertical structures of the local atmospheric circulation (Sooraj et al., 2019).

15.3   Monsoon influence over hot subtropical deserts

The idea that the existence of subtropical deserts is related to the ISM is dated back to the mid-1990s. Rodwell and Hoskins (1996) suggested a monsoon–desert mechanism whereby, in a linear modeling framework, remote diabatic heating in the Asian monsoon region induces a Rossby-wave pattern to the west, interacting with the southern flank of the mid-latitude westerlies and causing descent over eastern Sahara and the Mediterranean (Figs 15.2 and 15.3). The associated atmospheric response has the form of an anticyclone at upper levels and a cyclone at lower levels, with a deepening of the isentropic surfaces due to the mid-tropospheric warming (Fig. 15.3A). When this far poleward thermal structure interacts with the southern flank of the mid-latitude westerlies, the air moves down the isentropes on their western side (Wang et al., 2012), as evident from Fig. 15.3B where the northerly component of the westerly flow crosses the isentropic surfaces. Further, according to Rodwell and Hoskins (1996), the subsiding air is also of mid-latitude origin as revealed by their trajectory analysis. Regions of adiabatic descent are strengthened by "local diabatic enhancement"

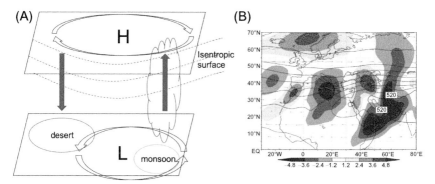

FIG. 15.3 (A) Schematic diagram of the monsoon desert-mechanism with descent over the desert region (orange) induced by the heating over the monsoon convective region (cyan) interacting with local westerlies. (Adapted by permission from Wang et al., 2012). (B) June to August mean wind (m/s, green vectors), meridional wind velocity (m/s, shaded), and pressure levels (hPa, contours with 50-hPa contour interval) on isentropic surfaces at 325K using NCEP/NCAR reanalysis dataset (Kalnay et al., 1996). The climatology is computed for the period 1958–2019.

and longitudinal mountain chains induce a blocking of the westerly flow (Rodwell and Hoskins, 2001). The idea of the monsoon–desert mechanism linking the ISM to the Mediterranean has been then easily extended to other regions with Mediterranean-type of climates, like California and Chile, remotely forced by monsoon regions to the east (Rodwell and Hoskins, 2001). Overall, the theoretical framework based on these pioneering works is well established in terms of the origin of subtropical highs in the summer hemispheres (Cherchi et al., 2018).

In agreement with idealized model simulations, results-focused on ERA40 reanalysis (Uppala et al., 2005) confirmed strong mid- and upper-level warming spreading westward and causing significant depression of the isentropes, which are linked to Rossby wave activity away from the diabatic heating sources (Tyrlis et al., 2013). Over the eastern Mediterranean and Middle East, the subsidence and northerly flow (Etesian winds) are recognized as manifestations of the Rossby wave structure triggered by the ISM convection (Tyrlis et al., 2013; Rizou et al., 2018). In a similar framework, processes at work in the monsoon–desert mechanism have been analyzed in 20th-century simulations from coupled GCMs participating to the fifth Coupled Model Intercomparison Project (CMIP5; Taylor et al., 2012), showing how few of them are able to simulate the mechanism for the correct reasons (Cherchi et al., 2014). CMIP5 coupled GCMs tend to underestimate (overestimate) ISM-related diabatic heating at upper (lower) levels, resulting in a weaker forced response (Cherchi et al., 2014). Moreover, CMIP5 coupled GCMs with a severe dry bias over the ISM region depict the weakest mid-latitude winds and minimum descent over the Mediterranean, as well as little vertical variation in diabatic heating (Cherchi et al., 2014). On the other hand, when simulated precipitation over India improves in the coupled GCMs, the extratropical teleconnection with the Mediterranean region improves as well (Jin et al., 2019). In other words, the uncertainty associated with monsoon simulations needs to be considered in future climate projections even outside the monsoon domain. Interestingly, air–sea coupling seems also to be important to correctly represent the monsoon–desert teleconnections, as in coupled model simulations the divergence field associated with the ISM is more favorable for inducing westward Rossby wave propagation than in uncoupled simulations (Osso' et al., 2019).

Within boreal summer, descent enhancement over the Mediterranean and east Sahara has been observed during the onset of the ISM (Rodwell and Hoskins, 1996). In early July, when convection migrates over northern continental India, the Rossby wave structure amplifies impacting the circulation over the eastern Mediterranean and Middle East (Tyrlis et al., 2013). In the peak of the monsoon season, the combined diabatic heating pattern over the Arabian Sea and Bay of Bengal regions exerts the largest descent over the eastern Mediterranean (Cherchi et al., 2014). As the monsoon activity is associated with active and breaks spells, it could be argued that successive Rossby wave pulses are released, inducing a cumulative effect over the region until the peak of the monsoon activity is reached in July (Tyrlis et al., 2013).

The processes recognized as part of the monsoon–desert mechanism vary at interannual timescales as well, with imprints of severe weak and strong ISMs noticeable over the Mediterranean (Cherchi et al., 2014). Interestingly, the monsoon forcing is more significant during strong ISMs, with enhanced subsidence over the Mediterranean region, than during weak ISMs (see Fig. 15.4). During

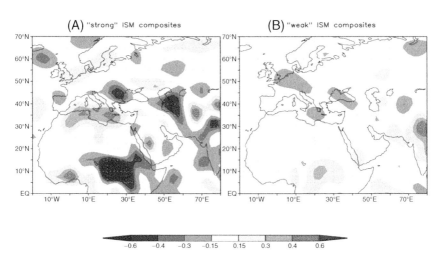

FIG. 15.4   Composite of omega at 500 hPa (units hPa/hr) for (A) strong and (B) weak monsoon years for the boreal summer monsoon period (June–September), using NCEP/NCAR reanalysis dataset (Kalnay et al., 1996) for the period 1958–2019. Strong years are 1959, 1961, 1970, 1975, 1983, 1988, 1994, and weak years are 1965, 1966, 1968, 1972, 1974, 1979, 1982, 1985, 1986, 1987, 2002, 2004, 2009, 2014, 2015 (based on normalized All India Rainfall (AIR) index exceeding 1 standard deviation). The AIR index is taken from https://www.tropmet.res.in/~kolli/MOL/Monsoon/Historical/air.html.

a strong monsoon, upper tropospheric heating expands westward reaching a maximum over the northern Arabian Peninsula modulating thermodynamic and dynamic patterns there and likely governing the occurrence of extremes over the region (Attada et al., 2019). ISM rainfall is significantly and negatively corre-lated with precipitation over the Black Sea/Balkans region (Osso' et al., 2019). This kind of relationship suggests the possibility of using the ISM characteris-tics as potential predictors for the summer conditions over the Mediterranean region. At interannual to interdecadal timescales, these dynamics are part of a larger picture involving North Africa (Liu et al., 2001; Wu et al., 2009; He et al., 2017) and portions of Eurasia via the so-called "silk road pattern" and "circum-global teleconnection" (Lu et al., 2002; Enomoto et al., 2003; Ding and Wang, 2005; Saeed et al., 2014; Wang et al., 2017; Stephan et al., 2019).

The monsoon–desert mechanism as first established using a linear modeling framework, is also supported by relating prehistoric lake-levels to Milankovitch-monsoon forcing (Rodwell and Hoskins, 1996) and extrapolated to the existence of "mega-deserts" and "mega-monsoon" areas in specific periods (Wang et al., 2014). Evidence of moist events over east Sahara have been observed (Gaven et al., 1981; Kowalski et al., 1989), consistent with shutting off of the mon-soon–desert mechanism (Quade et al., 2018). Some of those wet events have been dated to correspond to minimum NH summertime insolation, linked with a minimum intensity of ISM (Clemens et al., 1991). Some of these arguments have been confirmed with GCM studies (Claussen, 1994; Perez-Sanz et al., 2014).

As the ISM rainfall is projected to increase, the monsoon–desert telecon-nection is consistent with the changes in 21st-century projections over the Mediterranean region, at least for its central part corresponding to the largest increase in subsidence (Cherchi et al., 2016). However, other factors at play, i.e., the warming and moistening of the troposphere under climate change, which enhances significantly the greenhouse effect over the deserts (Zhou, 2016; Wei et al., 2017), the North Atlantic Oscillation (Blade et al., 2012; Kalimeris and Kolios, 2019) or the changes in frequency/intensity of blocking systems (Masato et al., 2013; Tyrlis et al., 2015), may influence and shape the response in future climate projections. In future projections, the local atmospheric dynamics con-tribute to maintaining local temperature and precipitation balance over eastern and southern Mediterranean regions, suggesting a stronger influence of land surface warming on local atmospheric circulation and progressing desertifica-tion (Zhou, 2016; Wei et al., 2017; Barcikowska et al., 2020). From a climate change perspective, the influence of ISM on the Mediterranean region may have consequences also on the sea surface characteristics in terms of temperature and ecosystems (Kim et al., 2019).

15.4 Potential role of deserts in modulating ISM

The "monsoon–desert mechanism" described in the previous section, perceives the subtropical deserts of the NH to be a rather "passive" recipient in the rela-tionship. However, from a different perspective, it is also argued that heat lows, changes of surface heating over the subtropical deserts, and dry air intrusions (whether natural or anthropogenic) from arid regions into the monsoon domain can exert significant control on the WAM and ISM systems, at both the intra-seasonal and seasonal time scales (Charney, 1975; Charney et al., 1977; Shukla and Mintz, 1982; Sud and Fennessy, 1982; Sud et al., 1988; Claussen, 1997; Bonfils et al., 2000; Douville et al., 2001; Xue et al., 2004; Yasunari et al., 2006; Xue et al., 2010; Krishnamurti et al., 2010; Bollasina and Nigam, 2011a, 2011b; Bollasina and Ming, 2013; Parker et al., 2016; Terray et al., 2018; Sooraj et al., 2019). For example, Sahelian Heat Low plays a key role in the WAM evolution through its low-level cyclonic circulation and induced moisture convergence, and this is well established in observations, reanalyzes, and GCM simulations (Haarsma et al., 2005; Biasutti et al., 2009; Lavaysse et al., 2009; Shekhar and Boos, 2017). On similar lines, ISM, especially its onset and early part, is known to be influenced by the atmospheric variability over the subtropical desert regions (i.e., northwest part of India and west-central Asia; Ramage, 1966; Smith, 1986a, 1986b; Mooley and Paolino 1988; Parthasarathy et al., 1992; Bollasina and Nigam, 2011b; Bollasina and Ming, 2013; Vinoj et al., 2014; Rai et al., 2015; Sooraj et al., 2019). During boreal summer, the monsoon depressions formed over the Bay of Bengal move northwestward across the Indian landmass, and eventually merge and dissipate in the heat-low region (Wang, 2006; Krishnamurti et al., 2013). The relationship between this seasonal heat low and ISM has been examined in perpetual boreal spring atmospheric GCM (i.e., AGCM) experiments by Bollasina and Ming, (2013). Remarkably, they found that the surface heating (i.e., with no seasonal variation of insolation in their perpetual experiments) over northwestern semi-arid areas is able to uniquely control the northwestward migration of ISM rainfall at seasonal time scale and of the monsoon depressions at intra-seasonal time scale.

Furthermore, recent modeling studies show increasing evidence of the impact of dust aerosols on the ISM at various time scales (i.e., from weekly to decadal) through aerosol radiative effects (Ramanathan et al., 2005; Lau et al., 2006; Meehl et al., 2008; Wang et al., 2009; Nigam and Bollasina, 2010; Bollasina et al., 2011; Vinoj et al., 2014; Jin et al., 2014, 2016a, 2016b; Solmon et al., 2015). For example, Vinoj et al. (2014) using sensitivity AGCM experiments argue that the presence of dust loading over North Africa and West Asian arid regions induced atmospheric heating over there, thus promoting abundant moisture transport into the ISM region, with enhanced associated rainfall (e.g., over Monsoon Core Zone, MCZ). Thus, their study basically showed that the arid regions can also act as a dust-induced atmospheric heat source (at least in a relative sense) during boreal summer and modulate ISM rainfall at the synoptic time scale. This is consistent with earlier results (Smith 1986a, 1986b; Mohalfi et al., 1998) and also in line with other recent studies focusing on the impact of Middle East dust aerosols on ISM rainfall (Jin et al., 2014, 2015, 2016a; Solmon et al., 2015). However, the rainfall responses in each of these studies display heterogeneous distributions both in space and time (Solmon et al., 2015; Jin et al., 2016b; Sanap and Pandithurai, 2015). For example, Solmon et al. (2015) conducted a study (using a regional climate model) to examine the interaction between Middle East dust and ISM. Similar to Vinoj et al. (2014), they found an increase in rainfall restricted to the southern part of India with other parts, say central and northern India, showing a significant decrease. Recently, Kumar and Arora (2019) also claimed that enhanced dust forcing and associated warming over the Arabian Sea are unlikely to create positive feedback on ISM rainfall, because of its limited spatial extent. To sum up, it is not altogether clear whether this part of the theory focusing on the role of Middle East dust on ISM is in definitive form.

This framework based on the relative heat source concept has also been extended to other time scales and the ISM onset (Rai et al., 2015; Chakraborty and Agrawal, 2017; Samson et al., 2017; Terray et al., 2018; Sooraj et al., 2019). In particular, Samson et al. (2017) and Terray et al. (2018) demonstrated the sensitivity of NH monsoon regions (in particular African–Asian monsoon) to the land surface thermal forcing over these arid regions. Using regional and global coupled GCMs, respectively, they modified the land surface albedo using up-to-date satellite estimates. According to them, the persistent tropical rainfall errors in current coupled GCMs are partly associated with insufficient surface thermal forcing and incorrect representation of the surface albedo over the NH continents. Improving the parameterization of the land albedo in a regional coupled model (Samson et al., 2017) and two global coupled GCMs (i.e., SINTEX; Masson et al., 2012 and CFSv2; Saha et al., 2014; Terray et al., 2018), leads to a significant reduction of the model systematic dry bias over land. They further showed that African–Asian monsoon circulation is, partly, a response to the large-scale pressure gradient between the hot NH subtropical deserts and the relatively cooler oceans to the south. A concept, which may have implications for the ISM response in the context of climate change as desertification has been identified as a robust feature of global warming (Zhou, 2016; Wei et al., 2017).

In a companion modeling study, Sooraj et al. (2019) found that ISM evolution and intensity are significantly affected with opposite polarity to prescribed negative and positive surface albedo perturbations over the whole hot subtropical desert lying to the west of ISM, including the remote Sahara Desert. This is consistent with the hypothesis that the arid regions can also act as a relative heat source and modulate the ISM, but also the whole tropical climate during boreal summer (Terray et al., 2018). The darkening of the deserts (negative albedo perturbations) leads to the advancement of the ISM onset by one month, with a rapid northward propagation of the rainfall band over the Indian domain (see Fig. 15.5). The brightening of the deserts (i.e., positive albedo perturbation) shows a nonlinear response in ISM rainfall and circulation with significantly larger amplitude (figure not shown).

Whilst the processes highlighted above (i.e., surface thermal forcing over arid regions) are mostly demonstrated at a seasonal time scale, a recent observational study shows the evidence at the daily timescale as well (Sooraj et al.,

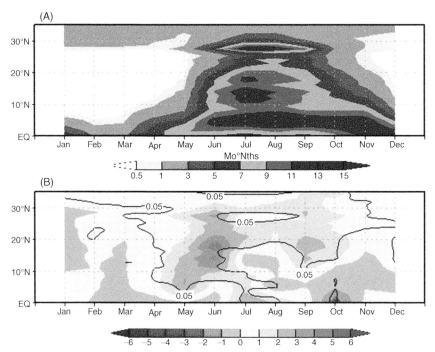

FIG. 15.5 (A) Time-latitude evolution of rainfall climatology (mm/day) averaged along 70°–90°E (over both ocean and land points), from a control simulation performed with the SINTEX-F2 coupled model (Masson et al., 2012). (B) Time-latitude evolution of anomalous rainfall response averaged along 70°–90°E from "Desert_m20" sensitivity experiment (see below). Note that here the anomalies are computed against the control rainfall climatology. In (B), the responses that are above the 95% confidence level according to a permutation procedure with 9999 shuffles are encircled. Desert_m20 is a sensitivity experiment performed with the SITEX-F2 coupled model in which the background land albedo has been artificially decreased by −20% over the whole hot subtropical desert extending up to the Sahara in the west (see highlighted box in Fig. 15.1) (Adpated by permission from Sooraj et al., 2019).

2020). While examining the ISM rainfall extremes over the MCZ region at a daily time scale, these authors found that a similar surface thermal signature over Indo-Pakistan Arid Region (IPAR) is among the best precursors of the ISM wet daily extremes at longer lead times (e.g., one or two weeks in advance), outperforming Sea Surface Temperature precursors based on the Monsoon Intra-Seasonal Oscillation (e.g., Wang et al., 2005, Roxy et al., 2017). With the additional role of moist processes, the study revealed the surprising existence of a strong water vapor positive feedback over the drier IPAR, at intraseasonal and daily time scales, in line with the strong longwave greenhouse effect as found over arid regions in the context of global warming (Zhou, 2016; Wei et al., 2017). Further, according to their study, the enhanced surface warming and amplified water vapor feedbacks maybe another significant contributor to the recent increasing trend in rainfall wet extremes over Indian landmass. In the backdrop of global warming, the results from Sooraj et al. (2020) may have larger implications concerning the changes in ISM rainfall extremes as well as their potential predictability.

15.5 Summary and future perspectives

This chapter presents a comprehensive review on the monsoon–desert system focusing on the mutual relationships between these two contrasting climates, which, surprisingly, coexist at the same latitude of the NH, and highlights the underlying mechanisms. Such a review assumes significance, as recently there is a renewed interest to understand the close relationships between the two systems since both of them are expected to be severely affected by anthropogenic climate change in the forthcoming decades (as described in Section 15.1).

Firstly, the literature focusing on the monsoon influence on subtropical deserts is reviewed by highlighting the monsoon–desert mechanism whereby convection over ISM (and also the neighboring Bay of Bengal) induces strong descending motions over the NH subtropical deserts (i.e., to the west of ISM region), with the local subsidence and northerly flow (Etesian winds) being manifestations of the Rossby wave structure associated with ISM convection (Fig. 15.3). These different processes are further summarized in Fig. 15.6. Hence, the theoretical framework, as suggested by earlier pioneering works (Rodwell and Hoskins, 1996, 2001), is now well established from more recent observational and modeling studies (e.g., Uppala et al., 2005; Tyrlis et al., 2013; Cherchi et al., 2014; Cherchi et al., 2018; Jin et al., 2019). Additional evidences are also presented to show that the monsoon–desert mechanism and the

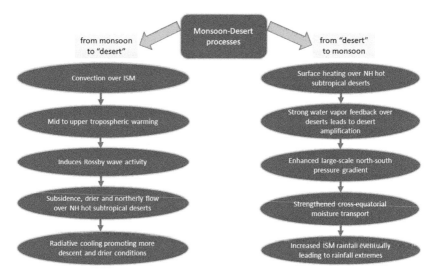

FIG. 15.6 Schematic diagram summarizing the various interactive processes and mechanisms in the monsoon–desert system over the African–Asian continent (i.e., from the monsoon to the "deserts" on the left and from the "deserts" to the monsoon on the right).

underlying processes operate at interannual timescales as well (Cherchi et al., 2014). However, most of the state-of-the-art coupled GCMs are unable to simulate this monsoon–desert relationship, partly due to the systematic monsoon dry bias affecting many current GCMs. For example, CMIP5 coupled GCMs with a severe dry bias over the ISM region depict the minimum descent over the subtropical deserts (Cherchi et al., 2014). In future projections, the monsoon–desert relationship remains elusive with ISM affecting mostly the central part of the Mediterranean region (Cherchi et al., 2016), despite that desertification and ISM rainfall are both projected to increase during the 21st century (Zhou, 2016; Wei et al., 2017; Wang et al., 2020). But, assessing the robustness of this counter-intuitive signal is challenging for several reasons. For example, it is not known if this paradox is related to the large uncertainty affecting ISM simulations (i.e., large biases in ISM mean characteristics as mentioned above) in current coupled GCMs or if it is the sign of the increasing role of radiative or surface land processes in shaping the climate of the deserts under global warming. Furthermore, the subtropical deserts are also affected by large temperature and radiation biases in current coupled GCMs (e.g., Samson et al., 2016; Terray et al., 2018). To sum up, both significant modeling developments, and an exhaustive analysis of all the external forcing at play (Blade et al., 2012; Kalimeris and Kolios, 2019; Masato et al., 2013; Tyrlis et al., 2015) are required to assess the fate of the monsoon–desert paradigm in our future world.

Secondly, we offered a different perspective by emphasizing the potential role of deserts in modulating ISM either through the changes of surface heating over the subtropical deserts or related to dry air intrusions from arid regions into the ISM domain (e.g., Krishnamurti et al., 2010; Bollasina and Nigam, 2011a, 2011b; Bollasina and Ming, 2013; Vinoj et al., 2014; Jin et al., 2014, 2016a; Solmon et al., 2015; Parker et al., 2016; Terray et al., 2018; Sooraj et al., 2019). Most of these studies, based on numerical modeling frameworks, show that surface land warming paves the way for enhanced ISM rainfall through increased southwesterly monsoon flow and hence moist transport into the Indian continent. The associated processes are further summarized in Fig. 15.6. For example, recent studies using fully coupled GCMs found that by darkening/brightening the subtropical deserts, the seasonality of ISM could be significantly modified (Fig 15.5; e.g., Samson et al., 2017; Terray et al., 2018; Sooraj et al., 2019). Furthermore, a recent observational study provided new insights into the role of subtropical deserts (e.g., surface thermal forcing over IPAR) on the subseasonal modulations of ISM, and the occurrence of ISM daily rainfall extremes (e.g., Sooraj et al., 2020). These new studies challenge both the passive role of the desert as described in the monsoon–desert paradigm and the view that the lower tropospheric thermal contrast has only a minor role in the ISM evolution at different time scales from days to decades (Dai et al., 2013).

The subtropical deserts are particularly vulnerable to climate change (i.e., desert amplification with enhanced warming signal especially over the Sahara, the Arabian Peninsula and Middle-East) as suggested by several recent studies (e.g., Cook and Vizy, 2015; Zhou, 2016). The warming and moistening of the troposphere under climate change may in turn influence the local monsoon atmospheric circulation and, thus, eventually, affect the monsoon–desert teleconnection in future climate. Further, the enhanced surface warming and amplified water vapor feedbacks over the arid regions adjacent to ISM (Sooraj et al., 2020) may have also significant implications for changes in frequency and intensity of ISM rainfall daily extremes under a global warming scenario. However, as for the monsoon–desert mechanism, elucidating in a more robust way the potential role of the subtropical deserts on ISM will require both additional analysis of observations and improved GCMs with reduced rainfall biases, more detailed parameterizations for aerosols, and improvements of the radiation codes embedded in these models.

Conflict of interest statement

The corresponding author states no conflict of interest on behalf of all authors.

Funding

Pascal Terray is funded by Institut de Recherche pour le Développement (IRD, France). The Centre for Climate Change Research (CCCR) at Indian Institute of Tropical Meteorology (IITM) is fully funded by the Ministry of Earth Sciences, Government of India.

References

Annamalai, H., Taguchi, B., Sperber., K.R., McCreary, J.P., Ravichandran, M., et al., 2015. Persistence of systematic errors in the Asian-Australian monsoon precipitation in climate models: a way forward. Clivar. Exchanges. 66. doi:10.2172/1178403.

Attada, R., Dasari, H.P., Parekh, A., Chowdary, J.S., Langodan, S., Knio, O., et al., 2019. The role of the Indian summer monsoon variability on Arabian Peninsula summer climate. Clim. Dyn. 52, 3389–3404. doi:10.1007/s00382-018-4333-x.

Barcikowska, M.J., Kapnick, S.B., Krishnamurty, L., Russo, S., Cherchi, A., Folland, C.K., 2020. Changes in the future summer Mediterranean climate: contribution of teleconnections and local factors. Earth Syst. Dyn 11, 161–181. doi:10.5194/esd-11-161-2020.

Biasutti, M., Sobel, A.H., Camargo, S.J., 2009. The role of the Sahara low in summertime Sahel rainfall variability and change in the CMIP3 models. J. Clim. 22, 5755–5771. doi:10.1175/2009JCLI2969.1.

Blade, I., Liebmann, B., Fortuny, D., van Oldenborgh, G.J., 2012. Observed and simulated impacts of the summer NAO in Europe: implications for projected drying in the Mediterranean region. Clim. Dyn. 39, 709–727. doi:10.1007/s00382-011-1195-x.

Bollasina, M., Ming, Y., 2013. The role of land-surface processes in modulating the Indian monsoon annual cycle. Clim. Dyn. 41 (9–10), 2497–2509. doi:10.1007/s00382-012-1634-3.

Bollasina, M., Nigam, S., 2011a. The summertime "heat" low over Pakistan/northwestern India: evolution and origin. Clim. Dyn. 37, 957–970.

Bollasina, M., Nigam, S., 2011b. Modeling of regional hydroclimate change over the Indian subcontinent: impact of the expanding Thar desert. J. Clim. 24, 3089–3106.

Bollasina, M.A., Ming, Y., Ramaswamy, V., 2011. Anthropogenic aerosols and the weakening of the South Asian summer monsoon. Science 334, 502–505. doi:10.1126/science.1204994.

Bonfils, C., Noblet-Ducoudré, N.de., Braconnot, P., Joussaume, S., 2000. Hot desert albedo and climate change: Mid-Holocene monsoon in North Africa. J Climate 14, 3724–3737.

Chakraborty, A., Agrawal, S., 2017. Role of west Asian surface pressure in summer monsoon onset over central India. Environ. Res. Lett. 12(7), 074002. https://doi.org/10. 1088/1748-9326/aa76ca.

Charney, J.G., 1975. Dynamics of deserts and drought in Sahel. Quart J Roy Meteor Soc 101, 193–202.

Charney, J., Quirk, W.J., Chow., S., Kornfield, J., 1977. A comparative study of the effects of albedo change on drought in semi-arid regions. J. Atmos. Sci. 34, 1366–1385.

Cherchi, A., Annamalai, H., Masina, S., Navarra, A., 2014. South Asian summer monsoon and the eastern Mediterranean climate: the monsoon-desert mechanism in CMIP5 simulations. J. Clim 27, 6877–6903. doi:10.1175/JCLI-D-13-00530.1.

Cherchi, A., Annamalai, H., Masina, S., Navarra, A., Alessandri, A., 2016. Twenty-first century projected summer mean climate in the Mediterranean interpreted through the monsoon-desert mechanism. Clim. Dyn. 47, 2361–2371. doi:10.1007/s00382-015-2968-4.

Cherchi, A., Ambrizzi, T., Behera, S., Freitas, A.C.V., Morioka, Y., Zhou, T., 2018. The response of subtropical highs to climate change. Curr. Clim. Ch. Rep. doi:10.1007/s40641-018-0114-1.

Claussen, M., 1994. On coupling global biome models with climate models. Clim. Res. 4, 203–221.

Claussen, M., 1997. Modeling biogeophysical feedback in the Africa and India monsoon region. Clim. Dyn. 13, 247–257.

Clemens, S., Prell, W., Murray, D., Shimmield, G., Weedon, G., 1991. Forcing mechanisms of the Indian Ocean monsoon. Nature 353, 720–725. doi:10.1038/353720a0.

Cook, K.H., Vizy, E.K., 2015. Detection and analysis of an amplified warming of the Sahara Desert. J. Clim. 28, 6560–6580. doi:10.1175/JCLI-D-14-00230.1.

Dai, A., Li, H., Sun, Y., Hong, L.-C., Ho, Lin., Chou, C., et al., 2013. The relative roles of upper and lower tropospheric thermal contrasts and tropical influences in driving Asian summer monsoons. J. Geophys. Res. Atmos. 118, 7024–7045. doi:10.1002/jgrd.50565.

Dee, D.P., Uppala, S.M., Simmons, A.J., Berrisford, P., Poli, P., Kobayashi, S., et al., 2011. The ERA-Interim reanalysis: Configuration 683 and performance of the data assimilation system. Quart. J. Roy. Meteorol. Soc. 137, 553–597.

Ding, Q.H., Wang, B., 2005. Circumglobal teleconnection in the northern hemisphere summer. J. Clim 18, 3483–3505. doi:10.1175/JCLI3473.

Douville, H., Chauvin, F., Broqua, H., 2001. Influence of soil moisture on the Asian and African monsoons. Part I: mean monsoon and daily precipitation. J. Clim. 14, 2381–2403.

Enomoto, T., Hoskins, B.J., Matsuda, Y., 2003. The formation mechanism of the Bonin high in August. Quart. J. Roy. Meteor. Soc 129, 157–178. doi:10.1256/qj.01.211.

Gaven, C., Hillaire-Marcel, C., Petit-Marie, N., 1981. A Pleistocene lacustrine episode in southeastern Libya. Nature 290, 131–133. doi:10.1038/290131a0.

Haarsma, R.J., Selten, F.M., Weber, S.L., Kliphuis, M., 2005. Sahel rainfall variability and response to greenhouse warming. Geophys Res Lett 32, L17702.

He, S., Yang, S., Li, Z., 2017. Influence of latent heating over the Asian and Western Pacific monsoon region on Sahel summer rainfall. Sci Rep 7, 7680. doi:10.1038/s41598-017-07971-6.

Haywood, J.M., Jones, A., Dunstone, N., Milton, S., Vellinga, M., Bodas-Salcedo, A., et al., 2016. The impact of equilibrating hemispheric albedos on tropical performance in the HadGEM2-ES coupled climate model. Geophys. Res. Lett. 43, 395–403. doi:10.1002/2015GL066903.

Hoskins, B.J., 1986. Diagnosis of forced and free variability in the atmosphere. In: Cattle, H. (Ed.), Atmospheric and Oceanic Variability. Royal Meteorological Society, Bracknell, pp. 57–63.

Huffman, G.J., Adler, R.F., Bolvin, D.T., Gu, G., 2009. Improving the global precipitation record: GPCP Version 2.1. Geophy. Res. Lett. 36, L17808. doi:10.1029/2009GL040000.

IPCC, 2013. Fifth assessment report of the intergovernmental panel on climate change, www.ipcc. ch/ipccreports/ar4-wg1.htm.

Jin, L., Zhang, H., Moise, A., Martin, G., Milton, S., Rodriguez, J., 2019. Australia-Asian monsoon in two versions of the UK Met Office Unified Model and their impacts on tropical-extratropical teleconnections. Clim. Dyn. 53, 4717–4741. doi:10.1007/s00382-019-04821-1.

Jin, Q., Wei, J., Yang, Z.L., 2014. Positive response of Indian summer rainfall to Middle East dust. Geophys. Res. Lett. 41, 4068–4074. doi:10.1002/2014GL059980.

Jin, Q., Wei, J., Yang, Z.L., Pu, B., Huang, J., 2015. Consistent response of Indian summer monsoon to Middle East dust in observations and simulations. Atmos. Chem. Phys. 15, 9897–9915. doi:10.5194/acp-15-9897-2015.

Jin, Q., Yang, Z.-L., Wei, J., 2016a. High sensitivity of Indian summer monsoon to Middle East dust absorptive properties. Sci. Rep. 6, 30690. doi:10.1038/srep30690.

Jin, Q., Yang, Z.-L., Wei, J., 2016b. Seasonal responses of Indian summer monsoon to dust aerosols in the Middle East, India, and China. J. Clim. 29, 6329–6349.

Kalimeris, A., Kolios, S., 2019. TRMM-based rainfall variability over the central Mediterranean and its relationship with atmospheric and oceanic climatic modes. Atm. Res. 230, 104649. doi:10.1016/j.atmosres.2019.104649.

Kalnay, E., Kanamitsu, M., Kirtler, R., Collins, W., Deaven, D., Gandin, L., et al., 1996. The NCEP/NCAR 40-year reanalysis project. Bull. Amer. Meteor. Soc. 77, 437–470.

Kato, S., Rose, F.G., Rutan, D.A., Thorsen, T.J., Loeb, N.G., Doelling, D.R., et al., 2018. Surface irradiances of Edition 4.0 Clouds and the Earth's Radiant Energy System (CERES) Energy Balanced and Filled (EBAF) data product. J. Clim. 31, 4501–4527. doi:10.1175/JCLI-D-17-0523.1.

Kim, G.-U., Seo, K.-H., Chen, D., 2019. Climate change over the Mediterranean and current destruction of marine ecosystem. Sci. Rep. 9, 18813. doi:10.1038/s41598-019-55303-7.

Kitoh, A., 2017. The Asian monsoon and its future change in climate models: a review. J. Meteor. Soc. Japan. 95, 7–33. doi:10.2151/jmsj.2017-002.

Kowalski, K., Van Neer, W., Bochenski, Z., Mlynarski, M., Rsebik-Kowalska, B., et al., 2009. Seasonal evolution of the West African heat low: a climatological perspective. Clim. Dyn. 33, 313–330.

Krishnamurti, T.N., Thomas, A., Simon, A., Kumar, V., 2010. Desert air incursions, an overlooked aspect, for the dry spells of the Indian summer monsoon. J. Atmos. Sci. 67, 3423–3441.

Krishnamurti, T.N., Stefanova, L., Misra, V., 2013. Tropical meteorology (pp 331). Springer, New York, NY.

Krishnan, R., Sabin, T.P., Vellore, R., Mujumdar, M., Sanjay, J., Goswami, B.N., et al., 2016. Deciphering the desiccation trend of the South Asian monsoon hydroclimate in a warming world. Clim. Dyn. 47 (3–4), 1007–1027. doi:10.1007/s00382-015-2886-5.

Kumar, S., Arora, A., 2019. On the connection between remote dust aerosol and Indian summer monsoon. Theor. Appl. Clim. 137, 929–940. doi:10.1007/s00704-018-2647-6.

Lau, K.M., Kim, M.K., Kim, K.M., 2006. Asian summer monsoon anomalies induced by aerosol direct forcing: the role of the Tibetan Plateau. Clim. Dyn. 26, 855–864. doi:10.1007/s00382-006-0114-z.

Lavaysse, C., Flamant, C., Janicot, S., Parker, D.J., Lafore, J.P., Sultan, B., et al., 2009. Seasonal evolution of the West African heat low: a climatological perspective. Clim. Dyn. 33, 313–330.

Levine, R.C., Turner, A.G., Marathayil, D., Martin, G.M., 2013. The role of northern Arabian Sea surface temperature biases in CMIP5 model simulations and future predictions of Indian summer monsoon rainfall. Clim. Dyn. 41, 155–172.

Liu, P., Wu, G., Sun, S., 2001. Local meridional circulation and deserts. Adv. Atmos. Sci. 18, 864–872. doi:10.1007/BF03403508.

Lu, R.Y., Oh, J.H., Kim, B.J., 2002. A teleconnection pattern in upper-level meridional wind over the North African and Eurasian continent in summer. Tellu 54A, 44–55. doi:10.3402/tellusa.v54i1.12122.

Masato, G., Hoskins, B.J., Woollings, T., 2013. Winter and summer northern hemisphere blocking in CMIP5 models. J. Clim 26, 7044–7059. doi:10.1175/JCLI-D-12-00466.1.

Masson, S., Terray, P., Madec, G., Luo, J.J., Yamagata, T., Takahashi, K., 2012. Impact of intra-daily SST variability on ENSO characteristics in a coupled model. Clim. Dyn. 39, 681–707.

Meehl, G.A., Arblaster, J.M., Collins, W.D., 2008. Effects of black carbon aerosols on the Indian monsoon. J. Climate. 21, 2869–2882. doi:10.1175/2007JCLI1777.1.

Mohalfi, S., Bedi, H.S., Krishnamurti, T.N., Cocke, S.D., 1998. Impact of shortwave radiative effects of dust aerosols on the summer season heat low over Saudi Arabia. Mon. Weather Rev. 126, 3153–3168.

Mooley, D.A., Paolino, D.A., 1988. A predictive monsoon signal in the surface level thermal fields over India. Mon. Weather Rev 116, 256–264.

Neelin, J.D., Held, I.M., 1987. Modeling tropical convergence based on the moist static energy budget. Mon. Weather Rev. 115, 3–12.

Nigam, S., Bollasina, M., 2010. "Elevated heat pump" hypothesis for the aerosol-monsoon hydroclimate link: "Grounded" in observations? J. Geophys. Res. 115, D16201. doi:10.1029/2009JD013800.

Osso, A., Shaffrey, L., Dong, B., Sutton, R., 2019. Impact of air-sea coupling on northern hemisphere summer climate and the monsoon-desert teleconnection. Clim. Dyn. 53, 5063–5078. doi:10.1007/s00382-019-04846-6.

Parker, D.J., Willetts, P., Birch, C., Turner, A.G., Marsham, J.H., Taylor, C.M., et al., 2016. The interaction of moist convection and mid-level dry air in the advance of the onset of the Indian monsoon. Quart. J. Roy. Meteor. Soc. 142, 2256–2272. doi:10.1002/qj.2815.

Parthasarathy, B., Rupakumar, K., Munot, A.A., 1992. Surface pressure and summer monsoon rainfall over India. Adv. Atmos. Sci 9, 359–366.

Perez-Sanz, A., Li, G., Gonzalez-Samperiz, P., Harrison, S.P., 2014. Evaluation of modern and mid-Holocene seasonal precipitation of the Mediterranean and northern Africa in the CMIP5 simulations. Clim. Past 10, 551–568. doi:10.5194/cp-10-551-2014.

Prodhomme, C., Terray, P., Masson, S., Izumo, T., Tozuka, T., Tamagata, T., 2014. Impacts of Indian Ocean SST biases on the Indian Monsoon as simulated in a global coupled model. Clim. Dyn. 42, 271–290.

Quade, J., Dente, E., Armon, M., Ben Dor, Y., Morin, E., Adam, O., et al., 2018. Megalakes in the Sahara? A review. Quart. Res. 90, 235–275. doi:10.1017/qua.2018.46.

Rai, A., Saha, S.K., Pokhrel, S., Sujith, K., Halder, S., 2015. Influence of pre-onset land-atmospheric conditions on the Indian summer monsoon rainfall variability. J. Geophy.s Res. Atmos. 120, 4551–4563.

Ramage, C.S., 1966. The summer atmospheric circulation over the Arabian Sea. J. Atmos. Sci. 23, 144–150.

Ramanathan, V., Chung, C., Kim, D., Bettge, T., Buja, L., Kiehl, J.T., et al., 2005. Atmospheric brown clouds: impacts on South Asian climate and hydrological cycle. Proc. Natl. Acad. Sci. USA 102, 5326–5333. doi:10.1073/pnas.0500656102.

Rizou, D., Flocas, H.A., Hatzaki, M., Bartzokas, A., 2018. A statistical investigation of the impact of the Indian monsoon on the eastern Mediterranean circulation. Atmosphere 9. doi:10.3390/atmos9030090.

Rodwell, M.J., Hoskins, B.J., 1996. Monsoons and the dynamics of deserts. Quart. J. Roy. Meteor. Soc 122, 1385–1404.

Rodwell, M.J., Hoskins, B.J., 2001. Subtropical anticyclones and summer monsoons. J. Clim. 14, 3192–3211.

Roehrig, R., Bouniol, D., Guichard, F., Hourdin, F., Redelsperger, J.L., 2013. The present and future of the West African monsoon: a process-oriented assessment of CMIP5 simulations along the AMMA Transect. J. Clim. 26, 6471–6505.

Roxy, M.K., Ghosh, S., Pathak, A., Athulya, R., Mujumdar, M., Murtugudde, R., et al., 2017. A threefold rise in widespread extreme rain events over central India. Nat. Commun., 1–11. doi:10.1038/s41467-017-00744-9.

Sabeerali, C.T., Rao, S.A., Dhakate, A.R., Salunke, K., Goswami, B.N., 2015. Why ensemble mean projection of south Asian monsoon rainfall by CMIP5 models is not reliable? Clim. Dyn. 45, 161–174.

Saeed, S., Lipzig, N.V., Muller, W.A., Saeed, F., Zanchettin, D., 2014. Influence of the circumglobal wave train on European summer precipitation. Clim. Dyn. 43, 503–515. doi:10.1007/s00382-013-1871-0.

Saha, S., Moorthi, S., Wu, X., Wang, J., Pan, H.-L., Wang, J., et al., 2014. The NCEP climate forecast system version 2. J. Clim. 27, 2185–2208.

Samson, G., Masson, S., Durand, F., Terray, P., Berthet, S., Jullien, S., 2017. Role of land surface albedo and horizontal resolution on the Indian Summer Monsoon biases in a coupled ocean-atmosphere tropical-channel model. Clim. Dyn. doi:10.1007/s00382-016-3161-0.

Sanap, S.D., Pandithurai, G., 2015. Inter-annual variability of aerosols and its relationship with regional climate over Indian subcontinent. Int. J. Climatol. 35, 1041–1053. doi:10.1002/joc.4037.

Sandeep, S., Ajayamohan, R., 2015. Origin of cold bias over the Arabian Sea in Climate Models. Sci. Rep. 4, 6403. doi:10.1038/srep06403.

Shekhar, R., Boos, W.R., 2017. Weakening and Shifting of the Saharan Heat Low Circulation During Wet Years of the West African Monsoon. J. Clim. 30, 7399–7422.

Shukla, J., Mintz, Y., 1982. Influence of land-surface evaporation on Earth's climate. Science 215, 1498–1501.

Sikka, D.R., 1997. Desert climate and its dynamics. Curr. Sci 72, 35–46.

Singh, R., AchutaRao, K., 2018. Quantifying uncertainty in twenty-first century climate change over India. Clim. Dyn. 52 (7–8), 3905–3928. doi:10.1007/s00382-018-4361-6.

Smith, E.A., 1986a. The structure of the Arabian heat low. Part I: surface energy budget. Mon. Weather Rev. 114, 1067–1083.

Smith, E.A., 1986b. The structure of the Arabian heat low. Part II: bulk tropospheric heat budget and implications. Mon. Weather Rev. 114, 1084–1102.

Solmon, F., Nair, V.S., Mallet, M., 2015. Increasing Arabian dust activity and the Indian summer monsoon. Atmos. Chem. Phys 15, 8051–8064. doi:10.5194/acp-15-8051-2015.

Sooraj, K.P., Terray, P., Mujumdar, M., 2015. Global warming and the weakening of the Asian summer monsoon circulation: assessments from the CMIP5 models. Clim. Dyn. 45, 233–252.

Sooraj, K.P., Terray, P., Masson, S., Cretat, J., 2019. Modulations of the Indian summer monsoon by the hot subtropical deserts: insights from coupled sensitivity experiments. Clim. Dyn. 52, 4527–4555. doi:10.1007/s00382-018-4396-8.

Sooraj, K.P., Terray, P., Shilin, A., Mujumdar, M., 2020. Dynamics of rainfall extremes over India: A new perspective. Int. J. Climatol 40, 5223–5245. doi:10.1002/joc.6516.

Sperber, K.R., Annamalai, H., Kang, I.S., Kitoh, A., Moise, A., Turner, A.G., et al., 2013. The Asian summer monsoon: an intercomparison of CMIP5 vs. CMIP3 simulations of the late 20th century. Clim. Dyn. 41, 2711–2744. doi:10.1007/s00382-012-1607-6.

Stephan, C.C., Klingaman, N.P., Turner, A.G., 2019. A mechanism for the recently increased interdecadal variability of the Silk Road Pattern. J. Clim. 32, 717–736. doi:10.1175/JCLI-D-18-0405.1.

Sud, Y.C., Fennessy, M., 1982. A study of the influence of surface albedo on July circulation in semi-arid regions using the GLAS GCM. Int. J. Climatol. (2), 105–125.

Sud, Y.C., Shukla, J., Mintz, Y., 1988. Influence of land surface roughness on atmospheric circulation and precipitation: A sensitivity study with a general circulation model. J. Appl. Meteor 27, 1036–1054.

Taylor, K., Stouffer, R., Meehl, G., 2012. An overview of CMIP5 and the experiment design. Bull. Am. Meteor. Soc 93, 485–498. doi:10.1175/BAMS-D-11-00094.1.

Terray, P., Sooraj, K.P., Masson, S., Krishna, R.P.M., Samson, G., Prajeesh, A.G., 2018. Towards a realistic simulation of boreal summer tropical rainfall climatology in state-of the-art climate coupled models. Clim. Dyn. 50, 3413–3439. doi:10.1007/s00382-017-3812-9.

Tyrlis, E., Lelieveld, J., Steil, B., 2013. The summer circulation over the eastern Mediterranean and the Middle East: influence of the South Asian monsoon. Clim. Dyn 40, 1103–1123. doi:10.1007/s00382-012-1528-4.

Tyrlis, E., Tymvios, F.S., Giannakopoulos, C., Lelieveld, J., 2015. The role of blocking in the summer 2014 collapse of Etesians over the eastern Mediterranean. J. Geophys. Res. Atm 120, 6777–6792. doi:10.1002/2015JD023543.

Uppala, S.M., Kallberg, P.W., Simmons, A.J., et al., 2005. The ERA-40 Reanalysis. Quart. J. Roy. Meteor. Soc. 131, 2961–3012. doi:10.1256/qj.04.176.

Vinoj, V., Rasch, P.J., Wang, H., Yoon, J.H., Ma, P.L., Landu, K., et al., 2014. Short term modulation of Indian summer monsoon rainfall by West Asian dust. Nat. Geo. Sci. doi:10.1038/NGEO2017.

Wang, B., 2006. The Asian Monsoon. Springer, Chichester.

Wang, B., Biasutti, M., Byrne, M., Castro, C., Chang, C.P., et al., 2020. Monsoons Climate Change Assessment. Bull. Amer. Meteor. Soc. doi:10.1175/BAMS-D-19-0335.1.

Wang, B., Liu, J., Kim, H.J., Webster, P.J., Yim, S.Y., 2012. Recent change of the global monsoon precipitation (1979-2008). Clim. Dyn. 39, 1123–1135. doi:10.1007/s00382-011-1266-z.

Wang, B., Webster, P.J., Teng, H., 2005. Antecedents and self-induction of active-break south Asian monsoon unraveled by satellites. Geophys. Res. Lett. 32, 4–7. doi:10.1029/2004GL020996.

Wang, C., Kim, D., Ekman, A.M.L., Barth, M.C., Rasch, P.J., 2009. Impact of anthropogenic aerosols on Indian summer monsoon. Geophys. Res. Lett. 36, L21704. doi:10.1029/2009GL040114.

Wang, L., Xu, P., Chen, W., Liu, Y., 2017. Interdecadal variations of the Silk Road Pattern. J. Clim. 30, 9915–9932. doi:10.1175/JCLI-D-17-0340.1.

Wang, P.X., Wang, B., Cheng, H., Fasullo, J., Guo, Z.T., Kiefer, T., et al., 2014. The global monsoon across timescales: coherent variability of regional monsoons. Clim. Past. 10, 2007–2052. doi:10.5194/cp-10-2007-2014.

Warner, T.T., 2004. Desert Meteorology. Cambridge University Press, London, p. 612.

Webster, P.J., 1994. The role of hydrological processes in ocean atmosphere interactions. Rev. Geophys. 32 (4), 427–476.

Wei, N., Zhou, L., Dai, Y., Xia, G., Hua, W., 2017. Observational evidence for desert amplification using multiple satellite datasets. Sci. Rep. 2043. doi:10.1038/s41598-017-02064-w.

Wu, G.X., Liu, Y., Zhu, X., Li, W., Ren, R., Duan, A., et al., 2009. Multi-scale forcing and the formation of subtropical desert and monsoon. Ann. Geophys. 27, 3631–3644. doi:10.5194/angeo-27-3631-2009.

Xue, Y., Juang, H.M.H., Li, W.P., Prince, S., Defries, R., Jiao, Y., et al., 2004. Role of land surface processes in monsoon development: East Asia and West Africa. J. Geophys. Res. (Atmos) 109, 03105–03128.

Xue, Y., De, Sales F., Vasic, R., Mechoso, C.R., Prince, S.D., Arakawa, A., 2010. Global and temporal characteristics of seasonal climate/vegetation biophysical process (VBP) interactions. J. Clim. 23, 1411–1433.

Yang, S., Webster, P.J., Dong, M., 1992. Longitudinal heating gradient: another possible factor influencing the intensity of the Asian summer monsoon circulation. Adv. Atmos. Sci. 9, 397–410.

Yasunari, T., Saito, K., Takata, K., 2006. Relative roles of large-scale orography and land surface processes in the global hydroclimate, Part I: Impacts on monsoon systems and the Tropics. J. Hydrometeor. 7, 626–641.

Zhou, L., 2016. Desert amplification in a warming climate. Sci. Rep. 6. doi:10.1038/srep31065.

Chapter 16

Interaction between South Asian high and Indian Summer Monsoon rainfall

Wei Wei[a,b], Song Yang[a,b]
[a]*School of Atmospheric Sciences, Sun Yat-sen University, and Southern Marine Science and Engineering Guangdong Laboratory (Zhuhai), Zhuhai, China,* [b]*Guangdong Province Key Laboratory for Climate Change and Natural Disaster Studies, Sun Yat-sen University, Zhuhai, China*

16.1   Introduction

The South Asian high (SAH) is the most intense and persistent upper-level high-pressure system in the Northern Hemisphere (Mason and Anderson, 1963; Tao and Zhu, 1964). It is a subtropical high located over southern Asia and extends from the western North Pacific to northern Africa during boreal summer (Fig. 16.1). The SAH is an important upper-level component of the Indian summer monsoon (ISM) system (Krishnamurti and Bhalme, 1976). Due to its stability, it can be considered as an important indicator for the Asian summer monsoon rainfall strength (e.g., Zhang et al, 2002).

During the spring-to-summer transition, the onset of monsoon in South Asia is closely related with the northwestward seasonal evolution of the SAH from the western Pacific to the South Asian plateaus (Wei et al., 2019a). The early northwestward shift of the SAH facilitates the advanced onset of the Asian summer monsoon. The upper-level divergence pumping effect associated with the SAH can trigger the monsoon onset vortex over the Bay of Bengal and the eastern Arabian Sea, which leads to the monsoon onset in South Asia (Liu et al., 2013, 2015; Wu et al., 2015).

In summer, the SAH is located stably over the South Asian plateaus. Its intensity represents the strength of the large-scale Asian monsoon (Zhang et al., 2005). On the one hand, the SAH intensity depends on the diabatic heating over the Tibetan Plateau (TP) and the surrounding monsoon regions (Krishnamurti et al., 1973; Liu et al., 2001; Wu and Liu, 2003; Wu et al., 2007, 2016). On the other hand, an intense SAH is accompanied by the enhanced upper-level easterly wind to its south, which intensifies the vertical wind shear over South Asia (Wang and Fan, 1999; Webster and Yang, 1992) and facilitates rainfall increase over the ISM region.

Indian Summer Monsoon Variability: El Niño-teleconnections and beyond.
DOI: https://doi.org/10.1016/B978-0-12-822402-1.00016-8
319

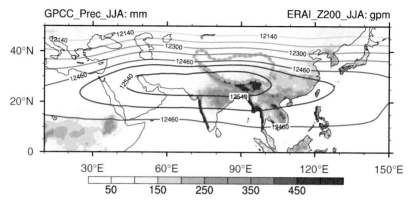

FIG. 16.1 Climatological summer rainfall (shaded; mm) and geopotential height at 200 hPa (contours; 12,140, 12,220, 12,300, 12,380, 12,460, 12,500, and 12,540 gpm) from June to August. The contours in green indicate the region with elevations exceeding 3000 m. The Global Precipitation Climatology Centre full data reanalysis (GPCC; Schneider et al., 2011) is applied in this figure.

One of the most obvious features is the zonal shift of the SAH. The SAH center is located frequently over the TP or the Iranian Plateau (IP) (Fig. 16.2A and B; Wei et al., 2014, 2019b; Zhang et al., 2002). When the SAH shifts eastward with the center over the TP, it is called the TP mode. When the SAH center is located in the west over the IP, it is referred to as the IP mode. Actually, the zonal shift of the SAH is always accompanied by a meridional movement (Fig. 16.3, Wei et al. 2015). Both the zonal shift and the meridional movement of the SAH are closely related with ISM rainfall variability (Wei et al., 2012, 2014, 2015).

In this chapter, we revisit the linkage between SAH and ISM rainfall variation during boreal summer from June to August (JJA). Focusing on the zonal shift of the SAH, we discuss the physical mechanism for the interaction between the SAH and ISM rainfall variations on intraseasonal time scales.

16.2 Methods and data

This study analyzes the daily and monthly data from the European Centre for Medium-range Weather Forecasts (ECMWF) reanalysis, namely the ERA-Interim Reanalysis (Dee et al., 2011). The Global Precipitation Climatology Project (GPCP) monthly precipitation (Adler et al., 2003) and the Multi-Source Weighted-Ensemble Precipitation (MSWEP) version 1.0 with a 0.25° × 0.25° horizontal resolution (Beck et al., 2017) are used to study the monthly and daily rainfall anomalies associated with the SAH, respectively. The MSWEP takes advantage of the complementary strengths of gauge, satellite, and reanalysis-based data (Beck et al., 2017).

On interannual time scales, the SAH indices are defined to quantify the intensity and the horizontal movements of the SAH by using the geopotential

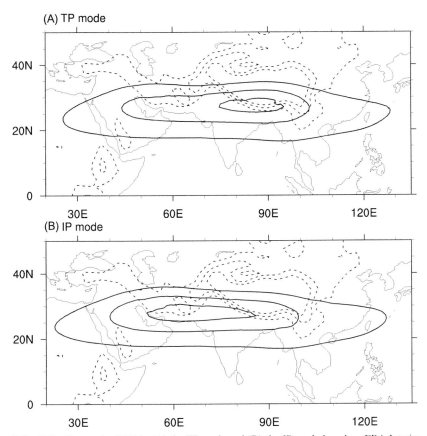

FIG. 16.2 Composite SAH in (A) the TP mode and (B) the IP mode based on ERA-Interim data (solid contours from outside to inside indicate the geopotential height of 12,500, 12,540, and 12,560 gpm, respectively). Black dashed contours from outside to inside indicate the plateau region with elevation of 1000, 2000, and 3000 m, respectively.

height at 200 hPa (Z200). The SAH intensity $I_{intensity}$ is defined as the domain average of Z200 in the SAH region (22.5–32.5°N, 50–100°E). The meridional shift can be well described by the index I_{NS} [Z200 (27.5–32.5°N, 50–100°E) – Z200 (22.5–27.5°N, 50–100°E)] (Wei et al., 2012), while the zonal shift by I_{WE} [Z200 (22.5–32.5°N, 55–75°E) – Z200 (22.5–32.5°N, 85–105°E)] (Wei et al., 2014), and the northwest-southeast movement by $I_{NW\text{-}SE}$ [Z200 (27.5–35°N, 50–80°E) – Z200 (20–27.5°N, 85–115°E)] (Wei et al., 2015). All indices are detrended and standardized.

On intraseasonal time scales, the zonal shift of the SAH is quantified by the South Asian high index (SAHI) [Z200 (22.5–32.5°N, 45–65°E) – Z200 (22.5–32.5°N, 82.5–102.5°E)] (Wei et al., 2019b). We applied the Butterworth bandpass filter (Butterworth, 1930) on all daily data to investigate the intraseasonal variation of SAH and the associated circulation and rainfall anomalies.

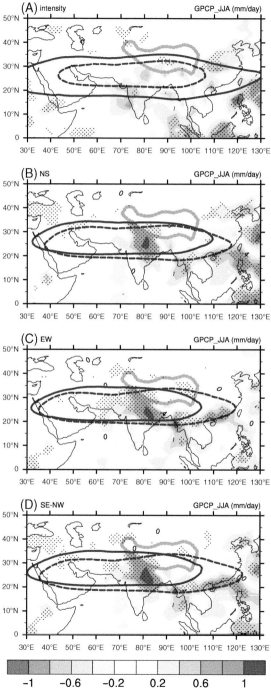

FIG. 16.3 Regressed GPCP rainfall anomalies against the SAH indices (A) $I_{intensity}$, (B) I_{NS}, (C) I_{WE}, and (D) I_{NW-SE} in JJA. Rainfall anomalies exceeding the 0.05 significance level are highlighted by dots. Blue contours indicate the composite SAH (isoline 12520 gpm) for the SAH indices higher than 1 (solid line) and lower than −1 (dashed line). The contours in green indicate the region with elevations exceeding 3000 m.

Before that, we subtracted the climatological mean of each day to remove the annual cycle of the original daily data. The 10- to 20-day period dominates the subseasonal zonal oscillation of the SAH (Wei et al., 2019b). Thus, the 10- to 20-day bandpass filter is applied to investigate the quasi-biweekly oscillation (QBWO) of the SAH and its interaction with ISM rainfalls. Based on the SAHI, we applied the composite analysis to find out the ISM rainfall and circulation anomalies in the extreme intraseasonal cases of the SAH. Day 0 in extreme cases is the day when the peak value of the filtered SAHI is higher than one standard deviation. Day-n and day n refer to n days before and after the peak days (day 0), respectively.

16.3   Interannual relationship between ISM rainfall and the SAH

On interannual time scales, the SAH intensity is positively related with ISM rainfall, especially the rainfall over west central India (Fig. 16.3A). However, more significant relationship can be observed between the horizontal movements of the SAH and the ISM rainfall (Fig. 16.3B–D). When the SAH shifts northward and westward, positive rainfall anomalies appear in a wider region over South Asia (Fig. 16.3B and C). Actually, the zonal shift of the SAH is always accompanied by its meridional movement (Fig. 16.3, Wei et al. 2015). The regressed rainfall anomalies against the northwest-southeast index I_{NW-SE} (Fig. 16.3D) are larger than those against the north-south index I_{NS} (Fig. 16.3B) and the west-east index I_{WE} (Fig. 16.3C), especially in the eastern parts of monsoon trough region.

The northwest-southeast movement is the dominant feature of the SAH movement on interannual time scales (Wei et al. 2015). When the SAH shifts northwestward, the anomalous 200-hPa anticyclone and cyclone are located over central Asia and southern China, respectively (Fig. 16.4A). The anomalous divergent northeasterly wind in upper level over the northern ISM region facilitates the enhancement of the positive vorticity anomaly at the middle and lower levels (Fig. 16.4B and C; Liu et al., 2013, 2015; Wu et al., 2015). The deepened monsoon trough leads to more rainfall in the monsoon trough region (Fig. 16.4C). In turn, the positive latent heat anomalies associated with the increased rainfall over the northern ISM region can excite anomalous anticyclone to its northwest and anomalous cyclone to its northeast in the upper troposphere, which result in the westward shift of the SAH (Wei et al., 2014, 2015).

Wei et al. (2015) proposed that the northwest-southeast movement is a result from the effects of the Indian and East Asian summer monsoon (EASM) rainfalls. The latent heat anomaly associated with the ISM rainfall is the main reason for the zonal shift of the SAH. To discuss the interaction between the ISM and the SAH, we focus on the zonal shift of the SAH on intraseasonal time scales in the next section.

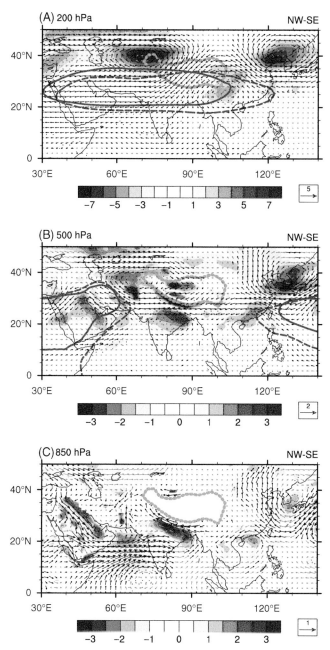

FIG. 16.4 Regressed vorticity anomalies (shaded; 10^{-6} s^{-1}) and the horizontal wind anomalies (vectors; m/s) at (A) 200 hPa, (B) 500 hPa, and (C) 850 hPa against the SAH index $I_{NW\text{-}SE}$ in JJA based on ERA-Interim data. Red contours indicate the composite values of (A) SAH at 200 hPa (isoline 12520 gpm) and (B) the subtropical high at 500 hPa (isoline 5870 gpm) for the $I_{NW\text{-}SE}$ higher than 1 (solid line) and lower than −1 (dashed line). Horizontal wind anomalies exceeding the 0.05 significance level are highlighted by dark vectors. The contours in green indicate the TP region with elevations exceeding 3000 m.

16.4   Intraseasonal relationship between the ISM rainfall and the SAH zonal shift

The QBWO is significant in both the ISM rainfall (Krishnamurti and Ardanuy, 1980; Krishnamurti and Bhalme, 1976) and the zonal shift of the SAH (Krishnamurti et al., 1973; Liu et al., 2007; Popovic and Plumb, 2001; Tao and Zhu, 1964; Wei et al., 2019b; Wu et al., 2015). In the ISM region, the 2-week alternation between active and break spells of the monsoon rainfall is possibly related with the movement of SAH (Ashfaq et al., 2009; Bansod et al., 2003; Wei et al., 2019b).

On quasi-biweekly time scales, a 13-day period exists in the zonal shift of the SAH between the IP mode (day 0) and the TP mode (day-6 and day 6) (Fig. 16.5). One day before the SAH moving to the western most position in the IP mode, the rainfall over the northern ISM region reaches its maximum value on day-1. The rainfall increase leads the westward shift of the SAH by 1 day. The result from the lead–lag correlation analysis also shows that the maximum correlation coefficient of 0.46 (exceeding the 0.001 significance level) between the west-east index SAHI and the rainfall over the northern ISM region appears on day-1, when the rainfall leads the zonal shift of SAH by 1 day. This one-day lead of the rainfall over the northern ISM region implies the possible influence of the ISM on the SAH (Wei et al., 2019b).

Besides, the rainfall anomaly over the northern ISM region turns from negative to positive on day-4 (Fig. 16.5), two days after the SAH turning to the TP mode, implying the possible impact of the SAH on the ISM rainfall. It also

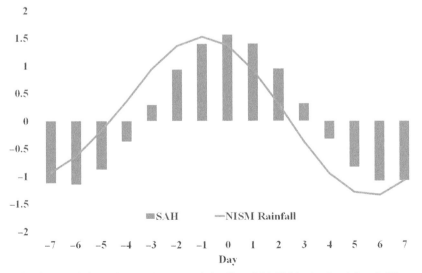

FIG. 16.5 Evolutions of composite 10- to 20-day filtered SAHI (blue bars) and the rainfall over the northern ISM region (gray curve; 23–32°N/70–100°E) from day -7 to day 7 based on the SAHI from ERA-Interim data. All series are standardized before the composite analysis.

suggests that the SAH can be an indicator for the rainfall over the northern ISM region according to its IP or TP status. Two days after the SAH moving to its easternmost position (TP mode), the above-normal rainfall may appear over the northern ISM region.

16.5 Interactive processes on quasi-biweekly time scales

Previous studies have shown that the QBWO in the SAH is associated with the forcing of diabatic heating (Liu et al., 2007; Murakami and Frydrych, 1974). In turn, the QBWO in the SAH can lead to the active and break monsoon in South Asia (Krishnamurti and Bhalme, 1976). The interactive processes between the SAH and ISM rainfall are proposed here on quasi-biweekly time scales.

16.5.1 Effect of the SAH on the ISM rainfall

Fig. 16.6 shows the circulation anomalies associated with the zonal shift of the SAH on the quasi-biweekly time scales. In the TP mode, a pair of cyclone–anticyclone anomalies appears in the SAH region at 200 hPa, with their centers at about 32.5°N on days-6 and 6 (Fig. 16.6). In the IP mode, the anticyclone and cyclone anomalies appear in the SAH region on day 0. In the westward moving process from the TP mode to the IP mode, the cyclone-anticyclone anomalies in the SAH region are replaced by the southward moving anticyclone-cyclone anomalies from the mid-latitudes from day-6 to day 0. From day 0 to day 6, the mid-latitude cyclone–anticyclone anomalies shift southward gradually into the SAH region. Then, the SAH shifts from the IP mode to the TP mode.

In the TP mode (day-6), a mid-latitude anticyclone is formed over central Asia, where the intense southerly anomalies exit between the cyclone and anticyclone anomalies (Fig. 16.6). This central Asia anticyclone is strengthened on day-5. In its southern flank, the easterly anomalies merge with the easterly anomalies in the southern flank of the cyclone over TP on day-5, and control the summer rainfall over the northern ISM region. As a result, the vertical shear is enhanced over this region, implying a strengthening ISM (Koteswaram, 1958; Webster and Yang, 1992). Accordingly, the rainfall anomaly over the northern ISM region increases from negative to positive on day-4 (Fig. 16.5), two days after the SAH turning to the TP mode.

In this process, the evolution of the central Asia anticyclone plays an important role in affecting the ISM rainfall. Ding and Wang (2007) pointed out that the anomalous high is well established over central Asia before the outbreak of ISM convection on intraseasonal time scales. The easterly anomalies in the southern flank of this anomalous central Asia anticyclone strengthen the vertical shear over the ISM region, favoring strong convection through the vertical shear mechanism (Wang and Xie, 1996; Xie and Wang, 1996). Besides, the strengthened easterly jet stream to the south of the anomalous anticyclone may induce a northerly ageostrophic flow and generate the upper-level divergence

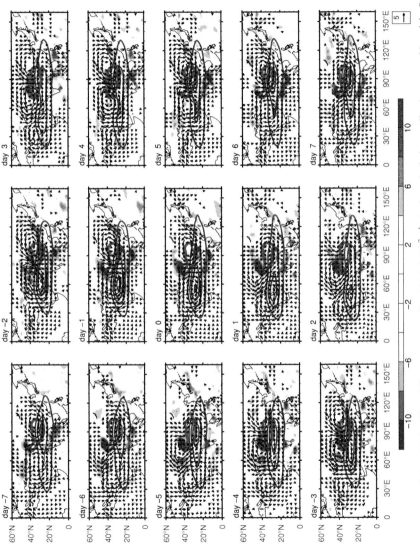

FIG. 16.6 Composite patterns of 10- to 20-day filtered divergence anomalies (shaded; 10^{-7} s^{-1}), horizontal wind anomalies (vectors; m/s), and the SAH (red contours from outside to inside indicate 12.500, 12.540, 12.560, and 12.575 gpm, respectively) at 200 hPa from day -7 to day 7 based on the SAHI from ERA-Interim data. The contours in magenta indicate the TP region with elevations exceeding 3000 m.

and strong upward motion over the northern ISM region (Hoskins and Wang, 2006). Karmakar et al. (2017) further pointed out the anomalous central Asia anticyclone as a crucial factor to trigger the ISM rainfall on quasi-biweekly time scales. Here we know that, this central Asia anticyclone, which is linked to the upper-level Eurasian wave train, may also be associated with the zonal oscillation of the SAH.

16.5.2   Effect of the ISM rainfall on the SAH

Associated with the positive ISM rainfall anomaly, positive latent heat anomaly appears over the northern ISM region on day-4, spreading westward gradually, and obtains the maximum value on day-1 (Fig. 16.7). The diabatic heating over this region can induce a baroclinic Rossby wave response to the west of the heating source (Gill, 1980; Ding and Wang, 2005; Liu et al., 2001; Rodwell and Hoskins, 1996). Accordingly, significant negative vorticity anomalies appear to the west of the ISM region on day-4, and then propagate westward, exerting an impact on the western SAH (Fig. 16.8A).

The negative vorticity anomalies in the west consist of two parts: the heating-induced negative vorticity anomalies over the IP (25–32.5°N/45–65°E) and the southward moving part from the mid-latitudes associated with the central Asia anticyclone (Fig. 16.8B). The heating-induced negative vorticity anomalies over the IP can intensify the western SAH directly. On the other hand, it

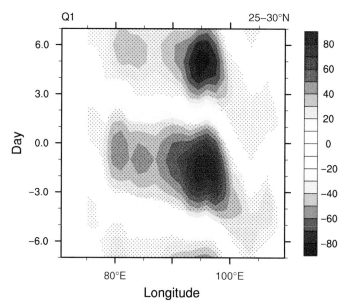

FIG. 16.7   Hovmöller diagram of composite 10- to 20-day filtered integrated apparent heat source Q1 anomalies (shaded; W/m^2) along 25–30°N from day -7 to day 7 based on the SAHI from ERA-Interim data. Heating anomalies exceeding the 0.05 significance level are highlighted by dots.

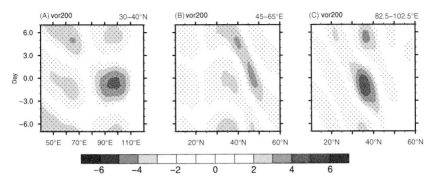

FIG. 16.8 Hovmöller diagram of composite 10- to 20-day filtered vorticity anomalies (shaded; 10^{-6} s^{-1}) at 200 hPa along (A) 30–40°N, (B) 45–65°E, and (C) 82.5–102.5°E from day -7 to day 7 based on the SAHI from ERA-Interim data. Vorticity anomalies exceeding the 0.05 significance level are highlighted by dots.

can play a role in drawing more intense perturbations from the mid-latitudes to the SAH region, which exerts a greater impact on the SAH (Wei et al., 2019b). After these two parts of negative vorticity anomalies merging over the IP on day-3 (Fig. 16.8B), significant southward shifts occur in both the central Asia anticyclone and the compensatory cyclone over northern China (Fig. 16.8B and C). These anticyclone–cyclone centers shift from 40°N to about 32.5°N into the SAH region from day-3 to day-0 (Fig. 16.6). As a result, the western SAH is strengthened and the eastern SAH is weakened, causing the SAH to move to the IP mode (Fig. 16.6).

Here we noted that the important factor for the zonal shift of the SAH is the significant southward shift of the mid-latitude anticyclone and cyclone anomalies to the south of 40°N. This significant southward movement is triggered by the ISM heating and lags the increase of the positive ISM rainfall anomalies by about 1 day. As a result, the zonal shift of the SAH lags the rainfall over the northern ISM region by about 1 day (Fig. 16.5).

16.6 Summary and Discussion

The ISM rainfall is more significantly related to the horizontal movements than the intensity of the SAH. Especially, the interaction between ISM rainfall and the SAH can be well depicted in the zonally moving process of the SAH on different time scales, which is uncorrelated with the SAH intensity. On interannual time scales, the westward and northward shifts of SAH indicate a rainfall increase over the northern ISM region, while the eastward and southward shifts imply less-than-normal summer rainfall over this region. On quasi-biweekly time scales, the rainfall increase in the northern ISM region leads the westward shift of the SAH by about 1 day, indicating the influence of ISM rainfall on the zonal shift of the SAH. In turn, 2 days after the SAH turning to the TP mode, rainfall anomaly over the northern ISM region turns from negative to positive, implying an impact of the SAH on the ISM rainfall.

During the zonal shift of the SAH from the TP mode to the IP mode on quasi-biweekly time scales, an anomalous anticyclone is formed over central Asia. The easterly anomalies in the southern flank of this central Asia anticyclone favor the enhancements of the vertical shear and the monsoon rainfall over the northern ISM region. The positive diabatic heating anomaly associated with the increased monsoon rainfall induces a baroclinic Rossby wave response to its west, leading to decreased negative vorticity anomalies over the IP. These heating-induced negative vorticity anomalies are merged with central Asia anticyclone, facilitating the southward movement of the mid-latitude anticyclone and cyclone anomalies to the SAH region. The SAH shifts to the IP mode. As a result, the rainfall increase over the northern ISM region leads the westward shift of the SAH by about 1 day.

The interaction between the ISM rainfall and the SAH is mainly reflected in the vorticity movement in the west of the SAH, which is different from the feature in the east (Fig. 16.8C). In the east of the SAH, the vorticity anomalies move further southward to the tropical region, which result from the positive feedback loop between the SAH and the EASM rainfall (Wei et al., 2019b; Yang and Li, 2016). Therefore, the zonal shift of the SAH can be considered as not only an indicator for the ISM rainfall, but also a bridge connecting the ISM rainfall and the EASM rainfall (Ha et al., 2018; Wei et al., 2014, 2015, 2019b; Wu, 2017).

The impact of the zonal shift of SAH mainly reflects the influencing signal from the mid-latitudes. In the tropics, El Niño/Southern Oscillation (ENSO) is a pronounced signal affecting the ISM rainfall. Hu et al. (2005) separated the ISM rainfall into ENSO-related and ENSO-unrelated variations. The ENSO-unrelated rainfall variation is associated with mid-latitude circulation anomalies (Fig. 4c in Hu et al, 2005), which resemble the circulation patterns associated with the zonal shift of the SAH. This feature indicates that the zonal shift of the SAH can explain the ENSO-unrelated part of ISM rainfall variability.

ENSO shows an insignificant relationship with the SAH zonal variation (Wei et al., 2014). However, it can influence the southern SAH by affecting the sea surface temperature (SST) in the tropical Indian Ocean (Huang et al. 2011; Wei et al. 2012; Xue et al. 2015, 2017, 2018; Yang et al. 2007). The positive SST anomalies in the tropical Indian Ocean in summer enhance the geopotential height in the southern SAH by exciting the Matsuno–Gill type response in the upper level (Matsuno 1966; Gill 1980). Associated with the southward extension of SAH, vertical wind shear is weakened in the core ISM region and the ISM rainfall is weaker than normal (Wei et al., 2012, 2015).

It is well known that the SAH plays an important role in the ISM rainfall variations, and that there is a connection between the ISM rainfall and the EASM rainfall. However, several issues need to be further addressed in the future. Will the SAH-ISM rainfall relationship be affected by external forcing?

Will this relationship change in the warming climate in future? What is the role of the SAH in the interdecadal change in the relationship between the ISM and the EASM? What is the predictability of the SAH on seasonal-to-subseasonal time scales? Improving the understanding of the SAH and the model representation of the interactions between the SAH and monsoon rainfall will benefit the monsoon rainfall prediction on seasonal-to-subseasonal time scales.

Acknowledgment

This research was supported by the National Natural Science Foundation of China (Grants 42088101, 41975074), Innovation Group Project of Southern Marine Science and Engineering Guangdong Laboratory (Zhuhai) (311020001), and the Guangdong Province Key Laboratory for Climate Change and Natural Disaster Studies (Grant 2020B1212060025).

References

Adler, R.F., Huffman, G.J., Chang, A., et al., 2003. The version 2 Global Precipitation Climatology Project (GPCP) monthly precipitation analysis (1979–present). J. Hydrometeor. 4 (6), 1147–1167.

Ashfaq, M., Shi, Y., Tung, W.W., Trapp, R.J., Gao, X., Pal, J.S., Diffenbaugh, N.S., 2009. Suppression of South Asian summer monsoon precipitation in the 21st century. Geophys. Res. Lett. 36, L01704. doi:10.1029/02008GL036500.

Bansod, S.D., Yin, Z.Y., Lin, Z., Zhang, X., 2003. Thermal field over Tibetan Plateau and Indian summer monsoon rainfall. Int. J. Climatol. 23 (13), 1589–1605.

Beck, H.E., Dijk, A.I.J.M.v., Levizzani, V., et al., 2017. Mswep: 3-hourly 0.25° global gridded precipitation (1979–2015) by merging gauge, satellite, and reanalysis data. Hydrol. Earth Syst. Sci. 21 (1), 589–615.

Butterworth, S. 1930. On the theory of filter amplifiers. Wirel. Eng. 7(6), 536-541.

Dee, D.P., Uppala, S.M., Simmons, A.J., et al., 2011. The ERA-interim reanalysis: Configuration and performance of the data assimilation system. Quart. J. Roy. Meteor. Soc. 137 (656), 553–597.

Ding, Q., Wang, B., 2005. Circumglobal teleconnection in the Northern Hemisphere summer. J. Clim. 18 (17), 3483–3505.

Ding, Q., Wang, B., 2007. Intraseasonal teleconnection between the summer Eurasian wave train and the Indian monsoon. J. Clim. 20 (15), 3751–3767.

Gill, A.E., 1980. Some simple solutions for heat-induced tropical circulation. Quart. J. Roy. Meteor. Soc. 106 (449), 447–462.

Ha, K.-J., Seo, Y.-W., Lee, J.-Y., Kripalani, R.H., Yun, K.-S., 2018. Linkages between the South and East Asian summer monsoons: A review and revisit. Clim. Dyn. 51 (11-12), 4207–4227.

Hoskins, B., Wang, B., 2006. Large-Scale Atmospheric Dynamics, The Asian monsoon. Springer, New York, pp. 357–415.

Hu, Z.-Z., Wu, R., Kinter, J.L., Yang, S., 2005. Connection of summer rainfall variations in South and East Asia: Role of ENSO. Int. J. Climatol. 25, 1279–1289.

Huang, G., Qu, X., Hu, K., 2011. The impact of the tropical Indian Ocean on South Asian high in boreal summer. Adv. Atmos. Sci. 28, 421–432.

Karmakar, N., Chakraborty, A., Nanjundiah, R.S., 2017. Space–time evolution of the low- and high-frequency intraseasonal modes of the Indian summer monsoon. Mon. Weather Rev. 145 (2), 413–435.

Koteswaram, P., 1958. The easterly jet stream in the tropics. Tellus 10 (1), 43–57.

Krishnamurti, T.N., Ardanuy, P., 1980. The 10 to 20-day westward propagating mode and "breaks in the monsoons". Tellus 32 (1), 15–26.

Krishnamurti, T.N., Bhalme, H., 1976. Oscillations of a monsoon system. Part I. Observational aspects. J. Atmos. Sci. 33, 1937–1954.

Krishnamurti, T.N., Daggupaty, S.M., Fein, J., Kanamitsu, M., Lee, J.D., 1973. Tibetan high and upper tropospheric tropical circulation during northern summer. Bull. Amer. Meteor. Soc. 54 (12), 1234–1249.

Liu, B., Wu, G., Mao, J., He, J., 2013. Genesis of the South Asian high and its impact on the Asian summer monsoon onset. J. Clim. 26 (9), 2976–2991.

Liu, B., Wu, G., Ren, R., 2015. Influences of ENSO on the vertical coupling of atmospheric circulation during the onset of South Asian summer monsoon. Clim. Dyn. 45 (7-8), 1859–1875.

Liu, Y., Hoskins, B., Blackburn, M., 2007. Impact of Tibetan orography and heating on the summer flow over. Asia. J. Meteor. Soc. Japan 85B, 1–19.

Liu, Y.M., Wu, G.X., Liu, H., Liu, P., 2001. Condensation heating of the Asian summer monsoon and the subtropical anticyclone in Eastern Hemisphere. Clim. Dyn. 17 (4), 327–338.

Matsuno, T., 1966. Quasi-geostrophic motions in the equatorial area. J. Meteor. Soc. Japan 44 (1), 25–43.

Mason, R.B., Anderson, C.E., 1963. The development and decay of the 100 mb summertime anticyclone over southern. Asia. Mon. Wea. Rev. 91 (1), 3–12.

Murakami, T., Frydrych, M., 1974. On the preferred period of upper wind fluctuations during the summer monsoon. J. Atmos. Sci. 31 (6), 1549–1555.

Popovic, J.M., Plumb, R.A., 2001. Eddy shedding from the upper-tropospheric Asian monsoon anticyclone. J. Atmos. Sci. 58 (1), 93–105.

Rodwell, M.J., Hoskins, B.J., 1996. Monsoons and the dynamics of deserts. Quart. J. Roy. Meteor. Soc. 122 (534), 1385–1404.

Schneider, U., Becker, A., Finger, P., Meyer-Christoffer, A., Rudolf, B., Ziese, M., 2011. GPCC full data reanalysis version 6.0 at 0.58: Monthly land-surface precipitation from rain-gauges built on GTS-based and historic data. GPCC Data Rep. doi:10.5676/DWD_GPCC/FD_M_V6_050.

Tao, S., Zhu, F., 1964. The 100-mb flow patterns in southern Asia in summer and its relation to the advance and retreat of the West Pacific subtropical anticyclone over the Far East. Acta Meteorol. Sin. 34, 385–396 (in Chinese).

Wang, B., Fan, Z., 1999. Choice of South Asian summer monsoon indices. Bull. Amer. Meteor. Soc. 80 (4), 629–638.

Wang, B., Xie, X., 1996. Low-frequency equatorial waves in vertically sheared zonal flow. Part I: Stable waves. J. Atmos. Sci. 53 (3), 449–467.

Webster, P.J., Yang, S., 1992. Monsoon and ENSO: Selectively interactive systems. Quart. J. Roy. Meteor. Soc. 118 (507), 877–926.

Wei, W., Wu, Y., Yang, S., Zhou, W., 2019a. Role of the South Asian high in the onset process of the Asian summer monsoon during spring-to-summer transition. Atmosphere 10 (5), 239.

Wei, W., Zhang, R., Wen, M., 2012. Meridional variation of South Asian high and its relationship with the summer precipitation over. China J. Appl. Meteorol. Sci. (in Chinese) 23, 650–659.

Wei, W., Zhang, R., Wen, M., Kim, B.-J., Nam, J.-C., 2015. Interannual variation of the South Asian high and its relation with Indian and East Asian summer monsoon rainfall. J. Clim. 28 (7), 2623–2634.

Wei, W., Zhang, R., Wen, M., Rong, X., Li, T., 2014. Impact of Indian summer monsoon on the South Asian high and its influence on summer rainfall over China. Clim. Dyn. 43 (6-7), 1257–1269.

Wei, W., Zhang, R., Yang, S., Li, W., Wen, M., 2019b. Quasi-biweekly oscillation of the South Asian high and its role in connecting the Indian and East Asian summer rainfalls. Geophys. Res. Lett. 46 (24), 14742–14750.

Wu, G., Duan, A., Liu, Y., Mao, J., Ren, R., Bao, Q., He, B., Liu, B., Hu, W., 2015. Tibetan Plateau climate dynamics: Recent research progress and outlook. Natl. Sci. Rev. 2 (1), 100–116.

Wu, G., Liu, Y., 2003. Summertime quadruplet heating pattern in the subtropics and the associated atmospheric circulation. Geophys. Res. Lett. 30 (5), 1201. doi:10.1029/2002GL016209.

Wu, G., Liu, Y., Zhang, Q., Duan, A., Wang, T., Wan, R., Liu, X., Li, W., Wang, Z., Liang, X., 2007. The influence of mechanical and thermal forcing by the Tibetan Plateau on Asian climate. J. Hydrometeor. 8 (4), 770–789.

Wu, G., Zhuo, H., Wang, Z., Liu, Y., 2016. Two types of summertime heating over the Asian large-scale orography and excitation of potential-vorticity forcing I. Over Tibetan Plateau. Sci. China Earth Sci. 59 (10), 1996–2008.

Wu, R., 2017. Relationship between Indian and East Asian summer rainfall variations. Adv. Atmos. Sci. 34 (1), 4–15.

Xie, X., Wang, B., 1996. Low-frequency equatorial waves in vertically sheared zonal flow. Part II: Unstable waves. J. Atmos. Sci. 53 (23), 3589–3605.

Xue, X, Chen, W., Chen, S., Zhou, D., 2015. Modulation of the connection between boreal winter ENSO and the South Asian High in the following summer by the stratospheric Quasi-Biennial Oscillation. J. Geophys. Res. Atmos. 120, 7393–7411.

Xue, X., Chen, W., Chen, S., 2017. The climatology and interannual variability of the South Asia high and its relationship with ENSO in CMIP5 models. Clim. Dyn. 48, 3507–3528.

Xue, X., Chen, W., Chen, S., Feng, J., 2018. PDO modulation of the ENSO impact on the summer South Asian high. Clim. Dyn. 50, 1393–1411.

Yang, J., Liu, Q., Xie, S., Liu, Z., Wu, L., 2007. Impact of the basin-wide mode of the Indian Ocean SST basin mode on the Asian summer monsoon. Geophys. Res. Lett. 24, L02708. doi:10.1 029/2006GL028571.

Yang, S., Li, T., 2016. Zonal shift of the South Asian high on the subseasonal time-scale and its relation to the summer rainfall anomaly in China. Quart. J. Roy. Meteor. Soc. 142 (699), 2324–2335.

Zhang, P., Yang, S., Kousky, V.E., 2005. South Asian high and Asian-Pacific-American climate teleconnection. Adv. Atmos. Sci. 22 (6), 915–923.

Zhang, Q., Wu, G.X., Qian, Y.F., 2002. The bimodality of the 100 hpa South Asia high and its relationship to the climate anomaly over East Asia in summer. J. Meteor. Soc. Japan 80 (4), 733–744.

Chapter 17

Southern annular mode teleconnections to Indian summer monsoon

C. Gnanaseelan[a], Sebastian Anila[a,b]

[a]*Indian Institute of Tropical Meteorology, Ministry of Earth Sciences, Pune, Maharashtra, India*
[b]*Savitribai Phule Pune University, Pune, Maharashtra, India*

17.1 Introduction

Indian summer monsoon (ISM) is a massive climatic phenomenon in the south Asian region characterized by rainfall over most of the country during the summer season (June–September). The Indian summer monsoon rainfall (ISMR) undergoes variability in different time scales including submonthly, intraseasonal, interannual, and interdecadal. ISMR variability is closely related to the Eurasian winter snow cover, land surface temperature, equatorial Pacific Sea Surface Temperature (SST), North Atlantic SST, equatorial Indian Ocean SST, winds and convection, equatorial Atlantic SST, Himalayan snow cover, etc. (e.g. Rasmusson and Carpenter, 1983; Rajeevan, 2002; Gadgil, 2003, and references therein). The interannual variability of ISMR is one of the most debated and challenging scientific problems in meteorology. ISM is closely linked to many of the climate modes such as El Niño Southern Oscillation (ENSO), North Atlantic Oscillation, Indian Ocean Dipole (IOD), equatorial Indian Ocean Oscillation (EQUINOO), etc., in the interannual time scale (e.g., Gadgil et al., 2004; Goswami et al., 2006; Krishnamurthy and Krishnamurthy, 2014). ENSO monsoon relationship in the interannual time scale has been studied extensively (e.g., Sikka, 1980; Pant and Parthasarathy, 1981; Rasmusson and Carpenter, 1983; Webster et al., 1998; Kumar et al., 1999). The external forcing which highly impact ISMR is ENSO, a coupled ocean-atmosphere variability over the Pacific Ocean (Webster and Yang, 1992; Wang and Picaut, 2004). ENSO exhibits an out of phase relation with ISMR in general, with the positive phase, El Niño, corresponding to a negative ISMR anomaly, and the negative phase, La Niña, corresponding to a positive ISMR anomaly (see Chapter 3). Several studies in the past have demonstrated ENSO as the most dominant external forcing for ISM variability mainly through modulations in the large-scale Walker and Hadley circulations (e.g., Rasmusson

Indian Summer Monsoon Variability: El Niño-teleconnections and beyond.
DOI: https://doi.org/10.1016/B978-0-12-822402-1.00026-0

and Carpenter, 1983). ENSO monsoon relationship however weakened in the recent years (Kumar et al., 1999) and is subjected to epochal changes as well. The modulation of variability in the equatorial Indian Ocean SST, Atlantic circulations, strengthening of Mascarene high are responsible for such weakening of ENSO monsoon relationship (Ashok et al., 2001; Chang et al., 2001; Feba et al., 2019). The change in ENSO, ISMR relation is attributed to the increase in the frequency of IOD events in the recent decades due to Indian Ocean warming (Ashok et al., 2001; Ashok et al., 2004). So, evaluating the relationships between ISM and other major climate modes is important.

There are many efforts in the recent years to explore the predictability of ISMR. In this context, understanding the possibility of other climate modes affecting the ENSO monsoon relationship is very important, it is important to highlight the fact that most of the prediction models capture the ENSO monsoon relationship reasonably well. The hemispheric scale variabilities in the northern and southern hemisphere sea level pressure, zonal wind, geopotential height, and temperature called annular modes: Southern Annular Mode (SAM) and Northern Annular Mode, primarily determine the variance in the extratropical atmospheric flow (Thompson and Wallace, 2000). The role of SAM on ISMR variability is not adequately studied in the past to the extent it deserves mainly due to lack of long-term data.

The SAM is a large-scale oscillation of atmospheric mass and the mean sea level pressure (MSLP) between mid and high latitudes (Thompson and Wallace, 2000). The leading empirical orthogonal function (EOF) in sea level pressure, surface temperature, geopotential height, and zonal wind display the structure of SAM (Thompson and Wallace, 2000). A negative MSLP anomaly is present over the Antarctic region and a positive MSLP anomaly over mid-latitude during the positive SAM events and vice versa for negative SAM events (Reboita et al., 2009). The SAM index (SAMI) compares average zonal MSLP over 40°S and 65°S (Gong and Wang, 1999). SAM is characterized by zonally symmetric structures of geopotential height anomalies of opposite signs over the Antarctic circles and mid-latitude belt, around 45°S (Thompson and Wallace, 2000). It is also a north–south vacillation of strong westerly winds (e.g., Prabhu et al., 2016). During positive SAM events, there is a poleward shift of stronger westerlies, with strong westerlies around 60°S and weak westerlies around 40°S latitude belts, and is opposite for negative SAM events (Hartmann and Lo, 1998; Gillett et al., 2006; Prabhu et al., 2016). These are closely associated with the Ferrel cell structure and variability (Raphael et al., 2011). Studies have reported that after the 1970s, SAM shows a positive trend, with lowering surface pressure over the Antarctic circle compared to mid-latitudes. The positive phase of SAM results in cooling over Antarctica and Australia, significant warming over the Antarctic peninsula, Argentina, Tasmania, and south of New Zealand (Abram et al., 2014). This trend leads to warming in the Antarctic Peninsula and cooling over East Antarctica (Thompson and Solomon, 2002; Marshall, 2003; Abram et al., 2014; Gillett et al., 2006).

SAM is also found to impact the Northern Hemisphere (Nan and Li 2003) mostly by modulating the equatorial Pacific SST (e.g., L'Heureux and Thompson 2006; Ciasto and Thompson 2008) and tropical and southern subtropical Indian Ocean SST (Dou et al., 2017). The SAM modulates the large-scale meridional circulations such as Ferrel cell and Hadley cell (Raphael et al., 2011; L'Heureux and Thompson, 2006), thereby establishing the teleconnections with equatorial Pacific SST. Liu et al. (2018) speculated the possibility of May SAM affecting the south China Sea summer monsoon through coupled ocean-atmospheric bridge in the South Pacific and the associated south–north Pacific teleconnections. Prabhu et al. (2017) reported a positive correlation between May and June SAMI and rainfall over Korea Japan Peninsula. The impact of boreal spring SAM on the east Asian monsoon (Nan et al., 2009) and North American summer monsoon (Sun, 2010) further suggests its global teleconnections and extent.

In this chapter, we focus on how SAM influences ISMR variability, the teleconnections, and the possible pathways connecting SAM and ISM. Prabhu et al. (2016) reported that the FM SAM is positively correlated with the following ISMR. They suggested the mechanism connecting SAM and ISMR is through the Pacific Ocean channel since SAM affects the equatorial Pacific SST. The FM SAM is negatively correlated to ENSO in the following June–September (JJAS) season (L'Heureux and Thompson, 2006; Ciasto and Thompson, 2008; Lim et al., 2013; Prabhu et al., 2016). The delayed relationship between FM SAM and following ISMR is therefore most likely through the Pacific pathway. FM SAMI shows a high correlation with the following JJAS El Niño Modoki index (EMI) as well (Prabhu et al., 2016). It is important to examine further how does the FM SAM and ISMR relationship evolve, as there is lead time available for monsoon prediction. Here we also explore the Indian Ocean pathways for the SAM monsoon relationship. Winter SAMI is also found to be correlated with Somali Jet (Shi et al., 2016). Further, SAMI is positively correlated with Mascarene High, which is then positively correlated with Somali Jet and ISM (Xue et al., 2004; Mohapatra et al., 2020), thereby speculating an Indian Ocean pathway in the SAM ISMR teleconnection. Considering the lead-lag relationship, there is a possibility of ocean dynamics playing significant role in this teleconnection pathway through the Indian Ocean.

Dou et al. (2017) on the other hand, showed that the positive SAM in the month of May would increase June–July (JJ) rainfall over India. The JJ Indian Ocean SST anomalies correlated with May SAM revealed a north–south dipole structure with positive SST anomalies over $50°S–30°S$ and negative SST anomalies north of it during the positive phase of SAM (Dou et al., 2017). These negative SST anomalies induce large-scale anomalous subsidence over the southern tropical Indian Ocean centered around $10°S$ and anomalous convection over the Indian latitudes (Dou et al., 2017) and intensify surface southerlies in the southern tropical Indian Ocean and thereby intensifying the trade winds and cross equatorial flow. This in turn supports anomalous excess moisture transport toward the Indian mainland. In the similar direction, Viswambharan and Mohanakumar (2014) and Pal et al. (2017)

showed that positive SAM in June negatively impact July-August rainfall over India. However, it is essential to understand the overall influence of SAM on the ISMR, the lag relationship between these two will be useful for assessing the possible new predictors of ISM.

17.2 Data and methodology

The ISMR index from the Indian Institute of Tropical Meteorology website (http://www.tropmet.res.in/) for the period 1901–2018 and India Meteorological Department observed monthly rainfall data for the period 1958–2018 have been used in this study. The ISMR index is the area average JJAS all India rainfall from a well-distributed network of rain gauges (Parthasarathy et al., 1995). This is the most reliable ISMR data and is available for the period 1871–2019. The India Meteorological Department rainfall data is $0.25° × 0.25°$ gridded dataset (Pai et al. 2014; Rajeevan et al. 2006).

The SST data from met office Hadley Centre Global Sea Ice and SST has been utilized to calculate the indices of different modes of SST variability. In this study, the indices using SST anomaly (SSTA) are Niño 3.4 index (SSTA averaged over the region 170°W–120°W and 5°S–5°N), El Niño Modoki index (EMI = SSTA(A) −0.5 × SSTA(B) − 0.5 × SSTA(C), where the bracket is representing the area average and A, B, C in the bracket are representing the regions 165°E–140°W, 10°S–10°N, 110°W–70°W, 15°S–5°N, and 125°E–145°E, 10°S–20°N, respectively), dipole mode index (DMI) (the difference between SSTA averaged over the boxes 50°E–70°E, 10°S–10°N and 90°E–110°E, 10°S-Equator) and subtropical IOD (SIOD) index (SDMI) (the difference between SSTA averaged over 55°E–65°E, 37°S–27°S, and 90°E–100°E, 28°S–18°S). Optimum Interpolation SST (Reynolds et al., 2002) for the period of 1983–2018 is also used in the study.

Marshall (2003) reported that the NCEP–NCAR reanalysis data are having a spurious negative trend in the SH high-latitude MSLP. They used station data from six stations at both 40°S and 65°S to calculate SAMI, as the difference between zonally averaged MSLP between 40°S and 65°S. They also suggested that the ECMWF's reanalysis dataset ERA-40 is ideal for the studies about SAM in recent years. The ECMWF's reanalysis datasets ERA-20C and ERA-5 MSLP are therefore merged here to obtain SAMI for the period 1901–2018 since ERA-5 is an improved version of ERA-40 reanalysis data. ERA-20C dataset is ECMWF's first atmospheric reanalysis dataset of the 20th century, which is available from 1900 to 2010. ERA-5 is the latest reanalysis data of ECMWF with higher spatial and temporal resolution than other reanalysis products of ECMWF. It is available for the period 1979–2020. So ERA-20C is used for the period 1901–78 and ERA-5 is used for the period 1979–2018. Nan and Li (2003) used the modified SAMI that is the difference between normalized monthly zonally averaged MSLP at

40°S and 70°S because the negative correlation between zonal mean MSLP at 40°S and 70°S is greater than that between 40°S and 65°S. So the SAMI used in this study is the difference between normalized monthly zonally averaged MSLP over 40°S and 70°S latitude belts.

For surface wind, ERA-20C from 1901 to 1978 and ERA-5 reanalysis datasets from 1979 to 2018 are merged. The EQUINOO index (EQWIN) is calculated as the negative of the anomaly of surface zonal wind over the region 60°E–90°E and 2.5°S–2.5°N (Gadgil et al., 2004). For computing different indices, anomalies are calculated based on the monthly climatology of respective datasets for the period 1983–2018 (1958–2018) while considering the period of 1983–2018 (1901–2018). The 41 months sliding window correlation between ISMR and the indices are computed based on band-pass filtered datasets. The Lanczos band-pass filter is used to remove the signals of periodicities above 90 months and below 13 months as in Ashok et al. (2001). Twenty-one years sliding window correlation is also computed to understand the decadal modulations in the relationship between ISMR and different indices. Partial correlation analysis is also carried out to remove the influence of ENSO on the above relations with the Indian Ocean indices to examine and emphasize more on the role of local Indian Ocean processes. Correlation and composite analyses are extensively used to understand the relation between SAM and ISMR. The composites considered in the chapter are computed as the average of different events of similar character (say positive SAM years).

The tropical and subtropical Indian Ocean modes that are likely modulate the SAM ISM relationships are defined below. Ashok and Saji (2007) suggested that IOD significantly influences the rainfall over the monsoon trough region, parts of the southwestern coast of India, and over Pakistan, Afghanistan, and Iran. The atmospheric component of IOD is EQUINOO, an oscillation in the convection over the western and eastern tropical Indian Ocean. EQUINOO occurs independently in non-IOD years also (See Chapter 2). EQWIN is significantly correlated with ISMR, but the IOD, ISMR relationship is weaker (Gadgil et al., 2004). Another mode of tropical Indian Ocean subsurface temperature variability driven by the meridional shear in zonal wind anomalies over the tropical Indian Ocean is subsurface dipole mode (Sayantani and Gnanaseelan, 2015). Yuan et al. (2008) suggested that the subsurface may memorize the dipole signal more than the surface IOD and influence the following Asian summer monsoon. A mode of SST variability over the subtropical Indian Ocean that affects ISMR variability is SIOD. It is an oscillation between the SST anomalies over the eastern part (off Australia) and southwestern part (south of Madagascar) of the subtropical Indian Ocean (Behera and Yamagata, 2001; Qian et al., 2002; Suzuki et al., 2004). Positive SIOD is significantly correlated (negatively) with ISMR, which is mostly attributed to the modulation of Mascarene High following the positive SIOD peaks (Terray et al., 2003).

17.3   Southern Annular Mode influence on Indian summer monsoon rainfall

The annual cycle of difference in zonally averaged MSLP at 40°S and 70°S is shown in Fig. 17.1, and it ranges from 18 hPa to 28 hPa and peaks during August to October with the secondary peak during March to April. The seasonal variability of SAMI displays about 40% of its peak amplitude. The correlation between the bimonthly SAMI and the following ISMR reveals that the FM SAMI is highly correlated with ISMR compared to other bimonthly indices, and is significantly high during the recent decades (Fig. 17.2). This high correlation suggests the possibility of strengthening teleconnections between SAM and ISM. The questions such as what are the selected pathways through which SAM affects ISM are addressed in the chapter. Fig. 17.2B highlights the possible subseasonal variability and the relationships SAM can display and may also help to identify the pathways of teleconnection. It is important to note that the subseasonal relationship displays a contrasting picture with FM and May SAM positively correlating with JJ rainfall over India, while January and April SAM is correlating negatively to August–September (AS) rainfall over India. This could be the reason for the poor correlation between April–May SAMI and ISMR. It also suggests that the possibility of SAM affecting the ISM positively during the onset phase and negatively during the withdrawal phase. Viswambharan and Mohanakumar (2014) also showed that SAM is positively influencing the onset phase of monsoon.

Fig. 17.3 shows the difference of composite JJAS rainfall over Indian land region between the positive and negative FM SAM years (1989, 1990, 1994, 1998, 2013, 2015, 2016 are positive FM SAM years and 1986, 1992, 2001, 2004, 2010, 2017 are negative FM SAM years). The association between FM SAMI and ISMR (Fig. 17.3A) is found to be significant especially over the Central India and Western Ghats, the regions of maximum summer rainfall variability. Therefore, FM SAMI provides some vital information on the following

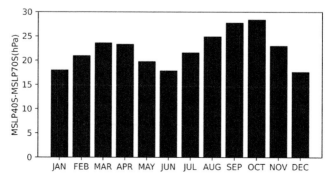

FIG. 17.1   Annual cycle of difference between zonally averaged MSLP at 40°S and 70°S (similar to Prabhu et al., 2016 but extended to the period 1901–2018). *MSLP*, mean sea level pressure.

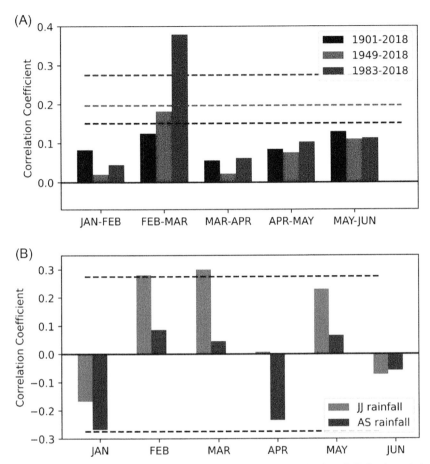

FIG. 17.2 Correlation between (A) bimonthly SAM index and the following ISMR for the period 1901–2018 (*black bars*), 1949–2018 (*red bars*), and 1983–2018 (*blue bars*). (B) Monthly (January–June) SAM index and the following June–July (*green bars*) and August–September (*pink bars*) rainfall over India for the period 1983–2018. The dashed lines are representing 90% confidence levels. *ISMR*, Indian summer monsoon rainfall; *SAM*, Southern Annular Mode.

ISM, suggesting some prediction values, as discussed by Prabhu et al. (2016). This, in the context of increasing correlation between FM SAMI and ISMR, especially in the recent years (shown below), highlights the urgent need of more observational and modeling studies on this important research topic. Fig. 17.3 suggests that the positive SAM years correspond to positive ISMR and vice-versa. In contrast, May SAMI is correlated mostly over the western part of India and Southern Peninsula. This has been demonstrated in Dou et al. (2017) as well. The correlation pattern shows similar evolution discussed in Chakravorty et al. (2016) as a response to Indian Ocean SST.

(A)   (B)

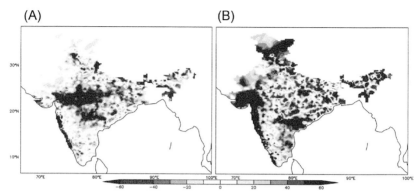

FIG. 17.3 Composite difference between (A) positive and negative FM SAM years JJAS rainfall anomaly (mm/month) (B) positive and negative May SAM years JJ rainfall anomaly (mm/month) over India for the period from 1983 to 2018. *FM*, Feb.–Mar.; *SAM*, Southern Annular Mode.

Some of the burning research questions, therefore, are whether the FM SAM influences the ISM by only through the equatorial Pacific pathways (as suggested by Prabhu et al., 2016) or multiple pathways including both the equatorial Pacific and Indian Ocean. FM SAMI and SIOD may have strong coupling between them. Close association between Spring time SAM and Indian Ocean SST has been reported by Nan et al. (2009) suggesting the possibility of pathways through the Indian Ocean. It is also important to examine whether the eastern equatorial Pacific pathways or central equatorial Pacific pathways are preferred by SAM-ISM teleconnections. It is also worth examining whether the good negative correlation between SAMI and equatorial Pacific SST suggests two ways feedback mechanism or SAM responds simply to Pacific SST as demonstrated by Zhou and You (2004) through atmospheric model experiments. If SAM is just responding to equatorial Pacific SST, then the Indian Ocean pathways connecting SAM and ISM need to be investigated and explored more.

The composite difference in SLP between the positive and negative FM SAM years and May SAM years is displayed in Fig. 17.4. The equatorward extension of mid-latitude SLP signals may suggest the possible teleconnections and pathways. In the case of FM SAM years, the equatorward extension of SLP is more predominant over the Australian longitudes with weaker midlatitude signals over Indian Ocean longitudes (Fig. 17.4A). On the other hand, the SLP signals in the midlatitude region are stronger over the Indian Ocean in the case of May SAM. The equatorward extension, however, is seen over the South American landmass and eastern Pacific region. This suggests the possible complex interaction between the different climate modes especially while considering the teleconnection between SAM and Northern Hemisphere climate.

To understand the pathways of FM SAM and May SAM on ISM variability, the correlation between the respective indices and Indo-Pacific SST

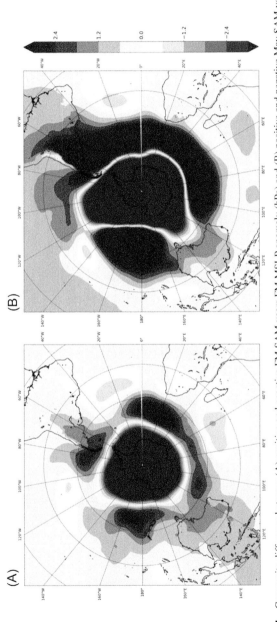

FIG. 17.4 Composite difference between (A) positive and negative FM SAM years FM MSLP anomaly (hPa) and (B) positive and negative May SAM years May MSLP anomaly for the period 1983–2018. *FM*, Feb.–Mar.; *MSLP*, mean sea level pressure; *SAM*, Southern Annular Mode.

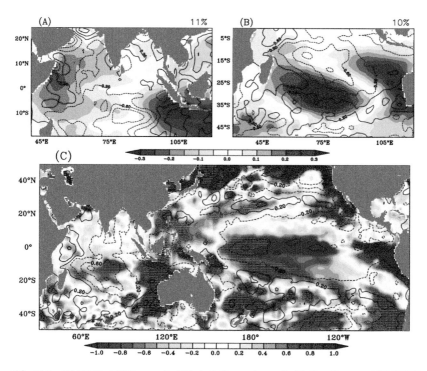

FIG. 17.5 (A) EOF2 of SST anomaly (°C, shaded) over the tropical Indian Ocean and JJAS SST regressed on FM SAM index (°C, contour), (B) EOF2 of SST anomaly over subtropical Indian Ocean (shaded) and JJ SST regressed on May SAM index (contour; °C), (C) JJAS SST regressed on FM SAM index (shaded) (°C; values significant at 90% confidence level are shown with dots) and JJ SST regressed on May SAM (contour) over the Indo-Pacific region for the period 1983–2018. *FM*, Feb.–Mar.; *SAM*, Southern Annular Mode; *SST*, Sea Surface Temperature.

is shown in Fig. 17.5. The leading modes of variability in the tropical and subtropical Indian Ocean, namely the IOD and SIOD, respectively, are also shown in Fig. 17.5A and B. The possible Indian Ocean pathways between FM SAM and ISM could be the western equatorial Indian Ocean and is weakly correlated with IOD. However, the May SAM is found to have strong correlation with both SIOD and subtropical Indian Ocean SST with the regression values exceeding 0.6°C. So the Indian Ocean pathways connecting May SAM are more evident in Fig. 17.5B, whereas evidence is weaker for FM SAM (Fig. 17.5A) except over southern tropical Indian Ocean. On the contrary both FM SAM and May SAM are found to have close association with central equatorial Pacific strongly suggesting the possibility of equatorial Pacific pathways.

Though the relationship between FM SAM and ISMR has been demonstrated to some extent, the convincing pathways connecting them are yet to be established. This, however, need careful analysis of long term observations and model simulations and sensitivity experiments which are beyond the scope of

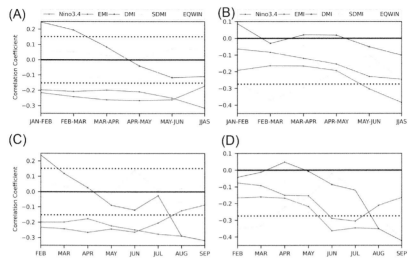

FIG 17.6   Correlation between FM SAM index and bimonthly Niño 3.4, El Niño Modoki index (*EMI*), dipole mode index (*DMI*), subtropical DMI (*SDMI*), and EQUINOO index (*EQWIN*) for the periods (A) 1901–2018 and (B) 1983–2018. The dotted lines are representing 90% confidence levels. (C) and (D) are the same as (A) and (B), respectively, but for monthly Niño 3.4, EMI, DMI, SDMI, and EQWIN. *SAM*, Southern Annular Mode.

this chapter. In line with exploring the possible pathways and teleconnections, we looked at the lead-lag correlation between FM SAMI and different climate indices of tropical Indo-Pacific and subtropical Indian Ocean in Fig. 17.6. It is found that both eastern equatorial Pacific (Niño3.4) and central Pacific (EMI) SSTs display significant correlations with FM SAMI.

It is also found that FM SAMI is significantly correlated (negatively) with both Niño3.4 and EMI for the subsequent 6 months or so and correlated (negatively) with DMI during the following AS (Fig. 17.6A and C). However, the relationships undergo changes in the recent years, say after the 1970s, this is clearly evident in Fig. 17.6B and D, with EMI showing significant correlation from JJAS and DMI during AS. This strongly suggests the possibility of central equatorial Pacific acting as the main pathway. It is important to note the negative correlation between FM SAMI and SDMI during FM and DMI during AS. The negative correlations between FM SAMI and Niño3.4 (EMI) strongly suggest that positive FM SAM favors ISMR through the Pacific pathways. On the other hand, negative correlation with DMI in the recent years compel us to examine whether the correlation between the tropical Indo Pacific climate modes and ISMR is stable over time. Moreover, the mechanisms connecting FM SAM and ISMR through Central Pacific pathway need to be explored more. It is important to note that the SAMI is not significantly correlated with EQWIN but is significantly correlated with FM SDMI and AS DMI, thereby speculating the possibility of Indian Ocean teleconnections from FM SAM through

SIOD. SIOD can leave strong footprints in the upper ocean heat content or sea level, as shown by Zhang et al. (2019), thereby impacting the tropical Indian Ocean subsurface temperature through oceanic Rossby waves (Sayantani and Gnanaseelan, 2015; Chakravorty et al., 2014). In such a scenario, the subsurface surface interaction in the southwestern TIO (Xie et al., 2002; Chowdary et al., 2009; Kakatkar et al., 2020) projects these in the surface, which can then be transmitted to North Indian Ocean through asymmetric warming, rainfall and wind pattern (Chakravorty et al. 2013). This, however, needs to be explored more and so is a potential future research area.

The 41 months sliding window correlation has been computed to understand the influence of ENSO, IOD, SIOD, and SAM on ISMR. Based on the 41 months sliding window correlation between Indian rainfall and ENSO, IOD indices for the period 1958–97, Ashok et al. (2001) suggested that when the correlation between DMI and ISMR increases, the relation between ENSO and Indian rainfall weakens. In this study, the 41 months sliding window correlation has been extended from 1958 to 2018 using the 13–90-month band-pass filtered datasets (Fig. 17.7). The relation between IOD and Indian rainfall and SIOD and Indian rainfall are also not stationary (oscillating between positive and negative correlation throughout without any clear periodicity). The Niño3.4 ISMR relationship is significantly negative most of the time and shows weak positive values in the late 1990s when both DMI and SAMI are positively correlated with ISMR. It is clear that the Niño3.4 and ISMR correlation is opposite to that of DMI, SDMI, and ISMR especially after the 1990s.

From the 21-year sliding window correlation, the decadal modulations of the correlation of ISMR with DMI, SDMI, SAMI, and EQWIN are examined for the period 1901–2018 (Fig. 17.8A) and a shorter period of 1958–2018 (Fig. 17.8B). All the above relations display decadal modulations. EQWIN showed a significant positive correlation with ISMR most of the time period. It is clear from the partial correlation (when ENSO influence has been removed), that EQWIN is significantly (positively) correlated with ISMR throughout the period (Fig. 17.5). The DMI–ISMR, SAMI–ISMR, and SDMI–ISMR correlations are varying mostly together. After removing the ENSO signal, IOD, ISMR

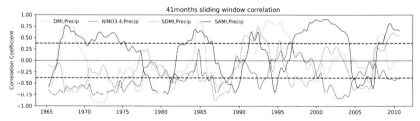

FIG. 17.7 Forty-one months sliding window correlation (simultaneous) between all India rainfall and DMI, Niño3.4, SDMI, and SAMI for the period 1958–2018. The black dashed lines representing the 90% confidence levels. *DMI*, dipole mode index; *SAMI*, Southern Annular Mode Index; *SDMI*, subtropical dipole mode index.

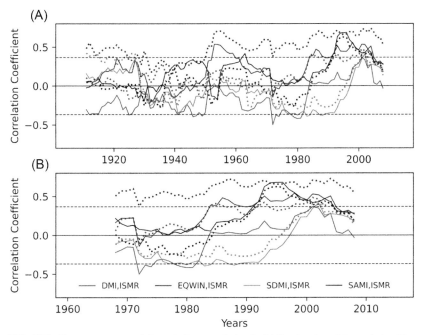

FIG. 17.8 Twenty-one years sliding window correlation (solid lines) and partial correlation (dotted lines) between DMI (JJAS), EQWIN (JJAS), SDMI (JFM), SAMI (FM), and ISMR after removing the ENSO signals (A) for the long period 1901–2018, (B) for the shorter period of 1958–2018. The black dashed lines are representing the 90% confidence levels. *DMI*, dipole mode index; *ENSO*, El Niño Southern Oscillation; *FM*, Feb.–Mar.; *ISMR*, Indian summer monsoon rainfall; *SAMI*, Southern Annular Mode Index; *SDMI*, subtropical dipole mode index.

relations become more positive, but the SDMI-ISMR and SAMI-ISMR correlations did not display any change. It suggests that the influence of ENSO on SIOD-ISMR and SAM-ISMR relation is weaker than IOD ISMR relation.

17.4   Summary and discussion

ENSO is the most significant external factor affecting the ISMR variability. However, the relationship between ISMR and ENSO weakened in the recent years and the relationship undergoes epochal modulations as well. In this context, understanding the teleconnections with the other major global climate modes and their time-varying relationships with ISMR remains important. The current coupled climate models used for monsoon prediction generally suffer from representing the ISMR variability and prediction. However, majority of them represent the SAM well (Zheng et al., 2017). So investigating the teleconnection between SAM and ISMR not only improves our understanding of Southern Ocean relation with ISMR but also provides predictive information. The FM SAM is better correlated with ISMR and the

Central India and Western Ghats rainfall, suggesting strong impact of SAM on the core monsoon regions. In the subseasonal scale FM and May SAM are found to have strong relationship with JJ rainfall over India. Previous studies have revealed that the pathways of teleconnections of FM SAM on ISMR is through equatorial Pacific, whereas the May SAM is through subtropical (southern) and tropical Indian Ocean. We speculate here the possibility of FM SAM affecting ISM through modulating the IOD and SIOD. These may require futher investigation for making any strong conclusion. An empirical model based on ENSO, SAM, and other climate modes that influence ISMR may serve as a better tool for hindcasting or predicting ISMR variations. The decadal modulations in the ISM during the recent years are closely related to SAM variability (Huang et al., 2021) thereby reinstating its importance in different time scales. Decadal changes in the correlation between SAM and precipitation over South America and Australia are also reported (e.g., Silvestri and Vera, 2003). The role of monthly SAM on subseasonal variability of ISMR has been discussed by Dou et al. (2017) and Pal et al. (2017) mainly through Indian Ocean (SST) pathways.

Acknowledgments

We acknowledge Director, Indian Institute of Tropical Meteorology and Ministry of Earth Sciences, Government of India for support. We also thank Drs. Amita Prabhu, Anant Parekh and Jasti S. Chowdary, and Sandeep Mohapatra for their suggestions to improve the chapter.

References

Abram, N.J., Mulvaney, R., Vimeux, F., Phipps, S.J., Turner, J., England, M.H., 2014. Evolution of the southern annular mode during the past millennium. Nat. Clim. Change 4 (7), 564–569. https://doi.org/10.1038/nclimate2235.

Ashok, K., Guan, Z., Saji, N.H., Yamagata, T., 2004. Individual and combined influences of ENSO and the Indian Ocean Dipole on the Indian summer monsoon. J. Clim. 17 (16), 3141–3155. https://doi.org/10.1175/1520-0442(2004)017<3141:IACIOE>2.0.CO;2.

Ashok, K., Guan, Z., Yamagata, T., 2001. Impact of the Indian Ocean Dipole on the relationship between the Indian Monsoon Rainfall and ENSO. Geophys. Res. Lett. 28 (23), 4499–4502. https://doi.org/10.1029/2001GL013294.

Ashok, K., Saji, N.H., 2007. On the impacts of ENSO and Indian Ocean Dipole events on sub-regional Indian summer monsoon rainfall. Nat. Hazards 42 (2), 273–285. https://doi.org/10.1007/s11069-006-9091-0.

Behera, S.K., Yamagata, T., 2001. Subtropical SST dipole events in the Southern Indian ocean. Geophys. Res. Lett. 28 (2), 327–330. https://doi.org/10.1029/2000GL011451.

Chakravorty, S., Chowdary, J.S., Gnanaseelan, C., 2013. Spring asymmetric mode in the tropical Indian Ocean: role of El Niño and IOD. Clim. Dyn. 40 (5–6), 1467–1481. https://doi.org/10.1007/s00382-012-1340-1.

Chakravorty, S., Gnanaseelan, C., Chowdary, J.S., Luo, J., 2014. Relative role of El Niño and IOD forcing on the southern tropical Indian Ocean Rossby waves. J. Geophys. Res. Oceans. doi: 10.1002/2013JC009713.

Chakravorty, S., Gnanaseelan, C., Pillai, P.A., 2016. Combined influence of remote and local SST forcing on Indian summer monsoon rainfall variability. Clim. Dyn. 47 (9–10), 2817–2831. https://doi.org/10.1007/s00382-016-2999-5.

Chang, C.P., Harr, P., Ju, J., 2001. Possible roles of Atlantic circulation on the weakening Indian monsoon rainfall-ENSO relationship. J. Clim. 14 (11), 2376–2380. https://doi.org/10.1175/1520-0442(2001)014<2376:PROACO>2.0.CO;2.

Chowdary, J.S., Gnanaseelan, C., Xie, S.P., 2009. Westward propagation of barrier layer formation in the 2006-07 rossby wave event over the tropical Southwest Indian ocean. Geophys. Res. Lett. 36 (4), 1–5. https://doi.org/10.1029/2008GL036642.

Ciasto, L.M., Thompson, D.W.J., 2008. Observations of large-scale ocean-atmosphere interaction in the southern hemisphere. J. Clim. 21 (6), 1244–1259. https://doi.org/10.1175/2007JCLI1809.1.

Dou, J., Wu, Z., Zhou, Y., 2017. Potential impact of the May southern hemisphere annular mode on the Indian summer monsoon rainfall. Clim. Dyn. 49 (4), 1257–1269. https://doi.org/10.1007/s00382-016-3380-4.

Feba, F., Ashok, K., Ravichandran, M., 2019. Role of changed Indo-Pacific atmospheric circulation in the recent disconnect between the Indian summer monsoon and ENSO. Clim. Dyn. 52 (3–4), 1461–1470. https://doi.org/10.1007/s00382-018-4207-2.

Gadgil, S., 2003. The Indian monsoon and its variability. Annu. Rev. Earth Planet. Sci. 31, 429–467. https://doi.org/10.1146/annurev.earth.31.100901.141251.

Gadgil, S., Vinayachandran, P.N., Francis, P.A., Gadgil, S., 2004. Extremes of the Indian summer monsoon rainfall, ENSO and Equatorial Indian Ocean Oscillation. Geophys. Res. Lett. 31 (12), 2–5. https://doi.org/10.1029/2004GL019733.

Gillett, N.P., Kell, T.D., Jones, P.D., 2006. Regional climate impacts of the southern annular mode. Geophys. Res. Lett. 33 (23), 1–4. https://doi.org/10.1029/2006GL027721.

Gong, D., Wang, S., 1999. Definition of Antarctic Oscillation Index. Geophys. Res. Lett. 26 (4), 459–462. https://doi.org/10.1029/1999GL900003.

Goswami, B.N., Madhusoodanan, M.S., Neema, C.P., Sengupta, D., 2006. A physical mechanism for North Atlantic SST influence on the Indian summer monsoon. Geophys. Res. Lett. 33 (2), 1–4. https://doi.org/10.1029/2005GL024803.

Hartmann, D.L., Lo, F., 1998. Wave-driven zonal flow vacillation in the southern hemisphere. J. Atmos. Sci. 55 (8), 1303–1315. https://doi.org/10.1175/1520-0469(1998)055<1303:WDZFVI>2.0.CO;2.

Huang, P.W., Lin, Y.F., Wu, C.R., 2021. Impact of the southern annular mode on extreme changes in Indian rainfall during the early 1990s. Sci. Rep. 11 (1), 1–8. https://doi.org/10.1038/s41598-021-82558-w.

Kakatkar, R., Gnanaseelan, C., Chowdary, J.S., 2020. Asymmetry in the Tropical Indian ocean subsurface temperature variability. Dyn. Atmos. Oceans 90, 101142. https://doi.org/10.1016/j.dynatmoce.2020.101142.

Krishnamurthy, L., Krishnamurthy, V., 2014. Influence of PDO on South Asian summer monsoon and monsoon-ENSO relation. Clim. Dyn. 42 (9–10), 2397–2410. https://doi.org/10.1007/s00382-013-1856-z.

Kumar, K.K., Rajagopalan, B., Cane, M.A., 1999. On the weakening relationship between the Indian monsoon and ENSO. Science 284 (5423), 2156–2159. https://doi.org/10.1126/science.284.5423.2156.

L'Heureux, M.L., Thompson, D.W.J., 2006. Observed relationships between the El-Niño-Southern oscillation and the extratropical zonal-mean circulation. J. Clim. 19 (1), 276–287. https://doi.org/10.1175/JCLI3617.1.

Lim, E.P., Hendon, H.H., Rashid, H., 2013. Seasonal predictability of the southern annular mode due to its association with ENSO. J. Clim. 26 (20), 8037–8054. https://doi.org/10.1175/JCLI-D-13-00006.1.

Liu, T., Li, J., Li, Y., Zhao, S., Zheng, F., Zheng, J., Yao, Z., et al., 2018. Influence of the May southern annular mode on the South China Sea summer monsoon. Clim. Dyn. 51 (11), 4095–4107.

Marshall, G.J., 2003. Trends in the southern annular mode from observations and reanalyses. J. Clim. 16 (24), 4134–4143. https://doi.org/10.1175/1520-0442(2003)016<4134:TITSAM>2.0.CO;2.

Mohapatra, S., Gnanaseelan, C., Deepa, J.S., 2020. Multidecadal to decadal variability in the equatorial Indian Ocean subsurface temperature and the forcing mechanisms. Clim. Dyn. 54 (7), 3475–3487.

Nan, S., Li, J., 2003. The relationship between the summer precipitation in the Yangtze River Valley and the Boreal Spring southern hemisphere annular mode. Geophys. Res. Lett. 30 (24), 1–4. https://doi.org/10.1029/2003GL018381.

Nan, S., Li, J., Yuan, X., Zhao, P., 2009. Boreal Spring southern hemisphere annular mode, Indian Ocean Sea Surface Temperature, and East Asian summer monsoon. J. Geophys. Res. Atmos. 114 (2), 1–13. https://doi.org/10.1029/2008JD010045.

Pai, D.S., Sridhar, L., Rajeevan, M., Sreejith, O.P., Satbhai, N.S., Mukhopadhyay, B., 2014. (1901-2010) Daily gridded rainfall data set over india and its comparison with existing data sets over the region. Mausam 1, 1–18.

Pal, J., Chaudhuri, S., Roychowdhury, A., Basu, D., 2017. An investigation of the influence of the southern annular mode on Indian summer monsoon rainfall. Meteorol. Appl. 24 (2), 172–179. https://doi.org/10.1002/met.1614.

Pant, G.B., Parthasarathy, S.B., 1981. Some aspects of an association between the southern oscillation and Indian summer monsoon. Arch. Meteorol. Geophys. Bioclimatol. Ser. B 29 (3), 245–252. https://doi.org/10.1007/BF02263246.

Parthasarathy, B., Yang, S., 1995. Relationships between regional Indian summer monsoon rainfall and Eurasian snow cover. Adv. Atmos. Sci. 12 (2), 143–150. https://doi.org/10.1007/BF02656828.

Prabhu, A., Kripalani, R.H., Preethi, B., Pandithurai, G., 2016. Potential role of the February–March southern annular mode on the Indian summer monsoon rainfall: a new perspective. Clim. Dyn. 47 (3–4), 1161–1179. https://doi.org/10.1007/s00382-015-2894-5.

Prabhu, A., Kripalani, R.H., Oh, J., Preethi, B., 2017. Can the southern annular mode influence the Korean summer monsoon rainfall? Asia-Pac. J. Atmos. Sci. 53 (2), 217–228.

Qian, W., Hu, H., Deng, Yi, Tian, J., 2002. Signals of interannual and interdecadal variability of air-sea interaction in the basin-wide Indian ocean. Atmosphere-Ocean 40 (3), 293–311. https://doi.org/10.3137/ao.400302.

Rajeevan, M., 2002. Winter surface pressure anomalies over Eurasia and Indian summer monsoon. Geophys. Res. Lett. 29 (10). 94-1–9-4-44. https://doi.org/10.1029/2001gl014363.

Rajeevan, M., Bhate, J., Kale, J.D., Lal, B., 2006. High resolution daily gridded rainfall data for the Indian region: analysis of break and active monsoon spells. Curr. Sci. 91 (3), 296–306.

Raphael, M.N., Hobbs, W., Wainer, I., 2011. The effect of Antarctic sea ice on the southern hemisphere atmosphere during the southern summer. Clim. Dyn. 36, 1403–1417.

Rasmusson, E.M., Carpenter, T.H., 1983. The relationship between eastern equatorial Pacific Sea surface temperatures and rainfall over India and Sri Lanka. Month. Weather Rev. 111 (3), 517–528. https://doi.org/10.1175/1520-0493(1983)111<0517:TRBEEP>2.0.CO;2.

Reynolds, R.W., Rayner, N.A., Smith, T.M., Stokes, D.C., Wang, W., 2002. An improved in situ and satellite SST analysis for climate. J. clim. 15 (13), 1609–1625. https://doi.org/10.1175/1520-0442(2002)015<1609:AIISAS>2.0.CO;2.

Reboita, M.S., Ambrizzi, T., Da Rocha, R.P., 2009. Relationship between the southern annular mode and southern hemisphere atmospheric systems. Revista Brasileira de Meteorologia 24 (1), 48–55.

Sayantani, O., Gnanaseelan, C., 2015. Tropical Indian Ocean subsurface temperature variability and the forcing mechanisms. Clim. Dyn. 44 (9–10), 2447–2462. doi:10.1007/s00382-014-2379-y.

Shi, W., Xiao, Z., Xue, J., 2016. Teleconnected influence of the Boreal Winter Antarctic Oscillation on the Somali Jet: bridging role of sea surface temperature in southern high and middle latitudes. Adv. Atmos. Sci. 33 (1), 47–57. https://doi.org/10.1007/s00376-015-5094-7.

Sikka, 1980. Some aspects of the large scale fluctuations of summer monsoon rainfall over India in relation to fluctuations in the planetary and regional scale circulation parameters. Proc. Ind. Acad. Sci. (Earth and Planetary Sciences) 89, 179–195.

Silvestri, G.E., Vera, C.S., 2003. Antarctic oscillation signal on precipitation anomalies over Southeastern South America. Geophys. Res. Lett. 30 (21), 1–4. https://doi.org/10.1029/2003 GL018277.

Sun, J., 2010. Possible impact of the boreal spring Antarctic oscillation on the North American summer monsoon. Atmos. Ocean. Sci. Lett. 3 (4), 232–236. https://doi.org/10.1080/16742 834.2010.11446870.

Suzuki, R., Behera, S.K., Iizuka, S., Yamagata, T., 2004. Indian ocean subtropical dipole simulated using a coupled general circulation model. J. Geophys. Res. C: Oceans 109 (9), 1–18. https://doi.org/10.1029/2003JC001974.

Terray, P., Delecluse, P., Labattu, S., Terray, L., 2003. Sea surface temperature associations with the late Indian Summer Monsoon. Clim. Dyn. 21 (7–8), 593–618. https://doi.org/10.1007/s00382-003-0354-0.

Thompson, D.W.J., Solomon, S., 2002. Interpretation of recent southern hemisphere climate change. Science 296 (5569), 895–899. https://doi.org/10.1126/science.1069270.

Thompson, D.W.J., Wallace, J.M., 2000. Annular modes in the extratropical circulation. Part I: month-to-month variability. J. Clim. 13 (5), 1000–1016. https://doi.org/10.1175/1520-0442(2000)013<1000:AMITEC>2.0.CO;2.

Viswambharan, N., Mohanakumar, K., 2014. Changes in the Walker and Hadley Circulations Associated with the Southern Annular Mode. Theor. Appl. Climatol. 117 (3–4), 535–547. https://doi.org/10.1007/s00704-013-1011-0.

Wang, C., Picaut, J., 2004. Understanding ENSO Physics—a review. Geophys. Monogr. Ser. 147, 21–48. https://doi.org/10.1029/147GM02.

Webster, P.J., Magana, V.O., Palmer, T.N., Shukla, J., Tomas, R.A., Yanai, M.U., Yasunari, T., 1998. Monsoons: Processes, predictability, and the prospects for prediction. J. Geophys. Res.: Oceans 103 (C7), 14451–14510. doi:10.1029/97JC02719.

Webster, P.J., Yang, S., 1992. Monsoon and Enso: selectively interactive systems. Quarterly J. Royal Meteorol. Soc. 118 (507), 877–926. https://doi.org/10.1002/qj.49711850705.

Xie, S.P., Annamalai, H., Schott, F.A., Julian, P.M., 2002. Structure and mechanisms of South Indian ocean climate variability. J. Clim. 15 (8), 864–878. https://doi.org/10.1175/1520-0442(2002)015<0864:SAMOSI>2.0.CO;2.

Xue, F., Wang, H., He, J., 2004. Interannual variability of Mascarene high and Australian high and their influences on East Asian Summer Monsoon. J. Meteorol. Soc. Japan 82 (4), 1173–1186. https://doi.org/10.2151/jmsj.2004.1173.

Yuan, Y., Yang, H., Zhou, W., Li, C., 2008. Influences of the Indian Ocean Dipole on the Asian Summer Monsoon in the following year. Int. J. Climatol. 28 (14), 1849–1859. https://doi.org/10.1002/joc.1678.

Zhang, L., Han, W., Li, Y., Lovenduski, N.S., 2019. Variability of sea level and upper-ocean heat content in the Indian ocean: effects of subtropical Indian Ocean dipole and Enso. J. Clim. 32 (21), 7227–7245. https://doi.org/10.1175/JCLI-D-19-0167.1.

Zheng, F., Li, J., Ding, R., 2017. Influence of the Preceding Austral Summer Southern Hemisphere annular mode on the amplitude of ENSO decay. Adv. Atmos. Sci. 34 (11), 1358–1379. https://doi.org/10.1007/s00376-017-6339-4.

Zhou, T., Yu, R., 2004. Sea-surface temperature induced variability of the Southern Annular Mode in an atmospheric general circulation model. Geophys. Res. Lett. 31 (24). https://doi.org /10.1029/2004GL021473.

Chapter 18

The Atlantic Multidecadal Oscillation and Indian summer monsoon variability: a revisit

Lea Svendsen

Geophysical Institute, University of Bergen and the Bjerknes Centre for Climate, Research, Bergen, Norway

18.1 Introduction

This chapter gives an overview of the present understanding of the observed link between Indian summer monsoon (ISM) rainfall and North Atlantic sea surface temperatures (SSTs) on multidecadal timescales. Observations of ISM rainfall covering the last one and a half century feature variability on a range of timescales from intraseasonal to multidecadal (Joseph, 2014; Maharana and Dimri, 2019; Parthasarathy et al., 1993). There is also large spatial variability in the rainfall distribution of the ISM (Guhathakurta et al., 2015). In total the ISM is responsible for around 70–80% of the annual precipitation over India (Parthasarathy et al., 1993,1994) and is crucial for India's economy and resources. The pronounced fluctuations of the ISM can therefore be calamitous. Wet years typically characterized by a total ISM rainfall exceeding 10% of the long-term average can give extensive flooding; dry years characterized by a total ISM rainfall below 90% of the long-term average can lead to severe droughts. The dry year of 2002 where rainfall was 19% below normal pulled the gross domestic product of India down by around 1% (Gadgil et al., 2003). The effect of two or more dry years in a row and longer multidecadal periods of below normal rainfall can be even more devastating.

Multidecadal variability in ISM rainfall can be seen in Fig. 18.1. The solid line is the normalized low-frequency filtered (13-year window) All-India monsoon rainfall index constructed from rain gauge data (Sontakke et al., 2008) updated with data from the India Meteorological Department (IMD) end of season reports. Across the twentieth century, ISM rainfall exhibits several 30-year epochs of dry and wet monsoons (Guhathakurta and Rajeevan, 2008; Joseph, 1976; 2014; Niranjan Kumar et al., 2013; Parthasarathy et al., 1993). The periods 1900–1930 and 1960–1990 were dry epochs with a higher frequency of dry years. The periods 1870–1900 and 1930–1960 were wet epochs with a lower

Indian Summer Monsoon Variability: El Niño-teleconnections and beyond.
DOI: https://doi.org/10.1016/B978-0-12-822402-1.00001-6

353

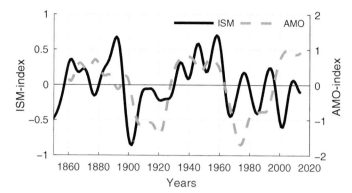

FIG. 18.1 Low-frequency filtered (low-pass Butterworth filter with cut-off window of 13 years) observed time series of ISM rainfall (Sontakke et al., 2008) (solid line) and the AMO-index (Enfield et al., 2001) (dashed line).

frequency of dry years (Joseph, 2014). This multidecadal variability in the ISM is in-phase with multidecadal variability in North Atlantic SSTs, referred to as the Atlantic multidecadal oscillation (AMO) (Fig. 18.1). The AMO–index (dashed line in Fig. 18.1) is the normalized linearly detrended low-frequency filtered (13-year window) annually averaged area-weighted North Atlantic SST anomalies using HadISST data (Rayner et al., 2003) from the equator to 60°N (Enfield et al., 2001; Schlesinger and Ramankutty, 1994). This in-phase variability suggests the possibility of a causal link between the ISM and the AMO.

18.1.1 The Atlantic multidecadal oscillation (AMO)

Multidecadal variability in North Atlantic SSTs is referred to as the AMO or the equivalent Atlantic multidecadal variability. The AMO has a period of around 60 years, and is characterized by basin-wide warming and cooling of the North Atlantic surface (Enfield et al., 2001; Schlesinger and Ramankutty, 1994). The temperature anomalies have a comma-shaped pattern with the maximum SST anomalies in the subpolar and tropical North Atlantic, and the minimum SST anomalies in the western extratropical North Atlantic (Fig. 18.2).

The AMO is thought to be a surface manifestation of the Atlantic meridional overturning circulation (Zhang et al., 2019 and references therein). However, to the first-order the AMO can be explained by the ocean response to stochastic atmospheric forcing (Clement et al., 2015; Frankignoul and Hasselmann, 1977; Hasselmann, 1976). Some studies have stressed the importance of natural external forcing for phasing the AMO (e.g., Otterå et al., 2010). Anthropogenic factors such as aerosols have also been suggested as a possible driver for the AMO (Booth et al., 2012), although debated (Zhang et al., 2013). Proxy reconstructions indicate that multidecadal variability in the Atlantic has been present for hundreds of years before significant anthropogenic impact (Gray et al., 2004;

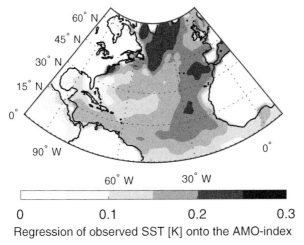

0   0.1   0.2   0.3
Regression of observed SST [K] onto the AMO-index

FIG. 18.2 AMO pattern defined as the regression of low-frequency filtered (low-pass Butterworth filter with cut-off window of 13 years) Atlantic SSTs in K using HadISST (Rayner et al., 2003) onto the normalized AMO-index.

Mann et al., 2009), suggesting that the AMO is part of the natural climate system. Currently, the AMO is seen as a combination of internal atmosphere and ocean variability and natural and anthropogenic external forcing. The relative role of these components is yet to be determined (Vecchi et al., 2017).

In the context of the ISM, there are at least two reasons for the interest in the AMO. First, the AMO could partly explain the observed multidecadal variability of the ISM. Studies have shown the vast impact that the AMO has on climate around the Atlantic, such as American and European temperatures and precipitation (Sutton and Hodson, 2007), Atlantic hurricane activity, and Sahel rainfall (Zhang and Delworth, 2006). But the AMO can also have far-reaching effects such as impacting El Niño–Southern Oscillation (ENSO) variability (Dong et al., 2006; Kang et al., 2014; Levine et al., 2017), Arctic sea ice cover (Miles et al., 2014), and the ISM (Goswami et al., 2006). Fig. 18.3 shows the regression of observed summer (June–September, JJAS) surface temperature from GISTEMP (Lenssen et al., 2019) and precipitation over land from the Climate Research Unit gridded Time Series (CRU TS) (Harris et al., 2014) onto the AMO and the ISM rainfall indices. When the AMO-index is positive, North America, Europe, and Eastern Asia is warmer (Fig. 18.3D), and there is enhanced precipitation over the Sahel and India (Fig. 18.3B). Similarly, when the ISM-index is positive, there is more ISM rainfall (Fig. 18.3A), and North Atlantic SSTs are higher (Fig. 18.3C). Thus, when the AMO was in a positive (negative) phase during the past 150 years, there was more (less) rain over India during the summer monsoon season.

Second, several studies have shown the possibility of decadal prediction of the AMO (Keenlyside et al., 2008; Kim et al., 2012; Yeager and Robson, 2017). Identifying a causal link between the AMO and ISM rainfall suggests

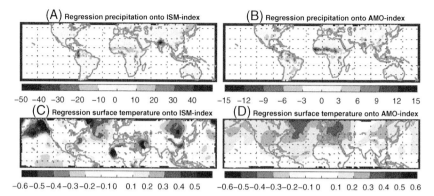

FIG. 18.3 Low-pass filtered precipitation over land (mm/month) (A, B) and surface temperature (K) (C,D) regressed onto the normalized observed ISM-index (A,C) and AMO-index (B,D). Filled contours indicate significant values at the 5% level. CRU TS (Harris et al., 2014) is used for precipitation, GISTEMP v4 (Lenssen et al., 2019) is used for surface temperature.

the possibility of decadal predictions of the ISM analogous to the seasonal monsoon forecasts that are well established and the weather forecasts that are performed across the world every day. Reliable decadal predictions of the ISM would improve multiyear planning for instance in water management and the agricultural and energy sectors.

In the following I will present the current understanding of the observed link between the AMO and the ISM. Section 18.2 introduces the observational basis as well as studies that have analyzed this link in past climate reconstructions. Section 18.3 presents the different mechanisms that have been proposed to explain the AMO–ISM link. Section 18.4 discusses how global climate models reproduce this link as models are our main tool to test mechanisms. Section 18.5 concludes and gives an overview of the current issues that need to be addressed to fully understand the AMO–ISM link.

18.2   The statistical AMO–ISM link in the past and present

Observations covering the past century and a half show a positive correlation between multidecadal variability in North Atlantic SSTs and ISM rainfall (Fig. 18.1). Decades of high (low) North Atlantic SSTs are associated with more (less) ISM rainfall. The correlation between the AMO and the multidecadal component (low-frequency filtered) of the All-India monsoon rainfall index is around 0.5. The exact value of the correlation depends on the smoothing window and the time period analyzed. Whether there is causality in these observations is unresolved and possible mechanisms for this AMO–ISM link are still being explored.

To shed light on the nature of the AMO–ISM link, proxy reconstructions can be used to investigate the robustness of the AMO–ISM relation prior to the instrumental records. The presence of an AMO–ISM link in proxy records can indicate that the link is part of the natural climate system and not due to either anthropogenic forcing or a statistical artifact of the specific time period when instrumental observations are available. Paleoclimate records provide evidence that there could have been low-frequency variability in Atlantic SSTs corresponding to variations in the ISM. For instance, there is evidence of changes in the Asian monsoon that correspond in timing to Dansgaard/Oeschger events (Burns et al., 2003), Heinrich events (Colin et al., 1998), and cooling ice-rafted debris events in the North Atlantic (Hong et al., 2003). For the ISM there are observational rainfall data from rain gauges going back all the way to 1813. The number of rain stations was limited for the first decade but steadily increased to a reliable number by 1844 (Sontakke et al., 2008). In addition to this extended instrumental record of ISM rainfall, there are several annual resolution proxy reconstructions of the ISM available. These records suggest multidecadal variability in the ISM with interchanging periods of more/less frequent dry years several hundreds of years back in time (Borgaonkar et al., 2010; Chakraborty et al., 2012; Gautam et al., 2019; Sinha et al., 2007; Yadava et al., 2004). The annual resolution proxy reconstructions of North Atlantic SSTs suggest that the AMO was also active prior to the instrumental era (e.g., Mann et al. 2009, Gray et al. 2004, Svendsen et al. 2014). On multidecadal timescales these North Atlantic SST reconstructions capture somewhat similar periods of low and high SSTs, although the amplitudes of the SST anomalies and the exact timing of AMO phase shifts do not always agree (Sankar et al., 2016; Svendsen et al., 2014).

Comparing the extended observed ISM rainfall data with the AMO reconstructions, the AMO–ISM relationship found in observations extends back to the beginning of the 1800s (Sankar et al., 2016), suggesting that the observed AMO–ISM relationship could be robust. The link between the ISM and the AMO has also been found in the past prior to the instrumental records using proxy reconstructions for both the AMO and the ISM (Berkelhammer et al., 2010; Feng and Hu, 2008; Shi et al., 2019). The observed AMO–ISM link might therefore be a consistent feature of the climate system and independent of anthropogenic climate forcing. However, most of these studies use single-proxy reconstructions for the ISM, and such ISM records are sensitive to local precipitation signals. Even though proxy reconstructions agree that there was decadal variability in the ISM prior to the instrumental data, there is little coherence between individual proxy reconstructions of ISM before 1750 (Sankar et al., 2016). The analysis of proxy reconstructions for the AMO–ISM link is therefore inconclusive so far. Combining a larger amount of records from a broad ISM core region could extract the large-scale ISM signal from more local variations.

18.3 Mechanisms of the AMO–ISM link

The ocean provides memory in the climate system and is a source of long-term variability in the atmosphere. The AMO is therefore seen as a possible driver of multidecadal variability in the ISM rather than the other way around. Several mechanisms for how the AMO can force the ISM have been suggested, and they can be organized into three categories: (1) meridional energy balance-related shifts of the intertropical convergence zone (ITCZ) (Section 18.3.1), (2) direct atmospheric teleconnections (section 18.3.2), and (3) air-sea interactions related to Atlantic-Pacific interbasin linkages (Section 18.3.3). In addition, external forcing can synchronize the AMO and the ISM (Section 18.3.4). The following subsections will summarize these mechanisms. Often, it is found that the combination of several of these mechanisms link the ISM to the AMO.

18.3.1   Meridional shifts of the ITCZ

A negative AMO phase is associated with an anomalously low North Atlantic SST and therefore induces a temperature difference anomaly to the South Atlantic. Anomalous cooling in the Northern Hemisphere extratropics can lead to a change in the meridional energy balance and a southward shift of the ITCZ (Broccoli et al., 2006; Frierson and Hwang, 2012; Kang et al., 2008). This is associated with an asymmetric response of the Hadley Circulation, a southward shift of tropical rainfall (Broccoli et al., 2006; Frierson and Hwang, 2012; Kang et al., 2008), and decreased rainfall over Southern Asia, including India (Broccoli et al., 2006; Mohtadi et al., 2016). For an anomalously warm North Atlantic, the opposite situation pervades with a northward shift of the ITCZ and enhanced ISM rainfall (Han et al., 2016; Knight et al., 2006; Monerie et al., 2019; Zhang and Delworth, 2006). This mechanism is suggested to be at work during past large climatic events in the Atlantic such as abrupt deglaciations (Zhang and Delworth, 2005), Dansgaard/Oeschger, and Heinrich events (Burns et al., 2003; Colin et al., 1998), but could also be responsible in less extreme events such as shifts in the AMO phase (Han et al., 2016; Knight et al., 2006; Monerie et al., 2019; Zhang and Delworth, 2006).

18.3.2   Direct atmospheric teleconnections

North Atlantic SST anomalies can also influence the ISM through atmospheric teleconnections. Fig. 18.4 shows schematically the identified direct atmospheric pathways presented in the following. During a positive AMO phase, the higher North Atlantic SSTs thermodynamically force low sea-level pressure (SLP) anomalies that can extend all the way to the Indian Ocean. Depending on the location of these wide-spread SLP anomalies, this can either lead to an enhanced low-level jet over the Arabian Sea (Wang et al., 2009) (Fig. 18.4A) or to a stronger Indian monsoon trough (Li et al., 2008) (Fig. 18.4B), both of which intensify ISM circulation and enhance rainfall.

The SST anomalies in the North Atlantic associated with a positive AMO can also induce a Rossby wave train across Eurasia (Li et al., 2008; Luo et al., 2011; Nath and Luo, 2019). The exact location of the troughs and crests of the Rossby wave train crossing Eurasia varies in different studies, but there is agreement that the Rossby wave enhances the upper-level South Asia high, heating the mid-to-upper troposphere there. This in turn increases the meridional

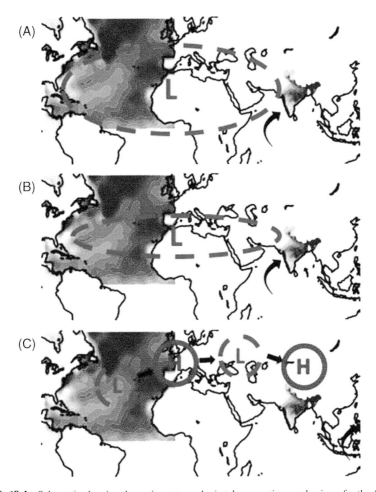

FIG. 18.4 Schematic showing the various atmospheric teleconnection mechanisms for the AMO influence on ISM rainfall described in Section 18.3.2. The observed AMO pattern from Fig. 18.2 is shown over the Atlantic. The mean ISM rainfall pattern based on the IMD gridded daily rainfall data (Rajeevan et al., 2006) is shown over India. The L (H) with dashed blue (solid red) circles around stand for negative (positive) pressure anomalies near the surface (A, B) or upper troposphere (C). Arrows over the Indian Ocean indicate circulation directions in the lower troposphere.

tropospheric temperature gradient between the Indian Ocean and the South Asian subcontinent. The following upper-level divergence over the Tibetan Plateau induces low-level convergence over India. The subsequent strengthening of the monsoon circulation can increase the ISM rainfall (Fig. 18.4C). Furthermore, the direct impact of North Atlantic SST anomalies on the upper-level Asian Jet and westerly advection could influence the meridional temperature gradient and affect the ISM (Wang et al., 2009).

In addition, Goswami et al. (2006) found that a positive (negative) AMO influenced the ISM by delaying (advancing) the withdrawal and enhancing (abating) rainfall. The ISM response was induced by tropospheric temperature anomalies over Eurasia through the North Atlantic Oscillation (NAO) acting as an atmospheric bridge. A positive summer NAO-index was associated with a similar SLP pattern as a positive AMO, and changes of the winds and storm tracks associated with NAO variability could be responsible for changes in the ISM. However, results from most subsequent studies (e.g., Joshi and Ha, 2019; Li et al., 2008; Luo et al., 2011; Monerie et al., 2019; Nath and Luo, 2019; Wang et al., 2009) do not concur with this mechanism involving the NAO.

18.3.3   Air-sea interactions and Atlantic-Pacific interbasin linkages

On interannual-to-decadal timescales, the Pacific Ocean constrains ISM variability (Joshi and Kucharski, 2017; Ju and Slingo, 1995). It is therefore apparent that the AMO could impact the ISM through interactions with the Pacific. A positive AMO leads to an interbasin zonal SST gradient because of the relative warming of the tropical Atlantic relative to the tropical Pacific. Anomalous convection in the tropical Atlantic and subsidence over the tropical Pacific reinforce La Niña–like circulation in the tropical Pacific, with positive SST anomalies in the tropical Indo-Pacific and anomalous easterly winds in the equatorial western Pacific. The global Walker circulation and the ISM circulation is modified and ISM rainfall increases (Chen et al., 2010; Lu et al., 2006; Luo et al., 2017; Monerie et al., 2019). The Atlantic can also change North Pacific climate (Zhang and Delworth, 2007) and in that way change the global meridional SST gradient. Changes in the meridional SST gradient in the Pacific can again impact the ISM (Joseph, 2014; Joseph et al., 2013; Joshi and Kucharski, 2017).

The AMO can also influence the ISM by modulating interannual variability in the tropical Atlantic and Pacific, and tropical interbasin interactions. ENSO is known to have an impact on the ISM (Ju and Slingo, 1995; Rasmusson and Carpenter, 1983; Walker, 1924; also see Chapters 2–7). The interannual variability in the tropical Atlantic, often referred to as the Atlantic Niño or Atlantic Zonal Mode, can influence the ISM as well (Kucharski et al., 2008; Pottapinjara et al., 2014; Pottapinjara et al., 2016). These tropical teleconnections with the ISM are covered in more detail in Chapter 11 of this book. ENSO and the Atlantic Niño are connected through interbasin interactions, with ENSO positively leading tropical Atlantic SST variability and the tropical Atlantic negatively leading ENSO (Rodríguez-Fonseca et al., 2009). North Atlantic temperatures have been shown to affect ENSO frequency and amplitude (Dong et al., 2006; Kang et al., 2014; Levine et al., 2017), as well as the coupling between ENSO and Atlantic Niños (Martín-Rey et al., 2014; Svendsen et al., 2013). The relative strength of the interaction between these tropical interannual modes of variability (Cai et al., 2019) can impact the interannual variability of ISM. The importance of these tropical interbasin linkages for the AMO–ISM link depends on whether the AMO is forcing a low-frequency signal in the ISM, or a low-frequency modulation of interannual variability such as longer periods with a higher or lower frequency of dry ISM years.

18.3.4 The role of external forcing

The observed records indicating a significant link between the AMO and ISM rainfall are influenced by external forcing, both natural and anthropogenic. External forcing works on the same timescales as the AMO, and it is therefore a complex exercise to separate internal variability and external forced signals in the AMO–ISM link. In recent years, as climate models become more sophisticated, evidence has emerged showing that both the AMO and ISM can be influenced by external forcing (Bellomo et al., 2018; Bollasina et al., 2011; Lau and Kim, 2017; Murphy et al., 2017; Turner and Annamalai, 2012; Vecchi et al., 2017; Wang et al., 2012).

External forcing can impact the AMO–ISM link in at least two ways. First, external radiative forcing can increase AMO variability (Bellomo et al., 2018; Murphy et al., 2017; Otto-Bliesner et al., 2016), especially in regions that are critical to induce AMO–ISM teleconnections, and therefore strengthen the AMO–ISM link (Luo et al., 2018a). Second, external radiative forcing is a common forcing for both the AMO and the ISM and could therefore lead to a statistical relationship between the AMO and ISM rainfall without a dynamical connection between the two regions.

External forcing that impacts the ISM on multidecadal timescales includes aerosol emissions (Bollasina et al., 2011) and land-use changes (Niyogi et al., 2010). Aerosols from industrialization and biomass burning have direct radiative effects on the ISM by reducing solar radiation over South Asia. This in turn can reduce the meridional tropospheric temperature gradient and cool the Indian Ocean, weakening the monsoon circulation (Lau and Kim, 2017; Ramanathan et al., 2005; Turner and Annamalai, 2012). At the same time, aerosols absorb heat in the atmosphere which can intensify ISM circulation and through aerosol–cloud interactions enhance deep convection and rainfall (Lau and Kim, 2017; Turner and Annamalai, 2012; see also Chapter 22). Furthermore, increased aerosol emissions can impact ISM rainfall by changing cloud lifetime and droplet characteristics (Turner and Annamalai, 2012). An increasing amount of anthropogenic aerosols in the atmosphere locally in South Asia could have resulted in a weak negative trend in the ISM rainfall since the 1950s (Bollasina et al., 2011), which coincided with a shift to a negative AMO.

Aerosol emissions, both anthropogenic and natural in relation to large volcanic eruptions can impact AMO strength and phasing (Otterå et al., 2010; Vecchi et al., 2017). Aerosols decrease the downward shortwave radiation through interactions with clouds and could lower North Atlantic SSTs. The North Atlantic SST response to anthropogenic aerosols in the twentieth century match the observed AMO-index in some climate models (Booth et al., 2012; Vecchi et al., 2017), but many of the other key elements of the AMO are not reproduced (Zhang et al., 2019). In addition, decadal hindcast simulations—where changes in external forcing are included—are not able to reproduce the AMO without initializing the ocean state (e.g., Kim et al., 2018; Matei et al., 2012), indicating that external forcing is not the primary driver of AMO (Zhang et al., 2019).

The concurrent responses of the AMO and the ISM to common external forcing such as aerosol emissions can lead to a correlation between the AMO and ISM indices without an actual dynamical link between the two phenomena. This would limit the prospect of skillful decadal predictions of the ISM that rely on AMO. However, model studies have identified dynamical mechanisms for the AMO–ISM link using sensitivity experiments and coupled atmosphere-ocean simulations with constant external forcing (Li et al., 2008; Lu et al., 2006, 2018a; Wang et al., 2009; Zhang and Delworth, 2006), meaning that the AMO–ISM link can exist without transient external forcing. The importance of external forcing for the AMO–ISM link remains unclear, especially because large uncertainties still exist with regard to the climate impact of aerosols (Boucher et al., 2013).

18.4   The model-simulated AMO–ISM link

Global climate models (or general circulation models) are the typical tools for investigating mechanisms of large-scale teleconnections. This section discusses how coupled climate models simulate the observed AMO–ISM link and what we can learn about the AMO–ISM link from these models. Preindustrial control simulations from the Coupled Model Intercomparison Project 5 (CMIP5) (Taylor et al., 2012) are used here for illustration (Table 18.1). For the CMIP5 models, the ISM-index is defined as rainfall over land in the region 10–30°N and 60–90°E. The AMO-index is defined as per observations. All model data have been filtered using a 10- to 70-year Butterworth bandpass filter to extract the multidecadal variability.

By using the preindustrial control simulations of CMIP5 with constant external forcing, internal variability is isolated and uncertainties with regard to external forcing are avoided. In preindustrial control simulation of CMIP5 only two of 38 models simulate significant positive correlations between the AMO and JJAS ISM rainfall (Fig. 18.5A). The correlations, although significant, are lower than the observed correlation of around 0.5. The majority of the CMIP5 models with constant forcing have no correlation between the AMO and ISM rainfall. A few more models (seven out of 38) show enhanced rainfall in the season September–November (SON) instead (Fig. 18.5B). For SON, the distribution of correlations across CMIP5 models is more skewed toward positive correlations compared to the season June-August (JJA) (Fig. 18.6). The enhanced correlation in SON rather than in JJA indicates that the AMO can be associated with a late withdrawal of the ISM (Goswami et al., 2006; Han et al., 2016). The spread in the strength of the AMO–ISM link in the CMIP5 models

TABLE 18.1 CMIP5 model names, modeling centers, and time span in years for each preindustrial control simulation.

Model ID	Modeling group	Time span (years)
ACCESS1-0	Commonwealth Scientific and Industrial Research Organization (CSIRO) and Bureau of Meteorology, (BOM), Australia	500
ACCESS1-3		500
bcc-csm1-1	Beijing Climate Center, China Meteorological Administration, China	500
bcc-csm1-1-m		400
BNU-ESM	College of Global Change and Earth System Science, Beijing Normal University, China	559
CanESM2	Canadian Centre for Climate Modelling and Analysis, Canada	996
CCSM4	National Center for Atmospheric Research, USA	501
CESM1-BGC		500
CESM1-CAM5		319
CESM1-FASTCHEM		222
CESM1-WACCM		200
CMCC-CESM	Centro Euro-Mediterraneo per I Cambiamenti Climatici, Italy	277
CMCC-CM		330
CMCC-CMS		500
CNRM-CM5	Centre National de Recherches Meteorologiques/ Centre Europeen de Recherche et Formation Avancees en Calcul Scientifique, France	850
CNRM-CM5-2		140
FIO-ESM	First Institute of Oceanography, China	800
GFDL-CM3	NOAA Geophysical Fluid Dynamics Laboratory, USA	500
GFDL-ESM2G		500
GFDL-ESM2M		500
GISS-E2-H	NASA Goddard Institute for Space Studies, USA	148
GISS-E2-R		850
HadGEM2-AO	Met Office Hadley Centre, UK	100
HadGEM2-CC		240
HadGEM2-ES		239
inmcm4	Institute of Numerical Mathematics, Russia	500

(continued)

TABLE 18.1 (Cont'd)

Model ID	Modeling group	Time span (years)
IPSL-CM5A-LR	Institute Pierre Simon Laplace, France	1000
IPSL-CM5A-MR		300
IPSL-CM5B-LR		300
MIROC-ESM	Japan Agency for Marine-Earth Science and Technology, Atmosphere and Ocean Research Institute (The University of Tokyo), and National Institute for Environmental Studies, Japan	531
MIROC-ESM-CHEM		255
MIROC5	Atmosphere and Ocean Research Institute (The University of Tokyo), National Institute for Environmental Studies, and Japan Agency for Marine-Earth Science and Technology, Japan	670
MPI-ESM-LR	Max Planck Institute for Meteorology, Germany	1000
MPI-ESM-MR		1000
MPI-ESM-P		1156
MRI-CGCM3	Meteorological Research Institute, Japan	500
NorESM1-M	Norwegian Climate Centre, Norway	501
NorESM1-ME		252

can also be seen in the regression maps for each model in Figs. 18.7 and 18.8. The earlier generation of models, CMIP3, shows similar results with a weak AMO–ISM relationship in the ensemble mean (Ting et al., 2011).

In contrast to the preindustrial control simulations, the historical CMIP5 simulations compare directly with observations because they include transient external forcing for the twentieth century. In around half the CMIP5 models, there is a positive correlation between the AMO and ISM rainfall when transient external forcing is included (Joshi and Ha, 2019; Luo et al., 2017). But only 15% (10 of 66 ensemble members from 22 different models) of these models have a significant positive correlation between the AMO and ISM rainfall that is comparable in strength to observations (Luo et al., 2017). Generally, CMIP5 models have difficulty reproducing the observed AMO–ISM link, but external forcing strengthens the link in some of the models.

Several issues with the coupled climate models have been identified to explain their failure to capture the observed AMO–ISM link. Many of the CMIP5 models, and the previous generation CMIP3, do not simulate the observed relationship between the AMO and the tropical Pacific (Joshi and Ha, 2019; Luo et al., 2017; Ting et al., 2011). In observations, positive North Atlantic SST anomalies are associated with negative equatorial Pacific SST anomalies, while in the historical CMIP5 ensemble an anomalously warm

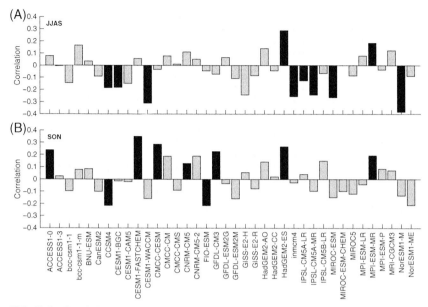

FIG. 18.5 Correlations between the AMO-index and the ISM-index in CMIP5 preindustrial control simulations for (A) JJAS and (B) SON. Black bars indicate that correlations are significant at the 90% confidence level. Indices have been linearly detrended and filtered with a 10- to 70-year third-order Butterworth bandpass filter.

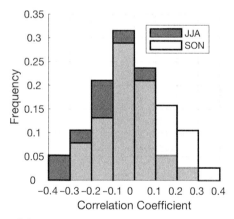

FIG. 18.6 Histogram of the correlations between the AMO-index and the ISM-index in CMIP5 preindustrial control simulations for seasons JJA (dark gray) and SON (light gray). Both indices have been linearly detrended and filtered with a 10- to 70-year third-order Butterworth bandpass filter.

North Atlantic is accompanied by an anomalously warm equatorial Pacific. These results indicate that the Pacific response to the Atlantic may be highly important for the response of the ISM to multidecadal changes in the Atlantic and that the air-sea interaction mechanisms presented in Section 18.3.3 may not be well reproduced in CMIP3 and CMIP5 models.

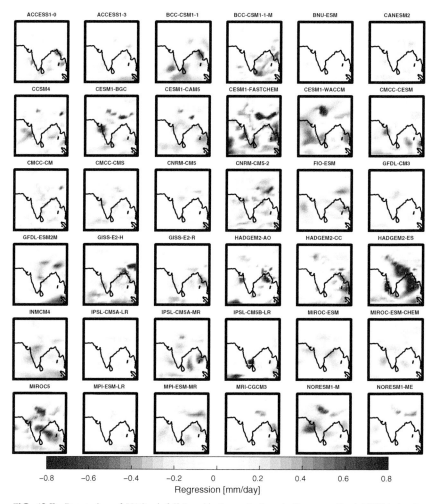

FIG. 18.7 Regression of JJAS rainfall (mm/day) over India onto the normalized AMO-index in CMIP5 preindustrial control simulations. Each panel represents one model. Values that are not significant at the 90% confidence level are set to zero. All data have been linearly detrended and filtered with a 10- to 70-year third-order Butterworth bandpass filter.

Coupled climate models also suffer from biases in their mean state (Wang et al., 2014) that could be relevant for the simulated AMO–ISM link. Tropical Atlantic SST and wind biases, and North Pacific SST biases could influence the global meridional SST gradient that has been identified as important for the AMO–ISM link (Joshi and Ha, 2019; Luo et al., 2017). In addition, most of the CMIP5 models do not reproduce the Rossby wave train response (Joshi and Ha, 2019; Luo et al., 2017) presented in Section 18.3.2 which links the AMO to the ISM, possibly due to atmospheric circulation biases. Furthermore,

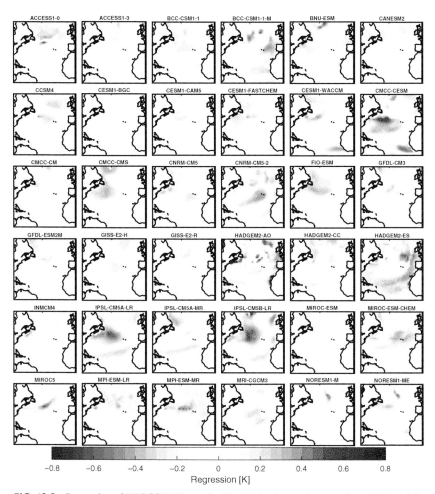

FIG. 18.8   Regression of JJAS SST (K) over the North Atlantic onto the normalized ISM rainfall-index in CMIP5 preindustrial control simulations. Each panel represents one model. Values that are not significant at the 90% confidence level are set to zero. All data have been linearly detrended and filtered with a 10- to 70-year third-order Butterworth bandpass filter.

the weak AMO–ISM link in CMIP5 and CMIP3 models could be impacted by errors in the simulation of aerosol forcing or the response to aerosol forcing in the models (Ting et al., 2011).

The spread in AMO patterns and amplitudes across models could also influence the simulation of the AMO–ISM link. AMO patterns and drivers of the AMO vary between models (Ba et al., 2014) and can change the simulated teleconnection patterns. If the AMO amplitude is too small, it might not be able to excite the far-reaching atmospheric response needed to impact the ISM. Studies also disagree about the relative importance of the tropical and the extratropical

North Atlantic part of the AMO pattern in influencing the ISM (Li et al., 2008; Sutton and Hodson, 2007). Some studies using atmosphere-only models found that the tropical North Atlantic-ISM link was the opposite of observations, where a cold (warm) tropical North Atlantic is associated with more (less) intense ISM rainfall (Li et al., 2008; Sutton and Hodson, 2007). If the tropical signal of the AMO is important for forcing the ISM response, then models need to realistically simulate the tropical coupled air-sea feedbacks that could reverse the sign of the ISM response. In addition, the tropical signal of the AMO pattern is weak in many coupled models (Joshi and Ha, 2019). The dependence of the AMO–ISM link on the AMO pattern alludes to the possibility that teleconnection patterns can vary as climate changes.

18.5 Summary and conclusions

Observations over the last century and a half show coinciding periods of positive (negative) AMO phases and more (less) ISM rainfall, resulting in a positive correlation between the AMO and the low-frequency filtered All-India monsoon rainfall index. As there is increasing evidence that the AMO could be predictable on decadal timescales, the identification of the AMO–ISM link implies that decadal predictions of the ISM might be possible.

Observational records are still very short relative to multidecadal timescales. Proxy reconstructions and model studies are therefore essential for increasing our understanding of climate variability on multidecadal timescales. In proxy reconstructions of past climate, large climatic events seem to connect the Atlantic and the ISM, but it is not clear yet if smaller multidecadal variations were related prior to the instrumental records. More proxy reconstructions of the ISM to extract large-scale climate variability from local rainfall signals will be beneficial to further our understanding of the past AMO–ISM link.

Our main tool to test mechanisms connecting the AMO to the ISM are climate models. Several mechanisms have been proposed, including shifts of the ITCZ, atmospheric teleconnections, and interbasin linkages between the Atlantic and the Pacific. Some climate models are able to simulate such mechanisms. Most coupled atmosphere-ocean models do not reproduce the observed AMO–ISM link, and different mechanisms are found to be dominating the link in the individual models. However, several of the proposed mechanisms connecting the AMO with the ISM implicate the Pacific and Indian Oceans. Taking into account interbasin interactions between the North Atlantic and the Indo-Pacific Oceans and improving the representation of these interactions in coupled climate models seem imperative for fully understanding the AMO–ISM link.

In addition, global climate models still have issues in simulating cumulus convection as it is not spatially resolved. A basic dilemma with teleconnection studies is the need for a global perspective while focusing on smaller scale regional climate. Global coupled climate models simulate a wide range of mean

ISM precipitation patterns explaining some of the spread in the modeling results (Jain et al., 2019; Mishra et al., 2018; Wang et al., 2018). As the horizontal and vertical resolution in global coupled climate models increases, we can envisage that the teleconnection between the AMO and the ISM will become clearer. Similarly, the role of external forcing in general, and aerosol forcing in particular, for the AMO–ISM link is still unclear. External forcing can strengthen the AMO–ISM link in some models; however the climate response and sensitivity to aerosol forcing still exhibits large uncertainties (Boucher et al., 2013) and leads to uncertainties in large-scale teleconnection patterns. Furthermore, these uncertainties open up for the possibility that the AMO–ISM link is a statistical link through the common driver of external radiative forcing rather than a dynamical link where the AMO forces multidecadal variability in the ISM.

   In recent decades, the observed AMO and ISM seem to have evolved independently (Luo et al., 2018b; Sankar et al., 2016). This calls into question whether there is any potential for decadal predictions of the ISM based on the AMO–ISM link. Mechanistical studies suggest that the AMO has potential to impact the ISM, but the AMO impact on the ISM may be competing with the anthropogenic control on the ISM (Maharana and Dimri, 2019) through aerosols and changes in the Walker and Hadley circulations due to global warming (Vecchi et al., 2006). Improving our comprehension of aerosol effects and the simulation of precipitation in global models, and taking into account interbasin interactions seem especially important to further our understanding of the AMO–ISM link. This is underscored by the prospect of skillful AMO predictions in the near future.

Acknowledgments

I thank Stephanie Gleixner and two anonymous reviewers for helpful and constructive comments improving this chapter. This work was funded by the Trond Mohn Foundation through the Bjerknes Climate Prediction Unit (Bjerknes CPU) project (number BFS2018TMT01). I acknowledge the World Climate Research Programme's Working Group on Coupled Modelling, which is responsible for CMIP, and I thank the climate modeling groups (listed in Table 18.1) for producing and making available their model output. The HadISST dataset was obtained from https://www.metoffice.gov.uk/hadobs/hadisst, the CRU precipitation data from the NCAS British Atmospheric Data Centre, and the GISTEMP data from https://data. giss.nasa.gov/gistemp. I also acknowledge the India Meteorological Department for the IMD gridded rainfall data and their publicly available end of season reports.

References

Ba, J., Keenlyside, N.S., Latif, M., Park, W., Ding, H., Lohmann, K., Mignot, J., Menary, M., Otterå, O.H., Wouters, B., Salas Y Melia, D., Oka, A., Bellucci, A., Volodin, E., 2014. A multi-model comparison of Atlantic multidecadal variability. Clim. Dyn. 43, 2333–2348.

Bellomo, K., Murphy, L.N., Cane, M.A., Clement, A.C., Polvani, L.M., 2018. Historical forcings as main drivers of the Atlantic multidecadal variability in the CESM large ensemble. Clim. Dyn. 50, 3687–3698.

Berkelhammer, M., Sinha, A., Mudelsee, M., Cheng, H., Edwards, R.L., Cannariato, K., 2010. Persistent multidecadal power of the Indian summer monsoon. Earth Planet. Sci. Lett. 290, 166–172.

Bollasina, M.A., Ming, Y., Ramaswamy, V., 2011. Anthropogenic aerosols and the weakening of the South Asian summer monsoon. Science 334, 502–505.

Booth, B.B.B., Dunstone, N.J., Halloran, P.R., Andrews, T., Bellouin, N., 2012. Aerosols implicated as a prime driver of twentieth-century North Atlantic climate variability. Nature 484, 228–232.

Borgaonkar, H.P., Sikder, A.B., Ram, S., Pant, G.B., 2010. El Niño and related monsoon drought signals in 523-year-long ring width records of teak (*Tectona grandis* L.F.) trees from south India. Palaeogeogr. Palaeoclimatol. Palaeoecol. 285, 74–84.

Boucher, O., Randall, D., Artaxo, P., Bretherton, C., Feingold, G., Forster, P., Kerminen, V.-M., Kondo, Y., Liao, H., Lohmann, U., Rasch, P., Satheesh, S.K., Sherwood, S., Stevens, B., Zhang, X.Y., 2013. Clouds and aerosols Climate Change. In: Stocker, T.F., Qin, D., Plattner, G.-K., Tignor, M., Allen, S.K., Boschung, J., Nauels, A., Xia, Y., Bex, V., Midgley, P.M. (Eds.), 2013:The Physical Science Basis. Contribution of Working Group I to the Fifth Assessment Report of the Intergovernmental Panel on Climate Change. Cambridge: Cambridge University Press.

Broccoli, A.J., Dahl, K.A., Stouffer, R.J., 2006. Response of the ITCZ to Northern Hemisphere cooling. Geophys. Res. Lett., 33.

Burns, S.J., Fleitmann, D., Matter, A., Kramers, J., Al-Subbary, A.A., 2003. Indian Ocean climate and an absolute chronology over Dansgaard/Oeschger events 9 to 13. Science 301, 1365–1367.

Cai, W., Wu, L., Lengaigne, M., Li, T., Mcgregor, S., Kug, J.-S., Yu, J.-Y., Stuecker, M.F., Santoso, A., Li, X., Ham, Y.-G., Chikamoto, Y., Ng, B., Mcphaden, M.J., Du, Y., Dommenget, D., Jia, F., Kajtar, J.B., Keenlyside, N., Lin, X., Luo, J.-J., Martín-Rey, M., Ruprich-Robert, Y., Wang, G., Xie, S.-P., Yang, Y., Kang, S.M., Choi, J.-Y., Gan, B., Kim, G.-I., Kim, C.-E., Kim, S., Kim, J.-H., Chang, P., 2019. Pantropical climate interactions. Science 363, eaav4236.

Chakraborty, S., Goswami, B.N., Dutta, K., 2012. Pacific coral oxygen isotope and the tropospheric temperature gradient over the Asian monsoon region: a tool to reconstruct past Indian summer monsoon rainfall. J. Quat. Sci. 27, 269–278.

Chen, W., Dong, B.W., Lu, R.Y., 2010. Impact of the Atlantic Ocean on the multidecadal fluctuation of El Nino-southern oscillation-South Asian monsoon relationship in a coupled general circulation model. J. Geophys. Res., 115. doi:10.1029/2009JD013596.

Clement, A., Bellomo, K., Murphy, L.N., Cane, M.A., Mauritsen, T., Rädel, G., Stevens, B., 2015. The Atlantic multidecadal oscillation without a role for ocean circulation. Science 350, 320–324.

Colin, C., Kissel, C., Blamart, D., Turpin, L., 1998. Magnetic properties of sediments in the Bay of Bengal and the Andaman Sea: impact of rapid North Atlantic Ocean climatic events on the strength of the Indian monsoon. Earth Planet. Sci. Lett. 160, 623–635.

Dong, B., Sutton, R.T., Scaife, A.A., 2006. Multidecadal modulation of El Niño–Southern Oscillation (ENSO) variance by Atlantic Ocean sea surface temperatures. Geophy. Res. Lett., 33. doi:10.1029/2006GL025766.

Enfield, D.B., Mestas-Nunez, A.M., Trimble, P.J., 2001. The Atlantic multidecadal oscillation and its relation to rainfall and river flows in the continental US. Geophys. Res. Lett. 28, 2077–2080.

Feng, S., Hu, Q., 2008. How the North Atlantic multidecadal oscillation may have influenced the Indian summer monsoon during the past two millennia. Geophys. Res. Lett., 35. doi:10.1029/2007GL032484.

Frankignoul, C., Hasselmann, K., 1977. Stochastic climate models, Part II Application to sea-surface temperature anomalies and thermocline variability. Tellus 29, 289–305.

Frierson, D.M.W., Hwang, Y.-T., 2012. Extratropical influence on ITCZ shifts in slab ocean simulations of global warming. J. Clim. 25, 720–733.

Gadgil, S., Vinayachandran, P.N., Francis, P.A., 2003. Droughts of the Indian summer monsoon: role of clouds over the Indian Ocean. Curr. Sci. 85, 1713–1719.

Gautam, P.K., Narayana, A.C., Band, S.T., Yadava, M.G., Ramesh, R., Wu, C.-C., Shen, C.-C., 2019. High-resolution reconstruction of Indian summer monsoon during the Bølling-Allerød from a central Indian stalagmite. Palaeogeogr. Palaeoclimatol. Palaeoecol. 514, 567–576.

Goswami, B.N., Madhusoodanan, M.S., Neema, C.P., Sengupta, D., 2006. A physical mechanism for North Atlantic SST influence on the Indian summer monsoon. Geophys. Res. Lett. 33. doi:10.1029/2005GL024803.

Gray, S.T., Graumlich, L.J., Betancourt, J.L., Pederson, G.T., 2004. A tree-ring based reconstruction of the Atlantic multidecadal oscillation since 1567 AD. Geophys. Res. Lett. 31. doi:10.102 9/2004GL019932.

Guhathakurta, P., Rajeevan, M., 2008. Trends in the rainfall pattern over India. Int. J. Climatol. 28, 1453–1469.

Guhathakurta, P., Rajeevan, M., Sikka, D.R., Tyagi, A., 2015. Observed changes in southwest monsoon rainfall over India during 1901–2011. Int. J. Climatol. 35, 1881–1898.

Han, Z., Luo, F., Li, S., Gao, Y., Furevik, T., Svendsen, L., 2016. Simulation by CMIP5 models of the Atlantic multidecadal oscillation and its climate impacts. Adv. Atmos. Sci. 33, 1329–1342.

Harris, I., Jones, P.D., Osborn, T.J., Lister, D.H., 2014. Updated high-resolution grids of monthly climatic observations – the CRU TS3. Dataset. Int. J. Climatol. 34, 623–642.

Hasselmann, K., 1976. Stochastic climate models Part I. Theory. Tellus 28, 473–485.

Hong, Y.T., Hong, B., Lin, Q.H., Zhu, Y.X., Shibata, Y., Hirota, M., Uchida, M., Leng, X.T., Jiang, H.B., Xu, H., Wang, H., Yi, L., 2003. Correlation between Indian Ocean summer monsoon and North Atlantic climate during the Holocene. Earth Planet. Sci. Lett. 211, 371–380.

Jain, S., Salunke, P., Mishra, S.K., Sahany, S., 2019. Performance of CMIP5 models in the simulation of Indian summer monsoon. Theor. Appl. Climatol. 137, 1429–1447.

Joseph, P.V., 1976. Climate change in monsoon and cyclones 1891 to 1974, Proc. Symp. 'Tropical Monsoon. Indian Institute of Tropical Meteorology, pp. 378–387.

Joseph, P.V., 2014. Role of ocean in the variability of Indian summer monsoon rainfall. Surv. Geophys. 35, 723–738.

Joseph, P.V., Gokulapalan, B., Nair, A., Sheela Wilson, S., 2013. Variability of summer monsoon rainfall in India on inter-annual and decadal time scales. Atmos. Oceanic Sci. Lett. 6, 398–403.

Joshi, M.K., Ha, K.-J., 2019. Fidelity of CMIP5-simulated teleconnection between Atlantic multidecadal oscillation and Indian summer monsoon rainfall. Clim. Dyn. 52, 4157–4176.

Joshi, M.K., Kucharski, F., 2017. Impact of interdecadal Pacific oscillation on Indian summer monsoon rainfall: an assessment from CMIP5 climate models. Clim. Dyn. 48, 2375–2391.

Ju, J., Slingo, J., 1995. The Asian summer monsoon and ENSO. Quart. J. R. Meteorol. Soc. 121, 1133–1168.

Kang, I.-S., No, H.-H., Kucharski, F., 2014. ENSO amplitude modulation associated with the mean SST changes in the Tropical Central Pacific induced by Atlantic multidecadal oscillation. J. Clim. 27, 7911–7920.

Kang, S.M., Held, I.M., Frierson, D.M.W., Zhao, M., 2008. The response of the ITCZ to extratropical thermal forcing: idealized slab-ocean experiments with a GCM. J. Clim. 21, 3521–3532.

Keenlyside, N.S., Latif, M., Jungclaus, J., Kornblueh, L., Roeckner, E., 2008. Advancing decadal-scale climate prediction in the North Atlantic sector. Nature 453, 84–88.

Kim, H.-M., Webster, P.J., Curry, J.A., 2012. Evaluation of short-term climate change prediction in multi-model CMIP5 decadal hindcasts. Geophys. Res. Lett. 39. doi:10.1029/2012GL051644.

Kim, W.M., Yeager, S.G., Danabasoglu, G., 2018. Key Role of internal ocean dynamics in Atlantic multidecadal variability during the last half century. Geophys. Res. Lett. 45 (13), 449-13,457.

Knight, J.R., Folland, C.K., Scaife, A.A., 2006. Climate impacts of the Aatlantic multidecadal oscillation. Geophys. Res. Lett. 33. doi:10.1029/2006GL026242.

Kucharski, F., Bracco, A., Yoo, J.H., Molteni, F., 2008. Atlantic forced component of the Indian monsoon interannual variability. Geophys. Res. Lett. 35. doi:10.1029/2007GL033037.

Lau, W.K.-M., Kim, K.-M., 2017. Competing influences of greenhouse warming and aerosols on Asian summer monsoon circulation and rainfall. Asia-Pac. J. Atmospheric Sci. 53, 181–194.

Lenssen, N.J.L., Schmidt, G.A., Hansen, J.E., Menne, M.J., Persin, A., Ruedy, R., Zyss, D., 2019. Improvements in the GISTEMP uncertainty model. J. Geophys. Res. 124, 6307–6326.

Levine, A.F.Z., Mcphaden, M.J., Frierson, D.M.W., 2017. The impact of the AMO on multidecadal ENSO variability. Geophys. Res. Lett. 44, 3877–3886.

Li, S.L., Perlwitz, J., Quan, X.W., Hoerling, M.P., 2008. Modelling the influence of North Atlantic multidecadal warmth on the Indian summer rainfall. Geophys. Res. Lett. 35. doi:10.1029/2 007GL032901.

Lu, R.Y., Dong, B.W., Ding, H., 2006. Impact of the Atlantic multidecadal oscillation on the Asian summer monsoon. Geophys. Res. Lett. 33. doi:10.1029/2006GL027655.

Luo, F., Li, S., Gao, Y., Keenlyside, N., Svendsen, L., Furevik, T., 2017. The connection between the Atlantic multidecadal oscillation and the Indian summer monsoon in CMIP5 models. Clim. Dyn 51, 3023–3039.

Luo, F., Li, S., Gao, Y., Svendsen, L., Furevik, T., Keenlyside, N., 2018a. The connection between the Atlantic multidecadal oscillation and the Indian summer monsoon since the industrial revolution is intrinsic to the climate system. Environ. Res. Lett. 13 094020.

Luo, F.-F., Li, S., Furevik, T., 2018b. Weaker connection between the Atlantic multidecadal oscillation and Indian summer rainfall since the mid-1990s. Atmos. Oceanic Sci. Lett. 11, 37–43.

Luo, F.F., Li, S.L., Furevik, T., 2011. The connection between the Atlantic multidecadal oscillation and the Indian summer monsoon in Bergen climate model version 2.0. J. Geophys. Res. 116. doi:10.1029/2011JD015848.

Maharana, P., Dimri, A.P., 2019. The Indian monsoon: past, present and future. Proc. Indian Natl. Sci. Acad. 85, 403–420.

Mann, M.E., Zhang, Z.H., Rutherford, S., Bradley, R.S., Hughes, M.K., Shindell, D., Ammann, C., Faluvegi, G., Ni, F.B., 2009. Global signatures and dynamical origins of the little ice age and medieval climate anomaly. Science 326, 1256–1260.

Martín-Rey, M., Rodríguez-Fonseca, B., Polo, I., Kucharski, F., 2014. On the Atlantic–Pacific Niños connection: a multidecadal modulated mode. Clim. Dyn. 43, 3163–3178.

Matei, D., Pohlmann, H., Jungclaus, J., Müller, W., Haak, H., Marotzke, J., 2012. Two tales of initializing decadal climate prediction experiments with the ECHAM5/MPI-OM model. J. Clim. 25, 8502–8523.

Miles, M.W., Divine, D.V., Furevik, T., Jansen, E., Moros, M., Ogilvie, A.E.J., 2014. A signal of persistent Atlantic multidecadal variability in Arctic sea ice. Geophys. Res. Lett. 41, 463–469.

Mishra, S.K., Sahany, S., Salunke, P., Kang, I.-S., Jain, S., 2018. Fidelity of CMIP5 multi-model mean in assessing Indian monsoon simulations. Clim. Atmos. Sci. 1, 39.

Mohtadi, M., Prange, M., Steinke, S., 2016. Palaeoclimatic insights into forcing and response of monsoon rainfall. Nature 533, 191–199.

Monerie, P.-A., Robson, J., Dong, B., Hodson, D.L.R., Klingaman, N.P., 2019. Effect of the Atlantic multidecadal variability on the global monsoon. Geophys. Res. Lett. 46, 1765–1775.

Murphy, L.N., Bellomo, K., Cane, M., Clement, A., 2017. The role of historical forcings in simulating the observed Atlantic multidecadal oscillation. Geophys. Res. Lett. 44, 2472–2480.

Nath, R., Luo, Y., 2019. Disentangling the influencing factors driving the cooling trend in boreal summer over Indo-Gangetic river basin, India: role of Atlantic multidecadal oscillation (AMO). Theor. Appl. Climatol. 138, 1–12.

Niranjan Kumar, K., Rajeevan, M., Pai, D.S., Srivastava, A.K., Preethi, B., 2013. On the observed variability of monsoon droughts over India. Weather Clim. Extremes 1, 42–50.

Niyogi, D., Kishtawal, C., Tripathi, S., Govindaraju, R.S., 2010. Observational evidence that agricultural intensification and land use change may be reducing the Indian summer monsoon rainfall. Water Resour. Res. 46. https://doi.org/10.1029/2008WR007082.

Otterå, O.H., Bentsen, M., Drange, H., Suo, L., 2010. External forcing as a metronome for Atlantic multidecadal variability. Nat. Geosci. 3, 688–694.

Otto-Bliesner, B.L., Brady, E.C., Fasullo, J., Jahn, A., Landrum, L., Stevenson, S., Rosenbloom, N., Mai, A., Strand, G., 2016. Climate variability and change since 850 CE: an ensemble approach with the community earth system model. Bull. Am. Meteorol. Soc. 97, 735–754.

Parthasarathy, B., Kumar, K.R., Munot, A.A., 1993. Homogeneous Indian monsoon rainfall: variability and prediction. Proc. Indian Acad. Sci. Earth Planet. Sci. 102, 121–155.

Parthasarathy, B., Munot, A.A., Kothawale, D.R., 1994. All-India monthly and seasonal rainfall series: 1871–1993. Theor. Appl. Climatol. 49, 217–224.

Pottapinjara, V., Girishkumar, M.S., Ravichandran, M., Murtugudde, R., 2014. Influence of the Atlantic zonal mode on monsoon depressions in the Bay of Bengal during boreal summer. J. Geophys. Res. 119, 6456–6469.

Pottapinjara, V., Girishkumar, M.S., Sivareddy, S., Ravichandran, M., Murtugudde, R., 2016. Relation between the upper ocean heat content in the equatorial Atlantic during boreal spring and the Indian monsoon rainfall during June–September. Int. J. Climatol. 36, 2469–2480.

Rajeevan, M., Bhate, J., Kale, J.A., Lal, B., 2006. High resolution daily gridded rainfall data for the Indian region: Analysis of break and active monsoon spells. Curr. Sci. 91, 296–306.

Ramanathan, V., Chung, C., Kim, D., Bettge, T., Buja, L., Kiehl, J.T., Washington, W.M., Fu, Q., Sikka, D.R., Wild, M., 2005. Atmospheric brown clouds: impacts on South Asian climate and hydrological cycleProc. Natl. Acad. Sci. USA102, 5326–5333.

Rasmusson, E.M., Carpenter, T.H., 1983. The relationship between eastern equatorial Pacific sea surface temperatures and rainfall over India and Sri Lanka. Mon. Weather Rev. 111, 517–528.

Rayner, N.A., Parker, D.E., Horton, E.B., Folland, C.K., Alexander, L.V., Rowell, D.P., Kent, E.C., Kaplan, A., 2003. Global analyses of sea surface temperature, sea ice, and night marine air temperature since the late nineteenth century. J. Geophys. Res. 108. https://doi.org/10.1029/2002JD002670.

Rodríguez-Fonseca, B., Polo, I., García-Serrano, J., Losada, T., Mohino, E., Mechoso, C.R., Kucharski, F., 2009. Are Atlantic Niños enhancing Pacific ENSO events in recent decades? Geophys. Res. Lett. 36. doi:10.1029/2009GL040048.

Sankar, S., Svendsen, L., Gokulapalan, B., Joseph, P.V., Johannessen, O.M., 2016. The relationship between Indian summer monsoon rainfall and Atlantic multidecadal variability over the last 500 years. Tellus A 68, 31717.

Schlesinger, M.E., Ramankutty, N., 1994. An oscillation in the global climate system of period 65-70 years. Nature 367, 723–726.

Shi, H., Wang, B., Liu, J., Liu, F., 2019. Decadal–multidecadal variations of Asian summer rainfall from the little ice age to the present. J. Clim. 32, 7663–7674.

Sinha, A., Cannariato, K.G., Stott, L.D., Cheng, H., Edwards, R.L., Yadava, M.G., Ramesh, R., Singh, I.B., 2007. A 900-year (600 to 1500 A.D.) record of the Indian summer monsoon precipitation from the core monsoon zone of India. Geophys. Res. Lett. 34, L16707.

Sontakke, N.A., Singh, N., Singh, H.N., 2008. Instrumental period rainfall series of the Indian region (AD 1813—2005): revised reconstruction, update and analysis. The Holocene 18, 1055–1066.

Sutton, R.T., Hodson, D.L.R., 2007. Climate response to basin-scale warming and cooling of the North Atlantic Ocean. J. Clim. 20, 891–907.

Svendsen, L., Hetzinger, S., Keenlyside, N., Gao, Y., 2014. Marine-based multiproxy reconstruction of Atlantic multidecadal variability. Geophys. Res. Lett. 41, 2013GL059076.

Svendsen, L., Kvamstø, N., Keenlyside, N., 2013. Weakening AMOC connects equatorial Atlantic and Pacific interannual variability. Clim. Dyn. 43, 2931–2941.

Taylor, K.E., Stouffer, R.J., Meehl, G.A., 2012. An overview of CMIP5 and the experiment design. Bull. Am. Meteorol. Soc. 93, 485–498.

Ting, M.F., Kushnir, Y., Seager, R., Li, C.H., 2011. Robust features of Atlantic multi-decadal variability and its climate impacts. Geophys. Res. Lett. 38. doi:10.1029/2011GL048712.

Turner, A.G., Annamalai, H., 2012. Climate change and the South Asian summer monsoon. Nat. Clim. Change 2, 587–595.

Vecchi, G.A., Delworth, T.L., Booth, B., 2017. Origins of Atlantic decadal swings. Nature 548, 284–285.

Vecchi, G.A., Soden, B.J., Wittenberg, A.T., Held, I.M., Leetmaa, A., Harrison, M.J., 2006. Weakening of tropical Pacific atmospheric circulation due to anthropogenic forcing. Nature 441, 73–76.

Walker, G.T., 1924. Correlation in seasonal variations of weather, IX. A further study of world weather. Mem. India Meteorol. Dept. 24, 275–333.

Wang, B., Liu, J., Kim, H.-J., Webster, P.J., Yim, S.-Y., 2012. Recent change of the global monsoon precipitation (1979–2008). Clim. Dyn. 39, 1123–1135.

Wang, C., Zhang, L., Lee, S.-K., Wu, L., Mechoso, C.R., 2014. A global perspective on CMIP5 climate model biases. Nat. Clim. Change 4, 201–205.

Wang, Y.M., Li, S.L., Luo, D.H., 2009. Seasonal response of Asian monsoonal climate to the Atlantic multidecadal oscillation. J. Geophys. Res. 114. doi:10.1029/2008JD010929.

Wang, Z., Li, G., Yang, S., 2018. Origin of Indian summer monsoon rainfall biases in CMIP5 multimodel ensemble. Clim. Dyn. 51, 755–768.

Yadava, M.G., Ramesh, R., Pant, G.B., 2004. Past monsoon rainfall variations in peninsular India recorded in a 331-year-old speleothem. The Holocene 14, 517–524.

Yeager, S.G., Robson, J.I., 2017. Recent progress in understanding and predicting Atlantic decadal climate variability. Curr. Clim. Change Rep. 3, 112–127.

Zhang, R., Delworth, T.L., 2005. Simulated tropical response to a substantial weakening of the Atlantic thermohaline circulation. J. Clim. 18, 1853–1860.

Zhang, R., Delworth, T.L., 2006. Impact of Atlantic multidecadal oscillations on India/Sahel rainfall and Atlantic hurricanes. Geophys. Res. Lett. 33. https://doi.org/10.1029/2006GL026267.

Zhang, R., Delworth, T.L., 2007. Impact of the Atlantic multidecadal oscillation on North Pacific climate variability. Geophys. Res. Lett. 34. https://doi.org/10.1029/2007GL031601.

Zhang, R., Delworth, T.L., Sutton, R., Hodson, D.L.R., Dixon, K.W., Held, I.M., Kushnir, Y., Marshall, J., Ming, Y., Msadek, R., Robson, J., Rosati, A.J., Ting, M., Vecchi, G.A., 2013. Have aerosols caused the observed Atlantic multidecadal variability? J. Atmos. Sci. 70, 1135–1144.

Zhang, R., Sutton, R., Danabasoglu, G., Kwon, Y.-O., Marsh, R., Yeager, S.G., Amrhein, D.E., Little, C.M., 2019. A review of the role of the Atlantic meridional overturning circulation in Atlantic multidecadal variability and associated climate impacts. Rev. Geophys. 57, 316–375.

Chapter 19

Indian summer monsoon and its teleconnection with Pacific decadal variability

Manish K. Joshi[a]**, Fred Kucharski**[b,c]**, Archana Rai**[a]**, Ashwini Kulkarni**[a]

[a]*Indian Institute of Tropical Meteorology, Ministry of Earth Sciences, Pune, Maharashtra, India,* [b]*Earth System Physics Section, Abdus Salam International Centre for Theoretical Physics, Trieste, Italy,* [c]*Center of Excellence for Climate Change Research/Department of Meteorology, King Abdulaziz University, Jeddah, Saudi Arabia*

19.1   Introduction

The Indian summer monsoon rainfall (ISMR) has a significant impact on the country's economy and agriculture. Therefore, the accurate prediction of the ISMR has immense importance that helps society to cope with monsoon's adverse effects (i.e., floods and droughts). Predicting monsoon rainfall is a challenging task, as it encompasses various climate forcings such as natural variability and anthropogenic changes. The natural variability is of two types: First is the internal variability, which is due to the natural internal processes within the climate system and the examples of internally generated variability on years to decadal timescales are the El Niño–Southern Oscillation (ENSO), the interdecadal Pacific oscillation (IPO)/Pacific decadal oscillation (PDO), and the Atlantic multidecadal oscillation (AMO) and on longer timescales is the thermohaline circulation. Second is the external variability, which is due to the forcing agents outside the climate system and the examples of such externally generated variability are orbital variations (i.e., the variations in the Earth's eccentricity, changes in the tilt angle of the Earth's axis of rotation, and precession of the Earth's axis), solar radiation, and volcanic eruptions. On the other hand, anthropogenic changes are caused by human beings, and the best known example is the increase in greenhouse gases (GHGs) emissions.

The Indian summer monsoon (ISM) fluctuates on various timescales, mainly intraseasonal, interannual, and interdecadal. The interannual variability of ISMR is mainly related to variations in sea surface temperatures (SSTs) over the equatorial Pacific (Sikka, 1980; Rasmusson and Carpenter, 1983; Krishna Kumar et al., 1999; Mujumdar et al., 2017), the Indian Ocean (Saji et al.,

Indian Summer Monsoon Variability: El Niño-teleconnections and beyond.
DOI: https://doi.org/10.1016/B978-0-12-822402-1.00018-1

1999), and the tropical South Atlantic (Kucharski and Joshi, 2017). So far, the predictability of ISMR mainly relied on ENSO and, to some extent, on Indian Ocean dipole and other factors. However, some predictability may arise from a better understanding of the relationship between other predictors such as AMO, PDO/IPO, and their incorporation into the forecast system.

In nature, the climate variability at decadal-to-multidecadal timescales has been recognized in the Atlantic (Kushnir, 1994; Schlesinger and Ramankutty, 1994, 1995; Kerr, 2000) as well as in the Pacific (Mantua et al., 1997; Power et al., 1998, 1999; Folland et al., 1999; Allan, 2000) Oceans. In the Atlantic Ocean, it has been referred to as the Atlantic multidecadal variability (see Chapter 18), whereas in the Pacific Ocean, it is termed as the Pacific decadal variability. Recent studies have furnished substantial evidence of Pacific decadal variability in causing persistent climate anomalies over different parts of the globe (Dai, 2013; Dong and Dai, 2015; Villamayor and Mohino, 2015; Joshi and Rai, 2015; Joshi and Kucharski, 2017).

Among different timescales, the ISM exhibits significant interdecadal variability. The ISMR shows alternate epochs of above- and below-normal rainfall on an interdecadal basis (see Fig. 19.1), each lasting for about three decades (Kripalani et al., 1997; Krishnamurthy and Goswami, 2000; Goswami, 2005; Joshi and Rai, 2015). However, different mechanisms for the epochal swings were proposed by previous studies. For example, the interdecadal variability of the ISM is strongly linked with the amplitude modulation of the ENSO (Krishnamurthy and Goswami, 2000), the Atlantic multidecadal variability (Goswami et al., 2006; Joshi and Pandey, 2011; Joshi and Rai, 2015; Joshi and Ha, 2019), and the Pacific decadal variability (Krishnan and Sugi, 2003; Joshi and Pandey, 2011; Krishnamurthy and Krishnamurthy, 2014; Joshi and Rai, 2015; Joshi and Kucharski, 2017).

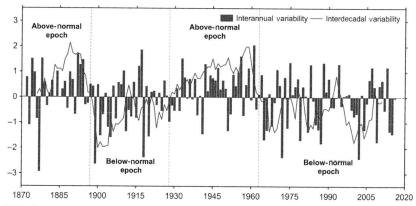

FIG. 19.1 Interannual and interdecadal (11-year moving averaged) variability of Indian summer monsoon rainfall (JJAS). Interannual and interdecadal anomalies are normalized by their own standard deviation.

In this chapter, an attempt has been made to understand the observed relationship between the ISM and Pacific decadal variability. Further, we will endeavor to unravel the plausible physical mechanism associated with ISM and Pacific decadal variability. A better understanding of the connection between Pacific decadal variability and ISM is essential, as it might help to improve the monsoon predictability, which would be crucial for policymakers to know the climate trajectory in the upcoming decades for the applications to water resources, agriculture, energy, and infrastructure development.

19.2   Data and methods

19.2.1   Data

To examine the spatial structure and temporal variability of IPO, the National Oceanic and Atmospheric Administration (NOAA) Extended Reconstructed Sea Surface Temperature version 5 (ERSST.v5; Huang et al., 2017) monthly data (resolution $2° \times 2°$) is obtained from https://www.ncdc.noaa.gov for the period 1891–2018. The $1° \times 1°$ gridded rainfall data from the India Meteorological Department (IMD; Rajeevan et al., 2006) is used for scrutinizing the IPO-related precipitation patterns for the period 1901–2008. Based on 1384 stations, this dataset minimizes the risk of generating temporal inhomogeneity due to varying station densities by considering only those stations that have at least a minimum of 70% of data availability. The multistage quality control is applied to observed station data before interpolating it into regular $1° \times 1°$ latitude/longitude grids (Rajeevan et al., 2006). For interpolating station rainfall data into regular grids, Shepard's interpolation scheme (Shepard, 1968) is used. The geographical area covered in this data is only within the India's political boundaries. For illustrating the epochal changes of ISMR, the longest rainfall dataset (Parthasarathy et al., 1994) for the period 1871–2016 is used.

To investigate the large-scale features allied with IPO, the monthly means of horizontal winds (i.e., zonal and meridional) and sea-level pressure (SLP) for the period 1948–2008 are obtained from the National Center for Environmental Prediction (NCEP)/National Center for Atmospheric Research (NCAR) atmospheric reanalysis V1 (Kalnay et al., 1996).

19.2.2   Methods

First, the annual mean SST anomalies (SSTAs) have been computed for the period 1891–2018. The computed annual mean SSTAs are then detrended to remove the global component of the anthropogenic forcing. As the focus of the present study is to scrutinize the IPO-ISMR teleconnection on a multidecadal basis; therefore, the obtained detrended annual mean SSTAs are filtered using Butterworth low-pass filter of order 4 and 21-year cut-off frequency. Due to the low-pass filter's end effects, there is an uncertainty in defining the low-pass filtered time series at the ends; therefore, the first and last 10 points of the filtered data are ignored in the analysis (Joshi and Rai, 2015). Thus, in the present study, IPO is defined as the pattern (Fig. 19.2A) and time series (Fig. 19.2B) of the first empirical orthogonal function (EOF-1) of low-pass filtered detrended annual mean SSTAs computed over the Pacific basin (45°S to 60°N, 140°E to 80°W) for the period 1901–2008. In many previous studies (e.g., Zhang et al., 1997; Power et al., 1999; Krishnan and Sugi, 2003; Meehl et al., 2013) the decadal (10–20 year) variations were often retained, but this chapter primarily focuses on the multidecadal variations from the entire Pacific.

The analysis is mainly based on regression maps, obtained by linearly regressing the field of interest onto the standardized index (Joshi and Kucharski, 2017; Kucharski and Joshi, 2017; Joshi and Ha, 2019; Joshi et al., 2020 Joshi et al., 2021). Thus, the units of regression maps are the same as that of the field used for regression. Before doing regression analysis, all

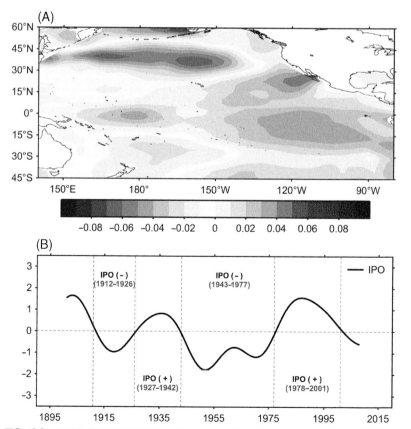

FIG. 19.2 (A) The first EOF (EOF-1) of the low-pass filtered detrended annual mean SSTAs computed over the Pacific basin (45°S to 60°N, 140°E to 80°W) and (B) the time series of the associated first principal component (PC-1).

fields are linearly detrended so that the trends in the data do not influence the results. The statistical significance of regression coefficients is assessed via a two-tailed Student's t-test.

19.3 Pacific decadal variability and Indian summer monsoon

19.3.1 Pacific decadal variability

The natural internal variability at decadal-to-multidecadal timescales in the Pacific Ocean is termed as the Pacific decadal variability, which is generally referred as the PDO (Mantua et al., 1997) for the North Pacific pattern, or the IPO (Power et al., 1998, 1999; Folland et al., 1999; Allan, 2000) for the basin-wide pattern.

The PDO is defined as the pattern and time series of leading EOF of monthly SSTAs over the North Pacific (north of 20°N), after removing the global mean SST anomaly (Mantua et al., 1997; Deser et al., 2004) and is characterized as cold SSTAs in the central North Pacific surrounded by warm anomalies along the west coast of North America. On the other hand, the IPO, whose precise definition varies (Dai, 2013; Dong and Dai, 2015; Villamayor and Mohino, 2015; Joshi and Kucharski, 2017), but is typically defined as the pattern and time series of the first EOF or EOF-1 of low-pass filtered detrended annual mean SSTAs computed over the Pacific basin (for details see Section 19.2.2) and is characterized as warm SSTAs in the tropical Pacific as well as in the north and south of the eastern part of the basin and cold SSTAs in the western part of the basin, poleward of 25°, and more prominent in the Northern Hemisphere. The time series of IPO is analogous to the PDO index of Mantua et al. (1997); for details, see Henley et al. (2015). According to the IPCC (2007), both the IPO and PDO are described as a long-lived El Niño–like pattern of Indo-Pacific climate variability or low-frequency residual of ENSO variability on multidecadal timescales.

The spatial signatures of IPO and ENSO are qualitatively similar, having large SSTAs in the tropical Pacific, anomalies of opposite polarity in the extratropical central North Pacific, and a high degree of equatorial symmetry (Zhang et al., 1997). On close observation, it can be seen that the positive SSTAs allied with the warm-equatorial phase of the IPO extends further west than that associated with El Niño (Figure not shown), and in the far eastern tropical Pacific, IPO has relatively weaker variability compared to ENSO. In the case of ENSO, the equatorial maximum in the eastern Pacific is more pronounced and more barely confined along the equator; whereas in case of IPO, the meridional scale of tropical anomalies is broader. Furthermore, unlike ENSO, the SST fluctuations in the extratropical central North Pacific are more prominent in IPO's case and are as large as those near the equator.

As discussed above, the PDO is defined over the North Pacific, whereas the IPO is seen as a broader Pacific basin phenomenon (Joshi and Kucharski, 2017; Henley, 2017), so here we have used the latter index as defined in Section 19.2.2.

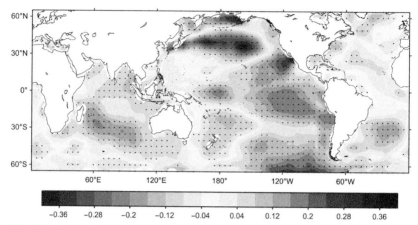

FIG. 19.3 Regression maps of annual SSTAs onto the standardized IPO index (units are °C per standard deviation). The black stippling indicates the grid points where the regression coefficients are statistically significant at the 90% confidence level, which is assessed via a two-tailed t test.

19.3.2 Teleconnection of ISMR with IPO

Herein, the EOF calculation for computing IPO index is restricted to the Pacific basin (45°S to 60°N, 140°E to 80°W) only (as shown in Fig. 19.2A). Therefore, to get further insight into the crucial elements in the observed global IPO pattern, the regression map, obtained by regressing annual SSTAs onto the standardized IPO index, is computed (see Fig. 19.3). The IPO-SST regression pattern illustrates a well-defined spatial pattern of IPO over the Pacific basin having positive SSTAs (i.e., warm loadings) in the tropical Pacific as well as in the north and south of the eastern part of the basin; and negative SSTAs in the western part of the basin, poleward of 25°, and more prominent in the Northern Hemisphere. Besides this, the regression pattern also shows warming over the Indian Ocean and parts of the Atlantic Ocean.

The IPO-SST regression pattern also exemplifies strong tropical-extratropical Pacific SST gradient, which is an important feature of the global SST pattern responsible for common multidecadal African and Indian rainfall variability (Feudale and Kucharski, 2013). Using the state-of-the-art climate models that were used in the fifth Coupled Model Intercomparison Project (CMIP5), Joshi and Kucharski (2017) also found a strong and statistically significant relationship (−0.66) between the tropical-extratropical SST gradient in the IPO regressions and the mean Indian land rainfall IPO regressions and reported that, in general, the models with strong tropical-extratropical IPO SST regression gradient produce larger negative Indian land rainfall responses.

To see the impact of IPO on Indian rainfall, the June-July-August-September (JJAS) precipitation anomalies are regressed onto the standardized IPO Index (see Fig. 19.4). The regression pattern reveals negative anomalies are seen over

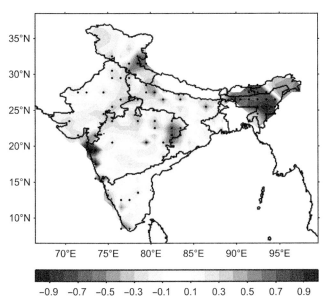

FIG. 19.4 Regression map of JJAS precipitation anomalies onto the standardized IPO index (units are mm/day per standard deviation). The black stippling indicates the grid points where the regression coefficients are statistically significant at the 90% confidence level, which is assessed via a two-tailed t test.

most parts of India (especially over the northwest and west-central India) and positive over the northeast region, whereas in the rest of the homogeneous monsoon regions, it is positive in some parts and negative in others. This signifies that during the warm phase of IPO, all-India will have below-normal rainfall, while the northeast region will have above-normal monsoon rainfall. Thus, the IPO precipitation pattern divulges that the influence of IPO on rainfall has spatial differences, and its effect is not homogeneous over all-India. This regression pattern is fairly consistent with Fig. 4a of Joshi and Rai (2015) and Fig. 5 of Joshi and Kucharski (2017) in which the gridded rainfall data from the IMD (Rajeevan et al., 2006) and climate research unit (CRU; Harris et al., 2014) are used for computing the correlation and regression maps, respectively between the low-pass filtered IPO and rainfall.

19.3.3 Mechanism of IPO-ISMR teleconnection

To scrutinize the atmospheric circulation pattern associated with the IPO, the JJAS horizontal (i.e., zonal and meridional) wind anomalies from NCEP/NCAR reanalysis at 850 hPa and the annual SSTAs are regressed onto the standardized IPO index for the period 1948 to 2008. The IPO SST regression pattern for the period 1948–2008 (figure not shown) is in close agreement with the one

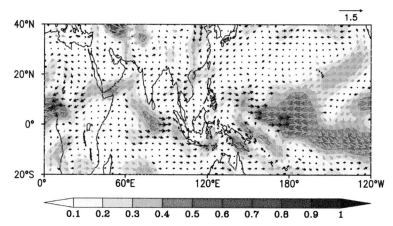

FIG. 19.5 Regression of JJAS seasonal anomalies of zonal and meridional winds at 850 hPa (plotted as vectors) from NCEP/NCAR reanalysis (1948–2008) onto the standardized IPO index. Magnitude of winds is represented by shaded color and vectors represent wind direction. Green vectors indicate wind anomalies that are statistically significant at 90% confidence level in at least one of the wind components (meridional or zonal). The unit of wind is m/s per standard deviation.

obtained for the period 1901–2008 (Fig. 19.3), that is, exhibiting warm SSTAs over the tropical Pacific and cold over the extratropics (poleward of 25°), especially in the northern Pacific sector, indicating that the wind circulation pattern shown in Fig. 19.5 is linked with the IPO's warm phase.

At the lower level, the IPO winds regression pattern depicts that the warm SSTAs in the tropical Pacific are allied with westerly anomalies along the equator, consistent with Joshi and Rai (2015). Joshi and Rai (2015) also reported that the easterlies in the equatorial Pacific are strengthened during the cold phase of IPO compared to its warm phase. The regression pattern also divulges that during the warm phase of IPO, the wind anomalies are blowing away from the Indian subcontinent into the Arabian Sea, reducing the moisture flow over India, which is consistent with the below-normal rainfall.

Wang et al. (2013) correlated this with mega-ENSO events and reported that the warming of the eastern and cooling of the western Pacific would lead to the weakening of the easterly trade winds, which causes the divergence of moisture from the Asian and African monsoon regions that, in turn, decreases the Northern Hemisphere summer monsoon rainfall and vice versa. Meehl and Hu (2006) had previously proposed that IPO's warm phase is characterized with relatively warm tropical SSTs, favorable convective precipitation and heating anomalies in the tropical Pacific, weak trade winds, and subtropical cells that will produce multidecadal drought-like conditions over the extended Indian monsoon region and anomalously wet conditions over the Great Basin region in the southwestern United States.

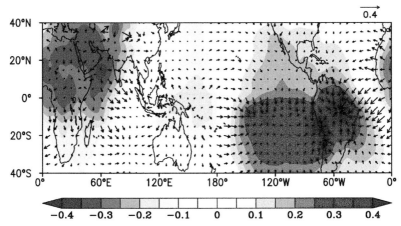

FIG. 19.6 Regression of JJAS anomaly of velocity potential at 850 hPa from NCEP/NCAR reanalysis (1948–2008) onto the standardized IPO index. The unit of velocity potential is 10^6 m^2/s per standard deviation. The vectors represent the divergent wind (m/s).

Generally, the convergence and divergence at the lower level typically coincide with divergence and convergence at the upper level, respectively, indicating the large-scale monsoonal overturning circulations. Velocity potential, a measure of divergent flow, is often used as a proxy for the Walker circulation. At the upper level, a region of negative (positive) potential has diverging (converging) winds that exemplify strong convection (subsidence), that is, rising (sinking) motion at a lower level.

The regression of the JJAS anomaly of velocity potential from NCEP/NCAR reanalysis (1948–2008) onto the standardized IPO index at lower and upper (i.e., 850 hPa and 150 hPa) levels is shown in Figs. 19.6 and 19.7, respectively. Fig. 19.6 depicts that the IPO's warm phase is allied with anomalous convergence (i.e., positive potential) over the central tropical Pacific as well as over the Southwest United States [in agreement with Dai (2013)] and divergence (i.e., negative potential) over West Africa [in accord with Villamayor and Mohino (2015)] as well as over the vast ISM region at lower levels and with anomalous divergence and convergence over the respective regions at higher levels (Fig. 19.7).

The anomalous convergence at lower levels over the central tropical Pacific is concordant with the eastern Pacific warming signal that illustrates anomalous Walker cell, that is, the weakening of zonal overturning circulation. Because of subsidence, the atmosphere over the western end of the Pacific is highly stable, which is unfavorable to and limits the occurrence of deep clouds and precipitation. In contrast, the atmosphere is unstable over the central-eastern tropical Pacific, and deep convective clouds and heavy rainfall occur frequently.

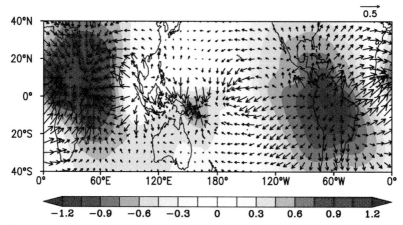

FIG. 19.7 Regression of JJAS anomaly of velocity potential at 150 hPa from NCEP/NCAR reanalysis (1948–2008) onto the standardized IPO index. The unit of velocity potential is 10^6 m^2/s per standard deviation. The vectors represent the divergent wind (m/s).

This anomalous circulation reduces the easterly trade winds across the tropical Pacific in the lower atmosphere, clearly distinct in the winds regression pattern at 850 hPa (Fig. 19.5). On the other hand, this anomalous circulation also weakens the westerly winds across the tropical Pacific in the upper atmosphere (figure not shown). This is in accord with Krishnamurthy and Krishnamurthy (2014), who proposed a mechanism that the warm phase of IPO-like variability affects the equatorial trade winds that in turn weakens the equatorial Walker circulation, which leads to enhanced ascending and descending motions in the central Pacific and Maritime Continent, respectively.

Fig. 19.8 illustrates the regression map of the JJAS SLP anomalies onto the standardized IPO index. In general, the Walker circulation is allied with low SLP over the Pacific's western end and high over the eastern end. This basin-wide pressure gradient is the key driving force for the low-level zonal winds, that is, the easterly trade winds of the Walker circulation.

During IPO's warm phase, the center of rising motion shifts east into the central-eastern Pacific, that is, away from the western end of the Pacific consistent with the eastern Pacific warming signal, as discussed earlier. This region is accompanied by negative SLP anomaly, while the western end of the Pacific will have positive SLP anomaly, which is seen in the IPO SLP anomaly regression pattern (Fig. 19.8). This basin-wide pressure gradient (i.e., low pressure over the central-eastern tropical Pacific and high pressure over the western end of the Pacific) is the main driving force for the low-level westerly anomalies over the equatorial Pacific (see Fig. 19.5). The IPO SLP anomaly regression pattern also divulges positive anomalies over South Asia and the Sahel, which is in accordant with below-normal rainfall, as discussed earlier.

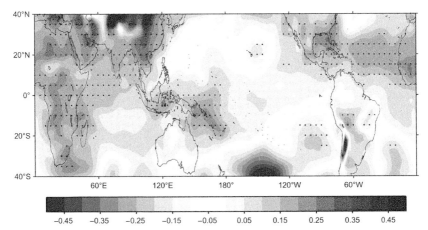

FIG. 19.8 Regression of JJAS seasonal anomaly of SLP from NCEP/NCAR reanalysis (1948–2008) onto the standardized IPO index. The black stippling indicates the grid points where the regression coefficients are statistically significant at the 90% confidence level, which is assessed via a two-tailed t test. The unit of SLP is hPa per standard deviation.

Overall the IPO-related atmospheric pattern indicates pronounced alteration of the Walker circulation, which is consistent with the fact that a larger zonal equatorial Pacific-Indian Ocean SST gradient also shows larger negative precipitation responses over India (Joshi and Kucharski, 2017). The tropical-extratropical Pacific SST gradient, a vital feature of the global IPO SST pattern, is strongly related to Indian precipitation (as discussed earlier), which may enhance the Walker circulation response in the Indian Ocean. The IPO SLP anomaly regression (Fig. 19.8) indicates negative SLP anomaly in the subtropical western north Pacific (around 20°N), which might be responsible for a westward shifted and strengthened Walker circulation response. It could be that the augmented baroclinicity induced by the stronger SST gradient in the western Pacific plays a role in this response. However, the details of the mechanism for the tropical-extratropical SST gradient influence on the South Asian monsoon are still unclear and may be further investigated using the idealized Atmospheric General Circulation Model simulations in a future study.

19.3.4   Modeling approach of IPO-ISMR teleconnection

In this section, we have discussed the modeling studies carried out on IPO-ISMR teleconnections. However, till date, only a few studies have scrutinized the simulated impact of IPO on ISMR in coupled general circulation models. Meehl and Hu (2006) observed large multidecadal variations in rainfall over India using a 1360-year control run from a global coupled climate model and stated that these variations are linked to multidecadal SST variations in the Pacific, which resemble IPO. Using observations and coupled model simulation, Krishnamurthy and

Krishnamurthy (2014) examined the linkage between ISMR and the IPO-like natural variability observed in the North Pacific Ocean and reported that the warm and cold phase of IPO-like natural variability is associated with the deficit and excess rainfall over India, respectively.

Recently, Joshi and Kucharski (2017) scrutinized the fidelity of 32 state-of-the-art climate models from the CMIP5 in evaluating the observed teleconnection of IPO with ISMR. The salient features of that study are as follows: (1) Of 32 models under consideration only two-thirds of them showed a well-defined spatial pattern of IPO over the Pacific basin, and most among these capture the IPO-ISMR teleconnection; (2) the models that fail to reproduce the observed IPO-ISMR teleconnection are the ones that are illustrating a poor spatial pattern of IPO, irrespective of the extent to which they simulate the precipitation climatology and seasonal cycle; (3) the results revealed that there is a strong relationship between the quality of reproducing the IPO pattern and its teleconnection with ISMR in the models, in particular with respect to the tropical-extratropical and the equatorial Pacific-Indian Ocean SST gradients during IPO phases; (4) the models that are able to reproduce the IPO-ISMR teleconnection reasonably, also simulates the atmospheric features allied with IPO; and (5) the physical mechanism for IPO-ISMR teleconnection is related to an alteration of the Walker circulation.

19.4 Discussion and summary

As discussed earlier, the climate system divulges various modes of internal natural variability. If these modes are successfully simulated and predicted, then the confidence in the coupled global climate models, used for climate predictions, can be increased. In this chapter, an attempt has been made to comprehend the relationship of Pacific decadal variability with ISMR and to explore the probable mechanism associated with it.

The IPO is negatively related to rainfall over most parts of all-India, except over the northeast region, signifying that during IPO's warm phase, most parts of all-India will experience below-normal rain, while the northeast region will have above normal. The physical mechanism for IPO-ISMR teleconnection is related to a modification of the Walker circulation. During the warm phase of IPO, the center of rising motion shifts east into the central-eastern Pacific, consistent with the eastern Pacific warming signal. So this region (i.e., central-eastern Pacific) is now accompanied by anomalously low SLP, compared to the western end of the Pacific having anomalously high SLP. This basin-wide pressure gradient is the key driving force for the low-level westerly anomalies over the equatorial Pacific. At lower levels, the anomalous convergence over the central tropical Pacific indicates anomalous Walker cell, that is, the weakening of zonal overturning circulation. Due to subsidence, the atmosphere over the western end of the Pacific is highly stable, which is unfavorable to and limits deep clouds and precipitation. In contrast, the atmosphere over the central-eastern tropical Pacific is unstable that favors deep convective clouds and precipitation to occur. This anomalous circulation reduces the easterly/westerly trade winds across the equatorial Pacific in the lower/upper atmosphere.

The decadal climate prediction is an emerging branch of climate science that fills the gap between seasonal climate forecasts and multidecadal-to-century projections of climate change and has two potential sources for the skill; first, the initialization, and second the external forcing. In terms of initialization, several phenomena in the Pacific (e.g., PDO/IPO and the North Pacific index) and the Atlantic (e.g., AMO and the Atlantic meridional overturning circulation) Oceans have been identified that can provide skill on decadal timescales (Meehl et al., 2009). In other words, decadal climate prediction depends on state-of-the-art climate models' skill to reliably reproduce these low-frequency climate variations (Henley et al., 2017). This is analogous to seasonal climate forecasts that derive much of their skill from the ENSO phenomenon. It is known that the regions that are influenced by ENSO are potentially benefited from seasonal forecasts. Similarly, the potential skill from decadal phenomena is also regionally dependent. For example, in India, the rainfall is modulated by IPO (as discussed in detail in this chapter as well as in previous studies, e.g., Joshi and Pandey, 2011; Krishnamurthy and Krishnamurthy, 2014; Joshi and Rai, 2015; Joshi and Kucharski, 2017) and AMO (Goswami et al., 2006; Joshi and Pandey, 2011; Joshi and Rai, 2015; Joshi and Ha, 2019). Generally, the decadal climate prediction focuses on time-evolving regional climatic conditions over the next 10–30 years that is a time period of interest to infrastructure planners, water resource managers, and others. So, the decadal predictions will help society in improving the plans to mitigate the adverse effect of monsoon floods or droughts. Thus, for a better understanding of decadal-to-multidecadal variability and improving the decadal predictions of rainfall over India and the regions whose climatic conditions are influenced by IPO, the models must simulate the IPO skillfully.

Acknowledgments

M.K.J., A.R., and A.K. gratefully acknowledge Director, IITM for support and encouragement. IITM is fully supported by the Ministry of Earth Sciences, Govt. of India. M.K.J. also thanks ICTP, Trieste, Italy for providing the facilities during his visits to the Centre under the Junior Associateship award received from ICTP.

References

Allan, R.J., 2000. ENSO and climatic variability in the last 150 years. In: Diaz, HF, Markgraf, V. (Eds.), El Niño and the Southern Oscillation: Multiscale Variability, Global and Regional Impacts. Cambridge University Press: Cambridge, UK, pp. 3–56.

Dai, A., 2013. The influence of the inter-decadal Pacific oscillation on US precipitation during 1923–2010. Clim. Dyn. 41 (3-4), 633–646.

Deser, C., Phillips, A.S., Hurrell, J.W., 2004. Pacific interdecadal climate variability: linkages between the tropics and the North Pacific during boreal winter since 1900. J. Clim. 17 (16), 3109–3124.

Dong, B., Dai, A., 2015. The influence of the interdecadal Pacific oscillation on temperature and precipitation over the globe. Clim. Dyn., 1–15.

Feudale, L., Kucharski, F., 2013. A common mode of variability of African and Indian monsoon rainfall at decadal timescale. Clim. Dyn 41 (2), 243–254.

Folland, C.K., Parker, D.E., Colman, A.W., Washington, R., 1999. Large scale modes of ocean surface temperature since the late nineteenth century. In: Navarra, A. (Ed.), Beyond El Niño: Decadal and Interdecadal Climate Variability. Springer, Berlin, pp. 73–102.

Goswami, B.N., 2005. Interdecadal variability. In: Wang, B. (Ed.), The Asian Monsoon. Praxis, Springer, Berlin Heidelberg, pp. 295–327.

Goswami, B.N., Madhusoodanan, M.S., Neema, C.P., Sengupta, D, 2006. A physical mechanism for North Atlantic SST influence on the Indian summer monsoon. Geophys. Res. Lett. 33 (2), L02706.

Harris, I., Jones, P.D., Osborn, T.J., Lister, D.H., 2014. Updated high-resolution grids of monthly climatic observations – the CRU TS3.10 dataset. Int. J. Climatol. 34 (3), 623–642.

Henley, B.J., 2017. Pacific decadal climate variability: indices, patterns and tropical-extratropical interactions. Global Planet. Change 155, 42–55.

Henley, B.J., Gergis, J., Karoly, D.J., Power, S., Kennedy, J., Folland, C.K., 2015. A tripole index for the interdecadal Pacific oscillation. Clim. Dyn. 45 (11), 3077–3090.

Henley, B.J., Meehl, G., Power, S.B., Folland, C.K., King, A.D., Brown, J.N., Karoly, D.J., Delage, F., Gallant, A.J.E., Freund, M., Neukom, R, 2017. Spatial and temporal agreement in climate model simulations of the interdecadal Pacific oscillation. Environ. Res. Lett. 12 (4), 044011.

Huang, B., Thorne, P.W., Banzon, V.F., Boyer, T., Chepurin, G., Lawrimore, J.H., Menne, M.J., Smith, T.M., Vose, R.S., Zhang, H.-M., 2017. Extended Reconstructed Sea Surface Temperature, Version 5 (ERSSTv5): upgrades, validations, and intercomparisons. J. Clim. 30 (20), 8179–8205.

IPCC, 2007. Summary for policymakersClimate Change 2007: The Physical Science Basis. Contribution of Working Group I to the Fourth Assessment Report of the Intergovernmental Panel on Climate Change. Cambridge University Press, Cambridge, United Kingdom and New York, NY, USA.

Joshi, M.K., Abid, M.A., Kucharski, F., 2021. The Role of an Indian Ocean Heating Dipole in the ENSO Teleconnection to the North Atlantic European Region in Early Winter during the Twentieth Century in Reanalysis and CMIP5 Simulations. J. Clim. 34 (3), , 1047–1060. doi:https://doi.org/10.1175/JCLI-D-20-0269.1.

Joshi, M.K., Ha, K.-J., 2019. Fidelity of CMIP5-simulated teleconnection between Atlantic multidecadal oscillation and Indian summer monsoon rainfall. Clim. Dyn. 52 (7), 4157–4176.

Joshi, M.K., Kucharski, F., 2017. Impact of Interdecadal Pacific oscillation on Indian summer monsoon rainfall: an assessment from CMIP5 climate models. Clim. Dyn. 48 (7), 2375–2391.

Joshi, M.K., Pandey, A.C., 2011. Trend and spectral analysis of rainfall over India during 1901–2000. J. Geophys. Res. 116 (D6).

Joshi, M.K., Rai, A., 2015. Combined interplay of the Atlantic multidecadal oscillation and the interdecadal Pacific oscillation on rainfall and its extremes over Indian subcontinent. Clim. Dyn 44 (11-12), 3339–3359.

Joshi, M.K., Rai, A., Kulkarni, A., Kucharski, F., 2020. Assessing changes in characteristics of hot extremes over India in a warming environment and their driving mechanisms. Sci. Rep. 10 (1), 2631. doi:https://doi.org/10.1038/s41598-020-59427-z.

Kalnay, E., Kanamitsu, M., Kistler, R., Collins, W., Deaven, D., Gandin, L., Iredell, M., Saha, S., White, G., Woollen, J., Zhu, Y., Leetmaa, A., Reynolds, R., Chelliah, M., Ebisuzaki, W., Higgins, W., Janowiak, J., Mo, K.C., Ropelewski, C., Wang, J., Jenne, R., Joseph,

D, 1996. The NCEP/NCAR 40-year reanalysis project. B. Am. Meteorol. Soc. 77 (3), 437–471.

Kerr, R.A., 2000. A North Atlantic climate pacemaker for the centuries. Science (New York, N.Y.) 288 (5473), 1984–1985.

Kripalani, R.H., Kulkarni, A., Singh, S.V., 1997. Association of the Indian summer monsoon with the northern hemisphere mid-latitude circulation. Int. J. Climatol. 17 (10), 1055–1067.

Krishna Kumar, K., Rajagopalan, B., Cane, M.A, 1999. On the weakening relationship between the Indian monsoon and ENSO. Science 284 (5423), 2156.

Krishnamurthy, L., Krishnamurthy, V., 2014. Influence of PDO on South Asian summer monsoon and monsoon–ENSO relation. Clim. Dyn. 42 (9-10), 2397–2410.

Krishnamurthy, V., Goswami, B.N., 2000. Indian monsoon–ENSO relationship on interdecadal timescale. J. Clim 13 (3), 579–595.

Krishnan, R., Sugi, M., 2003. Pacific decadal oscillation and variability of the Indian summer monsoon rainfall. Clim. Dyn. 21 (3-4), 233–242.

Kucharski, F., Joshi, M.K., 2017. Influence of tropical South Atlantic sea-surface temperatures on the Indian summer monsoon in CMIP5 models. Quart. J. R. Meteor. Soc. 143 (704), 1351–1363.

Kushnir, Y., 1994. Interdecadal variations in North Atlantic sea surface temperature and associated atmospheric conditions. J. Clim 7 (1), 141–157.

Mantua, N.J., Hare, S.R., Zhang, Y., Wallace, J.M., Francis, R.C., 1997. A Pacific Interdecadal climate oscillation with impacts on salmon production. B. Am. Meteorol. Soc. 78 (6), 1069–1079.

Meehl, G.A., Goddard, L., Murphy, J., Stouffer, R.J., Boer, G., Danabasoglu, G., Dixon, K., Giorgetta, M.A., Greene, A.M., Hawkins, E., Hegerl, G., Karoly, D., Keenlyside, N., Kimoto, M., Kirtman, B., Navarra, A., Pulwarty, R., Smith, D., Stammer, D., Stockdale, T., 2009. Decadal prediction: can it be skillful? B. Am. Meteorol. Soc. 90 (10), 1467–1486.

Meehl, G.A., Hu, A., 2006. Megadroughts in the Indian monsoon region and southwest North America and a mechanism for associated multidecadal Pacific sea surface temperature anomalies. J. Clim. 19 (9), 1605–1623.

Meehl, G.A., Hu, A., Arblaster, J.M., Fasullo, J., Trenberth, K.E., 2013. Externally forced and internally generated decadal climate variability associated with the interdecadal Pacific oscillation. J. Clim 26 (18), 7298–7310.

Mujumdar, M., Sooraj, K.P., Krishnan, R., Preethi, B., Joshi, M.K., Varikoden, H., Singh, B.B., Rajeevan, M., 2017. Anomalous convective activity over sub-tropical east Pacific during 2015 and associated boreal summer monsoon teleconnections. Clim. Dyn. 48 (11), 4081–4091.

Parthasarathy, B., Munot, A.A., Kothawale, D.R., 1994. All-India monthly and seasonal rainfall series: 1871–1993. Theor. Appl. Climatol. 49 (4), 217–224.

Power, S., Casey, T., Folland, C., Colman, A., Mehta, V., 1999. Inter-decadal modulation of the impact of ENSO on Australia. Clim. Dyn. 15 (5), 319–324.

Power, S., Tseitkin, F., Torok, S., Lavery, B., Dahni, R., McAvaney, B., 1998. Australian temperature, Australian rainfall and the Southern Oscillation, 1910-1992: coherent variability and recent changes. Aust. Meteorol. Mag. 47 (2), 85–101.

Rajeevan, M, Bhate, J., Kale, J.D, Lal, B., 2006. A high resolution daily gridded rainfall data for the Indian region: analysis of break and active monsoon spells. Curr. Sci. 91, 296–306.

Rasmusson, E.M., Carpenter, T.H., 1983. The relationship between eastern equatorial Pacific sea surface temperatures and rainfall over India and Sri Lanka. Mon. Weather Rev. 111 (3), 517–528.

Saji, N.H., Goswami, B.N., Vinayachandran, P.N., Yamagata, T., 1999. A dipole mode in the tropical Indian Ocean. Nature 401 (6751), 360–363.

Schlesinger, M.E., Ramankutty, N., 1994. An oscillation in the global climate system of period 65-70 years. Nature 367 (6465), 723–726.

Schlesinger, M.E., Ramankutty, N., 1995. Is the recently reported 65- to 70-year surface-temperature oscillation the result of climatic noise? J. Geophys. Res. 100 (D7), 13767–13774.

Shepard, D., 1968. A two-dimensional interpolation function for irregularly-spaced data, Proc. 23rd National Conf. ACM. Association of Computing Machinery., pp. 517–524.

Sikka, D.R., 1980. Some aspects of the large scale fluctuations of summer monsoon rainfall over India in relation to fluctuations in the planetary and regional scale circulation parameters. Proc. Indian Acad. Sci. Earth Planet. Sci. 89 (2), 179–195.

Villamayor, J., Mohino, E., 2015. Robust Sahel drought due to the Interdecadal Pacific oscillation in CMIP5 simulations. Geophys. Res. Lett. 42 (4), 1214–1222.

Wang, B., Liu, J., Kim, H.-J., Webster, P.J., Yim, S.-Y., Xiang, B., 2013. Northern Hemisphere summer monsoon intensified by mega-El Niño/southern oscillation and Atlantic multidecadal oscillation. Proc. Natl. Acad. Sci. USA 110 (14), 5347–5352.

Zhang, Y., Wallace, J.M., Battisti, D.S., 1997. ENSO-like interdecadal variability: 1900–93. J. Clim. 10 (5), 1004–1020.

Part IV

Climate change and Monsoon teleconnections

Chapter 20

Future changes of the ENSO–Indian summer monsoon teleconnection

June-Yi Lee[a,b], **Tamás Bódai**[a,c]

[a]*Institute for Basic Science (IBS), Center for Climate Physics (ICCP), Busan, South Korea,* [b]*Research Center for Climate Sciences and Department of Climate System, Pusan National University, Busan, South Korea,* [c]*Pusan National University, Busan, South Korea*

20.1   Introduction

The El Niño-Southern Oscillation (ENSO)–Indian summer monsoon (ISM) teleconnection has received tremendous attention due to its considerable socioeconomic impacts. During an El Niño developing summer, the equatorial convective system and the associated Walker circulation tend to shift eastward, leading to a decrease in the ISM rainfall (Kumar et al., 2006). It is evident that the so-called "ENSO–ISM inverse relationship on interannual timescales" (in other words the negative correlation between ENSO and ISM) is modulated on decadal timescales (Krishnamurthy and Goswami, 2000; Kumar et al., 1999). In particular, Kumar et al. (1999) claimed that there has been a statistically significant weakening of the relationship since the late 1970s (see Chapter 3).

What drove the decadal modulation of this ENSO–ISM teleconnection in the twentieth century has been a debated issue. On the one hand, many studies hypothesize that most or a considerable part of the change is deterministic, either as a result of low-frequency climate processes (Chen et al., 2010; Fan et al., 2017; Kucharski et al., 2009; Lu et al., 2006) or, concerning primarily the late twentieth century, as a response to anthropogenic greenhouse gas (GHG) or aerosol forcing (Azad and Rajeevan, 2016; Kumar et al., 1999; Wang et al., 2015). The low-frequency processes hypothesized to affect the ENSO–ISM teleconnection include the Atlantic multidecadal oscillation (AMO) (Chen et al., 2010; Kucharski et al., 2009; Lu et al., 2006), the Pacific decadal oscillation (PDO) (Krishnan and Sugi, 2003), the decadal variation of ENSO's center (Fan et al., 2017), the decadal shift in the tropical Indian Ocean (Feba et al., 2019; Mohapatra et al. 2020), and changes in the co-occurrence of Indian Ocean Dipole events with ENSO (Ashok et al., 2007; Hrudya et al., 2021;

Indian Summer Monsoon Variability: El Niño-teleconnections and beyond.
DOI: https://doi.org/10.1016/B978-0-12-822402-1.00007-7

Krishnaswamy et al., 2015). On the other hand, many studies (Cash et al., 2017; DelSole and Shukla, 2012; Yun and Timmermann, 2018) hypothesized that a trivial undersampling could be responsible for most of the decadal change.

Yun and Timmermann (2018), in particular, showed that both the standard deviation and the maximum of a short-term change of the observed decadal variation of the ENSO–ISM teleconnection during the past century, as test statistics, are consistent with a *stationary* linear regression model, as a null hypothesis, at the usual confidence level. However, they did not investigate when these two test statistics are sensitive to a considerable change of the relationship, for example, to a gradual complete vanishing of the correlations in the last 30 years, namely, whether the rejection rates are hardly different from those under the null hypothesis and in association with the adapted confidence level. If it is the case, it is meaningful to *carefully* explore the possible forced changes of the ENSO–ISM teleconnection and their physical mechanisms in modeling studies.

Several studies attributing the recent weakening of the ENSO–ISM teleconnection to GHG forcing also hypothesized that the weakening of the relationship will continue in a warmer world (Kripalani and Kulkarni, 1997; Kumar et al., 1999; Wang et al., 2015). However, there has been no consensus among modeling studies on either the past change or the projections of the future change of the teleconnection. By analyzing Coupled Model Intercomparison Project Phase 5 (CMIP5) model simulations, Li and Ting (2015) showed that natural variability plays a dominant role in the decadal ENSO–ISM teleconnection during the twentieth century. However, they reported a slightly weakened relationship in the twenty-first century, as a multimodel mean (MMM), under the representative concentration pathway 8.5 (RCP 8.5) scenario, resulting from enhanced ISM rainfall associated with SST warming. Roy et al. (2019), based on 35 CMIP5 models, suggested a tradeoff between changes in circulation acting to weaken the ENSO–ISM relationship and changes in moisture availability acting to strengthen it during the twenty-first century under the RCP 8.5. Contrary to Li and Ting (2015), Azad and Rajeevan (2016), based on 20 CMIP5 models, suggested that the inverse ENSO–ISM relationship will be stable but there will be more drought events in India due to more frequent El Niño events during the twenty-first century under the RCP 8.5 scenario.

Several conceptual issues in identifying the forced response as an MMM of running window temporal statistics have been recently discussed. First, a multimodel ensemble is the so-called "ensemble of opportunity", such that there is no underlying *objective* probability distribution that the member models sample. Therefore, the ensemble mean is not the forced response of anything, certainly not an estimate of that of the actual Earth system. Leith (1978) proposed that the climate is defined rather by an *initial condition* ensemble, borrowing the idea from statistical physics. Drótos et al. (2015) clarified that this ensemble owes its objectivity to the dissipative nature of the dynamical system, which gives rise to an attractor that supports a *unique* probability measure. Such an attractor is indeed well-defined also in the presence of external forcing, that is, under nonautonomous dynamics, and it is called the snapshot attractor (Romeiras et al., 1990). Therefore, the time evolution of the snapshot attractor and the probability measure, or any particular statistical quantity that is derived from it, including the mean, can be viewed to soundly represent climate change (Bódai and Tél, 2012; Tél et al., 2020).

Second, Drótos et al. (2016) argued and demonstrated that nonautonomous systems are generically nonergodic, which means that temporal statistics are biased, that is, their expectation value with respect to the objective probability measure supported by the snapshot attractor is not the same as the correct climatological ensemble-wise instantaneous statistics. This, of course, poses a problem even if not a multimodel ensemble but an initial condition ensemble is used, correctly recognizing at least that the forced response is strictly associated with individual models or systems. The analysis of the forced response should break with the tradition of evaluating temporal statistics —which clearly originates from the analysis of (single realization) *observational* data—before considering any variation across the ensemble; but, instead, one should evaluate ensemble-wise statistics before any temporal aspect is considered in association with forced climate change. Nevertheless, what makes estimations not trivial is that not every conceivable statistical quantity has an *unbiased* estimator, which means that the finiteness of the ensemble size will introduce biases also under the "ensemble-wise-statistics-first" approach. A case-dependent combination of temporal and ensemble-wise statistics should be sought to minimize the bias in question. The example of the sample correlation coefficient in association with the MCA analysis featuring a positive bias is showcased in Bódai et al., (2021).

An agreement between the majority of the members of a multimodel ensemble on the change would loosely suggest that the observable parameter of interest features a *robust* change in the sense that it is *not sensitive* to choices of model parameter or modeling errors. Moreover, the absence of such an intermodel agreement might just be due to the above sensitivity, rather than the lack of considerable change in most models. This can only be determined by analyzing a *multimodel ensemble of initial condition ensembles*. It is entirely possible that the actual Earth system is best represented by one particular model, whose forced response is considerable and is not close to the MMM. We leave such a decisive analysis for the future, and here we only show and compare results for two CMIP-class Earth system models, making use of the Max Plank Institute Grand Ensemble (MPI-GE) (Maher et al., 2019) and the Community Earth System Model version 1 Large Ensemble (CESM1-LE) (Kay et al., 2015) datasets. On the other hand, keeping close to the common analysis method of CMIP5 data, we evaluate the new CMIP6 data to see if the situation has changed with respect to CMIP5 intermodel (dis)agreement.

Section 20.2 provides a brief overview of the utilized data, including CMIP6 models, and methods, including the snapshot framework. In Section 20.3, we revisit the past and future changes of the ENSO–ISM teleconnection using the temporal method applied to CMIP6 simulations. Section 20.4 provides an analysis of the forced response of the relationship using the ensemble-wise method, reviewing Bódai et al. (2021) as for the MPI-GE and providing original results for the CEMS1-LE. Section 20.5 summarizes key points of this study.

20.2 Data and methods

20.2.1 Model data

This study analyzes the historical and future simulations by applying the two different methods. First, Section 20.3 uses the temporal method applied to 30-model simulations from CMIP6 (Eyring et al., 2016): the historical runs for the period of 1850–2014 and SSP5-8.5, and the high-emission scenario runs for the period of 2015–2100. One realization per model is used. For comparison, 50-member ensemble simulation by the Canadian Earth System Model version 5 (CanESM5) (Swart et al., 2019) is also utilized. The CMIP data can be obtained from https://esgf-node.llnl.gov. The SSP5-8.5 scenario has an identical radiative forcing level to RCP8.5 (i.e., 8.5 W/m^2 at 2100) but the scenario assumes accelerated globalization and rapid economic and social development in developing countries coupled with the exploitation of abundant fossil fuel resources (O'Neill et al., 2014). Second, Section 20.4 uses the ensemble-wise method applied to SMILE datasets obtained from the MPI-ESM and CESM1. Respective large ensemble (LE) monthly mean datasets, the MPI-GE and CESM1-LE, as presented in Maher et al. (2019) and Kay et al. (2015), respectively, are available from the Multi-model Large Ensemble Archive (Deser et al., 2020) http://www.cesm.ucar.edu/projects/community-projects/MMLEA/. We use MPI-GE ensemble members 19, 20, 40, …, 100 ($N = 63$), and CESM1-LE members 2, …, 33 ($N = 32$). Most results for the MPI-GE are preexisting and they were presented in Bódai et al. (2021), whereas the results for the CESM1-LE are new. Nevertheless, as for the latter, some comparison can be made with Haszpra et al. (2020a). The atmospheric model resolution is T63L47 for MPI-GE and 1.9° lat × 2.5° lon with 30 vertical levels for CESM1-LE.

20.2.2 Observations

Several observational datasets are used to compare with historical simulations for the ENSO–monsoon teleconnection in CMIP6 models. We use precipitation data from the Global Precipitation Climatology Centre (GPCC) ,V2018, dating from 1891 to 2020 (Ziese et al., 2018) and from the Climate Research Unit (CRU), version 4.04, dating from 1901 to 2019 (Harris et al., 2020). The GPCC precipitation data are obtained from https://psl.noaa/gov/ provided by the NOAA/OAR/ESRL PSL, Boulder, Colorado, USA, and the CRU precipitation data from https://crudata.uea.ac.uk/cru/data/hrg/. The observed SST data of NOAA Extended Reconstructed Sea Surface Temperature (ERSST), Version 5, dating from 1854 to 2020 (Huang et al., 2017) are obtained from https://psl.noaa/gov/. All observational data were interpolated to a common 1°lat × 1°lon grid.

20.2.3   The ensemble-wise method

We take the June-July-August (JJA) mean of the SST to represent ENSO, and the June-July-August-September (JJAS) mean of the rainfall to represent the Indian summer monsoon rain. That is, we study here only the quasi-synoptic teleconnection. Most simply, a spatial averaging can be applied to obtain a scalar quantity that would retain temporal variability only (for any individual ensemble member). As for the monsoon, to secure a correspondence with the so-called All India Summer Monsoon Rainfall (AISMR) rain gauge dataset (Parthasarathy et al., 1994), we consider the same area: Indian land mass, but excluding some States (see Fig. 1 of Parthasarathy et al. (1994)). Furthermore, one can characterize the variability by the first principal component (PC1) corresponding to the first empirical orthogonal function (EOF1) (Lund et al., 2000). Given that our principal interest is in the forced response, we must calculate the correlation coefficient *with respect to the ensemble members*, for any fixed year (Bódai et al., 2020; Herein et al., 2016; Yettella et al., 2018). Therefore, also the EOFs and PCs are calculated for fixed years, with respect to the ensemble-wise variability. This generalization of the EOF analysis is called a "snapshot" EOF (SEOF) analysis by Haszpra et al. (2020a, 2020b). As for possible representations or characteristics of the ENSO–ISM teleconnection, we will calculate correlation coefficients between (1) the AISMR and PC1 of ENSO as well as between (2) the PC1s of the ISM and ENSO. The SEOFs are evaluated with respect to the tropical Pacific box (30°S–30°N, 150°–295°E), and the box of (5°S, 40°N, 65°E, 100°E) encompassing the Indian subcontinent. As a further teleconnection characteristic, (3) we can consider spatial modes of variability derived with the requirement of maximal covariance between the data series belonging to those modes. Hence, it is called the maximal covariance analysis (MCA) (Lund et al., 2000), and we will refer to it as "snapshot" MCA (SMCA) in our analysis. We use the same boxes for the SMCA as for the SEOF.

We do not call the correlations between the PCs of the first SEOFs or the SMCA modes "representations" of the teleconnection, but only "characteristics," because the modes on the Indian rainfall do not provide or explain the majority of the variability or variance of the ISM rain. See a more comprehensive analysis in Bódai et al. (2020), which defines further representations of the teleconnection. It is also explained there how a temporal lumping of data in a short time window is used to estimate spatial modes more accurately and, subsequently, estimate the correlation coefficient conservatively. Without this, the correlation coefficient associated with the SMCA turns out to be *positively* biased—contrasting the situation with the negatively biased ordinary sample correlation coefficient associated with given time series. We also emphasize that correlation coefficients associated with the AISMR, on the one hand, or spatial modes, on the other, are not directly comparable, as the AISMR domain is very considerably smaller than the box domain of the modes, and it even excludes a major hotspot of variability: the foothills of the Himalayas.

Instead of trying to find the physical mechanisms responsible for a particular forced change signal in the teleconnection strength, here we pursue only a statistical attribution of that. In terms of a linear regression model, $\Psi = a\Phi + \xi$, underpinning the correlation coefficient, r, we can attribute changes in the teleconnection strength to three factors. We can identify these factors by considering the formula:

$$r = \frac{1}{\sqrt{1 + \left(\dfrac{\sigma_\xi / a}{\sigma_\Phi}\right)^2}}. \tag{20.1}$$

These factors are:

- ENSO variability $\sigma_\Phi = \text{std}[\Phi]$;
- ENSO–ISM "coupling" a being the regression coefficient;
- Noise strength $\sigma_\xi = \text{std}[\xi]$.

Note that, like r, also σ_Φ and σ_ξ are defined in terms of the variability with respect to the ensemble, for every year separately. That is, like $r(t)$, we have time series $\sigma_\Phi(t)$ and $\sigma_\xi(t)$, and also $a(t)$. Based on these, in a simple way, we can say that for example a change in $r(t)$ in a time period is due to a change in $\sigma_\Phi(t)$ if in that period no change is seen with respect to $a(t)$ and $\sigma_\xi(t)$. Or, perhaps we can also say that, if σ_ξ/a shows no change.

To check the hypothesis of the dominant role of ENSO variability, one can evaluate σ_Φ directly (i.e., compute the standard deviation of Φ over the ensemble), on the one hand, and σ_ξ/a by simply inverting Eq. (20.1), on the other. (The sign of a, indeterminate from this calculation, is recovered as being that of r). Anticipating that σ_ξ/a is not constant in time, we can evaluate a by, first, directly evaluating the ISM variability σ_Ψ, and, subsequently, using the textbook formula:

$$a = r\frac{\sigma_\Psi}{\sigma_\Phi}. \tag{20.2}$$

In turn, having now also a on hand, σ_ξ can be obtained by inverting Eq. (20.1).

20.3 Past and future changes estimated by the temporal method applied to CMIP6 models

We examine the ENSO–monsoon teleconnection in the past and the future using 30 CMIP6 models in comparison to the CMIP5 results shown in Figs. 1 and 2 of Li and Ting (2015). Fig. 20.1 indicates that the spatial patterns of the CMIP6 MMM match closely those of the correlation coefficients evaluated from observed time series of boreal summer rainfall over the Asian monsoon region and SST over the global ocean with Niño 3.4 SST index for the 50 years of 1965–2014. As Li and Ting (2015) claimed, the inverse ENSO–ISM relationship is prompted to be weakened in the latter half of the twenty-first century under the SSP5-8.5 scenario (Fig. 20.1E). However, the weakening of the relationship should be mainly

With Trend

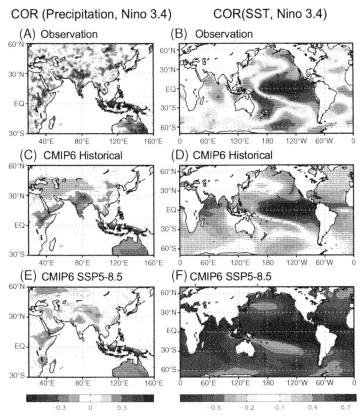

FIG. 20.1   Spatial patterns of sample correlation coefficients *including* long-term trends between JJA Niño 3.4 SST index and JJAS rainfall (left-hand side) and between JJA Niño 3.4 SST index and JJA SST (right-hand side) in observation (A, B) for 1965–2014, historical run (C, D) for 1965–2014, and SSP5-8.5 (E, F) for 2051–2100. For the historical and SSP5-8.5 run, the MMM of correlation coefficients from 30 models in CMIP6 is shown. Stippling denotes 5% significance based on two-sided Student's *t*-test in (A) and (B), and two-thirds model agreement on the sign in (C), (D), (E), and (F). The observed data are from GPCC V2018 for precipitation and ERSST V5 for SST.

due to the simultaneous increase of Niño 3.4 SST and AISMR. If we correctly remove the long-term trend, the MMM of the ENSO–ISM correlation on interannual timescale is rather stable as shown in Fig. 20.2C and E. (We also note that there is no robust change in the MMM of the global SST pattern associated with ENSO on interannual timescale under GHG warming (Fig. 20.2D and F). The simultaneous forced trends between two quantities should not be taken to represent any relationship between them (Bódai et al., 2021, footnote 2), which is an error made also by Pandey et al. (2020), leading them to find even a reversal of the sign of the associated correlation coefficient.

Without Trend

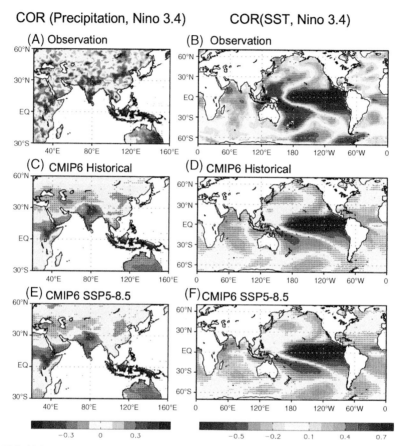

FIG. 20.2   Same as Fig. 20.1 except for having removed long-term linear trends.

In fact, Fig. 2 of Li and Ting (2015) showing the running-window sample correlation coefficients that naturally involve a detrending, contradicts their Fig. 1 retaining the trends. We construct this kind of diagram for the CMIP6 data and display it in Fig. 20.3. Just like the CMIP5 data used by Li and Ting (2015), it suggests no change in the ENSO–ISM correlation on interannual timescale across the twentieth and twenty-first century that could be viewed as robust in the sense that it is captured by most of the models. In this diagram we include also the time series for the observational data, which can be viewed as a visual representation of the "consistency test" done by Yun and Timmermann (2018). However, it should be noted that the standard deviations determined with respect to variations across the CMIP6 model ensemble members should

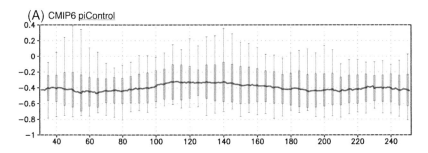

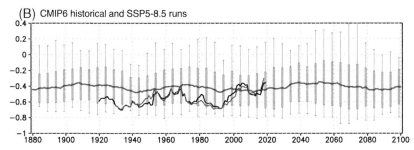

FIG. 20.3 Thirty-year backward running sample correlation coefficients between JJAS Indian rainfall (5°–30°N, 70°–90°E, land only) and JJA Niño 3.4 SST index obtained from CMIP6 (A) preindustrial control runs from year 30 to year 251 and (B) historical (1880–2014) and SSP5-8.5 (2015–2100) runs. The purple thick line indicates the MMM of individual models' correlation coefficients. The box edges give the 25th to 75th percentiles and the whiskers extend from minimum to maximum values. The value at 2100 is obtained for the 30 years (2071–2100). In (B), black and red solid lines indicate the sample correlation coefficients obtained from the GPCC rainfall and ERSST for 1920–2020 and from CRU rainfall and ERSST for 1930–2019, respectively.

be larger on average than that of the sample correlation coefficient, because the true correlation coefficients belonging to the different models should vary.

Fig. 20.4A shows that most of the individual models relying on a single realization tend to "project" insignificant change in the "relationship." Some models such as GISS-E2-1-G, MIROC6, and EC-Earth3-Veg project strengthening of the relationship, but somewhat more of the models such as CMCC-CM2-SR5, IPSL-CM6A-LR, and CanESM5 "project" a weakening of the relationship in future. The MMM of the sample correlation coefficient for individual model is −0.46 in the present (1965–2014) and −0.42 in the future (2051–2100). Fig. 20.4B displays the correlation coefficients in the present and future obtained from 50-member ensemble runs by the CanESM5, further indicating the undersampling problem for estimating future changes based on a single realization from one model. In Fig. 20.4A, the sample correlation coefficient for one ensemble member of CanESM5 is −0.36 in the present and −0.12 in the future, indicating the weakening of the ENSO–monsoon teleconnection. However, we note that the ensemble spread of correlation coefficients for a 50-member ensemble of CanESM5 is comparable to the intermodel spread across 30 models in Fig. 20.4B.

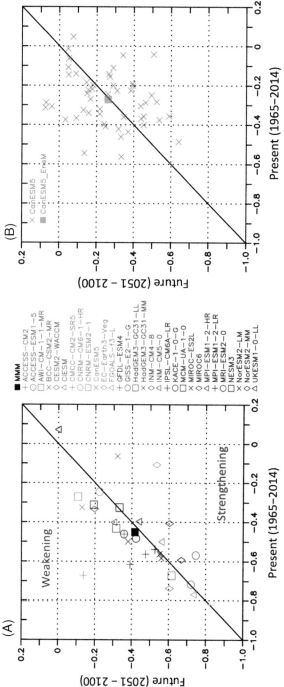

FIG. 20.4 The sample correlation coefficients between the JJAS Indian rainfall and JJA Niño 3.4 index during the present (1965–2014) and the future (2051–2100) obtained from (A) 3J CMIP6 models and (B) 50 ensemble runs by CanESM5. The MMM of the sample correlation coefficients of individual models is also shown by a black closed square in (A). The ensemble mean (EnsM) of 50 ensemble runs using CanESM5 is shown by a green closed square in (B).

To sum up, there is no *robust* change in the ENSO–ISM teleconnection under GHG warming estimated by the conventional multimodel temporal approach using CMIP6 model simulations; that is, such an analysis is *inconclusive*.

20.4 The forced response estimated by the ensemble-wise method for two SMILEs

As we discussed in Section 20.1 and 20.2.3, the conventional multimodel temporal approaches used in Section 20.3 have a limitation and other issues in capturing the forced response. In principle, the forced response of spatial aspects of the ENSO, such as those of the SEOFs introduced in Section 20.2.3, can have an impact on its teleconnections (Fan et al., 2017). The change of the first SEOFs for the MPI-GE and CESM1-LE is shown in Fig. 20.5, along with the first SMCA modes, averaged over consecutive 45-year periods. There is a somewhat stronger response for the CESM1, which corresponds to an eastward shift of the "center of action," while no such shift can be observed for the MPI-ESM. Otherwise, the SEOF1s of the two models are fairly similar, featuring a subtle east-west bimodality. It is interesting to observe that the SMCA1 modes are very similar to the SEOF1s in both models, with a slightly more accentuated western center of action. The agreement extends to the forced responses. This means that ENSO is the master in the examined relationship. We note that it is for this reason that we can apply the statistical attribution also to (3), interpreting σ_ϕ to represent autonomous ENSO characteristic.

On the side of the ISM, the EOF1 and MCA1 modes are similar only for the MPI-ESM; see Fig. 20.6. Furthermore, none of the EOF1 and MCA1 modes for the CESM1 is very similar to its counterpart belonging to the MPI-ESM. The EOF1 of CESM1 is extremely concentrated on a region in the west of the Tibetan plateau, on land, while the EOF1 of the MPI-ESM has a very accentuated dipole structure pertaining to the extreme west and east of the Bay of Bengal, and a third center of action in the Arabian sea directly adjacent to the subcontinent, all three over the sea. Furthermore, concerning the EOF1 of CESM1, in the majority of the area away from the center of action, the rain variability that belongs to this principal mode has an opposite phase (red vs. blue). The MCA1 mode is most different in this particular aspect: more of the land features the same phase (blue), for example, to include the whole of the Himalayan range; while areas of the opposite phase (red) have a slightly more importance for this mode. This is in fact an aspect with respect to which the MCA1 modes of the two models are more similar than their EOF1s.

Therefore, given that the correlation maps of both models are somewhat similar to the MCA1 modes of the respective models, the correlation maps of the two models also bear some resemblance. The differences are that for the CESM1 the areas of positive correlation are not restricted to the Bay of Bengal but "intrude into central India" and to Sri Lanka in the south; while the negative correlations are not as strong elsewhere. From another point of view, while the

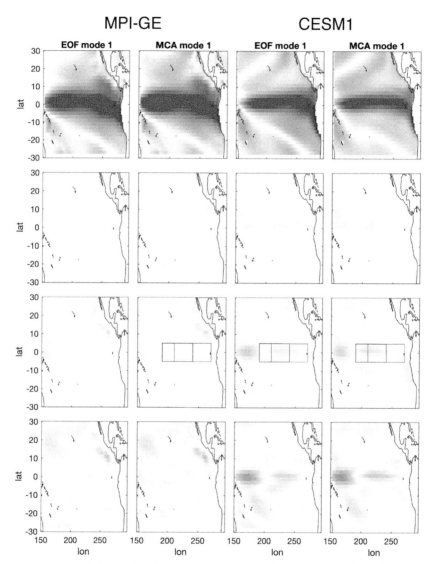

FIG. 20.5 Forced change of the first modes of JJA-mean SST variability in the Equatorial Pacific in the MPI-ESM and CESM1. Left: SEOF, right: SMCA first mode. Temporal means are taken in four consecutive 45-year periods starting from 1921. Thus, the lowest panels exhibit results during 2056–2100. The top row displays the temporal mean in the first period, and the subsequent rows display the difference with respect to that in the following periods. Color bars are not included for the modes because they are unit vectors, that is, only the pattern is nontrivial. However, the same color scales are used for plotting the differences as for the respective modes in the top row to indicate the relative change.

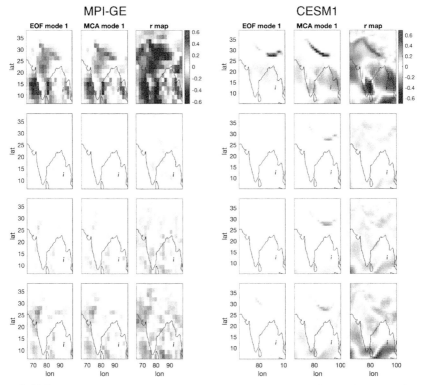

FIG. 20.6 Same as Fig. 20.5 with respect to the first two columns for each model but concerning the Indian summer monsoon. In the third column a correlation map and its changes are shown concerning the gridpoint-wise JJAS precipitation and the PC1 of the EOF1 of the SST in the box seen in Fig. 20.5. The color bars belong solely to the respective correlation maps.

relative levels of the correlations over small spatial scales match better between the models, being the source of similarity of the patterns, at a larger scale the relative levels do not match. In particular, for the MPI-ESM the negative correlation in northern India is as strong as in the south, unlike in the CEMS1.

The forced change of the MCA1 modes and correlation maps also show a considerable similarity for both models. As for the MPI-ESM, most of the areas of negative correlation over land become more negatively correlated, which foreshadows a strengthening correlation (in modulus) between the AISMR and the PC1 of ENSO. This strengthening pertains already to the historical period, but we give indication of any significance later below. As for the CESM1, some areas of positive and negative correlations see a negative and positive change, respectively, which makes any guess as above involving the AISMR not straightforward. We provide the actual results for the time evolution of the ensemble-wise correlation coefficient estimates in the top row of Fig. 20.7, where all three representations/characteristics (1)–(3) of the ENSO–ISM teleconnection are included.

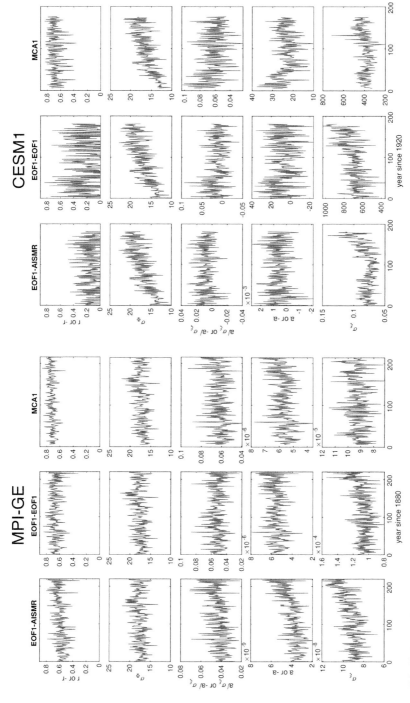

FIG. 20.7 The forced evolution of correlation coefficients $r(t)$ and the drivers of change; see the main text. The different columns correspond to different representations/characteristics of the ENSO–ISM teleconnection, labeled by (1)–(3) in the main text.

There are major differences between the two models regarding (1) and (2). First, in the CESM1 the level of correlations is a lot lower. In fact, this is an unrealistically low level considering the observational evidence as seen in Section 20.3, which issue was also documented by Bódai et al. (2020) and Ramu et al. (2018). Second, no obvious forced change is admitted for the CESM1 (see also Fig. 20.8D and E), which is hardly independent of the first issue. However, we should bear in mind that the part of CESM1-LE that we use is half the size of that of the MPI-GE. As for (3), in terms of the MCA1 modes, instead, both the correlation level and change are seemingly comparable. Diagrams of the Mann–Kendall test statistics Z_{MK} in Fig. 20.8C and F do indicate a long-term

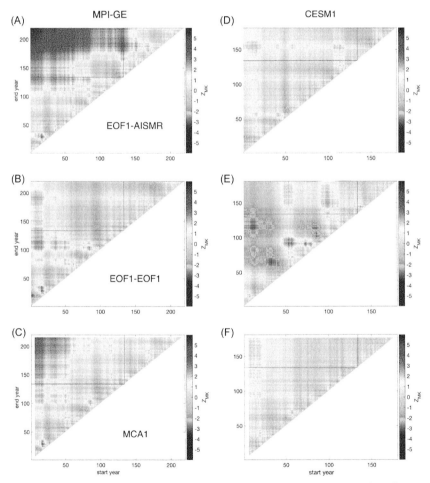

FIG. 20.8 Test statistics of the Mann–Kendall test for the stationarity of $r(t)$ or $-r(t)$, depending on which of these gives an overall positive value, for EOF1-AISMR (upper panels), EOF1-EOF1 (mid panels), and MCA1 (lower panels) using the MPI-GE (left panels) and CESM1 (right panels). Red and blue shades correspond to $p < 0.05$, that is, a detection of nonstationarity at that significance level.

increase in both models, although a much weaker one for the CESM1. It is conceivable though that with a comparable number of ensemble members, the significance would be comparable with that for the MPI-GE. This similarity is consistent with the observation of the similarity of correlation maps and their change between themselves and the respective MCA1 modes in Fig. 20.6. The vast difference of the correlation levels between for example representation (1) and characteristic (3) for CESM1, furthermore, is consistent with the very considerable difference between the EOF1 and MCA1 mode. However, it might be also due to the fact that the AISMR domain excludes the variability hotspot associated with the box domain.

Regarding the drivers of the evolution of $r(t)$ in terms of the regression model, Bódai et al. (2021) established for the MPI-GE that both the increasing ENSO variability, as seen in the second row of Fig. 20.7, and the increasing ENSO–ISM coupling given by the regression coefficient a (fourth row) take an effect, and typically the latter dominates, even outcompeting a typically increasing noise level σ_ξ (fifth row). The CESM1 model gives a very different picture. The ENSO variability σ_Φ features a much larger signal-to-noise ratio; not only the forced change signal is stronger, but the finite-N error fluctuations are comparable to that of the MPI-ESM despite the ensemble sizes differ by a factor of 2. The coupling, a, on the other hand, does not seem to feature much of a forced increase at all for (1) and (2), and shows a very different time dependence for (3). However, just like the ENSO variability σ_Φ, the noise level σ_ξ increases more in CESM1 compared to MPI-ESM, which seems to cancel the effect of the increasing σ_Φ.

It does seem, however, that in both models the coupling a features a non-monotonicity, at least for (2) and (3), a decrease around or before the turn of the century. This seems to be strong enough to cause a statistically significant decrease of r, as seen in Fig. 20.8 by the showing of blue patches. Yet, we do not think that this can lend support of any firmness to claims of a weakening ENSO–ISM teleconnection strength in the late twentieth century (Kumar et al., 1999), simply because our observation is made relying on information that is more than the historical record by at least as much as a factor of 30.

20.5   Summary

We investigate past and future changes in the ENSO–ISM relationship in response to GHG forcing based on two different approaches, the temporal and ensemble-wise methods. The temporal method estimates (biased) temporal statistics in single realizations of models and then takes the multimodel ensemble mean. This has been widely applied to model simulations contributed to different generations of the Coupled Model Intercomparison Project (CMIP). In this approach, the MMM of some temporal statistics seems to be commonly regarded as the forced response (signal). We argued that this is a flawed concept. The ensemble-wise method seeks to determine the time evolution of some ensemble-wise statistics, instead. In this conceptually sound approach, the forced response of teleconnections can be quantified by a time evolution of the ensemble-wise Pearson correlation coefficient.

The temporal method applied to historical (1850–2014) and SSP5-8.5 (2015–2100) runs of 30 CMIP6 models suggests that the negative correlation between ENSO and ISM will be weakened in response to GHG forcing in the latter half of the twenty-first century only if the long-term trend is incorrectly retained. The considerable *positive* trends of the AISMR and Niño 3.4 SST index in the SSP5-8.5 tend to reduce the negative correlation between them. If the long-term trend is removed as part of the calculation, the ENSO–ISM relationship during the twentieth and twenty-first century shows no robust change in the sense that most of the models agree on the sign of change, leaving us with inconclusive results. The MMM of the sample correlation coefficients of the individual models is −0.46 in the present (1965–2014) and −0.42 in the future (2051–2100).

In contrast, the ensemble-wise method applied to the MPI-GE does reveal a statistically significant strengthening of the negative ENSO–ISM correlation in response to GHG forcing. It is mainly due to the increase in the regression coefficient (Bódai et al., 2021) which can outcompete even an increasing noise strength. Although an increasing ENSO variability also plays a comparable role. On the other hand, the CESM1-LE shows only a slight strengthening of the relationship—which is also very weak in the first place—and only when the MCA is used to characterize the ENSO–ISM teleconnection. The main driver for the strengthening of the relationship in the CESM1-LE, the increase in ENSO variability under GHG warming, is compensated by the increasing noise strength while the coupling, a, does not show much increase. Beside these very different behaviors of the two ESMs, spatial characteristics including their forced response, such as the MCA modes and the maps of the local correlation coefficients on the Indian side, show a striking similarity.

Future forced changes of the ENSO–ISM teleconnection can be analyzed only in models. Even if any of the available models inaccurately represents some natural processes, at least an initial condition ensemble can be generated with them, which allows for a sound analysis of the forced response (Tél et al., 2020). Yet, in the case a statistical estimator is not unbiased, a *combination* of the temporal and ensemble-wise approaches should be sought to minimize biases (Bódai et al., 2021). Furthermore, a multimodel ensemble of initial condition ensembles should be analyzed to check if the forced change of an observable of interest is *robust or*, to the contrary, *sensitive* to parameter tuning and model errors.

Acknowledgment

JYL and TB were supported by the Institute for Basic Science (IBS), under IBS-R028-D1. TB would like to acknowledge the many discussions and collaboration with Gábor Drótos, who was the first to conceive the concept of SEOF, possibly jointly with János Márfy, as well as SMCA.

References

Ashok, K., Behera, S.K., Rao, S.A., Weng, H., Yamagata, T., 2007. El Niño Modoki and its possible teleconnection. J. Geophys. Res. 112, C11007.

Azad, S., Rajeevan, M., 2016. Possible shift in the ENSO-Indian monsoon rainfall relationship under future global warming. Sci. Rep. 6, 20145.

Bódai, T., Drótos, G., Ha, K.-J., Lee, J.-Y., Chung, E.-S., 2021: Nonlinear forced change and nonergodicity: The case of ENSO-Indian monsoon and global precipitation teleconnection. Front. Earth Sci. 8:599785. doi: 10.3389/feart.2020.599785.

Bódai, T., Drótos, G., Herein, M., Lunkeit, F., Lucarini, V., 2020. The forced response of the El Niño–Southern Oscillation–Indian monsoon teleconnection in ensembles of earth system models. J. Clim. 33, 2163–2182.

Bódai, T., Tél, T., 2012. Annual variability in a conceptual climate model: snapshot attractors, hysteresis in extreme events, and climate sensitivity. Chaos Interdiscip. J. Nonlinear Sci. 22, 023110.

Cash, B.A., Barimalala, R., Kinter, J.L., Altshuler, E.L., Fennessy, M.J., Manganello, J.V., Molteni, F., Towers, P., Vitart, F., 2017. Sampling variability and the changing ENSO–monsoon relationship. Clim. Dyn. 48, 4071–4079.

Chen, W., Dong, B., Lu, R., 2010. Impact of the Atlantic Ocean on the multidecadal fluctuation of El Niño–Southern Oscillation–South Asian monsoon relationship in a coupled general circulation model. J. Geophys. Res. 115, D17109.

DelSole, T., Shukla, J., 2012. Climate models produce skillful predictions of Indian summer monsoon rainfall. Geophys. Res. Lett. 39. doi:10.1029/2012GL051279.

Deser, C., Lehner, F., Rodgers, K.B., Ault, T., Delworth, T.L., DiNezio, P.N., Fiore, A., Frankignoul, C., Fyfe, J.C., Horton, D.E., Kay, J.E., Knutti, R., Lovenduski, N.S., Marotzke, J., McKinnon, K.A., Minobe, S., Randerson, J., Screen, J.A., Simpson, I.R., Ting, M., 2020. Insights from Earth system model initial-condition large ensembles and future prospects. Nat. Clim. Chang. 10, 277–286.

Drótos, G., Bódai, T., Tél, T., 2015. Probabilistic concepts in a changing climate: A snapshot attractor picture. J. Clim. 28, 3275–3288.

Drótos, G., Bódai, T., Tél, T., 2016. Quantifying nonergodicity in nonautonomous dissipative dynamical systems: an application to climate change. Phys. Rev. E 94, 022214.

Eyring, V., Bony, S., Meehl, G.A., Senior, C.A., Stevens, B., Stouffer, R.J., Taylor, K.E., 2016. Overview of the Coupled Model Intercomparison Project Phase 6 (CMIP6) experimental design and organization. Geosci. Model Dev. 9, 1937–1958.

Fan, F., Dong, X., Fang, X., Xue, F., Zheng, F., Zhu, J., 2017. Revisiting the relationship between the South Asian summer monsoon drought and El Niño warming pattern. Atmos. Sci. Lett. 18, 175–182.

Feba, F., Ashok, K., Ravichandran, M., 2019. Role of changed Indo-Pacific atmospheric circulation in the recent disconnect between the Indian summer monsoon and ENSO. Clim. Dyn. 52, 1461–1470.

Harris, I., Osborn, T.J., Jones, P., Lister, D., 2020. Version 4 of the CRU TS monthly high-resolution gridded multivariate climate dataset. Sci. Data 7, 109.

Haszpra, T., Herein, M., Bódai, T., 2020a. Investigating ENSO and its teleconnections under climate change in an ensemble view – a new perspective. Earth Syst. Dyn. 11, 267–280.

Haszpra, T., Topál, D., Herein, M., 2020b. On the time evolution of the Arctic oscillation and related wintertime phenomena under different forcing scenarios in an ensemble approach. J. Clim. 33, 3107–3124.

Herein, M., Márfy, J., Drótos, G., Tél, T., 2016. Probabilistic concepts in intermediate-complexity climate models: a snapshot attractor picture. J. Clim. 29, 259–272.

Hrudya, P.P.V.H., Varikoden, H., Vishnu, R.N., 2021. Changes in the relationship between Indian Ocean dipole and Indian summer monsoon rainfall in early and recent multidecadal epochs during different phases of monsoon. Int. J. Climatol. 41. doi:10.1002/joc.6685.

Huang, B., Thorne, P.W., Banzon, V.F., Boyer, T., Chepurin, G., Lawrimore, J.H., Menne, M.J., Smith, T.M., Vose, R.S., Zhang, H.-M., 2017. Extended Reconstructed Sea Surface Temperature, Version 5 (ERSSTv5): upgrades, validations, and intercomparisons. J. Clim. 30, 8179–8205.

Kay, J.E., Deser, C., Phillips, A., Mai, A., Hannay, C., Strand, G., Arblaster, J.M., Bates, S.C., Danabasoglu, G., Edwards, J., Holland, M., Kushner, P., Lamarque, J.F., Lawrence, D., Lindsay, K., Middleton, A., Munoz, E., Neale, R., Oleson, K., Polvani, L., Vertenstein, M., 2015. The community earth system model (CESM) large ensemble project : a community resource for studying climate change in the presence of internal climate variability. Bull. Am. Meteorol. Soc. 96 (8), 1333–1349.

Kripalani, R.H., Kulkarni, A., 1997. Climatic impact of El Niño/La Niña on the Indian monsoon: a new perspective. Weather 52, 39–46.

Krishnamurthy, V., Goswami, B.N., 2000. Indian monsoon-ENSO relationship on interdecadal timescale. J. Clim. 21 (3), 579–595.

Krishnan, R., Sugi, M., 2003. Pacific decadal oscillation and variability of the Indian summer monsoon rainfall. Clim. Dyn. 21, 233–242.

Krishnaswamy, J., Vaidyanathan, S., Rajagopalan, B., Bonell, M., Sankaran, M., Bhalla, R.S., Badiger, S., 2015. Non-stationary and non-linear influence of ENSO and Indian Ocean dipole on the variability of Indian monsoon rainfall and extreme rain events. Clim. Dyn. 45, 175–184.

Kucharski, F., Bracco, A., Yoo, J.H., Tompkins, A.M., Feudale, L., Ruti, P., Dell'Aquila, A., 2009. A Gill-Matsuno-type mechanism explains the tropical Atlantic influence on African and Indian monsoon rainfall. Quart. J. R. Meteorol. Soc. 135, 569–579.

Kumar, K.K., Rajagopalan, B., Cane, M.A., 1999. On the weakening relationship between the Indian monsoon and ENSO. Science 284, 2156–2159.

Kumar, K.K., Rajagopalan, B., Hoerling, M., Bates, G., Cane, M., 2006. Unraveling the mystery of Indian monsoon failure during El Nino. Science 314, 115–119.

Leith, C.E., 1978. Predictability of climate. Nature 276, 352–355.

Li, X., Ting, M., 2015. Recent and future changes in the Asian monsoon-ENSO relationship: natural or FORCED? Geophys. Res. Lett. 42, 3502–3512.

Lu, R., Dong, B., Ding, H., 2006. Impact of the Atlantic multidecadal oscillation on the Asian summer monsoon. Geophys. Res. Lett. 33, L24701.

Lund, R.B., von Storch, H., Zwiers, F.W., 2000. Statistical analysis in climate research. J. Am. Stat. Assoc. 95, 1375.

Maher, N., Milinski, S., Suarez-Gutierrez, L., Botzet, M., Dobrynin, M., Kornblueh, L., Kröger, J., Takano, Y., Ghosh, R., Hedemann, C., Li, C., Li, H., Manzini, E., Notz, D., Putrasahan, D., Boysen, L., Claussen, M., Ilyina, T., Olonscheck, D., Raddatz, T., Stevens, B., Marotzke, J., 2019. The Max Planck Institute Grand Ensemble: enabling the exploration of climate system variability. J. Adv. Model. Earth Syst. 11, 2050–2069.

Mohapatra, S., Gnanaseelan, C, Deepa, J.S., 2020. Multidecadal to decadal variability in the equatorial Indian Ocean subsurface temperature and the forcing mechanisms. Clim. Dyn 54, 3475–3487. https://doi.org/10.1007/s00382-020-05185-7.

O'Neill, B.C., Kriegler, E., Riahi, K., Ebi, K.L., Hallegatte, S., Carter, T.R., Mathur, R., van Vuuren, D.P., 2014. A new scenario framework for climate change research: the concept of shared socioeconomic pathways. Clim. Change 122, 387–400.

Parthasarathy, B., Munot, A.A., Kothawale, D.R., 1994. All-India monthly and seasonal rainfall series: 1871?1993. Theor. Appl. Climatol. 49, 217–224.

Ramu, D.A., Chowdary, J.S., Ramakrishna, S.S.V.S., Kumar, O.S.R.U.B., 2018. Diversity in the representation of large-scale circulation associated with ENSO-Indian summer monsoon teleconnections in CMIP5 models. Theor. Appl. Climatol. 132, 465–478.

Romeiras, F.J., Grebogi, C., Ott, E., 1990. Multifractal properties of snapshot attractors of random maps. Phys. Rev. A 41, 784–799.

Roy, I., Tedeschi, R.G., Collins, M., 2019. ENSO teleconnections to the Indian summer monsoon under changing climate. Int. J. Climatol. 39, 3031–3042.

Swart, N.C., Cole, J.N.S., Kharin, V.V., Lazare, M., Scinocca, J.F., Gillett, N.P., Anstey, J., Arora, V., Christian, J.R., Hanna, S., Jiao, Y., Lee, W.G., Majaess, F., Saenko, O.A., Seiler, C., Seinen, C., Shao, A., Sigmond, M., Solheim, L., von Salzen, K., Yang, D., Winter, B., 2019. The Canadian Earth System Model version (CanESM5.0.3). Geosci. Model Dev. 5 (12), 4823–4873.

Tél, T., et al., 2020. The theory of parallel climate realizations. J. Statis. Phys. 179, 1496–1530.

Wang, B., Xiang, B., Li, J., Webster, P.J., Rajeevan, M.N., Liu, J., Ha, K.-J., 2015. Rethinking Indian monsoon rainfall prediction in the context of recent global warming. Nat. Commun. 6, 7154.

Yettella, V., Weiss, J.B., Kay, J.E., Pendergrass, A.G., 2018. An ensemble covariance framework for quantifying forced climate variability and its time of emergence. J. Clim. 31, 4117–4133.

Yun, K.-S., Timmermann, A., 2018. Decadal monsoon-ENSO relationships reexamined. Geophys. Res. Lett. 45, 2014–2021.

Ziese, M., Rauthe-Schöch, A., Becker, A., Finger, P., Meyer-Christoffer, A., Schneider, U., 2018. GPCC Full Data Daily Version.2018 at 1.0°: daily land-surface precipitation from rain-gauges built on GTS-based and historic data. DOI: 10.5676/DWD_GPCC/FD_D_V2018_100.

Chapter 21

Response of the positive Indian Ocean dipole to climate change and impact on Indian summer monsoon rainfall

Wenju Cai[a,b], **Guojian Wang**[a,b], **Ziguang Li**[a], **Xiaotong Zheng**[a], **Kai Yang**[b,c], **Benjamin Ng**[b]

[a]*Key Laboratory of Physical Oceanography–Institute, for Advanced Ocean Studies, Ocean University of China, and Qingdao National Laboratory for Marine Science and Technology, Yushan Road, Qingdao, China,* [b]*Center for Southern Hemisphere Oceans Research (CSHOR), CSIRO Oceans and Atmosphere, Hobart, Australia,* [c]*State Key Laboratory of Numerical Modeling for Atmospheric Sciences and Geophysical Fluid Dynamics, Institute of Atmospheric Physics, Chinese Academy of Sciences, Beijing, China*

21.1 Introduction

The Indian Ocean dipole (IOD), a consequential mode of sea surface temperature (SST) variability of the tropical Indian Ocean, induces climate and weather extremes with devastating impacts (Abram et al., 2003; Ashok et al., 2003; Behera et al., 2005; Black et al., 2003; Cai et al., 2009; Murtugudde et al., 2000; Saji et al., 1999; Ueda and Matsumoto, 2000; Ummenhofer et al., 2009; Webster et al., 1999; Yu and Rienecker, 1999; Zubair et al., 2003). At its positive phase (pIOD), the observed SST is measured jointly by warm SST anomalies in the west pole (50°E–70°E and 10°S–10°N) and cold SST anomalies in the east pole (90°E–110°E and 10°S–0°S), described by an SST dipole mode index (DMI) (Saji et al., 1999). A pIOD event usually starts to develop in austral winter (June, July, August) and peaks in following spring (September, October, November, SON). Initially, cooling off Sumatra-Java suppresses local atmospheric convection, generating anomalous easterly winds. The easterlies promote upwelling and lift the thermocline that in turn reinforce the initial cooling (Saji et al., 1999; Webster et al., 1999), a process referred to as local Bjerknes feedback (Bjerknes 1969). The growth of cool anomalies causes a north-westward extension of the south-easterly trades, which flow over the southern tropical Indian Ocean in austral winter and spring, inducing anomalous easterlies along the equatorial

Indian Summer Monsoon Variability: El Niño-teleconnections and beyond.
DOI: https://doi.org/10.1016/B978-0-12-822402-1.00010-7

Indian Ocean where weak westerlies normally prevail (e.g., Schott et al., 2009). The anomalous equatorial easterly winds promote convergence, warm anomalies, and rainfall in the equatorial western Indian Ocean. The altered ocean-atmosphere circulations induce droughts, heat waves, and bushfires in Australia (Ashok et al., 2003; Cai et al., 2009; Ummenhofer et al., 2009), floods in East Africa (Behera et al., 2005; Black et al., 2003), coral reef death across western Sumatra (Abram et al., 2003), and malaria outbreaks in East Africa (Hashizume et al., 2012).

During the extreme 1997 pIOD, the IOD-induced anomalous circulation was strong (Murtugudde et al., 2000; Webster et al., 1999; Yu and Rienecker, 1999); floods in the African countries were particularly devastating, leading to several thousand deaths and hundreds of thousands of people displaced, whereas in Indonesia wild fires burnt for many months, affected the global carbon budget (Page et al., 2002) and the lives of tens of millions of people (Schott et al., 2009). During the 2019 pIOD, similar devastating impacts were seen (Wang et al., 2020a;(Cai et al., 2021)).

Development phase of a pIOD coincides with the Indian summer monsoon rainfall (ISMR) peak season, from June to September (JJAS). The ISMR derives its moisture from vapor transported by southwesterlies, supporting convection over the Indian subcontinent (Gadgil, 2003; Krishnan et al., 2012). Interannual fluctuations of ISMR often manifest as floods and droughts, affecting rainfall-dependent lives, food security, and the livelihood of more than one-sixth of the world's population (Gadgil and Gadgil, 2006; Kshiagar et al., 2006; Singh et al., 2014). In addition to impacts from the El Niño-Southern Oscillation (ENSO), the ISMR is also affected by the IOD. Although an El Niño event leads to anomalously low ISMR by suppressing convection over the Indian subcontinent through anomalous high pressure embedded in the Southern Oscillation (Ropelewski and Halpert 1987; Ju and Slingo, 1995; Webster and Yang, 1992), a strong pIOD tends to favor anomalously high ISMR, because the cold anomalies in the eastern Indian Ocean are conducive to vapor transport toward the continent as a result of the anomalous meridional circulation (Ashok et al., 2001; Ashok et al., 2004; Behera et al., 1999). The vapor and the warmer temperatures to the north delay the seasonal southward movement of the Indian Ocean Intertropical Convergence Zone, prolonging and enhancing the ISMR. Thus, a pIOD when concurrent with an El Niño offsets the impact of an El Niño on the ISMR, and vice versa. During 1997, the extreme pIOD occurred in conjunction with the extreme 1997 El Niño, leading to an almost neutral condition in terms of ISMR. During 2019, the extreme pIOD was responsible for the intense ISMR that lasted longer into the September, when the Pacific saw only a mild warm condition (Wang et al., 2020a). There is also evidence for a positive feedback between the pIOD and ISMR, in which the enhanced ISMR, which peaks earlier than the pIOD, in turn favors growth of the pIOD (Kulkarni et al. 2007).

Because of these impacts, determining whether and how pIOD events may change under greenhouse warming is one of the most important issues in climate science. However, for several model generations, there has been no intermodel consensus on its future change, owing to a lack of intermodel consensus on its future change using historical DMI (Cai et al., 2013; Hui and Zheng, 2018; Zheng et al., 2013). The lack of consensus in turn hinders progress on the issue of how the IOD's impact on the ISMR may change in a warming climate. Recent studies (Cai et al. 2014b, 2018b; Wang et al. 2017) show that in terms of a rainfall metrics, the frequency of extreme pIOD increases under greenhouse warming. In this chapter, we review the recent progress in the IOD response and assess the impact on ISMR from the potential change of pIOD under greenhouse warming.

The rest of this chapter is organized as follows. In Section 21.2, we describe data and models used; in Section 21.3.1, we review the recent progress in the response of pIOD to greenhouse warming, in particular, we identify extreme and moderate pIOD events for assessing their impact on the ISMR. In Section 21.3.2, we compare modeled and observed pIOD impact on ISMR, before examining its possible change. The chapter is concluded in Section 21.4.

21.2 Methods and data

To investigate observed influences of the IOD on ISMR, we use SST data from Hadley Centre Sea Ice and Sea Surface Temperature dataset version 1 (HadISST v1.1; Rayner et al., 2003) based on the 1900–1999 period and rainfall data from Global Precipitation Climatology Project monthly precipitation analysis (GPCP; Adler et al., 2003) from 1979 onwards. To assess the influence of greenhouse warming, we use SST and rainfall data from 34 models participating in the Coupled Model Intercomparison Project phase 5 (CMIP5; Taylor et al., 2012) under historical greenhouse gas and natural forcing and the Representative Concentration Pathway 8.5 (RCP8.5) emission scenario, each covering 200 years from 1900. We separate the coverage into two periods, that is, 1900–1999 as the "present-day" and 2000–2099 as the "future" period.

We use the SST data to calculate the traditional DMI (Saji et al., 1999), measured by an SST anomaly difference between the west (50°E–70°E, 10°S–10°N) and east (90°E–110°E, 10°S–0°) equatorial Indian Ocean. However, the DMI cannot capture the essential processes associated with extreme pIOD events (Cai et al., 2014b). Therefore, we utilize rainfall metrics following Cai et al. (2014b) by applying empirical orthogonal function (EOF) analysis to SON rainfall anomalies in the tropical Indian Ocean (40°E–100°E, 10°S–10°N) to define an extreme pIOD event. The first principal pattern is equivalent to the traditional dipole structure and the second principal pattern highlights the westward extension of dry anomalies from the eastern tropical Indian Ocean (see Figs. 21a and 21b in Cai et al., 2014b). An extreme pIOD event is defined as when the first principal component (PC1) ≥ 1 standard deviation (SD) and the second principal component (PC2) ≥ 0.5 SD. Not all models are able to generate extreme pIOD events. A total of 23 models are selected based on their ability to generate the observed nonlinear relationship between PC1 and PC2 (Cai et al., 2014b;

see also Fig. 21.3A for the observations). We use the DMI to define a moderate pIOD event as when the DMI is greater than 0.75 SD but not an extreme pIOD.

To separate the El Niño's influence on ISMR, we use the Niño3.4 SST index (5°N–5°S, 170°W–120°W) to define an El Niño event as when Niño3.4 averaged over austral summer (December, January, February, DJF) is greater than 0.5 SD. We then stratify extreme and moderate pIOD event into two subsets, that is, concurrent extreme (moderate) pIOD events with a developing El Niño event that peaks in the following DJF season, and independent extreme (moderate) pIOD events that do not occur with an El Niño event. Projected changes are calculated as the difference between the 2000–2099 and the 1900–1999 period, appropriately scaled by global warming in each model.

We deploy a Poisson distribution (Ahrens and Dieter, 1982) to evaluate the probability of pIOD occurrences, which is a discrete probability distribution expressing the probability of a given number of events occurring in a fixed interval of time. To test whether two composites are statistically significant, we apply a Student's t-test.

21.3 The Indian Ocean dipole in a warming climate

21.3.1 Response of the IOD to greenhouse warming

As discussed in the Introduction, based on the DMI, there is no definitive change in the amplitude of the IOD under greenhouse warming. This is confirmed in Fig. 21.1A, which compares the SD of the SON DMI in the 1900–1999 (present-day) and 2000–2099 (future) climate, based on CMIP5 models. Although a total of 21 of 34 models produce a decrease, the multimodel ensemble mean change is not statistically significant. This lack of a consistent change also translates to the frequency of extreme pIOD events defined as when the DMI is greater than 1.5 SD (Fig. 21.1B).

The lack of agreement is despite a strong intermodel consensus on the mean state change in the tropical Indian Ocean, which features an easterly wind trend along the equator and a faster warming rate in the west than in the east, faster warming in the northern than the southern tropical Indian Ocean (Fig. 21.2A), a shallowing equatorial thermocline (Cai et al., 2013; Vecchi et al., 2006; Zheng et al., 2010), and certain characteristics of pIOD extremes (Cai et al., 2014b). First, a projected faster warming in the ocean surface layer than at depth increases the role of the relatively cold subsurface water in the eastern Indian Ocean through a greater influence on eastern Indian Ocean SST, and a greater response of the thermocline to winds, but a weaker response of winds over the eastern Indian Ocean to SST anomalies because the lower atmosphere is more stable under greenhouse warming (Zheng et al., 2013). As such, the net change from the two opposing components to the strength of the positive feedback varies by models, giving rise to the lack of consensus. Second, the enhanced north-minus-south SST gradient and the weakened Walker Circulation facilitate the Intertropical Convergence Zone in the tropical eastern Indian Ocean to

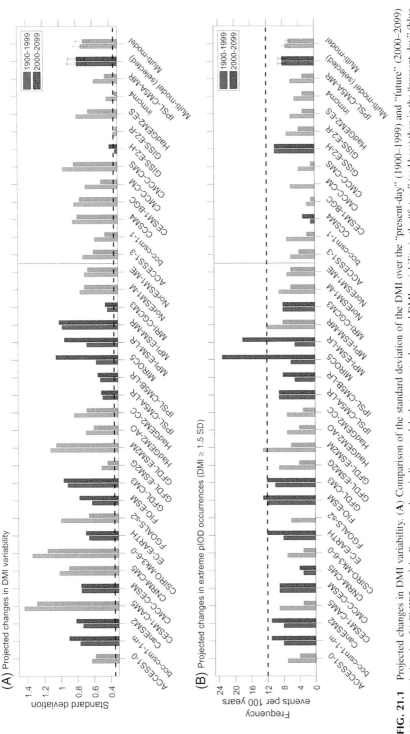

FIG. 21.1 Projected changes in DMI variability. (A) Comparison of the standard deviation of the DMI over the "present-day" (1900–1999) and "future" (2000–2099) 100-year periods using 34 CMIP5 models. Bars grayed out indicate models that generate decreased DMI variability in the "future" (red bars) than in the "present-day" (blue bars) climate. (B) Number of extreme pIOD occurrences, defined as when the DMI is greater than 1.5 SD that occurred in the two 100-year periods. Bars grayed out indicate models that generate decreased extreme pIOD frequency in the "future" (red bars) than in the "present-day" (blue bars) climate. Observed values are indicated by the horizontal dashed lines in A and B. The first 23 models (on the left side of the vertical line) are selected based on their ability to simulate the nonlinear relationship between rainfall EOF1 and EOF2 (Methods and Data). Multimodel ensemble values for the 23 selected models and all 34 models are indicated at the end of each bar plot. The error bars in (A) correspond to the one standard deviation of the multimodel DMI variance; the error bars in (B) correspond to the 90% confidence interval based on a Poisson distribution (Methods and Data). Results are based on the SON season when an IOD event matures. Observed DMI is calculated using HadISST over the 1900–1999 period.

stay north (Weller and Cai, 2014) closer to the Indian continent, or an anomalous northward shift, during a pIOD, particularly during strong events; in the zonal direction, the eastern Indian Ocean convection moves to the west. Third, a faster warming in the west than the east equatorial Indian Ocean facilitates an increased frequency of westward shifts of the atmospheric convergence zone from the east, even if SST variability does not change (Cai et al., 2014b; Wang et al. 2017). This leads to increased mean rainfall to the north under greenhouse warming (Fig. 21.2B), and an increase in extreme pIOD using rainfall metrics, as detailed next.

The assessment using variability of the DMI discussed above does not separate the response of moderate and extreme pIOD. Further, the DMI is not able

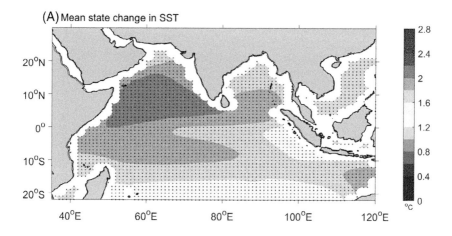

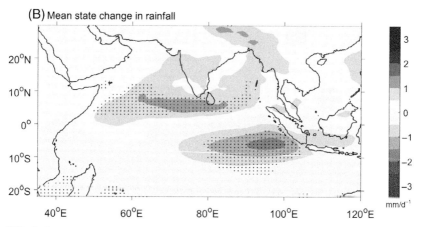

FIG. 21.2 Projected changes in the mean state of (A) JJAS SST (°C), and (B) JJAS rainfall (mm/day) between 2000–2099 and 1900–1999 based on 23 selected models (Methods and Data). Dotted areas indicate where the difference between the two periods is significant above the 90% confidence level based on a Student's *t*-test. Results are based on JJAS, the peak ISMR season.

to differentiate extreme from moderate pIOD events. In fact, the extremity of pIOD impacts needs to involve a combination of the first and second principal component of an EOF analysis on rainfall anomalies in the tropical Indian Ocean (40°E–100°E, 10°S–10°N) (Cai et al., 2014b, 2018b; Wang et al., 2017). The EOF analysis yields modes each with a spatial pattern and a principal component scaled to unity. The first EOF depicts the traditional dipole structure, the associated principal component being strongly correlated with the DMI. The second EOF highlights a westward extension of the dry anomalies from the eastern Indian Ocean along much of the equatorial Indian Ocean but an enhanced convergence over the equatorial western Indian Ocean and eastern Africa (See Fig. 2c and 2d of Cai et al., 2014b).

During an observed moderate pIOD, cold and dry anomalies are concentrated off Sumatra-Java, as indicated by the generally small values of EOF1 (green and cyan stars in Fig. 21.3A). In some pIOD events, the cold anomalies expand toward the equator, triggering additional equatorial processes in which equatorial easterlies over the eastern equatorial Indian Ocean drive equatorial ocean anomalous zonal currents, upwelling, and zonal SST gradients, initiating nonlinear zonal and vertical advection feedbacks resulting an extreme pIOD events, in which the equatorial cold and dry anomalies grow rapidly and expand westward, leading to strong negative rainfall skewness over the equatorial eastern Pacific (Cai et al., 2014b). In this way, the PC2 can be described in terms of the first component by a quadratic fit of $PC2(t) = \alpha[PC1(t)]^2 + \beta PC1(t) + \gamma$. The 1994, 1997, and 2019 pIOD events are extreme pIOD in terms of the DMI but also in terms of their impacts as depicted by the rainfall EOF analysis (red and purple stars in Fig. 21.3A). Moderate pIOD events are defined as when the DMI is greater than 0.75 SD but are not an extreme pIOD.

Using the ability to simulate negative rainfall skewness and the nonlinear relationship between EOF1 and EOF2 to benchmark CMIP5 models as in Cai et al. 2014b, we select 23 CMIP5 models, and apply EOF analysis as described above. The DMI in the selected models still shows no statistically significant change (the first 23 models in Fig. 21.1B). In aggregation, these models generate an increase in extreme pIOD frequency (red and purple dots, Fig. 21.3A and B) from 5.8 events per 100 years in the "present-day" climate to 16.0 events per 100 years in the "future" climate, with 21 of the 23 selected models showing an increase, that is, a strong intermodel consensus (Fig. 21.3C). On the contrary, there is a reduction in the frequency of moderate pIODs, from 17.3 events per 100 years to 12.5 events per 100 years (green and cyan dots, Fig. 21.3), similarly with a strong intermodel consensus.

Given that extreme El Niño and central Pacific El Niño are projected to increase (Cai et al., 2014a, 2018a; Wang et al., 2020b), it is relevant to address whether these changes in pIOD are in part forced by ENSO. A previous study (Cai et al., 2014b) has shown that the increase in extreme pIOD frequency is not associated with an increased frequency of extreme El Niño, which occurs due to a faster warming in the equatorial eastern Pacific under greenhouse warming.

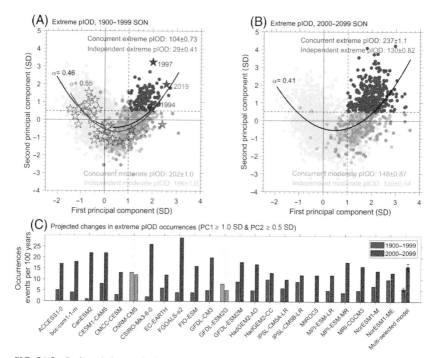

FIG. 21.3 Projected changes in the occurrence of extreme pIOD events for the (A) "present-day" (1900–1999) and (B) the "future" (2000–2099) period, as indicated by a and b, respectively. The red dots indicate extreme pIOD events as when PC1 is greater than 1 SD and PC2 is greater than 0.5 SD (Cai et al., 2014b) that occur concurrently with an El Niño event in the following DJF season defined as when Niño3.4 is greater than 0.5 SD; the purple dots indicate extreme pIOD events that are not followed by an El Niño event in the following DJF season. The moderate pIOD events are defined as when the DMI is greater than 0.75 SD but not an extreme pIOD. When it occurs with a developing El Niño that matures in the following DJF season, it is recorded as a concurrent moderate pIOD event by a green dot; otherwise, it is recorded as an independent moderate pIOD event by a cyan dot. All other years are indicated by gray dots. The total occurrences of each type of pIOD event are also indicated in each panel with the 90% confidence interval based on a Poisson distribution. Observed values are based on the 1979–2019 using GPCP rainfall data (Adler et al., 2003) and they are shown by pentagrams with 1994 and 1997 as concurrent extreme pIOD events; 2019 as an independent extreme pIOD event; 1982, 2002, and 2006 as concurrent moderate pIOD events; 1987 and 2018 as independent moderate pIOD events. The black and pink curves indicate a quadratic fit for model results and observations, respectively, following $PC2(t) = \alpha[PC1(t)]^2 + \beta PC1(t) + \gamma$. α indicates the nonlinearity of the curve. Results are based on the SON season. (C) Projected change in extreme pIOD occurrences defined as when PC1 is greater than 1 SD and PC2 is greater than 0.5 SD using rainfall anomalies (Cai et al., 2014b). The error bars in C correspond to the 90% confidence interval based on a Poisson distribution (Methods and Data).

Here we have examined if a relationship exists between the projected change in pIOD frequency and frequency of El Niño, defined as when Niño3.4 > 0.5 SD. We find that neither the reduction in the moderate pIOD nor the increase in extreme pIOD is associated with a change in the El Niño (Fig. 21.4A and B).

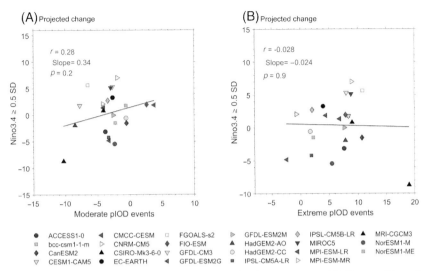

FIG. 21.4 Multimodel statistics between El Niño and pIOD occurrences. (A) Changes ("future" minus "present-day" period) in the number of occurrences per one hundred years of moderate pIOD events (DMI is greater than 0.75 SD, but not an extreme pIOD) and El Niño events (Niño3.4 is greater than 0.5 SD), normalized by the global warming in the corresponding model, by using the selected 23 models. (B) The same as (**A**), but for changes in the number of occurrences of extreme pIOD events and El Niño events. The intermodel correlation, slope, and statistical significance are indicated in each panel. Metrics for pIOD use data averaged over the SON season; Niño3.4 index is averaged over the following DJF season.

21.3.2   Impact of pIOD on the ISMR in a warming climate

Available observations show that a moderate pIOD tends to have a small impact (Fig. 21.5A and C; see Chapter-9). It is the strong pIOD (Fig. 21.5B) Fig. 21.5C that leads to an increase in ISMR (Fig. 21.5D). However this classification of the pIOD events might mix the impact of pIOD with that of El Niño because pIOD and El Niño tend to have opposite influences (Ashok et al., 2004; Li et al., 2017). Further stratification into pIOD events concurrent with or independent of an El Niño finds that for the moderate pIOD, the impact in either catalogue is similarly small (Fig. 21.6A and C). Although this stratification may be sensitive to the definition, for example, 1987 could be classified as an El Niño year, it would not change our conclusion, because there is little difference in the impacts between the two types (Fig. 21.6A and B). In both catalogues, an extreme pIOD favors anomalously high ISMR, as reported by previous studies (Ashok et al., 2004, 2001; Behera et al., 1999), but for the extreme pIOD concurrent with an El Niño, because of their opposing impacts, the increase in ISMR is smaller than that for the independent events (Fig. 21.6B and D), though there is only one such independent event as in 2019.

Climate model simulation of pIOD's impact on ISMR suffers from several deficiencies. First, the modeled pIOD impact on ISMR is overly weak,

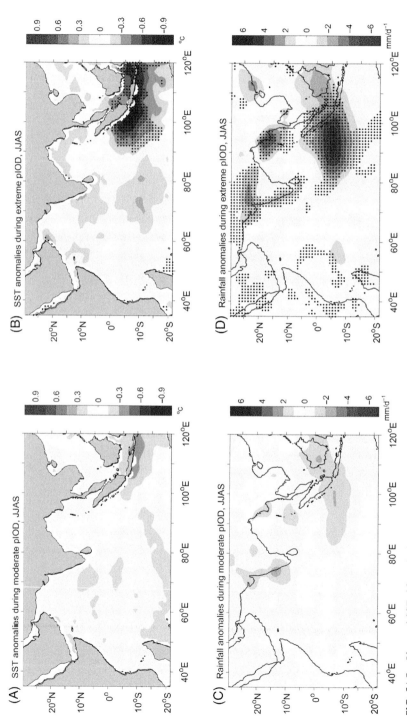

FIG. 21.5 Observed SST (°C) and rainfall (mm/day) distributions during moderate pIOD events and extreme pIOD events over the JJAS season. (A and B) SST composites during moderate pIOD events (green and cyan pentagrams in Fig. 21.3A) and extreme pIOD events (red and purple pentagrams in Fig. 21.3A), respectively. Dotted areas in B indicate where the difference between A and B is significant above the 90% confidence level based on a Student's t-test. (C and D) Same as A and B, but using rainfall anomalies. SST and rainfall anomalies are quadratically detrended before composites.

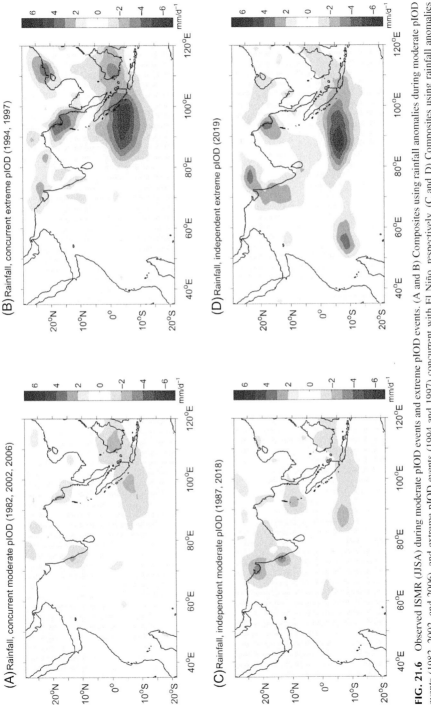

FIG. 21.6 Observed ISMR (JJSA) during moderate pIOD events and extreme pIOD events. (A and B) Composites using rainfall anomalies during moderate pIOD events (1982, 2002, and 2006), and extreme pIOD events (1994 and 1997) concurrent with El Niño, respectively. (C and D) Composites using rainfall anomalies during moderate pIOD events (1987 and 2018) and extreme pIOD event (2019) independent of El Niño, respectively. Rainfall anomalies are quadratically detrended before composites.

manifesting as a shorter time fraction in which a pIOD causes an increase in ISMR, leading to an overly weak IOD-ISMR positive correlation (Li et al., 2016). This is at least in part associated with the fact that both the IOD and the ISMR are highly controlled by the ENSO (Li et al., 2017). Thus, we will assess the potential change in the impact from pIOD using events independent from El Niño. Second, the impact from pIOD tends to be located south of the observed for both the moderate and extreme pIOD events (Fig. 21.7A and B), but shows little difference in the patterns between the moderate and the extreme pIOD. Taking this into account, there is a stronger impact from the extreme pIOD than from the moderate pIOD using detrended anomalies (Fig. 21.7A and B). In terms of detrending rainfall anomalies, there is little difference in the impact of a pIOD between the present-day and the future climate for both moderate (comparing Fig. 21.7A and C) and extreme pIOD (comparing Fig. 21.7B and D) events.

However, consistent with the increased frequency of extreme pIOD events (Fig. 21.3), the frequency of independent pIOD events increases dramatically by as much as 300% with a total of 19 of 23 models producing an increase (Fig. 21.3 and column 4 of Table 21.1). This suggests that the frequency of extreme ISMR as seen in 2019 is likely to increase, against a backdrop of a decreasing frequency of events with a mild impact, as the frequency of moderate pIOD events decreases with a 18 of 23 models showing a reduction (Fig. 21.3 and column 7 of Table 21.1).

To investigate if detrending removes some of the rainfall trend that is potentially part of the intensified impact of the pIOD under greenhouse warming, we carry out the same composite analysis as for Fig. 21.7 but using raw rainfall anomalies. In a similar context, a previous study has found that an increase in extreme El Niño frequency will contribute to more than 50% of the mean rainfall increase in the eastern equatorial Pacific under greenhouse warming (Cai et al., 2017). In terms of raw rainfall anomalies, there is an intensified impact on ISMR for either moderate or extreme pIOD events (compare Fig. 21.8A and C, and Fig. 21.8B and D). Importantly, the intensity increase is greater for extreme pIOD than for moderate pIOD. Further, compared with the mean state rainfall change (Fig. 21.2B), the ISMR increase is greater during the extreme pIOD than the increase in the mean, suggesting that the increased ISMR under greenhouse warming during an extreme positive pIOD is *not* due to the mean increase. On the contrary, the increased ISMR as a result of the extreme pIOD contributes to the mean. Thus under greenhouse warming, not only are extreme pIOD events more frequent, their impacts are also stronger.

21.4 Conclusions

Under greenhouse warming, a faster warming in the west and north of the equatorial Indian Ocean, relative to the eastern and southern, favors atmospheric convection in the west and convergence in the north, leading to an increased

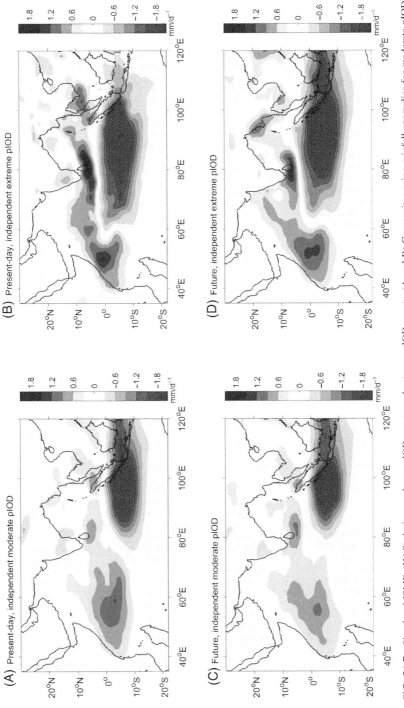

FIG. 21.7 Simulated ISMR (JJAS) during moderate pIOD events and extreme pIOD events. (A and B) Composites using rainfall anomalies for moderate pIOD events and extreme pIOD events independent of El Niño, respectively, in the "present-day" climate. (C and D) The same as A and B, but for the "future" climate. Rainfall anomalies are quadratically detrended before composites.

TABLE 21.1 Performance of 23 selected CMIP5 models forced under historical and climate change emission scenario RCP8.5. An extreme pIOD event is defined as when PC1 is greater than 1 SD and PC2 is greater than 0.5 SD. A moderate pIOD event is defined as when DMI is greater than 0.75 SD but not extreme. Metrics for pIOD is based on SON season. Niño3.4 index is utilized to evaluate if extreme and moderate pIOD events are followed by an El Niño event in the following DJF season defined as when Niño3.4 is greater than 0.5 SD.

CMIP5 models	Extreme pIOD 1900–1999 /2000–2099	Extreme pIOD with Niño3.4 ≥ 0.5 SD. 1900–1999 /2000–2099	Extreme pIOD with Niño3.4 < 0.5 SD. 1900–1999 /2000–2099	Moderate pIOD 1900–1999 /2000–2099	Moderate pIOD with Niño3.4 ≥ 0.5 SD. 1900–1999 /2000–2099	Moderate pIOD with Niño3.4 < 0.5 SD. 1900–1999 /2000–2099
ACCESS1-0	5/17	4/10	1/7	18/12	9/4	9/8
bcc-csm1-1-m	4/18	3/12	1/6	16/15	9/9	7/6
CanESM2	1/22	0/9	1/13	18/17	8/4	10/13
CESM1-CAM5	8/22	5/19	3/3	21/8	14/4	7/4
CMCC-CESM	3/13	1/11	2/2	18/12	10/5	8/7
CNRM-CM5	13/12	12/11	1/1	18/12	13/11	5/1
CSIRO-Mk3-6-0	2/26	2/11	0/15	24/11	9/2	15/9
EC-EARTH	6/12	3/8	3/4	18/14	7/8	11/6
FGOALS-s2	4/29	4/18	0/11	22/7	9/5	13/2
FIO-ESM	4/16	4/9	0/7	13/17	10/11	3/6
GFDL-CM3	5/20	5/17	0/3	17/12	11/5	6/7

GFDL-ESM2G	8/5	8/3	0/2	12/8	5/4	7/4
GFDL-ESM2M	9/18	9/12	0/6	14/11	9/7	5/4
HadGEM2-AO	5/17	3/10	2/7	22/9	6/4	16/5
HadGEM2-CC	10/13	8/5	2/8	13/12	6/3	7/9
IPSL-CM5A-LR	6/10	4/5	2/5	19/12	9/6	10/6
IPSL-CM5B-LR	9/12	6/8	3/4	19/14	9/8	10/6
MIROC5	0/12	0/10	0/2	19/15	13/13	6/2
MPI-ESM-LR	5/12	4/9	1/3	12/18	5/11	7/7
MPI-ESM-MR	4/18	4/11	0/7	14/11	6/6	8/5
MRI-CGCM3	5/16	4/13	1/3	19/14	7/5	12/9
NorESM1-M	7/14	6/8	1/6	16/13	11/7	5/6
NorESM1-ME	10/13	5/8	5/5	16/13	7/6	9/7
Multimodel ensemble	**133/367**	**104/237**	**29/130**	**398/287**	**202/148**	**196/139**

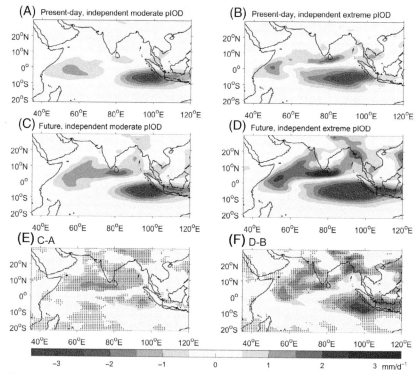

FIG. 21.8 Simulated ISMR (JJAS) during moderate and extreme pIOD events using total rainfall anomalies. (A and B) Composites using rainfall anomalies for moderate and extreme pIOD events independent of El Niño, respectively, in the "present-day" climate. (C and D) The same as A and B, but for the "future" climate. (E and F), Projected changes between the two periods. Dotted areas indicate where the differences are significant above the 90% confidence level based on a Student's *t*-test. Results are based on total rainfall anomalies before composites to highlight possible influence from the mean state changes.

frequency of extreme pIOD events, whereas frequency of moderate pIOD events decreases. Although these changes are not forced by changes in ENSO under greenhouse warming, pIOD events concurrent with developing El Niño events often occur. Because a strong pIOD tends to induce an increase in ISMR whereas a strong El Niño tends to decrease ISMR; and because the IOD and the ISMR are both overly dominated by ENSO, we focused on the impact of pIOD events independent of El Niño. We find that greenhouse warming leads to a dramatic increase in the frequency of independent extreme pIOD, whereas the frequency of independent moderate pIOD decreases. Therefore, a higher frequency of extreme ISMR is projected even if the impact of the pIOD does not change, but we further show that the impact of future extreme pIOD intensifies, suggesting that the impact of future extreme pIOD on ISMR is likely to be stronger in the future. Thus, our study suggests that more extreme ISMR is likely to occur more frequently as a result of the response of pIOD to greenhouse warming. However, these models still suffer from many biases. The extent to which the projected changes are impacted by these biases awaits examination.

Acknowledgments

This work is supported by National Key R&D Program of China (2018YFA0605700) and the Strategic Priority Research Program of Chinese Academy of Sciences, Grant No. XDB 40030000. Z. L. is supported by the National Natural Science Foundation of China (NSFC) Project (41975089). This work is also supported by the Centre for Southern Hemisphere Oceans Research, a joint research Centre between QNLM and CSIRO. W. C. and G. W. are supported by the Earth Systems and Climate Change Hub of the Australian Government's National Environmental Science Program.

Author Contributions

W.C. conceived the study and wrote the initial draft of the paper. G.W. performed the observed and model output analysis. All authors contributed to interpreting results, discussion of the associated dynamics and improvement of this paper.

References

Abram, N.J., Gagan, M.K., McCulloch, M.T., Chappell, J., Hantoro, W.S., 2003. Coral reef death during the 1997 Indian Ocean dipole linked to Indonesian wildfires. Science 301, 952–955.

Adler, R.F., Huffman, G.J., Chang, A., Ferraro, R., Xie, P.P., Janowiak, J., Rudolf, B., Schneider, U., Curtis, S., Bolvin, D., Gruber, A., Susskind, J., Arkin, P., Nelkin, E., 2003. The version-2 global precipitation climatology project (GPCP) monthly precipitation analysis (1979-present). J. Hydrometeorol. 4, 1147–1167.

Ahrens, J.H., Dieter, U., 1982. Computer generation of Poisson deviates. ACM Trans. Math. Softw. 8, 163–179.

Ashok, K., Guan, Z.Y., Saji, N.H., Yamagata, T., 2004. Individual and combined influences of ENSO and the Indian Ocean dipole on the Indian summer monsoon. J. Clim. 17, 3141–3155.

Ashok, K., Guan, Z.Y., Yamagata, T., 2001. Impact of the Indian Ocean dipole on the relationship between the Indian monsoon rainfall and ENSO. Geophys. Res. Lett. 28, 4499–4502.

Ashok, K., Guan, Z.Y., Yamagata, T., 2003. Influence of the Indian Ocean Dipole on the Australian winter rainfall. Geophys. Res. Lett., 30.

Behera, S.K., Krishnan, R., Yamagata, T., 1999. Unusual ocean-atmosphere conditions in the tropical Indian Ocean during 1994. Geophys. Res. Lett. 26, 3001–3004.

Behera, S.K., Luo, J.J., Masson, S., Delecluse, P., Gualdi, S., Navarra, A., Yamagata, T., 2005. Paramount impact of the Indian Ocean dipole on the East African short rains: a CGCM study. J. Clim. 18, 4514–4530.

Bjerknes, J., 1969. Atmospheric teleconnections from the equatorial Pacific. Mon. Wea. Rev. 97, 163–172. doi:10.1175/1520-0493(1969)097<0163:ATFTEP>2.3.CO;2.

Black, E., Slingo, J., Sperber, K.R., 2003. An observational study of the relationship between excessively strong short rains in coastal East Africa and Indian Ocean SST. Mon. Weather Rev. 131, 74–94.

Cai, W., Cowan, T., Raupach, M., 2009. Positive Indian Ocean dipole events precondition southeast Australia bushfires. Geophys. Res. Lett., 36.

Cai, W., Borlace, S., Lengaigne, M., van Rensch, P., Collins, M., Vecchi, G., Timmermann, A., Santoso, A., McPhaden, M.J., Wu, L., England, M.H., Wang, G., Guilyardi, E., Jin, F.-F., 2014a. Increasing frequency of extreme El Niño events due to greenhouse warming. Nat. Clim. Change 4, 111–116.

Cai, W., Santoso, A., Wang, G., Weller, E., Wu, L., Ashok, K., Masumoto, Y., Yamagata, T., 2014b. Increased frequency of extreme Indian Ocean dipole events due to greenhouse warming. Nature 510, 254.

Cai, W., Wang, G., Dewitte, B., Wu, L., Santoso, A., Takahashi, K., Yang, Y., Carreric, A., McPhaden, M.J., 2018a. Increased variability of eastern Pacific El Niño under greenhouse warming. Nature 564, 201.

Cai, W., Wang, G., Gan, B., Wu, L., Santoso, A., Lin, X., Chen, Z., Jia, F., Yamagata, T., 2018b. Stabilised frequency of extreme positive Indian Ocean dipole under 1.5 degrees C warming. Nat. Commun. 9.

Cai, W., Wang, G., Santoso, A., Lin, X., Wu, L., 2017. Definition of extreme El Niño and its impact on projected increase in extreme El Niño frequency. Geophys. Res. Lett. 44, 11184–11190.

Cai, W., Yang, K., Wu, L., Santoso, A., Benjamin, N., Wang, G., Yamagata, T., 2021. Opposite response of strong and moderate positive Indian Ocean Dipole to global warming. Nature Climate Change 11 (1), 27–32. doi: https://doi.org/10.1038/s41558-020-00943-1.

Cai, W., Zheng, X., Weller, E., Collins, M., Cowan, T., Lengaigne, M., Yu, W., Yamagata, T., 2013. Projected response of the Indian Ocean dipole to greenhouse warming. Nat. Geosci. 6, 999–1007.

Gadgil, S., 2003. The Indian monsoon and its variability. Annu. Rev. Earth Planet. Sci. 31, 429–467.

Gadgil, S., Gadgil, S., 2006. The Indian monsoon, GDP and agriculture. Econ. Polit. Wkly., 4887–4895.

Hashizume, M., Chaves, L.F., Minakawa, N., 2012. Indian Ocean dipole drives malaria resurgence in East African highlands. Sci. Rep. 2.

Hui, C., Zheng, X.T., 2018. Uncertainty in Indian Ocean dipole response to global warming: the role of internal variability. Clim. Dyn. 51, 3597–3611.

Ju, J.H., Slingo, J., 1995. The Asian summer monsoon and ENSO. Quart. J. R. Meteorol. Soc. 121, 1133–1168.

Krishnan, R., Sabin, T.P., Ayantika, D.C., Kitoh, A., Sugi, M., Murakami, H., Turner, A.G., Slingo, J.M., Rajendran, K., 2012. Will the South Asian monsoon overturning circulation stabilize any further? Clim. Dyn. 40, 187–211.

Kshirsagar, N., Shinde, R., Mehta, S., 2006. Floods in Mumbai: impact of public health service by hospital staff and medical students. J. Postgrad. Med. 52, 312.

Kulkarni, A, Sabade, S.S., Kripalani, R.H., 2007. Association between the extreme monsoons and the dipole mode over the Indian subcontinent. Meteorol. Atmos. Phys. 95, 255–268.

Li, Z.G., Cai, W.J., Lin, X.P., 2016. Dynamics of changing impacts of tropical Indo-Pacific variability on Indian and Australian rainfall. Sci. Rep. 6.

Li, Z.G., Lin, X.P., Cai, W.J., 2017. Realism of modelled Indian summer monsoon correlation with the tropical Indo-Pacific affects projected monsoon changes. Sci. Rep. 7.

Murtugudde, R., McCreary, J.P., Busalacchi, A.J., 2000. Oceanic processes associated with anomalous events in the Indian Ocean with relevance to 1997-1998. J. Geophys. Res. 105, 3295–3306.

Page, S.E., et al., 2002. The amount of carbon released from peat and forest fires in Indonesia in 1997. Nature 420, 61–65.

Rayner, N.A., Parker, D.E., Horton, E.B., Folland, C.K., Alexander, L.V., Rowell, D.P., Kent, E.C., Kaplan, A., 2003. Global analyses of sea surface temperature, sea ice, and night marine air temperature since the late nineteenth century. J. Geophys. Res. 108.

Ropelewski, C.F., Halpert, M.S., 1987. Global and regional scale precipitation patterns associated with the El Niño/Southern oscillation. Mon. Weather Rev. 115, 1606–1626.

Saji, N.H., Goswami, B.N., Vinayachandran, P.N., Yamagata, T., 1999. A dipole mode in the tropical Indian Ocean. Nature 401, 360–363.

Schott, F.A., Xie, S.P., McCreary, J.P., 2009. Indian Ocean circulation and climate variability. Rev. Geophys. 47.

Singh, D., Tsiang, M., Rajaratnam, B., Diffenbaugh, N.S., 2014. Observed changes in extreme wet and dry spells during the South Asian summer monsoon season. Nat. Clim. Change 4, 456–461.

Taylor, K.E., Stouffer, R.J., Meehl, G.A., 2012. An overview of CMIP5 and the experiment design. Bull. Am. Meteorol. Soc. 93, 485–498.

Ueda, H., Matsumoto, J., 2000. A possible triggering process of East-West asymmetric anomalies over the Indian Ocean in relation to 1997/98 El Niño. J. Meteorol. Soc. Jpn 78, 803–818.

Ummenhofer, C.C., England, M.H., McIntosh, P.C., Meyers, G.A., Pook, M.J., Risbey, J.S., Gupta, A.S., Taschetto, A.S., 2009. What causes southeast Australia's worst droughts? Geophys. Res. Lett., 36.

Vecchi, G.A., Soden, B.J., Wittenberg, A.T., Held, I.M., Leetmaa, A., Harrison, M.J., 2006. Weakening of tropical Pacific atmospheric circulation due to anthropogenic forcing. Nature 441, 73–76.

Wang, G., Cai, W., Santoso, A., 2017. Assessing the impact of model biases on the projected increase in frequency of extreme positive Indian Ocean dipole events. J. Clim. 30, 2757–2767.

Wang, G., Cai, W., Yang, K., Santoso, A., Yamagata, T., 2020a. A unique feature of the 2019 extreme positive Indian Ocean Dipole event. Geophys. Res. Lett. under review.

Wang, G., Cai, W., Santoso, A., 2020b. Stronger Increase in the frequency of extreme convective than extreme warm El Niño events under greenhouse warming. J. Clim. 33, 675–690.

Webster, P.J., Moore, A.M., Loschnigg, J.P., Leben, R.R., 1999. Coupled ocean-atmosphere dynamics in the Indian Ocean during 1997-98. Nature 401, 356–360.

Webster, P.J., Yang, S., 1992. Monsoon and ENSO—selectively interactive systems. Quart. J. R. Meteorol. Soc. 118, 877–926.

Weller, E., Cai, W.J., 2014. Meridional variability of atmospheric convection associated with the Indian Ocean Dipole mode. Sci. Rep. 4.

Yu, L.S., Rienecker, M.M., 1999. Mechanisms for the Indian Ocean warming during the 1997-98 El Niño. Geophys. Res. Lett. 26, 735–738.

Zheng, X.T., Xie, S.P., Du, Y., Liu, L., Huang, G., Liu, Q.Y., 2013. Indian Ocean dipole response to global warming in the CMIP5 multimodel ensemble. J. Clim. 26, 6067–6080.

Zheng, X.T., Xie, S.P., Vecchi, G.A., Liu, Q.Y., Hafner, J., 2010. Indian Ocean dipole response to global warming: analysis of ocean-atmospheric feedbacks in a Coupled Model. J. Clim. 23, 1240–1253.

Zubair, L., Rao, S.A., Yamagata, T., 2003. Modulation of Sri Lankan Maha rainfall by the Indian Ocean dipole. Geophys. Res. Lett. 30.

Chapter 22

South Asian summer monsoon response to anthropogenic aerosol forcing

Hai Wang

Department of Marine Meteorology, College of Oceanic and Atmospheric Sciences, Ocean University of China, Qingdao, China

22.1 Introduction

Anthropogenic aerosols increased dramatically due to industrial revolution and population growth in the past century, especially over the South Asia (Fig. 22.1). Distinct from well-mixed greenhouse gases (GHGs), the spatial distributions of anthropogenic aerosols are strong due to the inhomogeneous emission sources and short lifetimes. Understanding the dynamical mechanism of regional climate response to the changing radiative forcing due to anthropogenic aerosols has important scientific and societal implications.

The South Asian summer monsoon (SASM) has tremendous impacts on water resources, agriculture, economies, and ecosystem by providing up to 80% of the annual mean precipitation for much of the India (Webster et al. 1998). The South Asian rainfall has been documented as a persistent summertime drying trend during the latter half of the twentieth century (Fig. 22.2A and E) in observations (Bollasina et al. 2011; Harris et al. 2014; Roxy et al. 2015; Kulkarni et al. 2020). However, the climate models from the latest Coupled Model Intercomparison Project Phase 6 (CMIP6; Table 22.1; Erying et al. 2016; Pascoe et al. 2020) cannot capture this feature very well (Fig. 22.2B and E). Was this observed change of the SASM precipitation caused by natural variability or human influence? If the latter, what were the relative contributions of aerosols and GHGs, the two major anthropogenic climate forcing agents? Results from CMIP6 historical single-forcing experiments indicate that the anthropogenic aerosol forcing may be the key factor dominating the weakened SASM and the corresponding drying trend over the South Asia in the second half of the twentieth century (Fig. 22.2).

The potential impacts of anthropogenic aerosols on the Asian summer monsoon have drawn much attention in recent years (Li et al. 2016). Previous studies suggested both the local and nonlocal effects of anthropogenic aerosols can

Indian Summer Monsoon Variability: El Niño-teleconnections and beyond.
DOI: https://doi.org/10.1016/B978-0-12-822402-1.00006-5
433

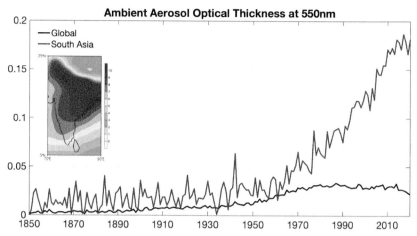

FIG. 22.1 Global and South Asia mean AOD (ambient aerosol optical thickness at 550 nm) change from the preindustrial to 2014 in CMIP6 multimodel ensemble.

weaken the SASM (Bollasina et al. 2011; Ganguly et al. 2012a, b; Li et al. 2016; Lau and Kim, 2017). Although there is a general recognition that the anthropogenic aerosols weaken the SASM, the detailed mechanisms remain controversial. The aerosol-induced climate change can be decomposed into a direct atmospheric response to radiative forcing without sea surface temperature (SST) change, and a SST-mediated ocean-atmosphere coupled interaction response.

In the direct atmospheric response, aerosols affect precipitation and atmospheric circulation by modifying radiation and cloud physics (Lau et al. 2006; Rosenfeld et al. 2008; Bollasina et al. 2013). The SST-mediated response refers to the atmospheric circulation change regulated by the anthropogenic aerosol-induced SST change via coupled ocean-atmosphere interaction processes (Xie et al. 2013; Wang et al. 2016a, b; Wang et al. 2019; Li et al. 2020). Till now, it is unclear whether the direct atmospheric response or the SST-mediated response dominates the weakened SASM and the corresponding drying trend over the South Asia in the second half of the twentieth century.

This chapter aims to review the recent progresses in understanding the SASM response to anthropogenic aerosol forcing from the direct atmospheric response and the ocean-atmospheric interaction perspectives, respectively. In the "Direct atmospheric response" section, how the regional atmospheric circulation and the corresponding SASM precipitation are regulated by the aerosol-induced radiative forcing without mediation from the ocean processes is presented. The "Ocean-atmosphere interaction response" section discusses the role of the aerosol-induced SST cooling pattern in shaping the SASM response. It is shown that the SST-mediated change dominates the aerosol-induced weakened SASM,

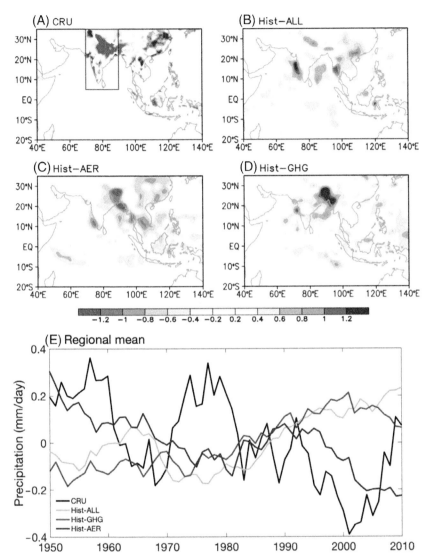

FIG. 22.2   Changes of June-July-August (JJA) precipitation during 1950–2010 (unit: mm/day) in (A) observations, (B) historical all-forcing simulation, (C) historical anthropogenic aerosol single-forcing simulation, and (D) historical GHG single-forcing simulation in CMIP6 multimodel ensemble. (E) Nine-year running mean of JJA average precipitation anomalies over South Asia (marked in red box in (A), 5°N–35°N; 70°E–90°E). Anomalies are calculated as deviations from the 1950 to 2010 climatology. The black line is based on the CRU observational dataset. The green, red, and blue lines are for the ensemble mean historical all-forcing, GHG single-forcing, and anthropogenic aerosol single-forcing simulations in CMIP6, respectively.

TABLE 22.1 List of CMIP5 and CMIP6 models used in this study.

CMIP6 model acronym	Institution
BCC-CSM2-MR	Beijing Climate Center, China
CanESM5*	Canadian Centre for Climate Modelling and Analysis, Environment and Climate Change Canada, Canada
CNRM-CM6-1*	Centre National de Recherches Meteorologiques; Centre Europeen de Recherche et de Formation Avancee en Calcul Scientifique, France
GISS-E2-1-G	Goddard Institute for Space Studies, USA
HadGEM3-GC31-LL*	Met Office Hadley Centre, UK
IPSL-CM6A-LR*	Institut Pierre Simon Laplace, France
MIROC6	Japan Agency for Marine-Earth Science and Technology; Atmosphere and Ocean Research Institute, The University of Tokyo; National Institute for Environmental Studies; RIKEN Center for Computational Science, Japan
MRI-ESM2-0	Meteorological Research Institute, Japan
NorESM2-LM*	NorESM Climate modeling Consortium consisting of CICERO, MET-Norway, NERSC, NILU, UiB, UiO and UNI, Norway
CMIP5 Model acronym	**Institution**
CanESM2	CCCma (Canadian Centre for Climate Modelling and Analysis), Canada
CSIRO Mk3.6.0	Australian Commonwealth Scientific and Industrial Research Organization (CSIRO) Marine and Atmospheric Research in collaboration with the Queensland Climate Change Centre of Excellence (QCCCE), Australia
GFDL CM3	NOAA (National Oceanic and Atmospheric Administration)/GFDL (Geophysical Fluid Dynamics Laboratory),USA
IPSL-CM5A-LR	Institut Pierre Simon Laplace, France
NorESM1-M	Norwegian Climate Centre, Norway

*CMIP6 AOD data.

with contributions from both the north-south interhemispheric SST gradient and the local SST cooling pattern over the tropical Indian Ocean. The progresses are summarized and the uncertainties in climate models calling for more research in understanding the forced SASM change are discussed in the "Summary and discussion" section.

22.2 Direct atmospheric response

How does the atmospheric circulation and precipitation respond to anthropogenic aerosol forcing directly without mediation from the ocean processes? To
detect the direct SASM response to anthropogenic aerosol forcing, the multimodel ensemble mean of five Coupled Model Intercomparison Project Phase 5
(CMIP5; Table 22.1; Taylor et al. 2012) models are used that performed all the
following experiments. (1) The historical anthropogenic aerosol single-forcing
simulation with the anthropogenic aerosols as the only time-varying forcing
agent, while with all the other external radiative forcings fixed at the preindustrial level. (2) The diagnostic Atmospheric General Circulation Model (AGCM)
experiments. The control run (sstClim) is an atmospheric-only run driven by
preindustrial climatological SST and sea ice concentration. The perturbed run
(sstClimAerosol) is forced with aerosols specified from year 2000 of the historical anthropogenic aerosol single-forcing simulation in CMIP5, but with SST
fixed at the preindustrial level. Thus, the difference between the perturbed run
and the control run isolates the direct atmospheric responses to anthropogenic
aerosol forcing.

In the direct atmospheric response, the wind anomalies at 850 hPa show an
easterly flow in South Asia around 20°N (Fig. 22.3A), which largely weaken
the climatological southwest monsoon circulation over the South Asia. The
weakened SASM circulation suppresses the moisture transport from the tropical Indian Ocean to the South Asia land regions and therefore leads to the rainfall decrease on land north of 20°N (Fig. 22.3A). Over the southwestern India,
Arabian Sea and Bay of Bengal, precipitation increases in a direct atmospheric
response due to the anomalous westerlies, inconsistent with the response pattern in coupled historical anthropogenic aerosol single-forcing simulations
(Fig. 22.2C). The changes in SASM and its corresponding precipitation are also
characterized by a high surface pressure anomaly over the Indian continent and
low sea level pressure (SLP) anomaly over the north Indian Ocean (Fig. 22.3B).
Besides the circulation anomaly, surface air temperature (SAT) decreases over
the Indian Continent (Fig. 22.3C), with little change over the Indian Ocean.
Effective radiative forcing at the top of the atmosphere resembles the SAT
change and the aerosol concentration very well (Fig. 22.3D). By reflecting and
scattering the solar radiation, aerosols cool the land, thereby reduce the land-
sea thermal contrast between the Indian Continent and the adjacent north Indian
Ocean (Fig. 22.3C).

Previous studies showed that anthropogenic aerosols from inside and outside South Asia both contributed to the overall reduction in precipitation, but
the dominant contribution comes from the local aerosol effect (Bollasina et al.
2014). The first explanation concerning the aerosol-induced direct atmospheric
response in SASM is the "solar dimming" effect (Ramanathan et al. 2005). It
showed that the reflection and scattering by anthropogenic aerosols reduced
the surface solar radiation that cooled the northern Indian Ocean, which further

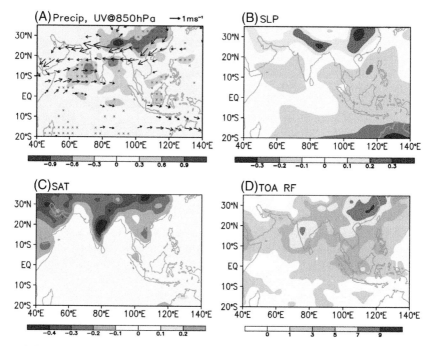

FIG. 22.3 Direct atmospheric response to aerosol changes from the preindustrial to 2000 in JJA in CMIP5: (A) precipitation (mm/day) and wind at 850 hPa (vectors with scale at top right, omitted where the magnitude < 0.2 m/s), (B) SLP (hPa), (C) SAT (°C), (D) TOA effective radiative forcing (W/m²). Based on multimodel mean of CMIP5 anthropogenic aerosol forcing AGCM with SST fixed at the preindustrial climatology. Hatched regions indicate precipitation changes exceed 90% statistical confidence. (Revised from Wang et al. 2019. © American Meteorological Society. Used with permission.)

reduced evaporation and caused a slowdown of the local atmospheric meridional overturning circulation, and weakened the SASM precipitation (Ramanathan et al. 2005). The subsequent studies have confirmed this pathway and further analysis claimed that the aerosol-induced SASM response via "solar dimming" effect is not only caused by direct effect of aerosols but also by their indirect effects on cloud physics (Chung and Ramanathan, 2006; Ganguly et al. 2012a). The aerosols, serving as cloud condensation nuclei or ice nuclei, can alter the atmospheric thermodynamic stability and convective potential of the lower atmosphere. Such aerosol indirect effects may lead to reduced temperatures, increased atmospheric stability, and weakened wind and atmospheric circulations in the South Asian monsoon region (Li et al. 2016).

A different pathway concerning the direct atmospheric response in SASM to aerosol forcing suggested that the radiative heating of absorbing aerosols over northern India and the Tibetan plateau may cause warm air to rise. The upper troposphere warm anomaly induced by the absorbing aerosols then acts as an "elevated heat pump" (Lau et al. 2006), which further leads to the anomalous rising motion in this region. By generating the anomalous meridional circulation, the "elevated heat pump" increases the moisture convergence over India from the north Indian Ocean in the lower troposphere, and causes the intensification of the SASM precipitation by enhancing the convection over northern India and the southern slope of the Tibetan plateau. Although being debated afterwards, modeling and observational studies have confirmed signatures of the "elevated heat pump" but found it relevant only in the premonsoon season due to its effect on convective instability and the wet scavenging on aerosol loading in June and July (Lau and Kim, 2006; Cowan and Cai, 2011; Ganguly et al. 2012a).

In either pathway, the aerosol-induced land-sea thermal contrast drives the anomalous regional atmospheric overturning circulation. Fig. 22.4 shows zonally averaged temperature and vertical motion changes in the South Asian sectors (60°E–100°E). Over the South Asian monsoon region, anthropogenic aerosols induce the surface cooling by reflecting and scattering the solar radiation (Fig. 22.3C), and the anomalous sinking motion develops north of 20°N (Fig. 22.4), resulting reduced precipitation and the anomalous easterlies north of 20°N is seen in the direct atmospheric response. To the south over southern India and the northern tropical Indian Ocean, strong increase in vertical motion in the mid-to-upper troposphere in 5°N–20°N (Fig. 22.4) can be linked to the

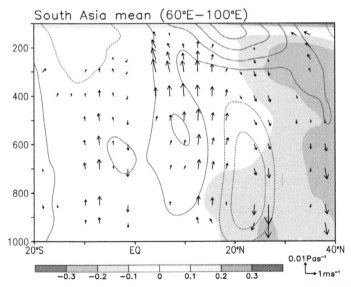

FIG. 22.4 Direct atmospheric response to aerosol changes from the preindustrial to 2000 in JJA in CMIP5. South Asia [60°E~100°E] zonal mean air temperature (shading in °C), zonal wind (contours at 0.2 m/s interval with 0 omitted, solid lines denote westerlies), meridional and vertical velocity (vectors with scale at bottom right, omitted where vertical motion response is weaker than 1×10^{-3} Pa/s). Based on multimodel mean of CMIP5 anthropogenic aerosol forcing AGCM with SST fixed at the preindustrial climatology. (Revised from Wang et al. 2019).

"elevated heat pump" effect. On the southern slope of the Tibetan Plateau, the heated air by absorbing aerosols leads to the strong upward motion in the upper troposphere, and draws the warm and moist low-level inflow from the north Indian Ocean, resulting in the intensified rainfall over the southwestern India, Arabian Sea and Bay of Bengal, and causing the anomalous westerly above.

The impacts on direct atmospheric adjustment in SASM due to anthropogenic aerosol forcing via radiative effects have been verified in climate models. However, due to the limited emission sources and short atmospheric residence time, the key dynamical process in a long-term SASM response to anthropogenic aerosol forcing may lie in the SST-mediated ocean-atmosphere coupled response process.

22.3   Ocean-atmosphere interaction response

Compared to the relatively uniform spatial distribution of well-mixed GHG, anthropogenic aerosol-induced radiative forcing change concentrates in the northern hemisphere and shows a pronounced interhemispheric asymmetry (Wang et al. 2016a, b). On global scale, the aerosol-induced interhemisphere asymmetric thermal forcing generates the interhemispheric SST difference and furthermore leads to the anomalous clockwise atmospheric overturning circulation over the equator due to the energy constrain (Kang et al. 2008; Chiang and Friedman, 2012; Wang et al. 2016a, b).

Using the CMIP5 control and perturbed AGCM simulations, the direct atmospheric response to the anthropogenic aerosol forcing from the preindustrial to 2000 is obtained. Furthermore, by excluding the results from the AGCM experiments in the coupled historical anthropogenic aerosol single-forcing simulations, the direct atmospheric response is linearly removed, and the atmospheric circulation response mediated by the aerosol-induced SST change is obtained (Wang et al. 2019). This decomposition method has been validated in the North Pacific and North Atlantic (Shaw and Voigt, 2015), and in the analysis of Asian summer monsoon response to aerosol forcing (Li et al. 2018).

On a regional scale, by eliminating the direct atmospheric response in the coupled simulations, the aerosol-induced SST feedback shows its importance in regulating the SASM response. Over tropical South Asia and the Indian Ocean, the SASM responses in coupled historical anthropogenic aerosol single-forcing simulation are dominated by the SST-mediated changes (Fig. 22.5). The SASM precipitation decreases on and north of the equator, while increases over the southeast tropical Indian Ocean (Fig. 22.5A and B). The aerosol-induced weakened precipitation centers over north India and Bay of Bengal, which corresponds well with the observations. In the coupled ocean-atmosphere interaction response process, the precipitation change is regulated by the SST pattern in response to aerosol forcing, which shows both zonal and meridional asymmetry with relatively cooling in the northwest and warming in the southeast tropical Indian Ocean (Fig. 22.5C and D). Corresponding to the SST pattern,

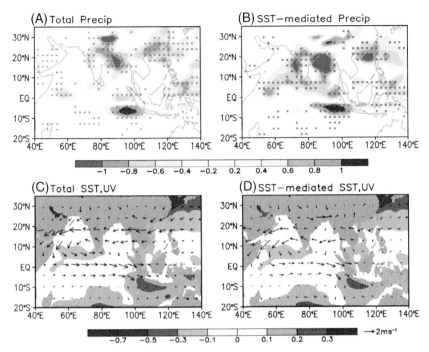

FIG. 22.5   Total (A), (C) and SST-mediated (B), (D) responses to aerosol changes from the prein-dustrial to 2000 in JJA in CMIP5: (A), (B) Precipitation (shading in mm/day), (C), (D) SST (shading in °C, tropical mean [25°N~25°S] has been removed), and wind at 850 hPa (vectors with scale at top right). Based on multimodel means of coupled and SST-mediated anthropogenic aerosol forcing simulations in CMIP5. Hatched regions indicate precipitation changes exceed 90% statistical confidence. (Revised from Wang et al. 2019). © American Meteorological Society. Used with permission.

an anomalous cyclonic circulation over the north Indian Ocean and south India weakens the climatological SASM circulation.

Through the SST changes, anthropogenic aerosols change the atmospheric temperature as well as shift the regional atmospheric meridional overturning circulation. Fig. 22.6 shows the global and South Asia (60°E–100°E) zonal mean air temperature and meridional circulation changes in coupled and SST-mediated response. Previous studies have pointed out that the spatial inhomogeneous anthropogenic aerosol forcing is much more efficient in causing circulation changes to compensate the interhemispheric energy imbalance (Wang et al. 2016a, b). The deep cooling structure over the Northern Hemisphere mid-latitudes has confirmed mediated by the SST feedback (Fig 22.6; Xu and Xie, 2015). From energy perspective (Kang et al. 2008; Chiang and Friedman, 2012), the boreal summer tropical meridional overturning circulation, of which the SASM is a major part (Trenberth et al. 2006; Bollasina et al. 2011), weakens to reduce the energy flow to the Southern Hemisphere and thus alleviates the interhemispheric asymmetry (Fig. 22.6). The anomalous sinking motion

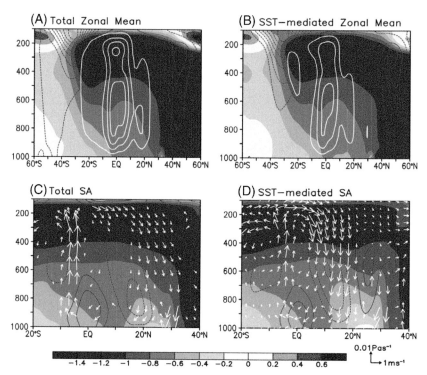

FIG. 22.6 Total (A), (C) and SST-mediated (B), (D) response to aerosol changes from the preindustrial to 2000 in JJA in CMIP5: (A), (B) Entire zonal mean, and (C), (D) South Asian (60°E~100°E) zonal mean air temperature (shading in °C), zonal wind (red contours at 0.2 m/s interval with 0 omitted, solid lines denote westerlies), meridional stream function (white contours in A, B at 3 × 10⁹ kg/s interval with 0 omitted, positive indicate clockwise circulation), and meridional and vertical velocity (vectors in C, D with scale at bottom right, omitted where vertical velocity response is weaker than 1 × 10⁻³ Pa/s). Based on multimodel means of coupled and SST-mediated anthropogenic aerosol forcing simulations in CMIP5. (Revised from Wang et al. 2019). © American Meteorological Society. Used with permission.

generated by the SST feedback reduces the ascent over the north Indian Ocean and South Asia, opposing the local Hadley circulation, and leads to the rainfall decrease over the tropical SASM region. The anomalous rising motion is centered over 10°S-Equator, contributing to the rainfall increase over the tropical southeast Indian Ocean.

Climate model results confirmed the important role of aerosol-induced SST pattern in shaping the SASM response via coupled ocean-atmosphere interaction process (Fig. 22.6), but it is still unclear which aspect of the SST change are most important: the uniform cooling, the global meridional temperature gradient, or the regional SST pattern. Recent studies based on AGCM experiments advanced our knowledge on the specific role of distinct aerosol-induced SST patterns in regulating the SASM response (Wang et al. 2019; Li et al. 2020). Following Wang et al. (2019), the SST-mediated atmospheric response

TABLE 22.2 AGCM experiments design (Revised from Wang et al. 2019. © American Meteorological Society. Used with permission.)

Name	Prescribed SST
Control	SST_{clim}
SST_global	SST_{clim} + SSTA[a] (global)
SST_uniform	SST_{clim} + (−0.51°C)[b]
SST_global_zm	SST_{clim} + SSTA (global zonal mean)
SST_tropical_IO	SST_{clim} + SSTA (tropical Indian Ocean[c]: 25°S–25°N, Indian Ocean sector)

[a]SSTA: changes in SST from the preindustrial to 2000 in 5 CMIP5 models ensemble.
[b]Tropical (25°S–25°N) annual mean SST change from the preindustrial to 2000 in 5 CMIP5 models ensemble.
[c]Tropical (25°S–25°N) mean has been removed from the tropical Indian Ocean SST anomaly.

to anthropogenic aerosol forcing is decomposed into a uniform cooling effect, interhemispheric asymmetry effect, and SST-patterned cooling effect over the tropical Indian Ocean via idealized AGCM experiments using the NOAA/GFDL Atmospheric Model, version 2.1 (AM2.1; Anderson et al. 2004). In the AGCM experiments (Table 22.2), the radiative forcing is fixed at the preindustrial level. The control simulation is run with the preindustrial climatological SST and sea ice concentration from the five CMIP5 models' ensemble mean. The forced experiments use the same radiative forcing and sea ice concentration as the control run, with different SST anomalies derived from the CMIP5 historical anthropogenic aerosol single-forcing simulations from the preindustrial to 2000. The differences between the forced experiments and the control run represent the SST feedback in the aerosol-induced SASM response.

The AGCM experiment forced by the CMIP5 aerosol-induced global SST anomaly resembles the result from the CMIP5 multimodel ensemble mean very well (Fig. 22.7A). Wang et al. (2019) showed that the north-south interhemispheric asymmetry in aerosol-induced SST response played a dominant role in causing the tropical atmospheric overturning circulation to weaken over the SASM region and led to the weakened monsoon circulation and the corresponding drying trend over South Asia since industrial revolution (Fig. 22.7C). Although with weaker magnitude, the local SST change in the tropical Indian Ocean induced similar precipitation and circulation response compared with the global SST anomaly and global zonal mean SST anomaly experiments. With the aerosol-induced SST cooling more in the northwest than the southeast tropical Indian Ocean, an anomalous cyclonic circulation is generated. Based on the regional ocean-atmosphere coupled response regulated by the Bjerknes feedback (Bjerknes, 1969), the wind anomalies furthermore shallow/deepen the thermocline in the tropical northwest/southeast Indian Ocean and enhance the

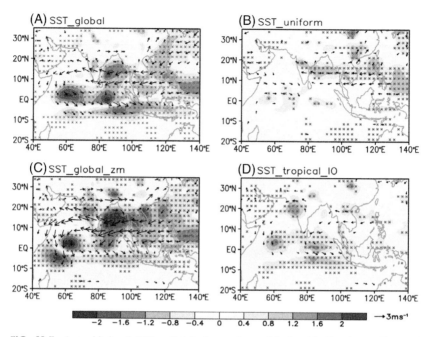

FIG. 22.7 Aerosol-induced SST-mediated changes in precipitation (shading in mm/day) and 850 hPa wind (vectors with wind speed < 0.3 m/s omitted) in JJA in GFDL AM2.1 experiments forced with different SST patterns in CMIP5 historical anthropogenic aerosol single-forcing simulation: (A) Global SST changes, (B) SST uniform cooling, (C) Global zonal mean SST changes, and (D) Tropical Indian Ocean SST changes with tropical mean SST change removed. Hatched regions indicate precipitation changes exceed 90% statistical confidence. (Revised from Wang et al. 2019. © American Meteorological Society. Used with permission.)

SST cooling pattern as well as its corresponding rainfall anomaly (Fig. 22.7D). The uniform cooling of SST mediated the precipitation change through the wet-get-wetter effect (Held and Soden, 2006) by decreasing the rainfall over 10°N–20°N, 80°E–120°E, where climatological convection is strong during boreal summer (Fig. 22.7B). Compared with the global zonal mean and Indian Ocean SST pattern, the uniform SST cooling effect on SASM wind and precipitation is feeble. Using different AGCM experiments, Li et al. (2020) showed that the merdional SST gradient in the Indian Ocean, South, and East China seas dominated the global interhemispheric SST asymmetry and determined the meridional overturning circulation response in South Asia.

Diagnostic analysis using distinct CMIP5 simulations showed that precipitation changes to anthropogenic aerosol forcing contradict each other in the direct atmospheric response and the SST-mediated response over the tropical SASM region (Wang et al. 2019). The AGCM results highlight the important role of aerosol-induced north-south interhemispheric SST gradient in altering the local atmospheric overturning circulation that weakens the SASM in the second half of the twentieth century.

22.4 Discussion and summary

Anthropogenic change in aerosols is a major driver of the twentieth century climate change, masking a considerable fraction of the GHG-induced warming since the industrial revolution. Rapid industrialization has resulted in a dramatic increase in emissions of anthropogenic aerosols globally until the 1990s, while the regional aerosol loading over the South Asia continues its fast increase (Fig. 22.1). The SASM shows robust weakness in the second half of the twentieth century corresponding with the significant drying trend in the observations. Climate model results attribute these changes to the anthropogenic aerosol forcing effect on the regional atmospheric circulation change. This chapter reviews recent findings in understanding the dynamical response of SASM to anthropogenic aerosol forcing in both direct atmospheric response and SST-mediated coupled ocean-atmosphere interaction response.

In the direct atmospheric response without ocean processes, anthropogenic aerosols weaken the SASM and the corresponding precipitation over South Asia north of 20°N by reflecting and scattering the solar radiation and therefore reduced the land-sea thermal contrast over South Asia. In the South Asia south of 20°N, the absorbing aerosols heated the troposphere and led to the rainfall increase via "elevated heat pump" mechanism. In either pathways, the direct atmospheric response to aerosol forcing shows adjustment in local atmospheric meridional overturning circulation that regulates the SASM north of 20°N. Considering the different species of aerosols, a more recent study reported that the impacts on optical characteristics were mostly due to sulfate aerosols, while the mass concentration was dominated by organic carbon over the Indo-Gangetic Basin (Srivastava et al. 2020). Furthermore, their study claimed that the elemental carbon showed relatively large impact on radiative forcing though their contribution to the AOD change was quite low. Therefore, evaluating the drastically different impacts of different aerosol species on atmospheric radiative forcing and detecting their distinct role in regulating the atmospheric circulation change require further exploration in better understanding the direct SASM response to anthropogenic aerosol changes.

In the SST-mediated ocean-atmosphere coupled interaction response, aerosol-induced SST pattern shows large north-south interhemispheric asymmetry, which generates the anomalous zonal mean cross-equatorial Hadley circulation due to the energy constrain. As a major part of the summertime tropical meridional overturning circulation, the SASM weakened in response to anthropogenic aerosol forcing due to the SST change. AGCM results confirmed the role of the interhemispheric SST difference in shaping the regional SASM change, with small contribution from the regional SST pattern in the tropical Indian Ocean which also led to the weakening of SASM through Bjerknes feedback.

Aerosols are the largest source of uncertainty in radiative forcing during the instrumental era because of insufficient understanding of indirect microphysical effects on cloud and precipitation (Boucher et al. 2013). Even though the changes in precipitation and SST in multimodel ensemble mean are robust, the

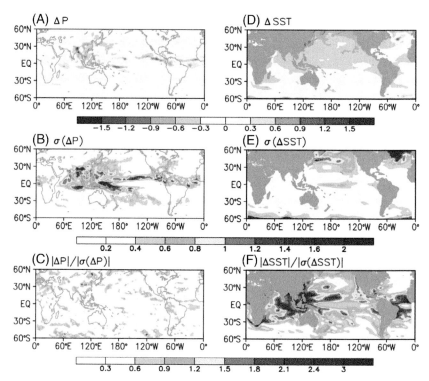

FIG. 22.8 Changes of JJA (A) precipitation (shading; unit: mm/day) and (D) SST (shading; unit: °C) during 1950–2010 in historical anthropogenic aerosol single-forcing simulation in CMIP6 multimodel ensemble. (B), (E) and (C), (F) are the intermodel spread and signal-to-noise ratio of the anthropogenic aerosol-induced precipitation and SST responses, respectively.

intermodel spreads are remarkable in precipitation response over the Asian monsoon regions in CMIP6 historical anthropogenic aerosol single-forcing simulations (Fig. 22.8 A–C). Previous studies linked the meridional shift of the tropical rain-belt to the aerosol-induced interhemispheric SST asymmetry, while large uncertainty in aerosol-induced north Pacific SST also leads to large uncertainty in rainfall response over the entire tropics (Fig. 22.8C,D). Signal-to-noise ratio indicates that over the broad Indo-western Pacific region, the aerosol-induced SST response is robust, while the precipitation change over the SASM region shows large intermodel spread compared with their robust ensemble mean changes (Fig. 22.8E,F). This points to the need to further investigate the large-scale coupled ocean-atmosphere response to anthropogenic aerosol forcing, including their effects on the regional climate change (Xie et al. 2015; Collins et al. 2018). Recent progress in large-ensemble climate model simulations (Deser et al. 2020) provides another useful metric to evaluate the internal versus forced response of regional climate change, which also calls for further investigations in understanding the SASM response to anthropogenic aerosol forcing.

Acknowledgment

This chapter is based on Wang et al. (2019) published in *Journal of Climate*. The author would also like to thank all the coauthors in that paper: Professor Shang-Ping Xie (Scripps Institution of Oceanography), Associate Professor Yu Kosaka (University of Tokyo), Professor Qinyu Liu (Ocean University of China), and Professor Yan Du (Chinese Academy of Science). This work is supported by the National Natural Science Foundation of China (Grant No. 41806006). I acknowledge the WCRP Working Group on Coupled Modeling, which is responsible for CMIP5 and CMIP6. The National Oceanic and Atmospheric Administration (NOAA)/Geophysical Fluid Dynamics Laboratory (GFDL) provided computer codes and related datasets of the AM2.1 model.

References

Anderson, J.L., et al., 2004. The new GFDL global atmosphere and land model AM2-LM2: evaluation with prescribed SST simulations. J. Clim. 17, 4641–4673.

Bjerknes, J., 1969. Atmospheric teleconnections from the equatorial Pacific. Mon. Weather Rev. 97, 163–172.

Bollasina, M.A., Ming, Y., Ramaswamy, V., 2011. Anthropogenic aerosols and the weakening of the South Asian summer monsoon. Science 334 (6055), 502–505.

Bollasina, M.A., Ming, Y., Ramaswamy, V., 2013. Earlier onset of the Indian monsoon in the late twentieth century: the role of anthropogenic aerosols. Geophys. Res. Lett. 40, 3715–3720.

Bollasina, M.A., Ming, Y., Ramaswamy, V., Schwarzkopf, M.D., Naik, V., 2014. Contribution of local and remote anthropogenic aerosols to the twentieth century weakening of the South Asian monsoon. Geophys. Res. Lett. 41, 680–687.

Boucher, O., et al., 2013. Clouds and aerosols. In: Stocker, T.F. et al (Ed.), Climate Change 2013: The Physical Science Basis. Cambridge University Press, Cambridge and New York, pp. 571–657.

Chiang, J.C., Friedman, A.R., 2012. Extratropical cooling, interhemispheric thermal gradients, and tropical climate change. Annu. Rev. Earth Planet. Sci. 40 (1), 383.

Chung, C., Ramanathan, V., 2006. Weakening of the North Indian SST gradients and the monsoon rainfall in India and the Sahel. J. Clim. 19, 2036–2045.

Collins, M., et al., 2018. Challenges and opportunities for improved understanding of regional climate dynamics. Nat. Clim. Chang. 8, 101–108.

Cowan, T., Cai, W., 2011. The impact of Asian and non-Asian anthropogenic aerosols on 20th century Asian summer monsoon. Geophys. Res. Lett., L11703.

Deser, C., et al., 2020. Isolating the evolving contributions of anthropogenic aerosols and greenhouse gases: A new CESM1 large Ensemble community Resource. J. Clim. 33 (18), 7835–7858.

Eyring, V., Bony, S., Meehl, G.A., Senior, C.A., Stevens, B., Stouffer, R.J., Taylor, K.E., 2016. Overview of the Coupled Model Intercomparison Project Phase 6 (CMIP6) experimental design and organization. Geosci. Model Dev. 9, 1937–1958.

Ganguly, D., Rasch, P.J., Wang, H., Yoon, J.-H., 2012a. Climate response of the South Asian monsoon system to anthropogenic aerosols. J. Geophys. Res. Atmos. 117 (D13), D13 209.

Ganguly, D., Rasch, P.J., Wang, H., Yoon, J.-H., 2012b. Fast and slow responses of the South Asian monsoon system to anthropogenic aerosols. Geophys. Res. Lett. 39 (18), L18804.

Harris, I., Jones, P.D., Osborn, T.J., Lister, D.H., 2014. Updated high-resolution grids of monthly climatic observations-the CRU TS3.10 dataset. Int. J. Climatol. 34 (3), 623–642.

Held, I.M., Soden, B.J., 2006. Robust responses of the hydrological cycle to global warming. J. Clim. 19, 5686–5699.

Kang, S.M., Held, I.M., Frierson, D.M., Zhao, M., 2008. The response of the ITCZ to extratropical thermal forcing: idealized slab-ocean experiments with a GCM. J. Clim. 21 (14), 3521–3532.

Kulkarni, A., Sabin, T.P., Chowdary, J.S., et al., 2020. Precipitation Changes in India. In Assessment of Climate Change over the Indian Region. Springer, Singapore, pp. 47–72.

Lau, K.M., Kim, K.M., 2006. Observational relationships between aerosol and Asian monsoon rainfall, and circulation. Geophys. Res. Lett. 33, L21810.

Lau, K.M., Kim, K.M., 2017. Competing influences of greenhouse warming and aerosols on Asian summer monsoon circulation and rainfall. Asia-Pac. J. Atmos. Sci. 53 (2), 181–194.

Lau, K.-M., Kim, M.-K., Kim, K.-M., 2006. Asian summer monsoon anomalies induced by aerosol direct forcing: the role of the Tibetan Plateau. Clim. Dyn. 26 (7-8), 855–864.

Li, X., Ting, M., Lee, D.E., 2018. Fast adjustment of the Asian summer monsoon to anthropogenic aerosols. Geophys. Res. Lett. 45, 1001–1010.

Li, X., Ting, M., You, Y., Lee, D.E., Westervelt, D.M., Ming, Y., 2020. South Asian summer monsoon response to aerosol-forced sea surface temperatures. Geophys. Res. Lett. 47, e2019GL085329.

Li, Z., et al., 2016. Aerosol and monsoon climate interactions over Asia. Rev. Geophys. 54, 866–929.

Pascoe, C., Lawrence, B.N., Guilyardi, E., Juckes, M., Taylor, K.E., 2020. Documenting numerical experiments in support of the Coupled Model Intercomparison Project Phase 6 (CMIP6). Geosci. Model Dev. 13, 2147–2167.

Ramanathan, V., et al., 2005. Atmospheric brown clouds: impacts on South Asian climate and hydrological cycle. Proc. Natl. Acad. Sci. 102 (15), 5326–5333.

Rosenfeld, D., et al., 2008. Flood or drought: how do aerosols affect precipitation? Science 321 (5894), 1309–1313.

Roxy, M.K., Ritika, K., Terray, P., et al., 2015. Drying of Indian subcontinent by rapid Indian Ocean warming and a weakening land–sea thermal gradient. Nat Commun 6, 7423.

Shaw, T., Voigt, A., 2015. Tug of war on summertime circulation between radiative forcing and sea surface warming. Nat. Geosci. 8, 560–566.

Srivastava, A.K., Mehrotra, B.J., Singh, A., Singh, V., Bisht, D.S., Tiwari, S., Srivastava, M.K., 2020. Implications of different aerosol species to direct radiative forcing and atmospheric heating rate. Atmos. Environ. 241, 117820.

Taylor, K.E., Stouffer, R.J., Meehl, G.A., 2012. An overview of CMIP5 and the experiment design. Bull. Am. Meteor. Soc. 93, 485–498.

Trenberth, K.E., Hurrell, J.W., Stepaniak, D.P., 2006. The Asian Monsoon: Global perspectives. In: Wang, B. (Ed.), The Asian Monsoon. Springer/Praxis Publishing, New York, pp. 67–87.

Wang, H., Xie, S.-P., Kosaka, Y., Liu, Q., Du, Y., 2019. Dynamics of Asian summer monsoon response to Anthropogenic aerosol forcing. J. Clim. 32 (2), 843–858.

Wang, H., Xie, S.-P., Liu, Q., 2016a. Comparison of climate response to anthropogenic aerosol versus greenhouse gas forcing: distinct patterns. J. Clim. 29 (14), 5175–5188.

Wang, H., Xie, S.-P., Tokinaga, H., Liu, Q., Kosaka, Y., 2016b. Detecting cross-equatorial wind change as a fingerprint of climate response to anthropogenic aerosol forcing. Geophys. Res. Lett. 43 (7), 3444–3450.

Webster, P.J., et al., 1998. Monsoons: processes, predictability, and the prospects for prediction. J. Geophys. Res. 103, 14451–14510.

Xie, S.-P., Lu, B., Xiang, B., 2013. Similar spatial patterns of climate responses to aerosol and greenhouse gas changes. Nat. Geosci. 6 (10), 828–832.

Xie, S.-P., et al., 2015. Towards predictive understanding of regional climate change. Nat. Clim. Chang. 5, 921–930.

Xu, Y., Xie, S.-P., 2015. Ocean mediation of tropospheric response to reflecting and absorbing aerosols. Atmos. Chem. Phys. 15 (10), 5827–5833.

Chapter 23

Moisture recycling over the Indian monsoon core region in response to global warming from CMIP5 models

T.V. Lakshmi Kumar[a], G. Purna Durga[a], K. Koteswara Rao[b], Harini Nagendra[b], R.K. Mall[c]

[a]*Atmospheric Science Research Laboratory, Department of Physics, SRM Institute of Science and Technology, Tamilnadu, India,* [b]*Centre for Climate Change and Sustainability, Azim Premji University, Bengaluru, India,* [c]*DST Mahamana Center of Excellence in Climate Change Research, Banaras Hindu University, Varanasi, India*

23.1 Introduction

Rainfall over any region is the resultant contribution of local evaporation recycled and the amount of advected water vapor into that region (Trenberth, 1999). The precipitated water component emerged from the evaporated moisture in the same region of the specified control volume is called as the moisture recycling (Brubaker et al, 1993). It is also reported that the precipitation recycling/moisture recycling ratio is a degree of measure of the land surface forcing (via evaporation) on the climate interactions (Brubakar et al., 1993). Moisture recycling plays an important role in explaining the land-atmosphere interactions (Savenije, 1995). It provides a key understanding of water resource engineering, hydrology, etc. by enhancing the knowledge on the coupling of land and atmospheric variables contributing to rainfall. Global moisture/precipitation recycling estimates infer less than 20% of the precipitation originates from evaporation for a mean space scale of 1000 km (Trenberth, 1999). Recent studies on precipitation recycling provided a deep understanding of its variability, sources, sinks of precipitation, the interaction between soil and precipitation, etc. The length scale of moisture recycling over tropical regions, temperate climates, and desert regions are found to be 500 to 2000 km, 3000 to 5000 km, and above 7000 km, respectively (van der Ent and Savenije, 2011). Except in desert areas, the time scales of moisture recycling are reported as 3 to 20 days. Studies by Keys et al. (2016) showed the percentage of precipitation triggered

Indian Summer Monsoon Variability: El Niño-teleconnections and beyond.
DOI: https://doi.org/10.1016/B978-0-12-822402-1.00008-9
449

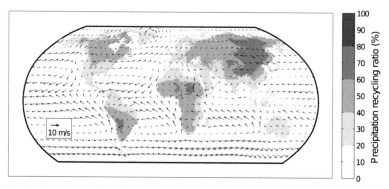

FIG. 23.1 Percentage of precipitation coming from the land evaporation (*Source:* Keys et al., 2016).

by land evaporation globally (Fig. 23.1) using the coupled land surface simple terrestrial evaporation to atmosphere model and atmospheric moisture budget water accounting model. It can be seen from Fig. 23.1 that the contribution of land evaporation to the monsoon rainfall is about less than 40%.

Humans' role in modifying water vapor entering the atmosphere through evaporation (Gordon et al. 2005) is a great concern. Deforestation and irrigation have significantly decreased and increased the water vapor flow of about 3000 km^3/year and 2600 km^3/year respectively in the atmosphere, leading to large spatiotemporal variations that affect the earth's hydrological cycle (Gordon et al. 2005). These changes, particularly concerning land around the Indian Ocean, will alter the Asian monsoon system behavior. As it is already witnessed that the increasing temperatures due to climate change cause the increasing moisture-holding capacity of the atmosphere that increases the atmospheric moisture together with the enhanced evapotranspiration, this will have large implications on the increasing risks of droughts and floods. Under these circumstances, studies on the quantification of rainfall from local evaporation and advection look essential. Kao et al. (2018) used the Coupled Model Intercomparison phase 5 (CMIP 5) models to study the moisture recycling phenomena and found that these models could capture the temporal variations of moisture recycling at long term scales. It is also found that the models are capable of simulating water vapor than precipitation and reported that the recycling rate of atmospheric moisture is a better indicator of climate change. Laine et al. (2014) concluded that evaporation changes to the global precipitation vary 11%–16% during the period of 2080 to 2099 under the intermediate GHG emission scenario of Representative Concentration Pathway (RCP 4.5) in CMIP5 models with more contribution from continents (32%–35%) than oceans (8–13%). It is also worth noting that the impact of global teleconnections on the regional moisture recycling ratios. The El Niño conditions could cause major fluctuations in atmospheric water vapor, particularly in the troposphere (Trenberth et al. 2005; and 2011). Moisture recycling estimated based on Chahine (1992) over India is found to be less than the normal during the El Niño Southern Oscillation (ENSO) years (Lakshmi Kumar et al. 2014). Sujith et al. (2017) found that the role of advection is strong while the local evaporative component is weak during the ENSO years over the Indian region. Pathak et al. (2014) also discussed the changes in local evaporation's contributions to rainfall during El Niño and La Niña years during the study period 1980–2010 and reported the lower recycling ratios during El Niño years. As it is known that the extreme El Niño events are responsible for disastrous events worldwide (Cai et al. 2014), these extreme El Niño events will be doubled in the future (Wang et al. 2017) which will have large bearing on the changes in the recycling ratios.

Taking into consideration of the above, the present chapter mainly focused on understanding the role of evapotranspiration in contributing to the regional rainfall over the Monsoon Core Region (MCR) during the southwest monsoon season. This objective has been studied using a moisture recycling method that accounts for the advected and evaporated components in the rainfall occurrence. Also, we focus on presenting the anomalies of recycling during the future El Niño conditions under two emission scenarios (RCP 4.5 and RCP 8.5). The changes in local evaporation contribution to rainfall over MCR spatially and temporally will provide a general understanding to identify the moisture sources and sinks over this region.

23.2 Data and methodology

To study the long-term variations of moisture recycling monthly simulations of 18 General Circulation Models (GCMs) data from a suite of the CMIP5 (Taylor et al. 2012) have been used. The models have been selected based on the availability of rainfall, evaporation and water vapor data for the historical and future emission scenarios. We estimated the moisture recycling in the historical period of 1976–2005 and also for the three future epochs 2011–2040, 2041–2070, and 2071–2100, which are treated as near, mid, and long term (Intergovernmental Panel on Climate Change, IPCC, 2013) under the two emission scenarios of RCP 4.5 and RCP 8.5. The recycling ratios of local evaporation (η) and advection (Ω) have been estimated for the historical period and for future scenarios. All the models used in the study are brought to a uniform spatial resolution of $0.25°$ $\times 0.25°$ using the bilinear interpolation method (Mishra et al., 2016; 2018) for obtaining the Multi-Model Mean (MMM) which will help to study the explicit diagnosis of moisture recycling on space and time scale. The list of models used and the details with grid resolutions are provided in Table 23.1.

To compare the MMM of rainfall and evaporation with the observed data, we have used India Meteorological Department (IMD) and European Centre for Medium range Weather Forecasting Reanalysis (ERA) data of rainfall and evaporation respectively for the historical period 1976 to 2005. The space scale resolutions of IMD rainfall and ERA evaporation are available at 0.25 × 0.25. The details of these data sets can be found in Pai et al. (2014) and Purnadurga et al. (2019).

TABLE 23.1 Description of the CMIP5 18 global climate models (Taylor et al. 2012).

S.No	Modeling Center (or Group)	Model Name	Grid Size
1.	Commonwealth Scientific and Industrial Research Organization (CSIRO)	ACCESS1.0	192 × 145
2.	Beijing Climate Center	BCC-CSM1-1	128 × 64
3.	Beijing Climate Center	BCC-CSM1-1-M	320 × 160
4.	College of Global Change and Earth System Science, Beijing, Normal University.	BNU-ESM	128 × 64
5.	Canadian Centre for Climate Modeling and Analysis	CanESM2	128 × 64
6.	Centre National de Recherches Météorologiques	CNRM-CM5	128 × 256
7.	Commonwealth Scientific and Industrial Research Organization.	CSIRO-Mk3.6.0	192 × 96
8.	NOAA Geophysical Fluid Dynamics Laboratory	GFDL-ESM2M	144 × 90
9.	Met Office Hadley Centre	HadGEM2-CC	192 × 144
10.	Met Office Hadley Centre	HadGEM2-ES	192 × 144
11.	Institute for Numerical Mathematics	INM-CM4	180 × 120
12.	Institut Pierre-Simon Laplace	IPSL-CM5A-LR	96 × 96
13.	Institut Pierre-Simon Laplace	IPSL-CM5A-MR	144 × 143
14.	Japan Agency for Marine-Earth Science and Technology, Atmosphere and Ocean Research Institute (The University of Tokyo)	MIROC-ESM	128 × 64
15.	Japan Agency for Marine-Earth Science and Technology, Atmosphere and Ocean Research Institute (The University of Tokyo)	MIROC-ESM-CHEM	128 × 64
16.	International Centre for Earth Simulation	MIROC5	256 × 128
17.	Meteorological Research Institute	MRI-CGCM3	320 × 160
18.	Norwegian Climate Centre	NorESM1-M	144 × 96

We identified the future El Niño years by using the CMIP5 Sea Surface Temperature (SST) anomalies over the Niño 3.4 region. The average SSTs of December–January–February have been obtained and based on the anomalies of the same (> 1.5°C), the future El Niño years have been recognized. Rao et al. (2019) has proposed this analysis, and this approach was able to reproduce the previous El Niño years, such as 1982–1983, 1987–1988, etc. The detailed approach of inferring the strength of El Niño based on the quantification of anomalies can be found in ggweather.com/enso/oni.htm. In this chapter, we considered the strong and very strong El Niño years for the different epochs and examined the spatiotemporal variations of moisture recycling.

The MCR of India is considered to be very important region during southwest monsoon season (Fig. 23.2). The MCR covers the major portions of central India and the Indo-Gangetic plains and this study is focused MCR region. The heat low characterizes the MCR before the arrival of the southwest monsoon. Thereafter, a tropical convergence zone will be established, which produces the rains (Gadgil, 2007; Lakshmi Kumar et al. 2019). The boundary between the lows of northwestern regions and the moist convective regime of the northeastern parts decides the MCR.

Moisture recycling is defined as the amount of moisture that has been recycled by the local evaporation before it precipitates (Brubaker et al. 1993). Moisture recycling provides a diagnostic measure to understand the role of local

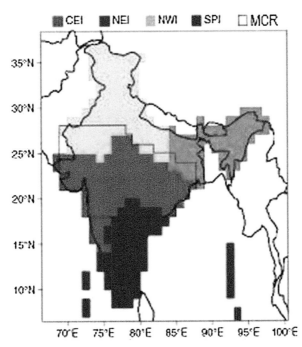

FIG. 23.2 Monsoon core region of India (MCR)—Solid black line drawn over India. (*Source:* https://nwp.imd.gov.in/ERF_Report_2017.pdf).

evaporation and advection in turning the moisture into rainfall over the region (Burde 2006; Fitzmaurice, 2007). Brubaker et al. (1993) proposed a methodology to estimate the rainfall components contributed by the local evaporation and advection, and the same is adopted in the present study.

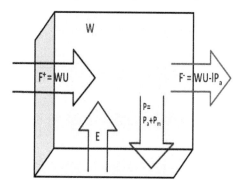

Total precipitation over a box region P = P_a (rainfall due to advection) + P_m (rainfall due to local evaporation)   (1)

Average Horizontal Flux of advected moisture: $Q_a = wu - \left(\dfrac{lPa}{2} \right)$ (2)

Average Horizontal Flux of locally exported moisture: $Q_m = \dfrac{1}{2}(E - P_m)$ (3)

Contribution from local evaporation (Pm/P): $Eta(\eta) = \left[1 + 2\dfrac{wu}{El} \right]^{-1}$ (4)

Contribution from advection (Pa/P): $Omega(\Omega) = \left(1 + \dfrac{El}{2wu} \right)^{-1}$ (5)

Where E is evapotranspiration in mm/day
P is precipitation in mm/day
w is precipitable water vapor in mm, u zonal wind speed in m/s and l is the length of the region

23.3   Results and discussion

23.3.1   Spatial scale variations of rainfall, evaporation, omega, and Eta for the historical period

The spatial variations of the seasonal mean rainfall (Fig. 23.3A) and evaporation (Fig. 23.3C) from CMIP5 MMM for the summer monsoon season for the historical period 1976–2005 are shown in Fig. 23.3. Also the space scale variations of observed data sets of rainfall and evaporation for the same study period are provided as Fig. 23.3 (B and D) respectively. The preponderant features

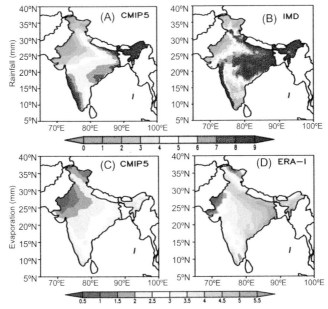

FIG. 23.3 Spatial distribution of seasonal mean (A) rainfall (mm/day) from CMIP5 MMM and (B) rainfall (mm/day) from IMD, (C) evaporation (mm/day) from CMIP5 MMM and (D) evaporation (mm/day) from ERA for the historical period 1976 to 2005.

of rainfall and evaporation are captured over India by the MMM data sets during the historical period when compared with the observed ones. Higher/lower rainfall associated with higher/lower evaporation features over Western Ghats/ Northwest portions have been observed over India. It is worth noting that several studies reported the agreement over the CMIP5 rainfall and the observational records in depicting the large-scale features of the southwest monsoon (Jena et al. 2016 and Mishra et al. 2018). The discrepancies in the quantification of rainfall/evaporation magnitudes do exist in the CMIP5 data sets from the IMD and ERA data sets. The spatial variations of CMIP5 MMM evaporation are analogous to the same obtained from the ERA data over India (Purnadurga et al. 2019). Time series analysis of seasonal rainfall and mean evaporation from the observed and MMM showed the increasing trends for the period 1976–2005 with a statistical significance at 0.05 level (fig not shown here). The seasonal mean daily all India rainfall from IMD (CMIP5) is 6.9 mm/day (5.4 mm/ day) and the seasonal mean daily all India evaporation from ERA (CMIP5) is 6.7 mm/day (2.4 mm/day) for the study period 1976–2005. The standard deviations of MMM rainfall and evaporation are 1.11 mm/day and 0.08 mm/day, indicating more interannual variability in rainfall than evaporation. The increase in evaporation due to increase in temperature causing an increase in the atmospheric moisture-holding capacity which modulates the rainfall pattern (Trenberth, 1998). Hence, in the present study, the increase in evaporation con-

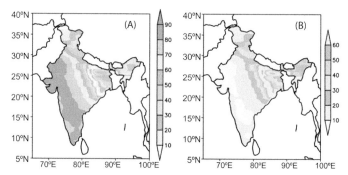

FIG. 23.4 Percentage in seasonal mean of (A) Omega (advected Component) and (B) Eta (evaporative component) from MMM of CMIP5 for the historical period 1976 to 2005.

tributes to an increase in the amount of precipitable water vapor, which has a large bearing on the rainfall amounts. However, the increasing rainfall trends and evaporation do not explicitly infer the evaporation's role in contributing to rainfall. For this purpose, we estimated the moisture recycling and found the changes in the local evaporative (η) and advective (Ω) components' contribution. The spatial variations of Omega (Ω) and Eta (η) (in %) estimated on a local scale over India are shown in Fig. 23.4 (A and B) for the historical period. It is observed that the Omega values vary spatially from 70%–90% over most of the India land region, with an exception to the parts of Northeast, Indo-Gangetic Plains, and Western Himalayan region (Fig. 23.4A). The contribution of the evaporative component, η, has also shown variations quite the opposite of Ω. The η values are high (up to 60%) over the Northeast parts and parts of the Ganges river basin, which show the contribution of evaporation in the rainfall is high over these regions (Fig. 23.4B). These results are similar to Pathak et al. (2014), where they reported that the recycled precipitation is high over Northeastern parts due to more vegetation, which triggers more recycling. They also reported that recycled rainfall estimated from the dynamic recycling model using NCEP Climate Forecast System Reanalysis data had shown a decreasing trend for the period 1980–2010 over northeastern regions of India. Also, studies of Tuineburg et al. (2012) pointed out the contribution of moisture recycling is up to 5% over the Ganges River basis when studied with ECMWF reanalysis data sets for the period 1990–2009. A similar type of study over the Australian monsoon also stressed that the land surface processes that drive the moisture recycling plays a significant role in the precipitation patterns and further can change the atmospheric circulation patterns (Xue et al. 2010). The inference of Fig. 23.4 provides a dominance of the advected component over the MCR except over the regions of northeastern parts during the southwest monsoon season. When recycling is estimated at a local scale, as expected, the contribution of advection dominates the contribution of evaporation, and the same is reflected in the present analysis. We may expect a considerable difference in these contributions when averaged for a region.

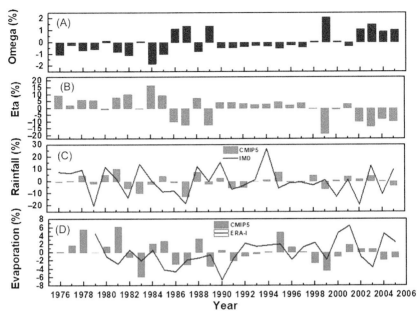

FIG. 23.5   (A-D) Time series of (A) Omega, (B) Eta, (C) rainfall and (D) evaporation anomalies (%) for the historical period 1976 to 2005 during southwest monsoon season.

Fig. 23.5 (A-D) shows the time series anomalies of evaporation, rainfall, Eta and Omega over the MCR for the historical period 1976–2005 obtained from the CMIP5 MMM data sets. The lines plotted in Fig.23.5 (C) and (D) are obtained from the IMD and ERA data sources for the respective variables. The agreement between the anomalies of observed and MMM rainfall and evaporation were found to be varied from year to year during the study period. It is worth recalling the results reported by Goswami et al. (2006) that rainfall over Central India has no trend over the study period 1951–2000. However, there are an increasing number of heavy rainfall events. Goswami et al. (2006) pointed out the decrease in the number of moderate rainfall days offset the increasing number of heavy rain days, thus leading to a steady state in the rainfall trend. However, observations show a declining trend of Indian summer monsoon rainfall from 1950–1999 due to global warming, aerosol loading, and deforestation and are recovered further (Huang et al. 2020). The estimated omega and eta also have shown the increasing and decreasing trends respectively during the study period. The increasing role of advection (denoted by omega) infers the increasing westerlies, from the Arabian Sea due to its warming, reaching the MCR and causing rainfall (Satyaban et al. 2014). Particularly the winds from the western Arabian Sea are reported to carry high moisture. The same is converged over the MCR, which acts as the sink of the moisture transported from the Arabian Sea (Pathak et al. 2014). The decreasing tendency in the evaporative component also complements the increasing role of advection. However, we also observed the interannual variability in Ω and η, as is observed in rainfall and evaporation. As mentioned by Trenberth (1999) that the advection dominates in some regions of Amazon, whereas the local evaporation shows dominance in southern parts of Amazon. However, during the annual cycle, it is reported that 34% of the moisture is recycled. Evaporation contributes to 1% of its total rainfall over the Mississippi basin (Benton and Estoque, 1954). The summer daytime rainfall over central North America was completely from local evaporation (Zangvil et al. 1993). The extent of recycling of moisture mainly depends on the study domain (Brubaker et al. 1993). If the selected domain is small, then a little amount of moisture gets recycled. Solander et al. (2020) reported that the El Niño conditions could cause a consistent decrease in soil moisture over the Amazon region due to less amplification of land — atmospheric interactions. Similarly, during El Niño years over the MCR, less moisture has been recycled due to low land-atmospheric interactions. The interannual variability of η shows that η is low during the El Niño years 1983, 1987, and 2002 compared to the previous years, which infer a low feedback mechanism between the land surface processes and the atmosphere. However, it is understood that external influences like El Niño can impact regional feedback, such as the drier than the normal conditions (Ropelewski and Halpert, 1987), decreasing the soil moisture (Solander et al. 2020). The lower values of recycling also depend on the rainfall pattern during the initial months of monsoon. If the monsoon is weak, there will be less moisture source for the recycling. Similarly, when the monsoon is strong during the initial phase, the El Niño offsets moisture recycling by showing the lesser values (Pathak et al. 2014).

23.4 Analysis of time and space scale moisture recycling – Future El Niño events

Fig. 23.6 shows the epochal trends of rainfall, evaporation, and η for RCP4.5 and RCP 8.5 scenarios. This analysis has been done mainly to examine the role of evaporation in contributing to the rainfall over the MCR. It is difficult to explain the role of evaporation in the rainfall just by analysing the trends and magnitudes of evaporation estimates as the locally evaporated water may transport and fall as rain in other regions. In such cases, the region of locally evaporative fluxes acts as the source, and the region where the fluxes converge are treated as the sinks of the moisture. However, as the MCR is very active during the southwest monsoon, where the cloud cover influences the local evaporation, quantifying evaporation's role is difficult, as reported by Hua et al. (2015). It is found from the Fig. 23.6 that rainfall and evaporation have shown statistically significant increasing trends (please refer Table 23.2 for significance levels) during all the epochs of RCP 4.5 and 8.5 scenarios. The trends of η have yielded decreasing trends during these epochs. The trends of the three variables are shown in Table 23.2, along with their statistical significance levels.

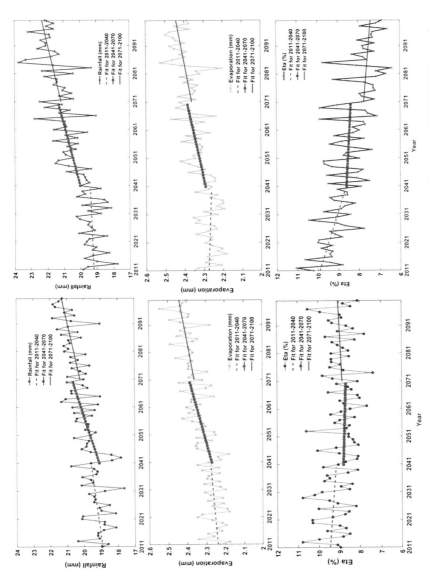

FIG. 23.6 Break trend analysis of rainfall, evaporation, and η for the period 2011–2100 under RCP4.5 (left panel) and RCP 8.5 (right) scenarios.

The MCR is characterized by more advection during the summer monsoon being proxy to the Arabian Sea. In such cases, one may expect the rainfall mainly contributed by the advection since the MCR falls under the humid seasonal climate category during the summer monsoon season. On the contrary, the advection is very low over the arid regions, and the local moisture will be recycled more, as reported by Li et al. (2016). In all three epochs, rainfall and evaporation have shown increasing trends with distinctive slopes. Rainfall variations are mimicked by the evaporation, which shows that the precipitation over MCR controls the local evaporation trends. The trends of η are exciting and not completely followed either rainfall or evaporation. The trend of η during the epochs 2011–2040 and 2041–2070 yielded slightly decreasing trends, and thereafter, it has shown the increasing trend during the RCP4.5 scenario. But in the case of the RCP8.5 scenario, though the rainfall and evaporation have shown increasing trends during all the epochs, the trends of η have undergone the decreasing trends during all the epochs. The overall analysis shows that though the evaporation increasing over the MCR, the local evaporation contribution in converting the moisture into rainfall decreases from 2011–2100 but changing when examined for different epochs. Analysis of recycling ratio estimated over Southeast China from 1979 –2010 discloses that local evaporation influence will be less but significant over the regions where the monsoon winds take major part (Hua et al. 2015). An increase in moisture recycling over Tibetian Plateau has been attributed to land-use changes and land cover associated with evapotranspiration, which intensifies moisture recycling. The extreme moisture recycling values are due to the large-scale atmospheric circulation and extreme weather events (An et al. 2017).

Spatiotemporal anomalies of η for the composite strong and very strong El Niño years over India, particularly in MCR, infer the changing role of evaporation in contributing to the MCR's rainfall during different epochs (Fig. 23.7). The anomalies of η during the composite El Niño years are decreased from the epoch 2011–2040 to 2041–2070 in the RCP4.5 scenario. But, during the epoch 2071–2100, the anomalies are slightly improved, indicating the lower role of evaporation over the western parts of MCR than the eastern part of MCR which includes the Indo-Gangetic Plains. The anomalies of η are varied from 9%–21% below normal during all the epochs in the RCP8.5 scenario, which infers the diminishing role of evaporation over the MCR in contributing to rainfall. Contribution of local evaporation has shown a considerable negative shift (more than 20% below normal) during the epoch 2071–2100 of the RCP8.5 scenario. A reduction of 6%–12% (2011–2040); 9%–21% (2041–2070 and 2071–2100) has been observed on local evaporation's contribution over most parts of the MCR in RCP8.5 scenario. The overall changes may be attributed to soil moisture interactions via evaporation with the circulation and associated surface heating mechanisms. Hence, further studies such as future land-atmospheric interactions, land surface feedback associated with Land Use Land Cover (LULC) changes are required to understand the changing contributions of local evaporation in

TABLE 23.2 Trends in Rainfall, evaporation, and Eta for different epochs of RCP4.5 and RCP8.5 future scenarios.

		Slope				Mean	
		Eta	Rainfall	Evaporation	Eta (%)	Rainfall (mm/day)	Evaporation (mm/day)
RCP4.5	2011–2040	−0.0103***	0.0222**	0.0013**	9.31	4.80	2.25
	2041–2070	−0.0041**	0.0532*	0.0039*	8.74	4.96	2.33
	2071–2100	0.0065**	0.0372*	0.0027*	8.95	5.19	2.40
RCP8.5	2011–2040	−0.0435*	0.0067***	−0.0005***	9.28	4.82	2.27
	2041–2070	−0.0078**	0.0452*	0.0032*	8.49	5.16	2.34
	2071–2100	−0.0116***	0.0398*	0.0028*	7.58	5.43	2.40

(*, ** and *** indicate 0.01, 0.05 and 0.10 levels of significance)

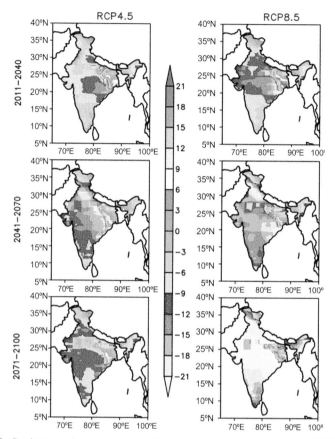

FIG 23.7 Spatiotemporal composite anomalies of η for the strong and very strong El Niño years for the epochs 2011–2040, 2041–2070, and 2071–2100 for RCP 4.5 (left pane) and RCP 8.5 (right panel) emission scenarios.

contributing to rainfall. With the increasing irrigation, changing vegetation, and anthropogenic activities, the changes in LULC may affect the regional feedback mechanisms (Gogoi et al. 2019). The moisture recycling needs to be revisited by considering these changes (Keys et al., 2016).

23.5 Conclusions

In the present chapter, an attempt has been made to understand the role of local evaporation and advection in contributing to the MCR rainfall using the CMIP5 MMM data sets. The necessity of using the CMIP5 for this purpose is to examine the future changes in response to global warming. Knowing the changes in moisture recycling in future El Niño events allows policymakers to enact the framework for water resource management. Our study reports that moisture recycling plays a key role in the eastern parts of MCR by contributing 50%–60% of the rainfall than in the other parts during the southwest monsoon season. During the El Niño years, moisture recycling is much less than normal. A gradual decrease in local evaporation contribution is observed in the two emission scenarios by the end of the century. The anomalies of moisture recycling during RCP8.5 scenario are declined up to 21% below normal from 2011–2100, which infers the diminishing role of evaporation over the MCR in contributing to rainfall. The spatiotemporal variations of moisture recycling in future strong and very strong El Niño years suggest the changing land-atmospheric interactions, which will have a great bearing on future LULC changes and moisture sources and sinks.

Acknowledgments

The authors would like to thank the Science Engineering and Research Board, Government of India, for sponsoring this work.

References

An, W., et al., 2017. Enhanced recent local moisture recycling on the northwestern tibetan plateau deduced from ice core deuterium excess records. J. Geophys. Res. Atmos. 122 (23), 541–556.

Benton, G., Estoque, A.M., 1954. A, "Water-vapor transfer over the North American continent. J. Atmos. Sci. 11 (6), 462–477.

Brubaker, K.L., Entekhabi, D., Eagleson, P.S., 1993. Estimation of continental precipitation recycling. J. Clim. 6, 1077–1089.

Burde, G.I., 2006. Bulk recycling models with incomplete vertical mixing, part I: Conceptual framework and models. J. Climate. 19, 1461–1472.

Cai, W., et al., 2014. Increasing frequency of extreme El Niño events due to greenhouse warming. Nat. Clim. Change. 4, 111–116.

Chahine, M., 1992. The hydrological cycle and its influence on climate. Nature 359, 373–380.

Fitzmaurice, Anne, J., 2007. A critical analysis of bulk precipitation recycling models Ph.D. thesis. Massachusetts Institute of Technology.

Gadgil, S., 2007. The Indian monsoon: 3. Physics of the monsoon. Resonance 12, 4–20.

Gogoi, P.P., et al., 2019. Land use and land cover change effect on surface temperature over Eastern India. Sci. Rep. 9, 8859.

Gordon, L.J., et al., 2005. Human modification of global water vapor flows from the land surface. Glob. Environ. Change. 102 (21), 7612–7622.

Goswami, B.N., Venugopal, V., Sengupta, D., Madhussodan, M.S, Xavier, PK, 2006. Increasing trend of extreme rain events over India in a warming environment. Science 314 (5804), 1442–1445.

Hua, L., Zhong, L., Ke, Z., 2017. Characteristics of the precipitation recycling ratio and its relationship with regional precipitation in China. Theor. Appl. Climatol. 127, 513–531.

Huang, X., et al., 2020. The recent decline and recovery of Indian summer monsoon rainfall: relative roles of external forcing and internal variability. J. Clim. 33, 5035–5060.

Jena, P., Azad, S., Rajeevan, M.N., 2016. CMIP5 projected changes in the annual cycle of Indian monsoon rainfall. Climate 4, 14.

Kao, A.X., et al., 2018. A comparative study of atmospheric moisture recycling rate between observations and models. J. Clim. 31, 2389–2398.

Keys, P.W., et al., 2016. Revealing invisible water: moisture recycling as an ecosystem service. PLoS ONE 11 (3), 1–16.

Laîné, A., Nakamura, H, Nishii, K., 2014. A diagnostic study of future evaporation changes projected in CMIP5 climate models. Clim. Dyn. 42, 2745–2761.

Lakshmi Kumar, T.V., Koteswara Rao, K., Uma, R., Aruna, K., 2014. The role of El Niño southern oscillation on the patterns of cycling rates observed over India during the monsoon season. J. Water Clim. Change. 5 (4), 696–706.

Lakshmi Kumar, T.V., Barbosa, H.A., Thakur, M.K., Trejo, F.P., 2019. Validation of Satellite (TMPA and IMERG) rainfall products with the IMD gridded data sets over monsoon core rgion of India. IntechOpen. doi:10.5772/intechopen.84999.

Li, R., Wang, C., Wu, D., 2016. Changes in precipitation recycling over arid regions in the Northern Hemisphere. Theor. Appl. Climatol. 131, 489–502.

Matthes, J.H., Goring, S., Williams, J.W., Dietze, M.C., 2016. Benchmarking historical CMIP5 plant functional types across the Upper Midwest and Northeastern United States. J. Geophys. Res. Biogeosci. 121, 523–535.

Mishra, S.K., et al., 2018. Fidelity of CMIP5 multi-model mean in assessing Indian monsoon simulations. NPJ Clim. Atmos. Sci. 1, 39.

Pathak, A., Ghosh, S., Kumar, P., 2014. Precipitation recycling in the Indian subcontinent during summer Monsoon. J. Hydrometeorol. 15 (5), 2050–2066.

Pai, D.S., Latha, S., Rajeevan, M., Sreejith, O.P., Satbhai, N.S., Mukhopadhyay, B., 2014. Development of a new high spatial resolution (0.25 × 0.25) long period (1901–2010) daily gridded rainfall data set over India and its comparison with existing data sets over the region. Mausam 65, 1–18.

Purnadurga, G., et al., 2019. Evaluation of evapotranspiration estimates from observed and reanalysis data sets over Indian region. Int. J. Climatol. 39, 1–10.

Rao, V.B., et al., 2019. Future increase in extreme El Nino events under greenhouse warming increases Zika virus incidence in South America. npj Clim Atmos Sci 2, 4.

Ropelewski, C.F., Halpert, M.S., 1987. Global and regional scale precipitation patterns associated with the El Niño/southern oscillation. Mon. Wea. Rev. 115, 1606–1626.

Satyaban, B.S., 2014. Moisture trend over the Arabian Sea and its influence on the Indian summer monsoon rainfall. Res. Papers Issue RP0225, 2014.

Savenije, H.H.G., 1995. New definitions for moisture recycling and the relationship with land-use changes in the Sahel. J. Hydrol. 167, 57–78.

Solander, K.C., et al., 2020. The pantropical response of soil moisture to El Niño. Hydrol. Earth Syst. Sci. 24, 2303–2322.

Sujith, K., Saha, S.K., Pokhrel, S., Hazra, A., Chaudhari, H.S., 2017. The dominant modes of recycled Monsoon rainfall over India. J. Hydrometeor. 18, 2647–2657.

Taylor, K.E., Ronald, S., Meehl, G.A., 2012. An overview of CMIP5 and the experiment design. Bull. Amer. Meteor. 93 (4), 485–498.

Trenberth, K.E., 1998. Atmospheric moisture residence times and cycling: Implications for rainfall rates and climate change. Clim. Change. 39, 667–694.

Trenberth, K.E., 1999. Atmospheric moisture recycling: Role of advection and local evaporation. J. Clim 12 (5), 1368–1381.

Trenberth, K.E., 2011. Changes in precipitation with climate change. Clim Res 47, 123–138.

Trenberth, K.E., Fasullo, J., Smith, L., 2005. Trends and variability in column-integrated atmospheric water vapor. Clim. Dyn. 24, 741–758.

Tuinenburg, O.A., Hutjes, R.W.A., Kaba, P., 2012. The fate of evaporated water from the Ganges basin. J. Geophys. Res. Atmos. 117, 1–17.

Van Der Ent, R.J, Savenije, H.H.G., 2011. Length and time scales of atmospheric moisture recycling. Atmos. Chem. Phys. 11 (5), 1853–1863.

Wang, G., et al., 2017. Continued increase of extreme El Niño frequency long after 1.5 °C warming stabilization. Nat. Clim. Change. 7, 568–572.

Xue, Y., et al., 2010. Global and seasonal assessment of interactions between climate and vegetation biophysical processes: a GCM study with different land-vegetation representations. J. Clim. 23, 1411–1433.

Zangvil, A., Portis, D.H., Lamb, P.J., 1993. Investigation of the large-scale atmospheric moisture field over the midwestern United States in relation to summer precipitation. Part II: Recycling of local evapotranspiration and association with soil moisture and crop yields. J. Clim. 17, 3283–3301.

Appendix

List of Reviewers

Dr. Krishna Achutarao, Indian Institute of Technology, Delhi, India

Dr. Anoop Ambili, Indian Institute of Science Education and Research, Mohali, India

Dr. Karumuri Ashok, University of Hyderabad, Hyderabad, India

Dr. Raju Attada, Indian Institute of Science Education and Research, Mohali, India

Dr. Swadhin Behera, Application Laboratory, Japan Agency for Marine-Earth Science and Technology, Yokohama, Japan

Dr. Preethi Bhaskar, Indian Institute of Tropical Meteorology, Pune, India.

Dr. Prasad K Bhaskaran, Indian Institute of Technology, Kharagpur, India

Dr. Mrinal Biswas, Developmental Testbed Center, Boulder, Colorado, USA

Dr. Naidu C.V, Andhra University, Visakhapatnam, India

Dr. Sanjib Kumar Deb, Space Applications Centre(SAC-ISRO), Ahmedabad, India

Dr. Somenath Dutta, India Meteorological Department, Pune, India

Dr. Pant G.B (Late), Indian Institute of Tropical Meteorology, Pune, India.

Dr. Yoo-Geun Ham, Chonnam National University, Gwangju, South Korea

Dr. Yoshiyuki Kajikawa, RIKEN Center for Computational Science, Kobe, Japan

Dr. Anand Karipot, Savitribai Phule Pune University, Pune, India.

Dr. Ashwini Kulkarni, Indian Institute of Tropical Meteorology, Pune, India

Dr. Rupa Kumar K, The International CLIVAR Monsoon Project Office, Indian Institute of Tropical Meteorology, Pune, India

Dr. Ravi Kumar Kunchala, Indian Institute of Techonology, Delhi, India

Dr. Jinbao Li, Department of Geography, University of Hong Kong, Hong Kong

Dr. Ramesh Kumar M. R, National Institute of Oceanography, Goa, India

Dr. Vikram Mehta, Center for Research on the Changing Earth System, Catonsville, Maryland, USA

Dr. Milind Mujumdar, Indian Institute of Tropical Meteorology, Pune, India

Dr. Francis P. A, Indian National Centre for Ocean Information Services, Hyderabad, India

Dr. Sunita P., Andhra University, Visakhapatnam, India

Dr. Joseph P. V, Cochin University of Science and Technology, Cochin, India

Dr. Amita Prabhu, Indian Institute of Tropical Meteorology, Pune, India

Dr Kriplani R. H., Indian Institute of Tropical Meteorology, Pune, India

Dr. Satyaban Bishoyi Ratna, School of Environmental Sciences, Climatic Research Unit, University of East Anglia, Norwich, UK

Dr. Ingo Richter, Application Laboratory, Japan Agency for Marine-Earth Science and Technology, Yokohama, Japan

Dr. Rajkumar Sharma, Space Applications Centre, ISRO, Ahmedabad, India

Dr. Atul Kumar Shrivastava, Indian Institute of Tropical Meteorology, Dehli, India

Dr. Pankaj Kumar Srivastava, Indian Institute of Science Education and Research (IISER) Bhopal, India

Dr. Hiroki Tokinaga, Kyushu University, Kasuga, Fukuoka, Japan

Dr. Umakanth Uppara, School of Environmental Sciences, University of East Anglia, Norwich, U.K.

Dr. Sathiyamoorthy V, Space Applications Centre (SAC-ISRO), Ahmedabad, India

Dr. Naresh Krishna Vissa, National Institute of Technology, Rourkela, India

Dr. Ramesh Kumar Yadav, Indian Institute of Tropical Meteorology, Pune, India.

Dr. Song Yang, Sun Yat-sen University, Guangzhou, China

Dr. Bin Yu, Climate Research Division, Environment and Climate Change Canada, Canada

Dr. Kyung-Sook Yun, Center for Climate Physics, Institute for Basic Science, Busan, South Korea

Dr. Xiao-Tong Zheng, Ocean University of China, Qingdao, China

Dr. Lei Zhou, Institute of Oceanography, Shanghai Jiao Tong University, Shanghai, China

Index

Page numbers followed by "*f*" and "*t*" indicate, figures and tables respectively.

Printed in the United States
by Baker & Taylor Publisher Services